ANNUAL PLANT REVIEWS
VOLUME 33

ANNUAL PLANT REVIEWS VOLUME 33

Intracellular Signaling in Plants

Edited by

Zhenbiao Yang

Professor of Plant Cell Biology
Center for Plant Cell Biology
University of California
Riverside, CA, USA

Annual Plant Reviews

WILEY-BLACKWELL

A John Wiley & Sons, Ltd., Publication

This edition first published 2008

Blackwell Publishing was acquired by John Wiley & Sons in February 2007. Blackwell's publishing programme has been merged with Wiley's global Scientific, Technical, and Medical business to form Wiley-Blackwell.
Registered office
John Wiley & Sons Ltd, The Atrium, Southern Gate, Chichester, West Sussex, PO19 8SQ, United Kingdom

Editorial office
9600 Garsington Road, Oxford, OX4 2DQ, United Kingdom
2121 State Avenue, Ames, Iowa 50014-8300, USA

For details of our global editorial offices, for customer services and for information about how to apply for permission to reuse the copyright material in this book please see our website at www.wiley.com/wiley-blackwell.

Library of Congress Cataloging-in-Publication Data

Intracellular signaling in plants / edited by Zhenbiao Yang.
p. cm. – (Annual plant reviews, volume 33)
Includes bibliographical references and index.
ISBN-13: 978-1-4051-6002-5 (hardback : alk. paper)
ISBN-10: 1-4051-6002-0 (hardback : alk. paper) 1. Plant cellular signal transduction.
I. Yang, Zhenbiao, 1961-
QK725.I584 2008
571.7'42–dc22
2007047501

A catalogue record for this book is available from the British Library

Annual plant reviews (Print) ISSN 1460-1494
Annual plant reviews (Online) ISSN 1756-9710

Set in 10/12 Palatino by Aptara Inc., New Delhi, India
Printed and bound in Singapore by Markono Print Media Pte Ltd
1 2008

Annual Plant Reviews

A series for researchers and postgraduates in the plant sciences. Each volume in this series focuses on a theme of topical importance and emphasis is placed on rapid publication.

Titles in the series:

1. **Arabidopsis**
   Edited by M. Anderson and J.A. Roberts
2. **Biochemistry of Plant Secondary Metabolism**
   Edited by M. Wink
3. **Functions of Plant Secondary Metabolites and Their Exploitation in Biotechnology**
   Edited by M. Wink
4. **Molecular Plant Pathology**
   Edited by M. Dickinson and J. Beynon
5. **Vacuolar Compartments**
   Edited by D.G. Robinson and J.C. Rogers
6. **Plant Reproduction**
   Edited by S.D. O'Neill and J.A. Roberts
7. **Protein–Protein Interactions in Plant Biology**
   Edited by M.T. McManus, W.A. Laing, and A.C. Allan
8. **The Plant Cell Wall**
   Edited by J.K.C. Rose
9. **The Golgi Apparatus and the Plant Secretory Pathway**
   Edited by D.G. Robinson
10. **The Plant Cytoskeleton in Cell Differentiation and Development**
    Edited by P.J. Hussey
11. **Plant–Pathogen Interactions**
    Edited by N.J. Talbot
12. **Polarity in Plants**
    Edited by K. Lindsey

13. **Plastids**
Edited by S.G. Moller
14. **Plant Pigments and Their Manipulation**
Edited by K.M. Davies
15. **Membrane Transport in Plants**
Edited by M.R. Blatt
16. **Intercellular Communication in Plants**
Edited by A.J. Fleming
17. **Plant Architecture and Its Manipulation**
Edited by C.G.N. Turnbull
18. **Plasmodeomata**
Edited by K.J. Oparka
19. **Plant Epigenetics**
Edited by P. Meyer
20. **Flowering and Its Manipulation**
Edited by C. Ainsworth
21. **Endogenous Plant Rhythms**
Edited by A. Hall and H. McWatters
22. **Control of Primary Metabolism in Plants**
Edited by W.C. Plaxton and M.T. McManus
23. **Biology of the Plant Cuticle**
Edited by M. Riederer
24. **Plant Hormone Signaling**
Edited by P. Hadden and S.G. Thomas
25. **Plant Cell Separation and Adhesion**
Edited by J.R. Roberts and Z. Gonzalez-Carranza
26. **Senescence Processes in Plants**
Edited by S. Gan
27. **Seed Development, Dormancy and Germination**
Edited by K.J. Bradford and H. Nonogaki
28. **Plant Proteomics**
Edited by C. Finnie
29. **Regulation of Transcription in Plants**
Edited by K. Grasser
30. **Light and Plant Development**
Edited by G. Whitelam and K.J. Halliday
31. **Plant Mitochondria**
Edited by David C. Logan
32. **Cell Cycle Control and Plant Development**
Edited by D. Inzé
33. **Intracellular Signaling in Plants**
Edited by Z Yang

CONTENTS

CONTRIBUTORS

Sarah M. Assmann
Biology Department
Penn State University
208, Mueller Laboratory
University Park, PA 16802
USA

Wendy F. Boss
Department of Plant Biology
North Carolina State University
Raleigh, NC 27695-7649
USA

Daniel A. Chamovitz
Department of Plant Sciences
Tel Aviv University
Tel Aviv 69978
Israel

Jin-Gui Chen
Department of Botany
University of British Columbia
Vancouver, BC V6T 1Z4
Canada

Viswanathan Chinnusamy
Water Technology Centre
Indian Agricultural Research Institute
New Delhi 110 012
India

Jesse Coutu
Department of Biochemistry and Molecular Biology
University of Nevada
Mail Stop 200
Reno, NV 89557
USA

Lei Ding
Biology Department
Penn State University
208 Mueller Laboratory
University Park, PA 16802
USA

Aurora Piñas Fernández
Department of Plant Physiology
Umeå Plant Science Centre
Umeå University
S-901 87 Umeå
Sweden

Ying Fu
State Key Laboratory of Plant Physiology and Biochemistry
College of Biological Sciences
National Center for Plant Gene Research (Beijing)
China Agricultural University
Beijing 100094
People's Republic of China

Simon Gilroy
Botany Department
University of Wisconsin
Birge Hall
430 Lincoln Drive
Madison, WI 53706
USA

Zhizhong Gong
State Key Laboratory of Plant Physiology and Biochemistry
College of Biological Sciences
National Center for Plant Gene Research (Beijing)
China Agricultural University
Beijing 100094
People's Republic of China

Yair Halimi
Department of Plant Sciences
Tel Aviv University
Tel Aviv 69978
Israel

Takashi Hashimoto
Graduate School of Biological Sciences,
Nara Institute of Science and Technology
Nara 630-0192
Japan

Patrick J. Hussey
The Integrative Cell Biology Laboratory
School of Biological and Biomedical Sciences
Durham University
South Road, Durham DH1 3LE
United Kingdom

Alan M. Jones
Departments of Biology and Pharmacology
University of North Carolina at Chapel Hill
Chapel Hill, NC 27599-3280
USA

Tsutomu Kawasaki
Laboratory of Plant Molecular Genetics
Nara Institute of Science and Technology
Takayama 8916-5, Ikoma
Nara 630-0192
Japan

Sheng Luan
Department of Plant and Microbial Biology
University of California
Berkeley, CA 94720
USA

Daniel V. Lynch
Department of Biology
Williams College
Williamstown, MA 01267
USA

Yasunori Machida
Division of Biological Science
Graduate School of Science
Nagoya University
Chikusa-ku, Nagoya 464-8602
Japan

Gad Miller
Department of Biochemistry and Molecular Biology
University of Nevada
Mail Stop 200
Reno, NV 89557
USA

Zhen-Ming Pei
Department of Biology
Duke University
Box 90338
Durham, NC 27708
USA

Ron Mittler
Department of Biochemistry and Molecular Biology
University of Nevada
Mail Stop 200
Reno, NV 89557
USA

Michiko Sasabe
Division of Biological Science
Graduate School of Science
Nagoya University
Chikusa-ku, Nagoya 464-8602
Japan

Ko Shimamoto
Laboratory of Plant Molecular Genetics
Nara Institute of Science and Technology
Takayama 8916-5, Ikoma
Nara 630-0192
Japan

Vladimir Shulaev
Virginia Bioinformatics Institute
Washington Street
Blacksburg, VA 24061
USA

Åsa Strand
Department of Plant Physiology
Umeå Plant Science Centre
Umeå University
S-901 87 Umeå
Sweden

Keiko U. Torii
Department of Biology
University of Washington
Seattle, WA 98195
USA

Xuemin Wang
Department of Biology
University of Missouri
St. Louis, MO 63121
USA
and
Donald Danforth Plant Science Center
St. Louis, MO 63132
USA

Yan Wu
Key Laboratory of Ministry of Education for Plant Developmental Biology
College of Life Sciences
Wuhan University
Wuhan 430072
People's Republic of China

Zhenbiao Yang
Department of Botany and Plant Sciences
University of California
Riverside, CA 92521
USA

Shuqun Zhang
Department of Biochemistry
371G Life Sciences Center
University of Missouri – Columbia
Columbia, MO 65211
USA

Jian-Kang Zhu
Department of Botany and Plant Sciences
Institute for Integrative Genome Biology
2150 Batchelor Hall
University of California
Riverside, CA 92521
USA

PREFACE

Sessile plants must rapidly respond to drastic environmental changes to survive and must timely adjust their growth and developmental behaviors in response to daily and seasonal environmental changes. Consequently, most differentiated plant cells are totipotent, and plants have flexible developmental programs that are highly adaptable to the environment. An intriguing and important question in our understanding of plant developmental program and responses to the environment is what kinds of strategies and mechanisms plant cells use for the transmission and the integration of various developmental and environmental signals. In recent years, we have witnessed an exponential increase in our knowledge of plant intracellular signaling mechanisms, pathways, and networks that plants utilize to monitor and process a specific extracellular signal and to modulate a given process. This rapid knowledge growth has been clearly aided by genetic and genomic approaches in model plant systems such as *Arabidopsis*. Many critical signaling components and pathways have been identified based on genetic mutations that affect a specific plant process. The availability of new biochemical, molecular, cell biological, and proteomic tools have also undoubtedly fueled our advancement in elucidating signaling mechanisms in plants.

The picture of intracellular signaling in plants that has emerged from these dramatic advances is such that plants use signaling mechanisms and networks that integrate ancient and universal intracellular signals and signaling mechanisms with their own inventions in essentially every plant signaling pathway/network known to date. Examples of conserved signaling mechanisms include the prokaryotic two-component regulatory systems, receptor and nonreceptor serine/threonine kinases, calcium, heterotrimeric G proteins, Rho family guanosinetriphosphatases (GTPases), and mitogen-activated protein kinase cascades, phospholipids, the cytoskeleton, and the ubiquitin-based protein turnover machinery. Many plant pioneer signaling components have also been uncovered, such as novel interactors of heterotrimeric G proteins and Rho GTPases, calcium-dependent protein kinases, and novel scaffolding proteins. The goal of this volume is to provide an in-depth discussion on many of these conserved and novel signaling mechanisms and to provide a few examples of how these conserved and novel signaling mechanisms are constructed into a signaling network that modulates a specific plant process. Obviously this volume is unable to cover several important topics in plant signaling either due to space limitation or due to significant coverage of these topics in other volumes. For example, the two-component systems have been discussed in the volume on plant hormones, and photoreceptors on the photomorphogenesis volume.

Despite the tremendous progress in recent years, many important questions and challenges remain in the field of plant intracellular signaling. It is likely that many new signaling components and mechanisms have yet to be unraveled. The roles of most plant "pioneer" proteins in signaling are unknown. The picture for many signaling pathways is still incomplete, and we are missing the knowledge of the nature and the mechanism of perception for many signals. For example, the vast majority of >400 receptor-like kinases have no known functions and corresponding ligands. The understanding of signaling networks and cross talks between signaling pathways/networks are just starting. Clearly, the conventional approaches such as genetics and biochemistry will continue to be valuable and *-omics* will also start to pay their dues. However, we are also in need of elucidating signaling networks at the systems level by using mathematical and computational approaches. It is my hope that this volume would provide a catalyst for propelling our understanding of plant intracellular signaling to the next level.

Zhenbiao Yang
Riverside
California
USA

ACKNOWLEDGMENTS

The editor of this volume thanks his wife, Yanping, for her support, patience, and understanding, as well as the members of his laboratory for stimulating discussions, which all together make this work possible. The work in Yang's laboratory is supported by grants from National Institute of Health, Department of Energy, and National Science Foundation.

Annual Plant Reviews (2008) **33**, 1–29
doi: 10.1111/b.9781405160025.2008.00001.x www.interscience.wiley.com

Chapter 1

TRANSMEMBRANE RECEPTORS IN PLANTS: RECEPTOR KINASES AND THEIR LIGANDS

Keiko U. Torii

Department of Biology, University of Washington, Seattle, WA 98195, USA

Abstract: Receptor-like kinases (RLKs) represent by far the largest family of cell surface receptors in plants. They mediate cell–cell signals regulating self-incompatibility, innate immunity, and a wide variety of developmental processes. The genetic dissection of paralogous RLKs illuminated their intricate redundancy, synergism, and antagonism. The nature of corresponding ligands for RLKs has been largely elusive. However, recent efforts using genetics, biochemistry, genome-wide functional genomics, and bioinformatics led to the discovery of small, secreted peptides and cysteine-rich secreted proteins as candidate ligands for RLKs. Studies of brassinosteroid signaling and innate immunity to bacterial flagellin peptide revealed a striking resemblance of the mechanism of plant RK activation to that of animal transforming growth factor β receptors.

Keywords: receptor kinases (receptor-like kinases); peptide ligands; brassinosteroids; pathogen-associated molecular patterns (PAMPs); leucine-rich repeats (LRRs); receptor endocycles

1.1 Introduction

As a sessile organism, a plant is equipped with mechanisms that detect its neighboring environment to adjust its developmental programs as well as to cope with numerous environmental challenges. It is therefore not surprising that plants possess a large number of genes predicted to encode cell surface receptors, transmembrane receptor kinases (RKs) and their variants, which are predicted to mediate cell–cell signaling. In addition, multiple two-component sensor histidine kinases, including members of the ethylene receptors and cytokinin receptors, and a small number of G-protein-coupled seven transmembrane receptors (GPCRs) are found in plants (Urao *et al.*, 2000; Jones and Assmann, 2004). The presence of a wide variety of membrane receptors highlights the intricacy of plant cell signaling.

A canonical transmembrane receptor kinase gene possesses a signal peptide, an extracytoplasmic ectodomain, a single-membrane-spanning region, and a C-terminal protein kinase domain. The paradigm of receptor-kinase signal transduction is that ligand binding to the extracellular domain triggers dimerization and activation of receptors, leading to intracellular signaling cascades via protein phosphorylation. Plant RKs comprise a monophyletic group related to animal RKs (Shiu and Bleecker, 2001a,b). Almost all plant RKs phosphorylate serine/threonine residues. Given that the vast majority of plant RKs remains as orphan receptors (i.e., their corresponding ligands are unknown), they are collectively called "receptor-like kinases (RLKs)." This chapter summarizes recent discoveries about intricate redundancy among RKs, identities of ligands for RKs, biological functions of ligand–RK pairs, and emerging mechanisms of receptor activation.

1.2 Classifications of the RLK superfamily

The RLK superfamily is classified into three major groups depending on the presence or absence of the receptor domain and the kinase domain. As described above, canonical RLKs possess both a receptor and a kinase domain. Those that lack the cytoplasmic kinase domain are collectively named "receptor-like proteins (RLPs)." Alternatively, kinases with no apparent signal peptides, those that lack extracellular domain, or those that lack both extracellular and transmembrane domains are collectively named "receptor-like cytoplasmic kinases (RLCKs)." Genetic and biochemical evidence suggest that RLKs, RLPs, and RLCKs may function at close molecular proximity to control signal transduction.

The RLKs are further divided into 17 subgroups based on the structural features of the predicted extracellular domain (Shiu and Bleecker, 2001a,b, 2003). Their brief descriptions are as follows.

1.2.1 Leucine-rich repeats (LRRs)

LRR-RLKs comprise by far the largest class of plant RLKs, with over two hundred members in *Arabidopsis*. LRRs are tandem repeats of approximately 24 amino acids with conserved leucine residues. The LRR motif is found in proteins with diverse functions in many organisms and is implicated in protein–protein interactions (Kobe and Deisenhofer, 1994). Several LRR-RLKs have been shown to regulate development, including ERECTA (ER: organ shape and plant growth), CLAVATA1 (CLV1: shoot meristem maintenance), EXCESS MICROSPOROCYTE/EXTRA SPOROGENOUS CELLS (EMS1/EXS: anther development), tomato LePRK1,2 (pollen tube elongation), BRASSINOSTEROID INSENSITIVE 1 (BRI1: brassinosteroid-mediated growth and development), BRI1-ASSOCIATED KINASE/SOMATIC EMBRYOGENESIS RECEPTOR KINASE (BAK/RK: brassinosteroid signaling/embryogenesis/microspore development), and tissue patterning, including ERECTA-LIKE 1 (ERL1) and ERL2 (stomatal patterning), SCRAMBLED/SUTRUBBELIG (SUB/SCM: ovule development and root epidermal patterning) (Torii *et al.*, 1996; Clark *et al.*, 1997; Li and Chory, 1997; Schmidt *et al.*, 1997; Canales *et al.*, 2002; Li *et al.*, 2002; Nam and Li, 2002; Zhao *et al.*, 2002; Chevalier *et al.*, 2005; Kwak *et al.*, 2005; Shpak *et al.*, 2005).

Other LRR-RLKs have been shown to mediate innate immunity and defense response. Those include Xa21 (resistant to *Xanthomonas oryzae* pv. *oryzae*), FLAGELLIN INSENSITIVE 2 (FLS2: innate immunity response to bacterial flagellin peptides), and PEP1R (defense response to endogenous elicitor peptide PEP1) (Song *et al.*, 1995; Gomez-Gomez and Boller, 2000; Yamaguchi *et al.*, 2006).

Interestingly, LRR-LRPs lacking the cytoplasmic kinase domain have also been shown to regulate both development and defense response. Those developmental regulators include CLV2 (shoot meristem maintenance) and TOO MANY MOUTHS (TMM: stomatal patterning), and those regulating defense responses include the tomato Cf9, Cf4, and Cf2 (race-specific resistance to *Cladosporium fulvum*) (Jones *et al.*, 1994; Thomas *et al.*, 1997; Jeong *et al.*, 1999; Nadeau and Sack, 2002). These LRR-RLPs may act in concert with partner LRR-RLKs.

1.2.2 S-domain

S-RLKs possess an extracellular S-domain, originally identified from S-glycoprotein (Nasrallah *et al.*, 1988). The S-domain consists of 12 conserved cysteine residues (10 of which are absolutely conserved) in a consensus $CX_5CX_5CX_7CXCX_nCX_7CX_nCX_3CX_3CXCX_nC$. In addition, the S-domain possesses the PTDT box, which has a conserved WQSFDXPTDTΦL sequence (X = nonconserved amino acid; Φ = aliphatic amino acid) (Walker, 1994). The S-RLKs are the female determinants for sporophytic self-incompatibility. In the *Brassica* self-incompatibility system, self-derived pollen will be recognized at the surface of stigma and its germination will be prevented (Nasrallah, 2005; Takayama and Isogai, 2005). Biochemically, S-RLK is a receptor for the pollen-derived ligand SCR/SP11 (Schopfer *et al.*, 1999; Takayama *et al.*, 2000; Kachroo *et al.*, 2001). Interestingly, the self-compatible plant species *Arabidopsis* possesses numerous (~40) S-RLK genes, some of which are expressed in vegetative tissues. They include ARK1, ARK2, ARK3, RLK1, and RLK4 from *Arabidopsis* (Tobias *et al.*, 1992; Walker, 1993). The presence of S-RLKs in self-compatible species implies their possible roles outside the self-incompatible response. Consistently, an S-RLK of *Brassica oleracea*, SFR2, may be involved in plant defense response signaling (Pastuglia *et al.*, 1997).

1.2.3 Others

There are many other RLKs with extracellular motifs distinct from the above two classes. Those include a tumor-necrosis factor receptor-like repeat motif, the epidermal growth factor-like repeat motif, the pathogenesis-related protein 5-like motif, and lectin-like motif (Becraft *et al.*, 1996; Harvé *et al.*, 1996; He *et al.*, 1996; Wang *et al.*, 1996; Shiu and Bleecker, 2001a). The structural diversity in extracellular receptor domains perhaps reflects their functional diversity. This chapter features LRR-RLKs and S-RLKs because of their importance in plant development and defense response, as well as of the wealth of knowledge about their ligands, modes of action, and signal transduction pathways.

1.3 Redundancy and antagonism among closely related RLKs

The completion of the *Arabidopsis* genome sequencing project accelerated efforts toward elucidating the biological functions of RLKs via reverse genetics. It has become routine to isolate T-DNA insertion lines of closely related RLK family members to uncover their collective function as a family. Such an approach is especially powerful when a single RLK gene knockout fails to confer a dramatic, visible phenotype. Collectively, several studies have revealed both conservation and uniqueness in closely related RLK paralogs arisen from gene duplication events. This highlights immense complexities in RLK signal transduction, given that these paralogous RLKs most likely perceive the same ligand molecules. Here, some examples are described.

1.3.1 Unequal Redundancy

1.3.1.1 ERECTA-family RLKs: synergistic interactions

The *Arabidopsis* accession Landsberg *erecta* (L*er*) carries a mutation in the *ERECTA* locus, which confers a compact inflorescence with short internodes, short pedicels, and short, blunt siliques (Torii *et al.*, 1996). *ERECTA* encodes an LRR-RLK and its promoter is highly active in the SAM and organ primordia (Torii *et al.*, 1996; Yokoyama *et al.*, 1998). ERECTA has two closely related paralogs, ERL1 and ERL2. The three ERECTA-family RLKs show synergistic interactions in regulating aboveground organ growth (Shpak *et al.*, 2004). While the *erecta* single loss-of-function mutation confers compact inflorescence, single and double loss-of-function mutations of other members display no growth phenotype. However, *erl1* and *erl2* mutations enhance the growth defects of *erecta* in a unique manner, and *erecta erl1 erl2* triple knockout plants displayed severe dwarfism due to reduced cell proliferation and abnormal floral patterning. The unequal contributions of the three *ERECTA*-family genes may be largely accounted for by their expression patterns rather than their functions as RLKs, given that both *ERL1* and *ERL2* are capable of rescuing *erecta* mutant phenotypes when driven by the native *ERECTA* promoter (Shpak *et al.*, 2004). Consistently, the *ERECTA*-family genes display intricate overlapping expression patterns during organ primordia growth. Successive loss of gene dosage revealed that *ERL2* is haplo-insufficient for ovule development and female fertility in the absence of *ERECTA* and *ERL1*, while *ERECTA* and *ERL1* are haplo-sufficient in the absence of other two members (Pillitteri *et al.*, 2007). Therefore, *ERL2* possesses the most minor role in promoting cell division and organ growth.

The dispensable roles of *ERLs* (especially *ERL2*) during organ growth lead to a question: how and why were these duplicated RLK genes maintained throughout the course of evolution? Briggs *et al.* (2005) proposes a hypothesis that the additional, specialized functions of *ERLs* in controlling stomatal differentiation may have allowed their retention (Shpak *et al.*, 2005; Briggs *et al.*, 2006).

1.3.1.2 BRI1 and its paralogs: specialized expression domains

Brassinosteroids (BRs) are growth-promoting steroid hormones in plants, and loss of their biosynthesis results in severely dwarfed plants and male sterility (Bishop and Koncz, 2002). BRI1 was identified as a cell surface receptor for BR, and *bri1* loss-of-function mutations result in severely dwarfed plants indistinguishable from BR-biosynthetic mutants (Li and Chory, 1997). *BRI1* encodes an LRR-RK with a 70-amino acid island domain intercepting the tandem repeats of LRRs (Li and Chory, 1997). *BRI1* belongs to a family of four closely related LRR-RLKs: BRI1, BRI1-LIKE 1 (BRL1), BRL2, and BRL3 (Cano-Delgado *et al.*, 2004). Unlike *BRI1*, *brl* mutations did not confer growth defects. Both BRL1 and BRL3, but not BRL2, bind BRs. Consistently, both *BRL1* and *BRL3*, but not BRL2, rescued *bri1* mutant phenotype when driven by the native *BRI1* promoter (Cano-Delgado *et al.*, 2004). These findings place BRL1 and BRL3 as redundant BR receptors.

Why do BRL1 and BRL3 have such a minor role in plant growth? Unlike ERECTA-family RLKs, BRI1-family RLKs appear to have adopted specialized expression patterns. While *BRI1* is ubiquitously expressed, *BRL1* and *BRL3* show highly specialized expression within a subset of vascular tissues (Cano-Delgado *et al.*, 2004). The *brl1* mutant plants indeed displayed vascular defects: increase in phloem tissues and concomitant decrease in xylem tissue, supporting the previous findings that BR stimulates xylem differentiation (Yamamoto *et al.*, 1997, 2001). Interestingly, *BRL2*, which does not bind BR, has been previously described as a gene regulating vascular cell-type differentiation, *VH1* (Clay and Nelson, 2002). It would be fascinating to address whether BRL2/VH1 is capable of binding sterols or other steroid-related molecules known to control plant growth and development (Jang *et al.*, 2000).

1.3.2 Redundancy

Simple redundancy, in which single mutations in given gene pairs fail to confer a visible phenotype, is commonly observed among closely related RLKs. A good example is SERK1 and SERK2 LRR-RLKs redundantly in the control of microspore development. Both SERK1 and SERK2 homo- and heterodimerize in plant protoplast, suggesting that both RLKs participate in receptor complex formation (Albrecht *et al.*, 2005; Colcombet *et al.*, 2005). In addition, two paralogous LRR-RLKs, PRK1 and TOAD2, were currently shown to redundantly regulate embryo patterning in *Arabidopsis* (Nodine *et al.*, 2007). Both PRK1 and TOAD2 show overlapping expression patterns in the developing embryo, though PRK1 turns on at earlier stage. The *rpk1 toad2* double loss-of-function mutant (precisely, a fraction of progenies from *rpk1 toad2/+*) fails to establish the protoderm at the early globular stage and results in a fluffy embryo with subprotodermal identity (Nodine *et al.*, 2007).

1.3.3 Antagonism among RLKs

1.3.3.1 CLV1 and its two paralogs: antagonistic actions of LRR-RLKs?

Genetic studies have shown that three *CLV* loci, *CLV1*, *CLV2*, and *CLV3*, act in a linear pathway to restrict undifferentiated stem cell populations within the shoot apical meristem (SAM), and the loss of function of each *CLV* gene confers SAM enlargement (Clark *et al.*, 1993, 1995; Kayes and Clark, 1998). The molecular identities of the three CLV gene products support the notion that they form a ligand–receptor complex: CLV1 is an LRR-RLK, CLV2 is an LRR-RLP without the cytoplasmic kinase domain, and CLV3 is a secreted peptide (Clark *et al.*, 1997; Fletcher *et al.*, 1999; Jeong *et al.*, 1999). A series of elegant genetic experiments revealed that the CLV signaling pathway along with a positive regulator of stem cells, WUSCHEL, constitutes a feedback loop, which maintains a stable population of stem cells within the SAM (Brand *et al.*, 2000; Schoof *et al.*, 2000).

There are three *CLV1* paralogs in the *Arabidopsis* genome: *BARELY ANY MERISTEM 1* (*BAM1*), *BAM2*, and *BAM3*. As opposed to *CLV1*, which negatively regulates the SAM, these three *CLV1* paralogs are required for maintenance of the SAM, since *bam1 bam2 bam3* triple loss-of-function mutation results in small plants with dramatically reduced SAM size (DeYoung *et al.*, 2006). Therefore, the three BAMs act antagonistically to CLV1. The mechanism for this antagonism is unclear: While CLV1 and BAMs have opposite function, the ectopic expression of *CLV1* and the *BAMs* in the entire shoot apex driven by the *ERECTA* promoter rescued *bam* and *clv1* defects, respectively. This reciprocal complementation implies that all four LRR-RLKs are capable of perceiving the same ligand molecules and triggering signal transduction. Perhaps CLV1 and the three BAM LRR-RLKs modulate the proliferation and differentiation of the SAM via competing for the ligands and/or receptor partners. Consistent with this hypothesis, *BAMs* are expressed in organ primordia, while *CLV1* is restricted within the inner central domain of the SAM (Clark and Schiefelbein, 1997; DeYoung *et al.*, 2006).

1.4 Ligands for RLKs

The presence of the plant cell walls prohibits direct contact of two transmembrane molecules in adjacent cells. Therefore, direct association of transmembrane ligands and corresponding receptors, such as Delta and Notch association, which control various patterning events in animals, will not likely occur in plants, and plant ligands need to be small, diffusible molecules. Despite the fact that plants possess numerous RLK multigene family members (~610 in total in *Arabidopsis*), the nature of their corresponding ligands remains poorly understood (Matsubayashi *et al.*, 2001; Shiu and Bleecker, 2001a; Torii, 2004). Three factors should be taken into account regarding the apparent discrepancy in the numbers of RLKs versus corresponding ligands: origin, size, and diversity. First, in regards to origin, ligands for plant RLKs may not be encoded by the plant genome. For example, pathogen-origin pathogen-associated molecular patterns (PAMPs) are recognized by plant RLKs to trigger defense response (Nurnberger *et al.*, 2004).

Second, the small size of peptide-ligand genes, as compared to the RLKs, puts significant biases against the prediction of open-reading frames (ORFs) by bioinformatics, and consequently, many ligand genes may have escaped from gene annotation. Third, a further complication arises from the possibility that many ligand molecules may not be peptides, and are not directly encoded by ORFs. BR, a well-known example is a steroid hormone, and other molecules, such as oligosaccharides, may act as ligands for RLKs as well.

To overcome the hurdle of small size, several approaches have been taken to uncover peptide ligand-like genes and further to identify their biological functions. For example, Olsen *et al.* (2002) performed bioinformatic analysis of previously annotated small secreted peptide genes and identified a potential, large *Arabidopsis* family of peptides closely related to tobacco rapid alkalization factors (RALF) (Pearce *et al.*, 2001). More recently, Lease and Walker (2006) developed algorithms to extract ligand-like features: ORFs encoding peptides and small proteins between 25 and 250 amino acids in length, the presence of an N-terminal signal peptide, the absence of transmembrane domains, and the absence of the ER retention sequence. Their analysis identified over 30,000 previously unannotated putative ORFs, which constitutes the Arabidopsis Unannotated Secreted Peptide Database (http://peptidome.missouri.edu). Among them approximately 1000 predicted ORFs satisfied the following categories: (1) they are expressed; (2) they comprise gene families; (3) they have homologs in rice. Through the database, Lease and Walker (2006) identified 12 additional RALF-like genes (Lease and Walker, 2006). Similarly, through bioinformatics, Silverstein *et al.* (2007) identified *in silico* ~13,000 plant genes encoding small, cysteine-rich peptides, including RALF-like peptides and defensins (Silverstein *et al.*, 2007). One-third of the genes identified were previously unannotated, although over half of the genes were expressed.

The *in silico* predictions of putative ligand genes provide the basis for systematic identification of their biological functions via functional genomics. For example, systematic overexpression of >100 small secreted ORFs in *Arabidopsis* led to the identification of *EPIDERMAL PATTERNING FACTOR 1* (*EPF1*), a gene controlling stomatal patterning (Hara *et al.*, 2007). Thus, functional-genomic approach, together with continued efforts in forward and reverse genetics, holds promise for advancing our knowledge of the small peptide world in plants. Some of the ligand molecules for plant RLKs with demonstrated biological functions are summarized in the following sections. See Table 1.1 for ligand–receptor pairs and their biological functions, and see Table 1.2 for the amino acid sequence of the ligand molecules.

Table 1.1 Receptor kinase–ligand pairs

Receptor-like kinases (RLKs)	Ligands (*candidates)	Biological function
BRI1 (LRR-RLK X)[a]	Brassinolide (BR)	BR plant growth
BRL1		
BRL3		
SR160 (LeBRI1)	Systemin*/BR	Defense signaling/plant growth
CLV1 (LRR-RLK XI)	MCLV3	Meristem development
BAM1		
BAM2		
BAM3		
CLV2 (LRR-RLP)		
PSKR (LRR-RLK X)	PSK	Cellular dedifferentiation Cell division in suspension culture
FLS2 (LRR-RLK XII)	flg22	Defense signaling
EFR (LRR-RLK XII)	elf18	
PEPR1 (LRR-RLK XI)	AtPep1	Defense signaling
EMS1/EXS1 (LRR-RLK X)	TPD1*	Microsporogenesis
HAESA (LRR-RLK XI)	IDA*	Organ shedding/senescence
TMM (LRR-RLP)	EPF1*	Stomatal patterning
ERECTA (LRR-RLK XIII)		Stomatal patterning/plant growth
ERL1		
ERL2		
Cf9 (LRR-RLP)	Avr9*	Defense signaling
Cf4	Avr4*	Defense signaling
LePRK1 (LRR-RLK)	Lat52/LeSTG1*	Pollen tube growth
LePRK2		
SRK (S-RLK)	SCR/SP11	Self-incompatibility

[a] Classification based on Shiu and Bleecker (2001b).

Table 1.2 Amino acid sequence of peptide ligands

Peptide ligands	Sequence
Peptides	
Systemin	AVQSKPPSKRDPPKMQTD
MCLV3	$\mathrm{RTVP^{h}SGP^{h}DPLHH}$
PSK	$\mathrm{Y^{(SO^{3}H)}IY^{(SO^{3}H)}TQ}$
flg22	QRLSSTGSRINSAKDDAAGLQIA
elf18	ac-SKEKFERTKPHVNVGTIG
AtPep1	ATKVKAKQRGKEKVSSGRPGQHN
Cysteine-patch proteins	
EPF1	MKSLLLLAFFLSFFFGSLLARHLPTSSHPSHHHVG MTGALKRQRRRPDTVQVAGSRLPD**C**SHA**C**GS**C**SP**C** RLVMVSFV**C**ASVEEAET**C**PMAYK**C**M**C**NNKSYPVP
Avr9	Y**C**NSS**C**TRAFD**C**LGQ**C**GR**C**DFHKLQ**C**VH
LeSTG1	MDFIILLIAILALSSTPITIISGSVTNHTYSTTNS YTNVALSARKVVFPPPRQLGKDNSDDDDLI**C**KT**C**K RLSEHRT**CC**FNYF**C**VDLFTNRFN**C**GS**C**GLV**C**IVGT R**CC**GGI**C**VDIKKDNGN**C**GK**C**NNV**C**SPGQN**C**SFGL**C** VSA
SCR/SP11	MRYATSIYTFLTNIHYLCFIFLILTYVQALDVG AWK**C**PEGIAYPSPISGR**C**FNSRSTE**C**KKHYEVE GHNVTN**C**R**C**DTYSMQNPARIT**C**Y**CC**KVKS
Others[b]	
TPD1[a]	MNRRRLLVSATLLSYLLYGMALVSVEASGGEKLRD NLDLTKTTTSSPSISHRKMLLLSPGTGKTERSVEP ERIGEKCKSTDIVVNQAVTEPMPNGIPGYMVEITN QCMSGCIISRIHINCGWFSSAKLINPRVFKRIHYD DCLVNNGKPLPFGSTLSFHYANTFPYHLSVAFVTCA
IDA[a]	MAPCRTMMVLLCFVLFLAASSSCVAAARIGATMEM KKNIKRLTFKNSHIFGYLPKGVPIPPSAPSKRHNS FVNSLPH

[a] Mature forms unknown.
[b] Mature forms unknown, may be small peptides. Underline, predicted signal peptide (SignalP 3.0); thick underline, conserved motif among closely related gene products; boldface, cysteines.

1.5 Small peptides

Many of the known signaling ligands are highly processed small peptides. These peptides may be modified by processes such as sulfonation, hydroxylation, and acetylation.

1.5.1 CLE family

CLV3 and maize ESR (*e*mbryo *s*urrounding *r*egion) proteins are founder members of the CLE (CLV3/ESR) family of peptide signaling molecules (Opsahl-Ferstad *et al.*, 1997; Fletcher *et al.*, 1999). CLV3 regulates stem cell population size within the SAM, and it acts as a ligand for CLV1/CLV2 LRR-RLK/LRR-RLP receptor heterodimers (Fletcher *et al.*, 1999; Brand *et al.*, 2000). The maize ESR proteins are proposed to act as a signal between the endosperm and the embryo during early maize development. CLV3 and ESR encode small precursor proteins with an N-terminal signal peptide, sharing a highly conserved 14-amino acid CLE motif at the C-terminus (Cock and McCormick, 2001). The *Arabidopsis* genome contains at least 31 *CLE* genes, of which 24 are expressed in various tissues (Cock and McCormick, 2001; Sharma *et al.*, 2003). Interestingly, the amino acid sequence of CLE proteins outside the C-terminal CLE motif is highly divergent, suggesting that the short CLE motif carries biological activities. Several researchers have performed deletion analysis as well as synthetic peptide application to address this hypothesis (Fiers *et al.*, 2006; Kondo *et al.*, 2006; Ni and Clark, 2006). For instance, Kondo *et al.* (2006) tested the biological activities of various synthetic CLV3 peptides. The direct application of the CLV3 peptide covering the CLE motif (CLV3L) to a shoot apex resulted in SAM differentiation, while deletion of a single arginine residue within the CLE motif abolished its activity (Fig. 1.1, Color plate 1). The mature form of the CLV3 peptide, MCLV3, was determined by in situ mass spectrometry (MALDI-TOF-MS) analysis using *Arabidopsis* callus tissues overexpressing CLV3 (Kondo *et al.*, 2006). MCLV3 is a dodeca peptide with two hyrdoxyproline residues (RTVP^{h}SGP^{h}DPLHH) (Fig. 1.1 (Color plate 1), Table 1.2), and is capable of terminating undifferentiated stem cell population when applied directly (Fig. 1.1, Color plate 1). Hydroxylation is not required for CLV3 function as a ligand; rather it affects stability of the peptide.

The large number of *CLE* genes in *Arabidopsis* and other plant species suggests that they act as ligands for additional LRR-RLKs regulating multiple developmental processes in plants. Consistent with this hypothesis, application of several CLE peptides triggered differentiation of the root meristem (Casamitjana-Martinez *et al.*, 2003; Fiers *et al.*, 2005). Furthermore, a CLE peptide was identified as a factor inhibiting xylem differentiation in *Zinnia* (TDIF: treachery element differentiation inhibitory factor) (Ito *et al.*, 2006).

Finally, the study of CLE peptides has revealed the fascinating possibility that a parasitic organism may have acquired a *CLE* gene (the gene *HgSYV46* from parasitic soybean nematode *Heterodera glycines*) during the course of evolution to modulate cell division and differentiation of a host plant (Wang *et al.*, 2005b). Further understanding of CLE peptides may therefore unravel exciting insight into development, evolution, and host–pathogen interaction.

1.5.2 PSK (phytosulfokine)

The peptide growth factor PSK is secreted into culture media and promotes cell proliferation of the cultured cells. The mature PSK is a sulfonated pentapeptide Y$^{(SO3H)}$IY$^{(SO3H)}$TQ (Table 1.2). The cloned rice *PSK* gene (*Os-PSK*) ORF is 276 bp, encoding an 89-amino acid precursor with an N-terminal signal sequence (Yang *et al.*, 1999). A C-terminal amino acid sequence (residues 80–84) corresponds to the mature PSK peptide.

PSK genes are found in asparagus, carrot, rice, *Zinnia*, and *Arabidopsis*, suggesting that PSK may act as a "universal" growth hormone in a wide variety of plant species (Yang *et al.*, 1999). There are five *PSK* genes in *Arabidopsis*, all of which possess the 100% identical PSK domain (MTDYIYTQ) with rice PSK (Yang *et al.*, 2001; Matsubayashi *et al.*, 2006). Outside this domain, PSKs among *Arabidopsis* and rice display very limited sequence similarity. The divergent, N-terminal regions of PSKs may reflect functional significance, such as target cleavage sites for distinct proteolytic enzymes or stability of precursor proteins.

The corresponding receptor for PSK was the first RLK identified via biochemical purification (Matsubayashi *et al.*, 2002). Using ligand-based affinity purification from microsomal fractions of carrot culture cells, Matsubayashi *et al.* purified a 120-kDa protein that specifically binds PSK. The protein has characteristic features of an LRR-RLK, with a 36-amino acid island domain intercepting LRR18 (Matsubayashi *et al.*, 2002). The *Arabidopsis* PSKR was also shown to bind PSK (Matsubayashi *et al.*, 2006). Somewhat surprisingly, the promoter activity of both AtPSK and AtPSKR is highest in mature leaves, suggesting that while PSK promotes cell proliferation in cell cultures, it likely possesses different function in plants (Matsubayashi *et al.*, 2006). Consistently, the loss of function of AtPSKR showed no defects in plant growth, but accelerated senescence. Conversely, AtPSKR overexpression conferred plants with increased longevity. On the basis of the expression patterns and the phenotype of mutants and overexpressors, Matsubayashi *et al.* proposed that PSK mediates signal transduction promoting cellular longevity (Matsubayashi *et al.*, 2006).

1.5.3 Flagellin peptides and other PAMPs

PAMPs are molecular signatures of pathogens that are recognized by the receptors of host plants eliciting innate immune response (Nurnberger *et al.*, 2004). Known PAMPs are derived from essential structural component of pathogens, such as fungal cell wall fragments and bacterial flagellin fragments (Gomez-Gomez and Boller, 2002). Using a simple bioassay (e.g., rapid alkalization assay of plant cell cultures), several peptide fragments have been identified as PAMPs. These include N-terminal fragment of bacterial flagellin (flg22: QRLSSTGSRINSAKDDAAGLQIA) (Felix *et al.*, 1999) as well as an acetylated 18-amino acid peptide from the N-terminus of elongation factor Tu (elf18: ac-SKEKFERTKPHVNVGTIG) (Table 1.2) (Kunze *et al.*, 2004). A receptor for flg22, FLS2 (FLAGELLIN INSENSITIVE 2), was identified from genetic screen for *Arabidopsis* mutants insensitive to flg22 (Gomez-Gomez and Boller, 2000). The elf18 receptor EFR was isolated through genome-wide expression profiling for genes rapidly induced by PAMPs (Zipfel *et al.*, 2006). Strikingly, FLS2 and EFR are closely related LRR-RKs, implying that the plant LRR-RLKs underwent duplication and specification to allow detection of diverse PAMPs.

1.5.4 Endogenous elicitors

In addition to the PAMPs, plants recognize their own peptide and oligosaccharide fragments as indicators of pathogen attack and trigger innate immunity.

Among such elicitors, tomato systemin induces plant wound response (McGurl *et al.*, 1992). Systemin is the first peptide hormone discovered in plants. The mature systemin peptide is 18-amino acid long and is processed from the C-terminus of a 200-amino acid precursor called prosystemin (McGurl *et al.*, 1992). Similar to the PAMPs, addition of synthetic systemin to culture media induces rapid extracellular alkalization. Interestingly, prosystemin does not possess an N-terminal signal peptide. Therefore, the mechanism of its secretion remains unclear.

The corresponding receptor of systemin was identified biochemically using photoaffinity-labeled systemin (Scheer and Ryan, 2002). Strikingly, the purified systemin receptor SR160 turned out to be an LRR-RLK orthologous to the BR receptor BRI1. The finding suggests the intriguing possibility that in Solanaceae, BRI1 may have acquired additional function as a systemin receptor. The molecular nature of the dual function of SR160 is unclear. It appears that BR and systemin bind to different domains of SR160/LeBRI1, as systemin does not compete for the BR binding (Scheer and Ryan, 2002). Recently, genetic studies revealed that the tomato *bri1* mutant *curl3* (*cu3*) exhibits normal response to systemin-induced wound signaling, while it is defective in BR response (Holton *et al.*, 2007). This leaves the biological relevance of SR160/LeBRI1 ambiguous.

AtPep1, an endogenous elicitor in *Arabidopsis*, was identified from *Arabidopsis* leaf extracts for its ability to cause rapid extracellular alkalization of cell cultures. The mature form of AtPep1 is a 23-amino acid peptide derived from the C-terminus of a 92-amino acid precursor protein, whose expression is upregulated by wounding or various stress hormones (Table 1.2) (Huffaker *et al.*, 2006). The corresponding AtPep1 receptor (PEPR1) was subsequently identified via biochemical purification using radiolabeled AtPep1 peptide (Yamaguchi *et al.*, 2006). PEPR1 is an LRR-RK with 26 extracellular LRRs, further emphasizing the central roles for LRR-RLKs in innate immunity response. A phylogenetic analysis by Ryan *et al.* (2007) revealed the presence of a closely related paralog of PEPR1 and PEPR2 (Ryan *et al.*, 2007). It would be interesting to address the biological function of PEPR2.

1.6 Cysteine-rich extracellular proteins

Several of known RLK ligands are cysteine-rich extracellular proteins. They appear to form a stable structure via intramolecular disulfide bonds. Consistently, bacterially produced recombinant cysteine-rich proteins have been shown to have biological activity.

1.6.1 LAT52 and LeSTG1

Both LAT52 and LeSTG1 were isolated as a candidate ligand molecule for the petunia LRR-RK LePRK2 by a yeast two-hybrid screen using the extracellular domain of LePRK2 as bait (Tang *et al.*, 2002, 2004) (Tables 1.1 and 1.2). Both LAT52 and LeSTG1 encode cysteine-rich proteins that are predicted to be secreted, glycosylated, and have a mature size of 16 and 13 kDa, respectively (Muschietti *et al.*, 1994; Tang *et al.*, 2004). While LAT52 is pollen-specific, LeSTG1 is expressed in stigmatic tissues in the pistils where pollen tube elongation occurs. In vitro binding assays revealed that LeSTG1 is capable of displacing LAT52 for binding to the ectodomain of LePRK2. On the basis of expression patterns and biochemical analyses, Tang *et al.* (2004) proposed a hypothesis that LePRK2 associates with LAT52 in nongerminated pollen grains, and when pollen grains land on stigma, LePRK2 changes its partner to LeSTG1 to promote pollen tube growth. Consistent with this hypothesis, the recombinant LeSTG1 protein promoted pollen tube growth in vitro when applied to the culture media (Tang *et al.*, 2004).

1.6.2 Avr4 and Avr9

Tomato disease-resistant proteins Cf9 and Cf4 confer race-specific resistance to the fungal pathogen *C. fulvum* that express the *Avr9* and *Avr4* genes, respectively (Jones and Jones, 1997). Fungal-produced Avr9 and Avr4 are small, secreted proteins, and both contain a patch of cysteine residues, which may participate in the formation of a stable tertiary structure via disulfide bonds (Table 1.2). In fact, ^{1}H NMR analysis revealed that the Avr9 peptide forms a compact structure containing three antiparallel β-sheets connected by three disulfide bridges (Vervoort *et al.*, 1997). Other than the cysteine patch, Avr9 and Avr4 display no similarity in primary sequence and differ significantly in size. The mature Avr9 and Avr4 proteins possess 28 and 86 amino acids, respectively (Van den Ackerveken *et al.*, 1992; Joosten *et al.*, 1997; Jones and Jones, 1997). This is rather surprising given that the extracellular domain of Cf9 and Cf4 share >90% amino acid sequence identity.

1.6.3 SCR/SP11

SCR (*S*-locus cysteine-rich proteins), also known as SP11 (S-protein 11), was identified as a small ORF within the *S*-locus (Schopfer *et al.*, 1999; Takayama *et al.*, 2000). The SCR/SP11 protein possesses patches of eight cysteines at conserved positions (Table 1.2). The spacer regions are highly divergent, consistent with its role as signaling ligand for self-incompatibility. The tertiary structure of SCR/SP11 from S_8 allele has been resolved by NMR crystallography (Mishima *et al.*, 2003). SCR/SP11 possesses an α/β sandwich structure stabilized with intramolecular disulfide bonds. Therefore, SCR/SP11 most likely exists as a stable secreted protein (as opposed to being processed into a small peptide). Consistent with this, a bacterial-produced recombinant SCR/SP11 is biologically active to trigger self-incompatible response (Kachroo *et al.*, 2001).

1.6.4 EPF1 (EPIDERMAL PATTERNING FACTOR 1)

EPF1 was identified through genome-wide overexpression studies for small, secreted proteins (Hara *et al.*, 2007). The constitutive overexpression of *EPF1* (*CaMV35S::EPF1*) conferred an epidermis solely composed of pavement cells, lacking any stomata. Conversely, *epf1* loss-of-function mutations disrupt proper stomatal patterning, with occasional stomatal clustering (Hara *et al.*, 2007). *EPF1* encodes a putative secretory protein with patches of eight cysteines (Table 1.2). Although EPF1 shares no sequence similarity with other cysteine-rich protein ligands, the positions of the cysteine patches are somewhat conserved. Thus, EPF1 may also adopt a stable structure with intramolecular disulfide bonds.

A previous genetic model suggests that STOMATAL DENSITY AND DISTRIBUTION 1 (SDD1), a putative subtilisin protease, process ligands for TOO MANY MOUTHS (TMM) LRR-RLP and ERECTA-family LRR-RLKs. Perception of the ligands by TMM and ERECTA-family RLKs triggers signal transduction controlling stomatal patterning (Ingram, 2005; Bergmann and Sack, 2007). The loss-of-function *sdd1*, *tmm*, and *erecta erl1 erl2* mutants exhibit an epidermis with increase numbers of stomata that are adjacent to each other or in clusters (Berger and Altmann, 2000; Nadeau and Sack, 2002; Shpak *et al.*, 2005). Inhibition of stomatal development by *EPF1* requires functional alleles of *TMM* and *ERECTA*-family genes, since CaMV35S::EPF1 failed to rescue stomatal cluster phenotype of *tmm* and *erecta erl1 erl2* (Hara *et al.*, 2007). This finding is consistent with the hypothesis that the secreted EPF peptides act upstream of TMM and ERECTA-family receptors. On the other hand, constitutive overexpression of *EPF* rescued the increased stomatal density by *sdd1* (Hara *et al.*, 2007). Therefore, SDD1 may function in a separate pathway from EPF, suggesting that it is not modified by the putative subtilisin-like protease SDD1. Alternatively, in high doses, EPF may be active regardless of whether or not it is processed by SDD1.

1.7 Other possible ligands and their corresponding receptors

1.7.1 IDA (INFLORESCENCE DEFICIENT IN ABSCISSION), a ligand for HAESA?

The recessive *Arabidopsis ida* mutation confers defects in floral organ abscission, resulting in sepals and petals, which remain attached to flowers long after fertilization (Butenko *et al.*, 2003). *IDA* encodes a 77-amino acid protein with an N-terminal signal sequence (Butenko *et al.*, 2003) (Table 1.2). Consistent with the predicted nature of the secreted protein, the IDA:GFP fusion protein gave a high signal at the cell periphery. The *IDA* gene is conserved in a wide variety of plant species, and *Arabidopsis* has five *IDA-LIKE* genes (*IDL1–IDL5*). Comparison of these *IDA*-family genes revealed a conserved 12-amino acid motif, named the PIP motif, at the C-terminus (Butenko *et al.*, 2003). It would be of special interest to test whether the mature form of IDA (and IDL) is a short peptide encompassing the PIP motif. Strikingly, the constitutive overexpression of IDA (CaMV35S::IDA) led to ectopic abscission zone formation in various places, such as at the base of pedicels (Stenvik *et al.*, 2006). Therefore, IDA is sufficient to trigger abscission and cell separation.

The LRR-RLK HAESA is an attractive candidate for the IDA receptor (Jinn *et al.*, 2000). The *HAESA* promoter is specifically active in the abscission zone, and antisense suppression of *HAESA* reduced floral organ abscission, a phenotype similar to that of *ida* (Jinn *et al.*, 2000). It would be therefore of special interest to test the genetic and biochemical interactions of *IDA* and *HAESA*.

1.7.2 TPD1 (TAPETUM DETERMINANT 1), a ligand for EMS1/EXS1?

The *tpd1* loss-of-function mutation in *Arabidopsis* confers male sterility due to the loss of tapetum and concomitant increase in microsporocytes (Yang *et al.*, 2003). *TPD1* is thus required for the specification of tapetal cell identity in the *Arabidopsis* anther, and in the absence of *TPD1*, more cells commit to adopt microsporocyte identity instead of differentiating into tapetal cells (Yang *et al.*, 2003). *TPD1* encodes a small protein of 176 amino acids with an N-terminal signal peptide (Table 1.2). The C-terminal region of TPD1 possesses a CLVNNG motif that is shared with other gene products of unknown function, suggesting that this portion may be retained in the mature peptide.

The phenotype of *tpd1* highly resembles that of *ems1/exs1* (*excess microsporocytes 1/extra sporogenous cells*). The *EMS1/EXS1* gene product is an LRR-RLK, and thus it would be exciting to postulate that TPD1 acts as a ligand for EMS1/EXS1. The anther phenotype of *tpd1 ems1/exs1* double mutant plants was indistinguishable from *tpd1* single mutants, suggesting that *TPD1* and *EMS1/EXS1* act in a linear pathway (Yang *et al.*, 2003). The constitutive overexpression of *TPD1* (*CaMV35S::TPD1*) conferred excessive cell division within carpels, and this phenotype was dependent on the functional *EMS1/EXS1* allele (Yang *et al.*, 2005). The results indicate that *TPD1* acts upstream of *EMS1/EXS1*, and TPD1 may indeed be the EMS1/EXS1 ligand.

1.8 Ligand–receptor interactions

In animals, it is widely accepted that ligand binding to corresponding RKs leads to conformational changes, which trigger activation of the cytoplasmic protein kinase domain and further signal transduction. Biochemical association of ligands to corresponding receptors has been demonstrated for a few plant ligand–receptor pairs, including BR to BRI1, PSK to PSKR, LAT52 and

LeSTG1 to LePRK2, fls22 to FLS2, AtPep1 to PEPR1, and SCR/SP11 to SRK (Kachroo *et al.*, 2001; Matsubayashi *et al.*, 2002; Tang *et al.*, 2002; Tang *et al.*, 2004; Kinoshita *et al.*, 2005; Chinchilla *et al.*, 2006; Yamaguchi *et al.*, 2006). In known cases, ligand binding to corresponding RKs does not require a cytoplasmic kinase domain. For instance, the association of LAT52 and LeSTG1 to the extracellular domain of LePRK2 was tested using a yeast two-hybrid approach (Tang *et al.*, 2002, 2004). The binding of BR to BRI1 was tested using the bacterially produced fragments within the extracellular domain of BRI1 (Kinoshita *et al.*, 2005). The binding of PSK to PSKR was even enhanced by removal of the cytoplasmic kinase domain, perhaps due to increased receptor stability (Shinohara *et al.*, 2007). The exact sites of ligand-binding pocket have been determined for two LRR-RKs, PSKR and BRI1 (Kinoshita *et al.*, 2005; Shinohara *et al.*, 2007). Interestingly, these two studies demonstrate that an extracellular "island" domain intercepting tandem repeats of LRRs is the key ligand-binding site. In the case of BRI1, a biotin-tagged photoaffinity castasterone as well as tritium-labeled brassinolide were shown to bind to the bacterially produced BRI1 fragment including 70-amino acid island domain and flanking LRR22 (ID-LRR22), indicating that this region is sufficient for BR binding (Kinoshita *et al.*, 2005). More specific analysis was performed to delineate the PSK binding site of PSKR (Shinohara *et al.*, 2007). Using a radiolabeled photoaffinity PSK analog, ^{125}I-[N$^{\varepsilon}$-(4azidosalicyl)Lys^{5}]PSK, Shinohara *et al.* (2007) successfully cross-linked the radiolabeled ligand to carrot PSKR and subsequently delineated the ligand-binding site via peptide fragmentation followed by mass spectrometry. The PSK binding site was mapped to the 15-amino acid fragment within the island domain. Consistently, the deletion of the island domain from PSKR abolished ligand binding.

Several LRR-RLPs, including CVL2 and Cf-family of tomato disease resistant receptors, possess an island domain (Jones *et al.*, 1994; Dixon *et al.*, 1998; Jeong *et al.*, 1999). Strikingly, the position of the island domain is conserved in all cases, between the fourth and fifth LRRs from the transmembrane domain, despite the fact that the primary sequence of the island domain shares no similarities (Torii, 2004). This intriguing observation implies that the island domain and its adjacent LRRs may form a structure favored for ligand binding and subsequent receptor activation.

Although the active protein kinase domain appears dispensable for ligand binding to RKs, the proper transmembrane anchoring of RKs may be critical for some cases of ligand–RK interaction. For instance, SCR/SP11 displays high–binding affinity to the naturally occurring, truncated SRK that possesses the extracellular domain and a transmembrane domain but lacks the cytoplasmic kinase domain (Shimosato *et al.*, 2007). The extracellular, soluble form of SRK, however, did not show high ligand-binding affinity (Shimosato *et al.*, 2007). Since *Brassica* self-incompatibility involves a set of three molecules, ligand (SCR/SP11), receptor (SRK), and extracellular glycoprotein with S-domain (SLG), the differential affinity of the ligands to soluble versus membrane-anchored receptors may reflect the complex regulation of signal transduction.

1.9 Early events in receptor kinase signaling: dynamics of receptor activation

After ligand binding, how do RKs activate their kinase domains and trigger downstream signal transduction, and how are they eventually downregulated to quench signaling? The early in vivo events after ligand–RK association have been extensively studied in two LRR-RKs, BRI1 and FLS2. Recent studies on these two RKs are beginning to unravel the conserved mechanism of plant RK activation, which is strikingly similar to that of animal transforming growth factor β (TGFβ) receptors. TGFβ receptors consist of two distinct receptors: type I receptors (i.e., transducers), which are unable to bind ligands in the absence of type II receptors, and type II receptors (i.e., primary receptors), which are unable to signal in the absence of type I receptors. Binding of the ligand, TGF, to the type II receptors promotes heteromultimeric (tetrameric) receptor complex formation and triggers subsequent signal transduction (Massagué, 1996). Both BRI1 and FLS2 appear to function as primary receptors, and intriguingly, they share the same transducer LRR-RLK, BAK1 (Li *et al.*, 2002; Nam and Li, 2002; Russinova *et al.*, 2004). The resemblance of the modes of activation of plant RKs to that of TGFβ receptors extends to subcellular receptor dynamics of compartmentalization/endocytosis.

1.9.1 Dynamics of BRI1 activation

Extensive studies by the group of Dr Joanne Chory revealed the modes of BRI1 activation (Fig. 1.2a). In the absence of ligands, BRI1 exists as inactive homodimers (Wang *et al.*, 2005a). The activity of BRI1 is suppressed by its own C-terminal domain as well as by BKI1 (BRI1 kinase inhibitor 1), a protein directly interacting with the kinase domain of BRI1 (Wang and Chory, 2006). Upon direct binding of BR, the cytoplasmic kinase domain of BRI1 will be activated, undergoing autophosphorylation as well as phosphorylating BKI1. The phosphorylated BKI1 dissociates from BRI1 and leaves from the plasma membrane. Subsequently, BRI1 forms a heterodimer with a second LRR-RLK, BAK1 (also known as SERK3). This triggers downstream events leading to the activation of key transcription factors BES1/BZR1 via dephosphorylation and regulation of BR-specific gene expression (Fig. 1.2a) (Belkhadir and Chory, 2006; Karlova and de Vries, 2006). Although the precise stoichiometry of the activated BRI1 receptor complex is not known, it has been proposed that, similar to TGFβ receptors, the active BRI1 complex would be heterotetramer with BRI1 analogous to the type II receptor and BAK1 to the type I receptor (Belkhadir and Chory, 2006; Karlova and de Vries, 2006).

In TGFβ receptor signaling, the activated receptors are compartmentalized into an endosome population, where they associate with an adaptor protein that mediates further signal transduction to the transcription factor Smad2 (Tsukazaki *et al.*, 1998; Raikhel and Hicks, 2007). While BRI1 homodimers are localized predominantly at the plasma membrane, BRI1–BAK1 heterodimers

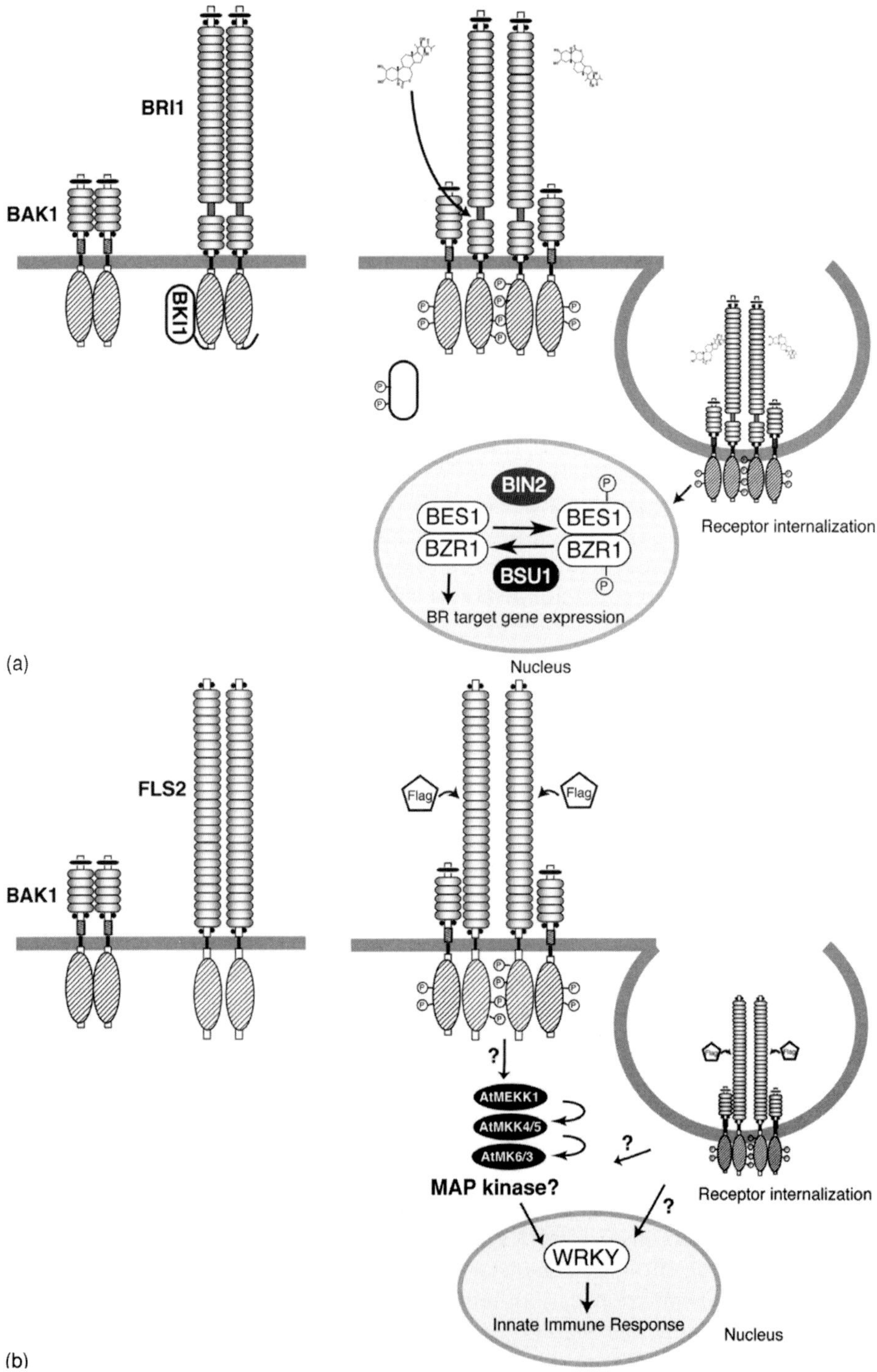
BRI1
BAK1
BKI1
BIN2
BES1
BZR1
BES1
BZR1
BSU1
BR target gene expression
Receptor internalization
(a)
Nucleus
FLS2
BAK1
Flag
Flag
?
AtMEKK1
AtMKK4/5
AtMK6/3
?
MAP kinase?
?
Receptor internalization
WRKY
Innate Immune Response
Nucleus
(b)

are mainly detected in the endosomes (Russinova *et al.*, 2004). Since internalization and endocytosis of BRI1 was observed constitutively, Russinova *et al.* (2004) postulated that BAK1 functions to redistribute BRI1 within the endomembrane system. Consistently, the pharmacological treatment of *Arabidopsis* cells revealed that endosomal population of BRI1 is indeed active in signal transduction. Geldner *et al.* (2007) discovered that application of an endosomal trafficking inhibitor, brefeldin A, to *Arabidopsis* roots increased the population of BRI1-containing endosomes, and promoted dephosphorylation of BES1 and reduced the BR early responsive gene expression (Geldner *et al.*, 2007). These studies place BAK1 as a critical effector that activates BRI1 signaling, and further highlight the importance of endomembrane trafficking for plant RK signaling.

1.9.2 Dynamics of FLS2 activation and internalization

FLS2 recognizes a bacterial flagellin peptide fragment (flg22) and triggers defense response (Gomez-Gomez and Boller, 2002). Recent studies by Robatzek *et al.* (2006) demonstrated that FLS2 is ubiquitously expressed and plasma-membrane localized in the absence of a ligand, but the addition of the ligand, flg22, rapidly induces internalization of FLS2 into endosomes (Fig. 1.2b). Therefore unlike BRI1, which exhibits constitutive endocycles, FLS2 is subjected to a ligand-induced endocytosis (Robatzek *et al.*, 2006). The internalized FLS2 is most likely targeted for destruction to quench the innate immune response (Robatzek *et al.*, 2006).

Two PAMP receptors, FLS2 and EFR, recognize unrelated ligands, flg22 and elf18, respectively, via direct binding (Chinchilla *et al.*, 2006; Zipfel *et al.*, 2006). Nevertheless, both FLS2 and EFR trigger a common set of target gene expression potentially via a downstream map kinase cascade (Zipfel *et al.*, 2006).

Figure 1.2 (a) Model of BRI1 activation. (Left) In the absence of BR, both BRI1 and BAK1 form homodimers. The BRI1 homodimer is inactive due to autoinhibition of its C-terminus as well as association with the negative regulator, BKI1. Both BRI1 and BAK1 are constitutively recycled into endomembranes. (Right) The BR binding to BRI1 leads to its activation (and dissociation of BKI1) via phosphorylation, and formation of receptor heterodimer (heterotetramer) with BAK1. The receptor heterodimers are internalized into endosomes, and they signals to inhibit BIN2 kinase via unknown mechanism. In the absence of BR signaling, BIN2 inhibits BES1/BZR1 activity via phosphorylation. Dephosphorylation of BES1/BZR1 transcription factors (mediated via BSU1 phosphatase) activates the target gene expression (b). Model of FLS2 activation. (Left) In the absence of PAMP (flg22), both FLS2 and BAK1 likely form homodimers. (Right) The flg22 binding to FLS2 leads to its activation via phosphorylation and formation of receptor heterodimer (heterotetramer) with BAK1. The active receptor complex is internalized rapidly via ligand-dependent manner. The signal may be transduced via mitogen-activated protein kinase (MAP kinase) cascades and eventually trigger innate immune response likely via WRKY-family transcription factors.

How can two distinct ligand–receptor pairs induce common downstream targets? Through a genome-wide transcriptome analysis of PAMP-induced genes and biochemical purification, Chinchilla *et al.* (2007) and Hesse *et al.* (2007) made an intriguing discovery that BAK1, a known heterodimeric partner of BRI1, in addition acts as a heterodimeric partner of FLS2 for defense signaling (Chinchilla *et al.*, 2007; Heese *et al.*, 2007) (Fig. 1.2b). Similar to *fls2*, *bak1* mutant seedlings are insensitive to flg22, indicating that BAK1 is required for flagellin perception in *Arabidopsis*. Application of flg22 rapidly induced FLS2–BAK1 receptor complex formation. These results point to the mechanism by which the recognition of a specific PAMP by FLS2 leads to association with BAK1 and active receptor complex formation, which in turn signals to downstream components. Similar to BRI1, FLS2 may exist as an inactive homodimer at the plasma membrane in the absence of its ligand. Consistently, the absence of *BAK1* has no effects on flg22 binding to FLS2 (Chinchilla *et al.*, 2007).

BAK1's role as a transducer of PAMP response goes beyond FLS2, given that BAK1 is required in restricting other bacterial and oomycetes infections (Heese *et al.*, 2007). Furthermore, BAK/SERK-family LRR-RLKs control multiple developmental processes, such as somatic embryogenesis and microsporocyte formation. Thus, BAK/SERK-family RLKs may form an active receptor heterodimers with numbers of different LRR-RLKs.

1.10 Early events in receptor kinase signaling: emerging link to small GTP-binding proteins

How do activated RKs transduce signals? Recent studies unraveled an exciting potential link between RKs and small GTP-binding proteins known as ROP (Rho of plants) or RACs (reviewed in Shichrur and Yalowsky, 2006; Yang and Fu, 2007). ROPs/RACs function as a switch during cellular morphogenesis, such as pollen tube tip growth, root hair elongation, and pavement cell intercalation, as well as for hormonal and environmental responses (Gu *et al.*, 2004; Yang and Fu, 2007; Chapter 3). ROPs were first implicated in RLK signaling through the biochemical analysis of the CLV1 receptor complex, which contained potential ROP proteins (Trotochaud *et al.*, 1999). However, the role of ROPs in CLV1 signaling remained elusive.

Similar to the small GTP-binding proteins in animals and yeast, the plant ROP/RACs shuttle between a GDP-bound inactive form and a GTP-bound active form. The activation of small GTP-binding proteins requires guanine-nucleotide exchange factors (GEFs). Recently, through a yeast two-hybrid approach using the dominant-negative form of ROP4, plant-specific RhoGEFs that possess specific activity toward ROPs have been identified (Berken *et al.*, 2005; Gu *et al.*, 2006). Interestingly, the plant-specific RhoGEFs do not share any primary sequence similarities to their animal/yeast counterparts, and the GDP dissociation from ROPs is mediated via a previously undescribed domain, now named PRONE (plant-specific Rop nucleotide exchanger) (Berken *et al.*, 2005).

The real surprise came from the discovery that a tomato homolog of the plant-specific RhoGEF interacts with two LRR-RLKs, LePRK1 and LePRK2, that are known to regulate pollen tube elongation (see Section 1.6.1; Wengier, *et al.*, 2003). The protein, named kinase partner protein (KPP), was isolated via yeast two-hybrid screen using the kinase domain of LePRK1 and LePRKs as bait. KPP was shown to be phosphorylated in pollen tubes and localized at the membrane periphery (i.e., attached to the plasma membrane from the cytoplasmic side) (Kaothien, *et al.*, 2006). The overexpression of KPP conferred severe swelling of pollen tube tips, a phenotype highly resembling that induced by RopGEF1 overexpression and the constitutive activation of a pollen-specific ROP, ROP1, in *Arabidopsis* (Li *et al.*, 1999; Kaothien, *et al.*, 2006; Gu *et al.*, 2006). These studies provide an intriguing link between RLKs and small GTPase signaling in nongenomic responses, such as polar cell elongation. The *Arabidopsis* genome encodes 14 members of RhoGEFs and 11 ROPs. It would be a daunting but exciting task to understand the specific interactions among RhoGEFs and ROPs, and further identify the upstream RLKs that initiate the signal.

1.11 Future perspectives

It has been 17 years since the first RLK was reported by Walker and Zhang (1990), and it has been over a decade since the first clear demonstrations of RLK function in plant development and defense against pathogens (Song *et al.*, 1995; Torii *et al.*, 1996). Similarly, the first plant peptide hormone, systemin, was discovered by Clarence Ryan's group in 1991. Recently, the ligand–receptor pairs have been identified and the molecular and biochemical mechanisms of signal perception and receptor activation are beginning to emerge. However, important questions still remain to be addressed: What is the structural basis of ligand–receptor interactions? How does the same transducer, BAK1, trigger developmental and defense response depending on its receptor pair? Which RLKs directly bind ligands and which RLKs act as transducers? Is there a conserved mechanism of RLK signaling? Fueled with a synergism of traditional molecular genetics and biochemistry as well as development of bioinformatics and structural modeling, studies of plant receptor-kinase signaling will continue to be hot and exciting.

Note added in proof

After the submission of the book chapter, two interesting reports came out. First, Ogawa *et al.* (2008) demonstrated that the MCLV3 peptide directly associates with the ectodomain of CLV1, thus concluding the long-standing debate on whether CLV1 is indeed a receptor for CLV3. It would be interesting to address whether CLV2 also binds MCLV3. In such a case, both CLV1 and CLV2 may independently act as receptors for CLV3. Second, Jia *et al.* (2008) reported that TPD1 and EMS1/EXR1 associate in vivo and in vitro, and TPD1 induces phosphorylation of EMS/EXR1 in vivo. Collectively, findings by Jia *et al.* (2008) strongly suggest that TPD1 is a ligand for EMS1/EXR1.

Acknowledgments

I thank Shannon Bemis and Lynn Pillitteri for commenting on this chapter, and Shinichiro Sawa for Fig. 1.1 (Color plate 1). Research in my laboratory is currently supported by NSF (IOB-0520548) and DOE (DE-FG02-03ER15448).

References

Albrecht, C., Russinova, E., Hecht, V., Baaijens, E. and De Vries, S. (2005) The *Arabidopsis thaliana* SOMATIC EMBRYOGENESIS RECEPTOR-LIKE KINASES1 and 2 control male sporogenesis. *Plant Cell*, **17**, 3337–3349.

Becraft, P.W., Stinard, P.S. and Mccarthy, D.R. (1996) *CRINKLY4*, a TNFR-like receptor kinase involved in maize epidermal differentiation. *Science*, **273**, 1406–1409.

Belkhadir, Y. and Chory, J. (2006) Brassinosteroid signaling: a paradigm for steroid hormone signaling from the cell surface. *Science*, **314**, 1410–1411.

Berger, D. and Altmann, T. (2000) A subtilisin-like serine protease involved in the regulation of stomatal density and distribution in *Arabidopsis thaliana*. *Genes Dev*, **14**, 1119–1131.

Bergmann, D.C. and Sack, F.D. (2007) Stomatal development. *Annu Rev Plant Biol*, **58**, 163–181.

Berken, A., Thomas, C. and Wittinghofer, A. (2005) A new family of RhoGEFs activates the Rop molecular switch in plants. *Nature*, **436**, 1176–1180.

Bishop, G.J. and Koncz, C. (2002) Brassinosteroids and plant steroid hormone signaling. *Plant Cell*, **14**, S97–S110.

Brand, U., Fletcher, J.C., Hobe, M., Meyerowitz, E.M. and Simon, R. (2000) Dependence of stem cell fate in *Arabidopsis* on a feedback loop regulated by CLV3 activity. *Science*, **289**, 617–619.

Briggs, G.C., Osmont, K.S., Shindo, C., Sibout, R. and Hardtke, C.S. (2006) Unequal genetic redundancies in *Arabidopsis*—a neglected phenomenon? *Trends Plant Sci*, **11**, 492–498.

Butenko, M.A., Patterson, S.E., Grini, P.E., Stenvik, G.E., Amundsen, S.S., Mandal, A. and Aalen, R.B. (2003) Inflorescence deficient in abscission controls floral organ abscission in *Arabidopsis* and identifies a novel family of putative ligands in plants. *Plant Cell*, **15**, 2296–2307.

Canales, C., Bhatt, A.M., Scott, R. and Dickinson, H. (2002) EXS, a putative LRR receptor kinase, regulates male germline cell number and tapetal identity and promotes seed development in *Arabidopsis*. *Curr Biol*, **12**, 1718–1727.

Cano-Delgado, A., Yin, Y., Yu, C., Vafeados, D., Mora-Garcia, S., Cheng, J.C., Nam, K.H., Li, J. and Chory, J. (2004) BRL1 and BRL3 are novel brassinosteroid receptors that function in vascular differentiation in *Arabidopsis*. *Development*, **131**, 5341–5351.

Casamitjana-Martinez, E., Hofhuis, H.F., Xu, J., Liu, C.M., Heidstra, R. and Scheres, B. (2003) Root-specific CLE19 overexpression and the sol1/2 suppressors implicate a CLV-like pathway in the control of *Arabidopsis* root meristem maintenance. *Curr Biol*, **13**, 1435–1441.

Chevalier, D., Batoux, M., Fulton, L., Pfister, K., Yadav, R.K., Schellenberg, M. and Schneitz, K. (2005) STRUBBELIG defines a receptor kinase-mediated signaling pathway regulating organ development in *Arabidopsis*. *Proc Natl Acad Sci USA*, **102**, 9074–9079.

Chinchilla, D., Bauer, Z., Regenass, M., Boller, T. and Felix, G. (2006) The *Arabidopsis* receptor kinase FLS2 binds flg22 and determines the specificity of flagellin perception. *Plant Cell*, **18**, 465–476.

Chinchilla, D., Zipfel, C., Robatzek, S., Kemmerling, B., Nurnberger, T., Jones, J.D., Felix, G. and Boller, T. (2007) A flagellin-induced complex of the receptor FLS2 and BAK1 initiates plant defence. *Nature*, **448**, 497–500.

Clark, S.E., Running, M.P. and Meyerowitz, E.M. (1993) *CLAVATA1*, a regulator of meristem and flower development in *Arabidopsis*. *Development*, **119**, 397–418.

Clark, S.E., Running, M.P. and Meyerowitz, E.M. (1995) *CLAVATA3* is a specific regulator of shoot and floral meristem development affecting the same processes as *CLAVATA1*. *Development*, **121**, 2057–2067.

Clark, S.E. and Schiefelbein, J.W. (1997) Expanding insights into the role of cell proliferation in plant development. *Trends Cell Biol*, **7**, 454–458.

Clark, S.E., Williams, R.W. and Meyerowitz, E.M. (1997) The *CLAVATA1* gene encodes a putative receptor kinase that controls shoot and floral meristem size in *Arabidopsis*. *Cell*, **89**, 575–585.

Clay, N.K. and Nelson, T. (2002) VH1, a provascular cell-specific receptor kinase that influences leaf cell patterns in *Arabidopsis*. *Plant Cell*, **14**, 2707–2722.

Cock, J.M. and Mccormick, S. (2001) A large family of genes that share homology with *CLAVATA3*. *Plant Physiol*, **126**, 939–942.

Colcombet, J., Boisson-Dernier, A., Ros-Palau, R., Vera, C.E. and Schroeder, J.I. (2005) *Arabidopsis* SOMATIC EMBRYOGENESIS RECEPTOR KINASES1 and 2 are essential for tapetum development and microspore maturation. *Plant Cell*, **17**, 3350–3361.

Deyoung, B.J., Bickle, K.L., Schrage, K.J., Muskett, P., Patel, K. and Clark, S.E. (2006) The CLAVATA1-related BAM1, BAM2 and BAM3 receptor kinase-like proteins are required for meristem function in *Arabidopsis*. *Plant J*, **45**, 1–16.

Dixon, M.S., Hatzixanthis, K., Jones, D.A., Harrison, K. and Jones, J.D. (1998) The tomato Cf-5 disease resistance gene and six homologs show pronounced allelic variation in leucine-rich repeat copy number. *Plant Cell*, **10**, 1915–1925.

Felix, G., Duran, J.D., Volko, S. and Boller, T. (1999) Plants recognize bacteria through the most conserved domain of flagellin. *Plant J*, **18**, 265–276.

Fiers, M., Golemiec, E., Van Der Schors, R., Van Der Geest, L., Li, K.W., Stiekema, W.J. and Liu, C.M. (2006) The CLAVATA3/ESR motif of CLAVATA3 is functionally independent from the nonconserved flanking sequences. *Plant Physiol*, **141**, 1284–1292.

Fiers, M., Golemiec, E., Xu, J., Van Der Geest, L., Heidstra, R., Stiekema, W. and Liu, C.M. (2005) The 14-amino acid CLV3, CLE19, and CLE40 peptides trigger consumption of the root meristem in *Arabidopsis* through a CLAVATA2-dependent pathway. *Plant Cell*, **17**, 2542–2553.

Fletcher, J.C., Brand, U., Running, M.P., Simon, R. and Meyerowitz, E.M. (1999) Signaling of cell fate decisions by *CLAVATA3* in *Arabidopsis* shoot meristems. *Science*, **283**, 1911–1914.

Geldner, N., Hyman, D.L., Wang, X., Schumacher, K. and Chory, J. (2007) Endosomal signaling of plant steroid receptor kinase BRI1. *Genes Dev*, **21**, 1598–1602.

Gomez-Gomez, L. and Boller, T. (2000) FLS2: an LRR receptor-like kinase involved in the perception of the bacterial elicitor flagellin in *Arabidopsis*. *Mol Cell*, **5**, 1155–1163.

Gomez-Gomez, L. and Boller, T. (2002) Flagellin perceptiron: a paradigm for innate immunity. *Trends Plant Sci*, **7**, 251–256.

Gu, Y., Li, S., Lord, E.M. and Yang, Z.B. (2006) Members of a novel class of *Arabidopsis* Rho guanine nucleotide exchange factors control Rho GTPase-dependent polar growth. *Plant Cell*, **18**, 366–381.

Gu, Y., Wang, Z. and Yang, Z.B. (2004) ROP/RAC GTPases: an old new master regulator of plant signaling. *Curr Opin Plant Biol*, **7**, 527–536.

Hara, K., Kajita, R., Torii, K.U., Bergmann, D.C. and Kakimoto, T. (2007) Secretory peptide gene EPF1 enforces the stomatal one-cell spacing rule. *Genes Dev*, **15**, 1720–1725.

Harvé, C., Dabos, P., Galaud, J.-P., Rougé, P. and Lescure, B. (1996) Characterization of an *Arabidopsis thaliana* gene that defines a new class of putative plant receptor kinases with an extracellular lectin-like domain. *J Mol Biol*, **258**, 778–788.

He, Z.-H., Fujiki, M. and Kohorn, B.D. (1996) A cell wall-associated, receptor-like protein kinase. *J Biol Chem*, **271**, 19789–19793.

Heese, A., Hann, D.R., Gimenez-Ibanez, S., Jones, A.M., He, K., Li, J., Schroeder, J.I., Peck, S.C. and Rathjen, J.P. (2007) The receptor-like kinase SERK3/BAK1 is a central regulator of innate immunity in plants. *Proc Natl Acad Sci USA*, **104**, 12217–12222.

Holton, N., Cano-Delgado, A., Harrison, K., Montoya, T., Chory, J. and Bishop, G.J. (2007) Tomato BRASSINOSTEROID INSENSITIVE1 is required for systemin-induced root elongation in *Solanum pimpinellifolium* but is not essential for wound signaling. *Plant Cell*, **19**, 1709–1717.

Huffaker, A., Pearce, G. and Ryan, C.A. (2006) An endogenous peptide signal in *Arabidopsis* activates components of the innate immune response. *Proc Natl Acad Sci USA*, **103**, 10098–10103.

Ingram, G.C. (2005) Plant development: spacing out stomatal pores. *Curr Biol*, **15**, R663–R665.

Ito, Y., Nakanomyo, I., Motose, H., Iwamoto, K., Sawa, S., Dohmae, N. and Fukuda, H. (2006) Dodeca-CLE peptides as suppressors of plant stem cell differentiation. *Science*, **313**, 842–845.

Jang, J.C., Fujioka, S., Tasaka, M., Seto, H., Takatsuto, S., Ishii, A., Aida, M., Yoshida, S. and Sheen, J. (2000) A critical role of sterols in embryonic patterning and meristem programming revealed by the fackel mutants of *Arabidopsis thaliana*. *Genes Dev*, **14**, 1485–1497.

Jeong, S., Trotochaud, A.E. and Clark, S.E. (1999) The *Arabidopsis CLAVATA2* gene encodes a receptor-like protein required for the stability of the CLAVATA1 receptor-like kinase. *Plant Cell*, **11**, 1925–1933.

Jai, G., Liu, X., Owen, H.A. and Zhao, D. (2008) Signaling of cell fate determination by the TPD1 small protein and EMS1 receptor kinase. *Proc Natl Acad Sci USA*, **105**, 2220–2225.

Jinn, T.L., Stone, J.M. and Walker, J.C. (2000) HAESA, an *Arabidopsis* leucine-rich repeat receptor kinase, controls floral organ abscission. *Genes Dev*, **11**, 108–117.

Jones, A.M. and Assmann, S.M. (2004) Plants: the latest model system for G-protein research. *EMBO Rep*, **5**, 572–578.

Jones, D.A. and Jones, J.D.G. (1997) The role of leucine-rich repeats in plant defences. *Advances in Botanical Research Incorporating Advances in Plant Pathology*, Vol. 24, pp. 90–167. Academic Press, London, UK.

Jones, D.A., Thomas, C.M., Hammond-Kosack, K.E., Balintkurti, P.J. and Jones, J.D.G. (1994) Isolation of the tomato *Cf-9* gene for resistance to *Cladosporium fulvum* by transposon tagging. *Science*, **266**, 789–793.

Joosten, M.H.A.J., Vogelsang, R., Cozijnsen, T.J., Verberne, M.C. and Dewit, P.J.G.M. (1997) The biotrophic fungus *Cladosporium fulvum* circumvents Cf-4-mediated resistance by producing unstable AVR4 elicitors. *Plant Cell*, **9**, 367–379.

Kachroo, A., Schopfer, C.R., Nasrallah, M.E. and Nasrallah, J.B. (2001) Allele-specific receptor-ligand interactions in *Brassica* self-incompatibility. *Science*, **293**, 1824–1826.

Kaothien, P., Ok, S.H., Shuai, B., Wengier, D., Cotter, R., Kelley, D., Kiriakopolos, S., Muschietti, J. and McCormick, S. (2005) Kinase partner protein interacts with the LePRK1 and LePRK2 receptor kinases and plays a role in polarized pollen tube growth. *Plant J*, **42**, 492–503.

Karlova, R. and De Vries, S.C. (2006) Advances in understanding brassinosteroid signaling. *Sci STKE*, **2006**, pe36.

Kayes, J.M. and Clark, S.E. (1998) *CLAVATA2*, a regulator of meristem and organ development in *Arabidopsis*. *Development*, **125**, 3843–3851.

Kinoshita, T., Cano-Delgado, A., Seto, H., Hiranuma, S., Fujioka, S., Yoshida, S. and Chory, J. (2005) Binding of brassinosteroids to the extracellular domain of plant receptor kinase BRI1. *Nature*, **433**, 167–171.

Kobe, B. and Deisenhofer, J. (1994) The leucine-rich repeat—a versatile binding motif. *Trends Biochem Sci*, **19**, 415–421.

Kondo, T., Sawa, S., Kinoshita, A., Mizuno, S., Kakimoto, T., Fukuda, H. and Sakagami, Y. (2006) A plant peptide encoded by CLV3 identified by in situ MALDI-TOF MS analysis. *Science*, **313**, 845–848.

Kunze, G., Zipfel, C., Robatzek, S., Niehaus, K., Boller, T. and Felix, G. (2004) The N terminus of bacterial elongation factor Tu elicits innate immunity in *Arabidopsis* plants. *Plant Cell*, **16**, 3496–3507.

Kwak, S.H., Shen, R. and Schiefelbein, J. (2005) Positional signaling mediated by a receptor-like kinase in *Arabidopsis*. *Science*, **307**, 1111–1113.

Lease, K.A. and Walker, J.C. (2006) The *Arabidopsis* unannotated secreted peptide database, a resource for plant peptidomics. *Plant Physiol*, **142**, 831–838.

Li, J. and Chory, J. (1997) A putative leucine-rich repeat receptor kinase involved in brassinosteroid signal transduction. *Cell*, **90**, 929–938.

Li, J., Wen, J., Lease, K.A., Doke, J.T., Tax, F.E. and Walker, J.C. (2002) BAK1, and *Arabidopsis* LRR receptor-like protein kinase, interacts with BRI1 and modulates brassinosteroid signaling. *Cell*, **110**, 213–222.

Liu, X., Yue, Y., Li, B., Nie, Y., Li, W., Wu, W.H. and Ma, L. (2007) A G protein-coupled receptor is a plasma membrane receptor for the plant hormone abscisic acid. *Science*, **315**, 1712–1716.

Massagué, J. (1996) TGFβ signaling: receptors, transducers and Mad proteins. *Cell*, **85**, 947–950.

Matsubayashi, Y., Ogawa, M., Kihara, H., Niwa, M. and Sakagami, Y. (2006) Disruption and overexpression of *Arabidopsis* phytosulfokine receptor gene affects cellular longevity and potential for growth. *Plant Physiol*, **142**, 45–53.

Matsubayashi, Y., Ogawa, M., Morita, A. and Sakagami, Y. (2002) An LRR receptor kinase involved in perception of a peptide plant hormone, phytosulfokine. *Science*, **296**, 1470–1472.

Matsubayashi, Y., Yang, H. and Sakagami, Y. (2001) Peptide signals and their receptors in higher plants. *Trends Plant Sci*, **6**, 573–577.

Mcgurl, B., Pearce, G., Orozco-Cardenas, M. and Ryan, C.A. (1992) Structure, expression, and antisense inhibition of the systemin precursor gene. *Science*, **255**, 1570–1573.

Mishima, M., Takayama, S., Sasaki, K., Jee, J.G., Kojima, C., Isogai, A. and Shirakawa, M. (2003) Structure of the male determinant factor for *Brassica* self-incompatibility. *J Biol Chem*, **278**, 36389–36395.

Muschietti, J., Dircks, L., Vancanneyt, G. and Mccormick, S. (1994) LAT52 protein is essential for tomato pollen development: pollen expressing antisense *LAT52* RNA hydrates and germinates abnormally and cannot achieve fertilization. *Plant J*, **6**, 321–338.

Nadeau, J.A. and Sack, F.D. (2002) Control of stomatal distribution on the *Arabidopsis* leaf surface. *Science*, **296**, 1697–1700.

Nam, K.H. and Li, J. (2002) BRI1/BAK1, a receptor kinase pair mediating brassinosteroid signaling. *Cell*, **110**, 203–212.

Nasrallah, J.B. (2005) Recognition and rejection of self in plant self-incompatibility: comparisons to animal histocompatibility. *Trends Immunol*, **26**, 412–418.

Nasrallah, J.B., Yu, S.-M. and Nasrallah, M.E. (1988) Self-incompatibility genes of *Brassica oleracea*: expression, isolation, and structure. *Proc Natl Acad Sci USA*, **85**, 5551–5555.

Ni, J. and Clark, S.E. (2006) Evidence for functional conservation, sufficiency, and proteolytic processing of the CLAVATA3 CLE domain. *Plant Physiol*, **140**, 726–733.

Nodine, M.D., Yadegari, R. and Tax, F.E. (2007) RPK1 and TOAD2 are two receptor-like kinases redundantly required for *Arabidopsis* embryonic pattern formation. *Dev Cell*, **12**, 943–956.

Nurnberger, T., Brunner, F., Kemmerling, B. and Piater, L. (2004) Innate immunity in plants and animals: striking similarities and obvious differences. *Immunol Rev*, **198**, 249–266.

Ogawa, M., Shinohara, H., Sakagami, Y. and Matsubayashi, Y. (2008) *Arabidopsis* CLV3 peptide directly binds CLV1 ectodomain. *Science*, **319**, 5861.

Opsahl-Ferstad, H.-G., Le Deunff, E., Dumas, C. and Rogowsky, P.M. (1997) *ZmEsr*, a novel endosperm-specific gene expressed in a restricted region around the maize embryo. *Plant J*, **12**, 235–246.

Pastuglia, M., Roby, D., Dumas, C. and Cock, J.M. (1997) Rapid induction by wounding and bacterial infection of an *S* gene family receptor-like kinase gene in *Brassica oleracea*. *Plant Cell*, **9**, 49–60.

Pearce, G., Moura, D.S., Stratmann, J. and Ryan, C.A., Jr. (2001) RALF, a 5-kDa ubiquitous polypeptide in plants, arrests root growth and development. *Proc Natl Acad Sci USA*, **98**, 12843–12847.

Pillitteri, L.J., Bemis, S.M., Shpak, E.D. and Torii, K. U. (2007) Haploinsufficiency after successive loss of signaling reveals a role for *ERECTA*-family genes in ovule development. *Development*, **134**, 3099–3109.

Raikhel, N. and Hicks, G. (2007) Signaling from plant endosomes: compartments with something to say. *Genes Dev*, **21**, 1578–1580.

Robatzek, S., Chinchilla, D. and Boller, T. (2006) Ligand-induced endocytosis of the pattern recognition receptor FLS2 in *Arabidopsis*. *Genes Dev*, **20**, 537–542.

Russinova, E., Borst, J.W., Kwaaitaal, M., Cano-Delgado, A., Yin, Y., Chory, J. and De Vries, S.C. (2004) Heterodimerization and endocytosis of *Arabidopsis* brassinosteroid receptors BRI1 and AtSERK3 (BAK1). *Plant Cell*, **16**, 3216–3229.

Ryan, C.A., Huffaker, A. and Yamaguchi, Y. (2007) New insights into innate immunity in *Arabidopsis*. *Cell Microbiol*, **9**, 1902–1908.

Scheer, J.M. and Ryan, C.A. (2002) The systemin receptor SR160 from *Lycopersicon peruvianum* is a member of the LRR receptor kinase family. *Proc Natl Acad Sci USA*, **99**, 9585–9590.

Schmidt, E.D.L., Guzzo, F., Toonen, M.A.J. and De Vries, S.C. (1997) A leucine-rich repeat containing receptor-like kinase marks somatic plant cells competent to form embryos. *Development*, **124**, 2049–2062.

Schoof, H., Lenhard, M., Haecker, A., Mayer, K.F.X., Jürgens, G. and Laux, T. (2000) The stem cell population of *Arabidopsis* shoot meristem is maintained by a regulatory loop between the *CLAVATA* and *WUSCHEL* genes. *Cell*, **100**, 635–644.

Schopfer, C., Nasrallah, M.E. and Nasrallah, J.B. (1999) The male determinant of self-incompatibility in *Brassica*. *Science*, **286**, 1697–1700.

Sharma, V.K., Ramirez, J. and Fletcher, J.C. (2003) The *Arabidopsis CLV3-like* (*CLE*) genes are expressed in diverse tissues and encode secreted proteins. *Plant Mol Biol*, **51**, 415–425.

Shichrur, K., and Yalowsky, S. (2006) Turning ON the switch-RhoGEFs in plants. *Trends Plant Sci*, **11**, 57–59.

Shimosato, H., Yokota, N., Shiba, H., Iwano, M., Entani, T., Che, F.S., Watanabe, M., Isogai, A. and Takayama, S. (2007) Characterization of the SP11/SCR high-affinity binding site involved in self/nonself recognition in *Brassica* self-incompatibility. *Plant Cell*, **19**, 107–117.

Shinohara, H., Ogawa, M., Sakagami, Y. and Matsubayashi, Y. (2007) Identification of ligand binding site of phytosulfokine receptor by on-column photoaffinity labeling. *J Biol Chem*, **282**, 124–131.

Shiu, S.H. and Bleecker, A.B. (2001a) Plant receptor-like kinase gene family: diversity, function, and signaling. *Sci STKE*, **2001**, RE22.

Shiu, S.H. and Bleecker, A.B. (2001b) Receptor-like kinases from *Arabidopsis* form a monophyletic gene family related to animal receptor kinases. *Proc Natl Acad Sci USA*, **98**, 10763–10768.

Shiu, S.H. and Bleecker, A.B. (2003) Expansion of the receptor-like kinase/Pelle gene family and receptor-like proteins in *Arabidopsis*. *Plant Physiol*, **132**, 530–543.

Shpak, E.D., Berthiaume, C.T., Hill, E.J. and Torii, K.U. (2004) Synergistic interaction of three ERECTA-family receptor-like kinases controls *Arabidopsis* organ growth and flower development by promoting cell proliferation. *Development*, **131**, 1491–1501.

Shpak, E.D., Mcabee, J.M., Pillitteri, L.J. and Torii, K.U. (2005) Stomatal patterning and differentiation by synergistic interactions of receptor kinases. *Science*, **309**, 290–293.

Silverstein, K.A., Moskal, W.A., Jr., Wu, H.C., Underwood, B.A., Graham, M.A., Town, C.D. and Vandenbosch, K.A. (2007) Small cysteine-rich peptides resembling antimicrobial peptides have been under-predicted in plants. *Plant J*, **51**, 262–280.

Song, W.-Y., Wang, G.-L., Chen, L.-L., Kim, H.-S., Pi, L.-Y., Holsten, T., Gardner, J., Wang, B., Zhai, W.-X., Zhu, L.-H., Fauquet, C. and Ronald, P. (1995) A receptor kinase-like protein encoded by the rice disease resistance gene, *Xa21*. *Science*, **270**, 1804–1806.

Stenvik, G.E., Butenko, M.A., Urbanowicz, B.R., Rose, J.K. and Aalen, R. B. (2006) Overexpression of INFLORESCENCE DEFICIENT IN ABSCISSION activates cell separation in vestigial abscission zones in *Arabidopsis*. *Plant Cell*, **18**, 1467–1476.

Takayama, S. and Isogai, A. (2005) Self-incompatibility in plants. *Annu Rev Plant Biol*, **56**, 467–489.

Takayama, S., Shiba, H., Iwano, M., Shimosato, H., Che, F.-S., Kai, N., Watanabe, M., Suzuki, G., Hinata, K. and Isogai, A. (2000) The pollen determinant of self-incompatibility in *Brassica campesteris*. *Proc Natl Acad Sci USA*, **97**, 1920–1925.

Tang, W., Ezcurra, I., Muschietti, J. and Maccormick, S. (2002) A cysteine-rich extracellular protein, LAT52, interacts with the extracellular domain of the pollen receptor kinase lePRK2. *Plant Cell*, **14**, 2277–2287.

Tang, W., Kelley, D., Ezcurra, I., Cotter, R. and Mccormick, S. (2004) LeSTIG1, an extracellular binding partner for the pollen receptor kinases LePRK1 and LePRK2, promotes pollen tube growth in vitro. *Plant J*, **39**, 343–353.

Thomas, C.M., Jones, D.A., Parniske, M., Harrison, K., Balint-Kurti, P.J., Hatzixanthis, K. and Jones, J.D.G. (1997) Characterization of the tomato *Cf-4* gene for resistance to *Cladosporium fulvum* identifies sequences that determine recognitional specificity in Cf-4 and Cf-9. *Plant Cell*, **9**, 2209–2224.

Tobias, C.M., Howlett, B. and Nasrallah, J.B. (1992) An *Arabidopsis thaliana* gene with sequence similarity to the *S*-locus receptor kinase of *Brassica oleracea*—sequence and expression. *Plant Physiol*, **99**, 284–290.

Torii, K.U. (2004) Leucine-rich repeat receptor kinases: structure, function, and signal transduction pathways. *Int Rev Cytol*, **234**, 1–46.

Torii, K.U., Mitsukawa, N., Oosumi, T., Matsuura, Y., Yokoyama, R., Whittier, R.F. and Komeda, Y. (1996) The *Arabidopsis ERECTA* gene encodes a putative receptor protein kinase with extracellular leucine-rich repeats. *Plant Cell*, **8**, 735–746.

Trotochaud, A.E., Hao, T., Wu,G., Yang, Z.B and Clark, S.E.(1999) The CLAVATA1 receptor-like kinase requires CLAVATA3 for its assembly into a signaling complex that includes KAPP and a Rho-related protein. *Plant Cell*, **11**, 393–406.

Tsukazaki, T., Chiang, T.A., Davison, A.F., Attisano, L. and Wrana, J.L. (1998) SARA, a FYVE domain protein that recruits Smad2 to the TGFβ receptor. *Cell*, **95**, 779–791.

Urao, T., Yamaguchi-Shinozaki, K. and Shinozaki, K. (2000) Two-component systems in plant signal transduction. *Trends Plant Sci*, **5**, 67–74.

Van Den Ackerveken, G.F.J.M., Van Kan, J.A.L. and De Wit, P.J.G.M. (1992) Molecular analysis of the avirulence gene *Avr9* of the fungal pathogen *Cladosporium fulvum* fully supports the gene-for-gene hypothesis. *Plant J*, **2**, 359–366.

Vervoort, J., Vandenhooven, H.W., Berg, A., Vossen, P., Vogelsang, R., Joosten, M.H.A. and Dewit, P.J.G.M. (1997) The race-specific elicitor AVR9 of the tomato pathogen *Cladosporium fulvum*: a cysteine knot protein sequence-specific H-1 NMR assignments, secondary structure and global fold of the protein. *FEBS Lett*, **404**, 153–158.

Walker, J.C. (1993) Receptor-like protein kinase genes of *Arabidopsis thaliana*. *Plant J*, **3**, 451–456.

Walker, J.C. (1994) Structure and function of the receptor-like kinases of higher plants. *Plant Mol Biol*, **26**, 1599–1609.

Wang, X. and Chory, J. (2006) Brassinosteroids regulate dissociation of BKI1, a negative regulator of BRI1 signaling, from the plasma membrane. *Science*, **313**, 1118–1122.

Wang, X., Li, X., Meisenhelder, J., Hunter, T., Yoshida, S., Asami, T. and Chory, J. (2005a) Autoregulation and homodimerization are involved in the activation of the plant steroid receptor BRI1. *Dev Cell*, **8**, 855–865.

Wang, X., Mithum, M., Gao, B., Li, C., Diab, H., Baum, T., Hussay, R. and Davis, E. (2005b) A parasitism gene from a plant-parasitic nematode with function similar to CLAVATA/ESR (CLE) of *Arabidopsis thaliana*. *Mol Plant Pathol*, **6**, 187–191.

Wang, X., Zafian, P., Choudhary, M. and Lawton, M. (1996) The PR5K receptor protein kinase from *Arabidopsis thaliana* is structurally related to a family of plant defense protein. *Proc Natl Acad Sci USA*, **93**, 2598–2602.

Yamaguchi, Y., Pearce, G. and Ryan, C.A. (2006) The cell surface leucine-rich repeat receptor for AtPep1, an endogenous peptide elicitor in *Arabidopsis*, is functional in transgenic tobacco cells. *Proc Natl Acad Sci USA*, **103**, 10104–10109.

Yamamoto, R., Demura, T. and Fukuda, H. (1997) Brassinosteroids induce entry into the final stage of tracheary element differentiation in cultured zinnia cells. *Plant Cell Physiol*, **38**, 980–983.

Yamamoto, R., Fujioka, S., Demura, T., Takatsuto, S., Yoshida, S. and Fukuda, H. (2001) Brassinosteroid levels increase drastically prior to morphogenesis of tracheary elements. *Plant Physiol*, **125**, 556–563.

Yang, H., Matsubayashi, Y., Nakamura, K. and Sakagami, Y. (1999) *Oryza sativa* PSK gene encodes a precursor of phytosulfokine-alpha, a sulfated peptide growth factor found in plants. *Proc Natl Acad Sci USA*, **96**, 13560–13565.

Yang, H., Matsubayashi, Y., Nakamura, K. and Sakagami, Y. (2001) Diversity of *Arabidopsis* genes encoding precursors for phytosulfokine, a peptide growth factor. *Plant Physiol*, **127**, 842–851.

Yang, S.L., Jiang, L., Puah, C.S., Xie, L.F., Zhang, X.Q., Chen, L.Q., Yang, W.C. and Ye, D. (2005) Overexpression of *TAPETUM DETERMINANT1* alters the cell fates in the *Arabidopsis* carpel and tapetum via genetic interaction with excess microsporocytes1/extra sporogenous cells. *Plant Physiol*, **139**, 186–191.

Yang, S.L., Xie, L.F., Mao, H.Z., Puah, C.S., Yang, W.C., Jiang, L., Sundaresan, V. and Ye, D. (2003) *TAPETUM DETERMINANT1* is required for cell specialization in the *Arabidopsis* anther. *Plant Cell*, **15**, 2792–2804.

Yang, Z and Fu, Y. (in press) ROP/RAC GTPase signaling. *Curr Opin Plant Biol*, **10**, 490–494.

Yokoyama, R., Takahashi, T., Kato, A., Torii, K.U. and Komeda, Y. (1998) The *Arabidopsis ERECTA* gene is expressed in the shoot apical meristem and organ primordia. *Plant J*, **15**, 301–310.

Zhao, D.-Z., Wang, G.-F., Speal, B. and Ma, H. (2002) The *EXCESS MICROSPOROCYTES1* gene encodes a putative leucine-rich repeat receptor protein kinase that controls somatic and reproductive cell fates in the *Arabidopsis* anther. *Genes Dev*, **16**, 2021–2031.

Zipfel, C., Kunze, G., Chinchilla, D., Caniard, A., Jones, J.D., Boller, T. and Felix, G. (2006) Perception of the bacterial PAMP EF-Tu by the receptor EFR restricts *Agrobacterium*-mediated transformation. *Cell*, **125**, 749–760.

Annual Plant Reviews (2008) **33**, 30–63
doi: 10.1111/b.9781405160025.2008.00002.x
www.interscience.wiley.com

Chapter 2

HETEROTRIMERIC G-PROTEIN-COUPLED SIGNALING IN HIGHER PLANTS

Lei Ding[1], Jin-Gui Chen[2], Alan M. Jones[3], and Sarah M. Assmann[1]

[1] *Biology Department, Penn State University, 208 Mueller Laboratory, University Park, PA 16802, USA*
[2] *Department of Botany, University of British Columbia, Vancouver BC V6T 1Z4, Canada*
[3] *Departments of Biology and Pharmacology, University of North Carolina at Chapel Hill, Chapel Hill, NC 27599-3280, USA*

Abstract: Heterotrimeric G proteins are key signaling elements in eukaryotes. The fundamental building blocks of this pathway, the Gα, Gβ, and Gγ subunits, are encoded in plant genomes, as are regulator of G-protein signaling (RGS) proteins, and candidate seven-transmembrane (7TM) G-protein-coupled receptors (GPCRs). However, plants are distinguished from other metazoans by having far fewer genes encoding these functions: for example, the genome of the model plant species *Arabidopsis thaliana* encodes single canonical Gα and Gβ subunits, two Gγ subunits, one RGS protein (which, unlike animal RGS proteins, contains a 7TM domain), and many fewer candidate GPCRs than mammalian genomes. Nevertheless, genetic approaches have demonstrated the importance of heterotrimeric G-protein signaling in a wide diversity of responses that are fundamental to plant growth and survival, including cell division, ion channel regulation, responses to most of the major plant hormones, and aspects of light signaling, oxidative stress, and pathogen response. These studies have also demonstrated that, similar to the situation in other eukaryotes, some responses are primarily mediated by the Gα subunit and others by the Gβ subunit (βγ dimer). The role that a given G-protein component plays in a given signaling process can differ between different plant cell types, as illustrated most thoroughly for regulation of cell division and hormonal response. These results imply that different plant cell types may employ different upstream and downstream proteins to couple with the heterotrimeric subunits. However, to date, only a few proteins have been shown to physically interact with plant G-protein subunits, and this is a fertile area for future research.

Keywords: Gα subunit; Gβ subunit; Gγ subunit; G-protein-coupled receptor (GPCR); heterotrimeric G protein; RGS protein

2.1 Introduction

GTP-binding proteins, also called G proteins, belong to a superfamliy of GTPase domain-containing proteins, which bind GTP and hydrolyze the bound GTP to GDP by their intrinsic GTPase activity; thus G proteins have two states (Bourne *et al.*, 1990, 1991). A subfamily called the heterotrimeric G proteins are composed of α, β, and γ subunits. G proteins function as molecular switches in the regulation of numerous biological processes. In the classical model of heterotrimeric G-protein action, signal transmission involves binding of a molecular signal (ligand) to a G-protein-coupled receptor (GPCR), activation of the heterotrimer by the ligand-bound receptor, and action of the Gα and/or the Gβγ dimer on target proteins, designated as effectors (Hepler and Gilman, 1992; Neer, 1995; Hamm, 1998).

Heterotrimeric G proteins of plants and the signaling coupled by them are the focus of this chapter. However, because the field of plant G-protein signaling has developed within the realm of a large and well-established body of work on metazoan G-protein signaling, an overview is first given of G-protein signaling models developed from information from nonplant systems.

2.2 Heterotrimeric G proteins in nonplant systems

Heterotrimeric G proteins are widely present in eukaryotes, and are intensively studied using model species such as *Saccharomyces cerevisiae, Neurospora crassa, Drosophila melanogaster, Dictyostelium discoideum* and *Caenorhabditis elegans*. Heterotrimeric G-protein-based signaling has recently been assessed using transgenic (knockout) mammals (Offermanns *et al.*, 1997), building upon a long history of their study using mammalian cell lines. Heterotrimeric G proteins also exist in protozoa and ancestral metazoans such as the ciliate *Stentor coeruleus* and the Anthozonan *Renilla koellikeri* (New and Wong, 1998).

In mammalian cells, stimuli transduced by G proteins include hormones, photons, odorants, certain taste ligands, neurotransmitters, phospholipids, and growth factors (Hepler and Gilman, 1992; Gutkind, 1998). Other G-protein-based responses include mating in yeast, visual signal transduction in *Drosophila*, aggregation of *Dictyostelium* in response to cAMP, and chemotaxis and thermotaxis of *C. elegans*. Underlying these physiological and behavioral responses are rapid cellular responses such as alterations in ion fluxes (Brown and Birnbaumer, 1990; Wickman and Clapham, 1995), transient changes in concentrations of secondary messages such as Ca^{2+}, cAMP, and inositol phosphates (Restrepo *et al.*, 1996; Rhee, 2001), and restructuring of cytoskeletal architecture (Luttrell, 2002; Zheng, 2004).

2.2.1 Structure of heterotrimeric G proteins

Among the heterotrimeric G proteins in eukaryotic organisms, those from mammalian systems have been studied most intensively. So far, at least 20 Gα, 5 Gβ, and 12 Gγ subunits have been identified in mammals (Hamm, 1998; McCudden *et al.*, 2005). Gα subunits are divided into four classes based on their sequence similarity: $G\alpha_i$, $G\alpha_s$, $G\alpha_q$, and $G\alpha_{12}$. Corresponding to the diversity of G-protein complexes, 800–1000 GPCRs exist in mammalian cells.

The α subunit of heterotrimeric G proteins contains two key structural domains: one is a GTPase domain (G domain; also known as a Ras domain); the other is a helical domain. The GTPase domain is a common structure shared by Gα subunits and the monomeric members of the GTPase superfamily. The basic Gα core contains a number of common functionalities, namely, nucleotide binding domains, e.g., switches that establish the basal and activated conformations, and protein interfaces such as for the Gβγ dimer, the receptor, the RGS proteins, the modulators, and the effectors. A motif called the P-loop is involved in nucleotide triphosphate binding, a motif called the DxxGQ motif is used for GTP hydrolysis, and the NKxD motif confers guanine recognition (Sprang, 1997). Upon GDP/GTP exchange, Gα subunits undergo a conformational change so as to switch on or off their signaling and GTPase activity. The helical domains in some Gα subunits interact with RGS proteins (Echeverria *et al.*, 2000; Brito *et al.*, 2002) or with effectors (Liu and Northup, 1998), as discussed further below.

All Gβ subunits contain seven WD40 repeat sequence motifs, in which a repeat of tryptophan (W) and aspartate (D) or similar residues are present about every 40 amino acids. The β subunits of heterotrimeric G proteins have a seven-bladed antiparallel β-propeller structure. Among the three G-protein subunits, the γ subunits are least conserved. The γ subunits are small (6–9 kDa) and contain a C-terminal cysteine-aliphatic-aliphatic-undefined amino acid (CAAX) motif required for isoprenylation as well as a N-terminal coiled-coil domain essential for interaction with Gβ subunits. Gγ subunits are always associated with Gβ subunits, forming a tightly bound Gβγ heterodimer that can only be separated under denaturing conditions. Both Gα and Gγ subunits tether to the plasma membrane via covalently bound lipids (Casey, 1994; Milligan and Grassie, 1997).

2.2.2 Signaling by heterotrimeric G proteins

In the classical model of G-protein signaling, the stimulus (ligand) first binds to an associated ligand-specific GPCR, which has a seven-transmembrane (7TM) domain structure and is located on the plasma membrane. Binding of the GPCR by a ligand triggers a conformational change in the receptor, which accordingly prompts a conformational change in the Gα subunit. The Gα subunit of the heterotrimer then releases bound GDP and the vacated site is available for GTP binding. Thus, ligand-bound GPCRs serve as guanine nucleotide exchange factors (GEF), accelerating GDP/GTP exchange. GTP binding on the Gα prompts a conformational change that triggers dissociation of the heterotrimer into Gα and the Gβγ dimer, each of which can interact with downstream effectors. Among the effectors for Gα in yeast and metazoans are second messenger generating enzymes such as adenylate cyclases, cGMP phosphodiesterase, and phospholipase C (Hamm, 1998; McCudden *et al.*, 2005). The Gβγ dimer can directly activate ion channels, such as K^+ and Ca^{2+} channels, and possibly Na^+ channels (Hamm, 1998). The Gβγ dimer can also activate Gβγ-responsive phosphoinositide 3-kinases (PI 3-kinases) (Tang and Downes, 1997). MAP kinases can be activated by the Gβγ dimer in a Ras-dependent pathway (Gutkind, 1998). Eventually, the intrinsic GTPase activity of the Gα subunit allows the bound GTP to be hydrolyzed into GDP and the G protein returns to an inactive state (Hepler and Gilman, 1992; Gutkind, 1998; Hamm, 1998).

2.2.3 Modifiers of G-protein signaling

As shown in Fig. 2.1 (Color plate 2), there are a number of mechanisms by which G-protein signaling can be modulated. For example, removal of GPCRs from the cell surface acts to desensitize signaling. Some GPCRs become phosphorylated when occupied by ligand, and after phosphorylation by G-protein-coupled receptor kinases (GRKs), these GPCRs are bound by arrestins that are cytosolic adaptor and scaffold proteins. This interaction leads to desensitization by promoting GPCR internalization (Gurevich and Gurevich, 2004; Moore *et al.*, 2007). It is now believed that besides their desensitizing role, arrestins recruit nonreceptor tyrosine kinases or MAP kinases to GPCRs in G-protein-independent GPCR signaling. Two arrestins and seven GRKs have been identified in mammalian cells. The intrinsic GTPase activity of Gα is up-regulated by RGS proteins that have GTPase activating protein (GAP) activity (Berman and Gilman, 1998; De Vries *et al.*, 2000b). RGS proteins promote return of active G proteins to their inactive state and often inhibit Gα-dependent signaling. Thirty-nine mammalian RGS proteins have been identified to date. Some of these are delimited to the plasma membrane by phospholipid binding through protein–protein interactions, but no mammalian RGS protein to date has membrane spanning domains.

Phosducin, a phosphoprotein that binds Gβγ dimers in the retina, triggers a conformational change in the Gβ subunit, preventing the Gβγ dimer from binding to its downstream effectors or associating with the Gα subunit (Gaudet *et al.*, 1996; Schulz, 2001). Other important modifiers such as activators of G-protein signaling (AGS proteins) and G-protein-signaling modifiers (HUGO, http://www.gene.ucl.ac.uk/nomenclature) also act on the activated complex or even are able to activate the complex in the absence of GPCRs

(Cismowski *et al.*, 2001; Lanier, 2004). AGS1, a member of the Ras superfamily of small GTPases, has GEF activity on some Gα subunit types (Cismowski *et al.*, 1999). AGS2 is identical to the mouse Tctex1, a component of the light chain of dynein, and specifically interacts with the Gβγ dimer (Takesono *et al.*, 1999). AGS3 is similar to human LGN (Takesono *et al.*, 1999), a protein involved in spindle positioning in mitotic cell division. AGS3 (now designated GPSM1) serves as a guanine dissociation inhibitor (GDI) of $G\alpha_i$ and $G\alpha_t$ (transductin, a member of the $G\alpha_i$ family) (Natochin *et al.*, 2000; De Vries *et al.*, 2000a), and modulates mitotic spindle orientation (Sanada and Tsai, 2005). AGS4 (now designated GPSM3), like AGS3, is a GDI of $G\alpha_i$ (Cao *et al.*, 2004). AGS proteins preferentially target GDP-bound Gα subunits not in the heterotrimeric complex; thus, they act by either prolonging signaling via Gbγ and/or by blocking activation of Gα. Activation of heterotrimeric G proteins by AGS proteins may represent a new mode of heterotrimeric G-protein signaling in response to external signals.

A novel role of GDP-bound Gαs in proper positioning of the mitotic spindle to effect asymmetric cell division has been uncovered in studies in *Drosophila* and *C. elegans* embryos, and there are hints that similar mechanisms function in mammalian cells as well (Hampoelz and Knoblich, 2004; Willard *et al.*, 2004; McCudden *et al.*, 2005). Similar to regulation by AGS proteins, these results also reflect a mechanism that is divergent from the classical model of heterotrimeric G-protein signaling.

2.3 Heterotrimeric G proteins in higher plants

As in other eukaryotes, the central components of the classical heterotrimeric G-protein signaling paradigm, including heterotrimeric G-protein subunits, candidate GPCRs, and an RGS protein have been found in higher plants, although plant GRKs and arrestins have not been identified by sequence or functional homology (Ma, 1994; Assmann, 2002; Jones and Assmann, 2004). Many of the mammalian effectors are also found in plants, but to date only a few have been assessed with regard to their coupling with heterotrimeric G-protein-based pathways.

2.3.1 Components of heterotrimeric G-protein signaling in plants

Although early experiments using ADP ribosylation, GTP-binding assays, and anti-Gα immunoassays implied the existence of heterotrimeric G-protein α subunits in higher plants, their presence in plants was verified when the first G-protein α subunit gene, *GPA1*, was cloned from *Arabidopsis* (Ma *et al.*, 1990). In striking contrast to metazoans, the repertoire of G-protein subunits is simple. Most plant species have one type each of the Gα and Gβ subunits and two Gγ subunits.

The *GPA1* gene encodes a 45-kDa protein that is 36% identical to bovine transducin and rat G_{i1}. The consensus sequences for guanine nucleotide binding in mammalian and yeast Gα subunits are also conserved in the GPA1 protein. *GPA1* is expressed in all tissues and organs at various developmental stages except mature pollen and mature seeds. *GPA1* is highly expressed in root and shoot meristems, lateral root primordia, and vascular tissues (Weiss *et al.*, 1993; Huang *et al.*, 1994; Pandey *et al.*, 2006). This broad expression pattern of *GPA1* may reflect the broad functions of heterotrimeric G proteins in plants (see below) despite the presence of only one prototypical Gα gene in the *Arabidopsis* genome. Subsequently, a tomato homolog of *Arabidopsis GPA1* (*TGA1*) was isolated using *Arabidopsis GPA1* as a probe (Ma *et al.*, 1991). As in *Arabidopsis*, there is only one copy of the Gα gene in tomato. Since these early studies, genes encoding heterotrimeric G-protein α subunits have been cloned from other plants, such as alfalfa, lotus, lupin, pea, rice, soybean, spinach, tobacco, and wild oat (Assmann, 2002). Among these G-protein α subunits, recombinant *Arabidopsis* GPA1 (Wise *et al.*, 1997; Chen *et al.*, 2003), rice G-protein α subunit (RGA1) (Seo *et al.*, 1997), and tomato TGA1 (Aharon *et al.*, 1998) have been shown to bind and hydrolyze GTP.

The first two genes identified to encode plant G-protein β subunits, *ZGB1* and *AGB1*, were isolated in maize by screening a subtracted maize tassel cDNA library and in *Arabidopsis* by screening a cDNA library with a partial *ZGB1* cDNA as a probe (Weiss *et al.*, 1994). The proteins encoded by *ZGB1* and *AGB1* share 76% identity with each other and each has 41% or more identity to human and *C. elegans* Gβ subunits. As for other known Gβ subunits, ZGB1 and AGB1 contain seven WD40 repeats and an α-helical amino terminal domain (Ullah *et al.* 2003), two structural features of G-protein β subunits. *AGB1* is ubiquitously expressed in vegetative organs and flowers (Weiss *et al.*, 1994; Ullah *et al.* 2003; Chen *et al.*, 2006d; Anderson and Botella, 2007). Consistent with its functions, the *AGB1* gene is expressed in vascular tissues, meristems, and guard cells. *AGB1* is also expressed in trichomes and root cap, suggestive of possible AGB1 signaling in these two locations (Anderson and Botella, 2007). Besides localization at the plasma membrane, AGB1 protein is also found in the nucleus (Anderson and Botella, 2007). Genes encoding Gβ subunits also have been isolated in alfalfa, rice, tobacco, and wild oat (Assmann, 2002).

Two Gγ subunit genes, *AGG1* and *AGG2*, were isolated by screening an *Arabidopsis* yeast two-hybrid library with a tobacco Gβ subunit (TGB1) as bait (Mason and Botella, 2000, 2001). In an in vitro binding assay, ^{35}S-labeled AGG1 and AGG2 strongly bound to AGB1. The AGG1 and AGG2 proteins maintain the features of mammalian Gγ subunits, such as small size (98 and 100 amino acids, respectively), a C-terminal CAAX motif required for isoprenyl modification, and an N-terminal coiled-coil domain crucial for interaction with the Gβ subunit. They also have three short regions conserved in mammalian Gγ subunits (Mason and Botella, 2000, 2001). *AGG1* and *AGG2* have mostly distinct (albeit partial overlapping) expression patterns, but these expression patterns in combination match those of *AGB1*. *AGG1* is mainly expressed in the apical meristem, hypocotyl, veins of rosette leaves, and stele of roots; *AGG2* is mainly expressed in root hairs, hydathodes of rosette leaves, guard cells, and cortex and epidermis of roots (Chen *et al.*, 2006d; Trusov *et al.*, 2007). Interestingly, AGB1 and AGG1 are reported to be dependent upon each other for plasma membrane localization, whereas AGG2 does not require AGB1 for plasma membrane targeting (Adjobo-Hermans *et al.*, 2006). Fluorescently tagged AGG2 localizes to the plasma membrane, whereas AGG1 localizes to both the plasma membrane and internal membranes (Zeng *et al.*, 2007). Nonetheless, both AGG1 and AGG2 could form heterotrimers with GPA1 and AGB1 (Adjobo-Hermans *et al.*, 2006).

Two rice Gγ subunits genes, *RGG1* and *RGG2*, were uncovered by homology-based searches of the rice databases (Kato *et al.*, 2004). In both yeast two-hybrid and gel filtration assays, RGG1 and RGG2 associate with RGB1 (Kato *et al.*, 2004), a rice Gβ subunit (Ishikawa *et al.*, 1996). The four G-protein subunits in rice, RGA1, RGB1, RGG1, and RGG2, make up two large G-protein complexes associated with the plasma membrane (Kato *et al.*, 2004). The molecular mass of each G-protein complex is ~400 kDa, indicating the presence of other components in the complex in addition to the three G-protein subunits.

Structural comparisons between plant and animal G-protein subunits have led to a number of interesting conclusions (Temple and Jones, 2007). First, the plant heterotrimeric complex most closely approximates the reconstructed ancestral G-protein complex. Second, tertiary structure involving the central core functions of GTP binding, GTP hydrolysis, and the interfaces between the subunits are conserved between metazoan and divergent plant G proteins. Finally, some surface residues that form interaction faces are shared with metazoan subunits and some are unique to plants suggesting that plants have a unique set of interaction partners (Temple and Jones, 2007).

To date, two *Arabidopsis* 7TM proteins, GCR1 and AtRGS1, have been shown to physically interact with a plant Gα subunit, satisfying one of the fundamental criteria for designating them as GPCRs. GCR1 shares weak similarity to the cyclic AMP receptor, CAR1, found in the slime mold and the Class B Secretin family GPCRs (Josefsson and Rask, 1997; Josefsson, 1999). It was determined by in vivo split-ubiquitin assay, in vitro binding, and co-immunoprecipitation (co-IP) that GCR1 directly interacts with *Arabidopsis* GPA1 (Pandey and Assmann, 2004). GCR1 has been shown to have both GPA1-dependent and GPA1-independent roles (Chen *et al.*, 2004a; Pandey *et al.*, 2006). AtRGS1 has been shown to interact with GPA1 genetically and physically in *Arabidopsis* and to complement the yeast RGS deletion mutant, sst2Δ (Chen *et al.*, 2003). Furthermore, *Atrgs1* loss-of-function phenotypes are recapitulated by overexpression of a GTPase-dead Gα, GPA1$^{(Q222L)}$, suggesting that the sole or main function of AtRGS1 is to regulate the active state of Gα. Because GPA1 has a very slow GTPase activity as compared to mammalian Gαs, in the absence of proteins such as AtRGS1 acting as GAPs, GPA1 (and presumably other plant

Gαs) would be predicted to exist in vivo primarily in the GTP-bound form (Johnston *et al.*, 2007a).

Liu *et al.* (2007) reported the identity of a new plant GPCR said to function as an ABA receptor, which they designate GCR2, but independent groups challenged their claim of 7TM receptor structure and function based on more rigorous modeling, which predicts that GCR2 lacks TM domains (Johnston *et al.*, 2007b), and on genetic analyses in which *gcr2* mutants did not consistently display insensitivity to ABA in seed germination and early seedling development (Gao *et al.*, 2007).

In metazoans, GPCRs constitute a superfamily of 7TM proteins subdivided into five subfamilies based on sequence; however, the overall conservation between the subfamilies, and to some extent even within the subfamilies, is low (Schioth and Fredriksson, 2005). Consequently, divergent GPCRs have been difficult to identify by homology-based approaches. Highly divergent *Arabidopsis* GPCR candidates have been identified through the use of non-alignment approaches (Moriyama *et al.*, 2006). This approach retrieved 394 *Arabidopsis* protein sequences, and the list contained GCR1, AtRGS1, and 15 proteins called MLOs (Devoto *et al.*, 2003). The MLO family of 7TM proteins was originally identified in barley and is widely present in plant species (Devoto *et al.*, 1999, 2003; Chen *et al.*, 2006d). While MLO-mediated response of barley to powdery mildew appears to be independent of heterotrimeric G proteins (Kim *et al.*, 2002), this does not preclude the possibility that MLO family members function as GPCRs (Jones, 2002). Restricting candidates in the list to those predicted to have strictly seven membrane spans and with the amino terminus located on the extracellular face of the membrane culls the candidate list to 54 proteins, including GCR1, AtRGS1, and 7 of the 15 MLOs (Moriyama *et al.*, 2006). To summarize, 7TM proteins, similar to the situation with the heterotrimer, are of low complexity in plants (Fredriksson and Schioth, 2005), and none to date have been shown to have GEF activity on a plant heterotrimeric G protein.

As illustrated in Fig. 2.2 (Color plate 3), four interacting proteins of GPA1, namely, AtPirin1, PLDα1, PD1, and THF1, are currently identified in *Arabidopsis* (Lapik and Kaufman, 2003; Zhao and Wang, 2004; Huang *et al.*, 2006; Warpeha *et al.*, 2006). These proteins are proposed candidate effectors of GPA1. AtPirin1, a member of the cupin protein superfamily, which has diverse biological functions, was isolated as an interacting protein of GPA1 in a yeast two-hybrid library screen (Lapik and Kaufman, 2003). The interaction between AtPirin1 and GPA1 was confirmed by an in vitro binding assay. Like the *gpa1* mutants, loss-of-function *AtPirin1* mutants display hypersensitive responses to abscisic acid (ABA) in seed germination and early seedling development. AtPirin1 was also implicated in a signaling chain that mediates both blue light and ABA responses in *Arabidopsis* (Warpeha *et al.*, 2007).

PLDα1 is a major isoform of phospholipase D (PLD) in *Arabidopsis*. In co-IP assays, both recombinant PLDα1 protein expressed in *Escherichia coli* and native PLDα1 protein extracted from leaves of *Arabidopsis* bound recombinant *Arabidopsis* Gα protein (GPA1) expressed in *E. coli* (Zhao and Wang, 2004). In an assay for PLDα1 activity, application of recombinant wild-type Gα decreased PLDα1 activity, while a mutant Gα that could not interact with PLDα1 did not have this inhibitory effect. Interestingly, binding of PLDα1 to the Gα stimulates GTP hydrolysis by the Gα.

PD1 is a cytosolic prephenate dehydratase discovered in a yeast two-hybrid screen for GPA1 interactors (Warpeha *et al.*, 2006). The interaction between PD1 and GPA1 was confirmed by an in vitro protein binding assay. It was proposed that PD1 lies in a signal transduction pathway with GPA1 and GCR1 for the blue light-mediated synthesis of phenylpyruvate and phenylalanine.

THF1 was also identified in a yeast two-hybrid library screen (Huang *et al.*, 2006). THF1 is a plastid protein localized to both the outer plastid membrane and the stroma, and does not share significant sequence with any known protein, although similar proteins can be found in a number of other plant species. The interaction between THF1 and GPA1 was confirmed by in vitro and in vivo co-IP, FRET analysis, and genetic epistatic analysis. Molecular and genetic analyses indicated that THF1 functions downstream of the plasma membrane-delimited GPA1 in a D-glucose signaling pathway (Huang *et al.*, 2006), suggesting a sugar signaling mechanism between plastids and the plasma membrane.

2.3.2 Genetic and physiological evidence for heterotrimeric G-protein-mediated signaling in plants

Study of both loss-of-function mutants and gain-of-function transgenic plants confirmed some of the biological functions of heterotrimeric G proteins implicated by early biochemical and pharmacological experiments (reviewed in Assmann, 2002), identified new functions regulated by heterotrimeric G proteins, and elucidated heterotrimeric G-protein-mediated signaling pathways in higher plants (Fujisawa *et al.*, 2001; Jones, 2002; Perfus-Barbeoch *et al.*, 2004). Table 2.1 summarizes the plant responses known to be modulated by heterotrimeric G proteins.

2.3.2.1 Cell division

Figure 2.3 (Color plate 4) illustrates the complex role that the heterotrimeric G-protein complex and AtRGS1 play in modulating cell division. Strong expression of GPA1 in shoot and root meristems and lateral root primordia prompted the hypothesis that GPA1 is involved in regulation of cell division (Ma, 1994). This hypothesis was confirmed when T-DNA insertional mutants of *Arabidopsis GPA1* were isolated and characterized (Ullah *et al.*, 2001). The *Arabidopsis gpa1* null mutants exhibit reduced cell division in several developmental stages. Dark-grown *gpa1* null mutants show shorter hypocotyls, caused by a decrease in cell number. The rosette leaves of the *gpa1* mutants

Table 2.1 Responses that are mediated by heterotrimeric G proteins in plants

Physiological responses	G-protein element involved	Plant species	Regulation by G proteins	Reference(s)
Cell division	GPA1	*Arabidopsis*	Stimulates cell division in hypocotyls, primary root meristem, and rosette leaves	Ullah *et al.*, 2001; Chen *et al.*, 2006a
			Stimulates advance of the cell cycle from G1 phase to G2 phase and promotes formation of nascent cell plates in tobacco BY-2 cells	
	AGB1	*Arabidopsis*	Stimulates axial cell division, but attenuates circumferential cell division in hypocotyls	Ullah *et al.*, 2003; Chen *et al.*, 2006a
			Attenuates cell division in pericycle founder cells for lateral root primordia	
			Attenuates cell division in the root apical meristem but this may require a functional Gα subunit	
	GCR1	*Arabidopsis*	Overexpression of *Arabidopsis* GCR1 promotes cell division in tobacco BY-2 cells	Colucci *et al.*, 2002; Apone *et al.*, 2003
	AtRGS1	*Arabidopsis*	Attenuates cell division in primary roots	Chen *et al.*, 2003, 2006a
Ion channel activity	GPA1	*Arabidopsis*	Mediates inhibition of inward K^+channels and activation of anion channels by ABA in guard cells	Wang *et al.*, 2001; Coursol *et al.*, 2003
	TGA1	Tomato	Recombinant TGA1 activates Ca^{2+} channels	Aharon *et al.*, 1998
Stomatal movement	GPA1	*Arabidopsis*	Promotes inhibition of stomatal opening by ABA and S1P	Wang *et al.*, 2001; Coursol *et al.*, 2003; Mishra *et al.*, 2006
	GCR1	*Arabidopsis*	Attenuates inhibition of stomatal opening and promotion of stomatal closure by ABA	Pandey and Assmann, 2004
Auxin responsiveness	GPA1	*Arabidopsis*	Positively modulates regulation of cell division by auxin	Ullah *et al.*, 2001, 2003
	AGB1	*Arabidopsis*	Negatively modulates auxin-inducible cell division	Ullah *et al.*, 2003
	AGG1	*Arabidopsis*	Negatively modulates acropetally transported auxin activity	Trusov *et al.*, 2007

(*continued*)

Table 2.1 *(continued)*

Physiological responses	G-protein element involved	Plant species	Regulation by G proteins	Reference(s)
	AGG2	*Arabidopsis*	Negatively modulates basipetally transported auxin activity	Trusov *et al.*, 2007
GA responsiveness	GPA1	*Arabidopsis*	Positively regulates GA-stimulated seed germination	Ullah *et al.*, 2002; Chen *et al.*, 2004a
	RGA1	Rice	Promotes α-amylase activity and *OsGAmyb* transcript induction by low concentration of GA (10^{-7} M or lower) and GA-induced internode elongation	Fujisawa *et al.*, 1999; Ueguchi-Tanaka *et al.*, 2000
	AGB1	*Arabidopsis*	Positively regulates GA-stimulated seed germination	Chen *et al.*, 2004a
	GCR1	*Arabidopsis*	Positively regulates GA-stimulated seed germination	Chen *et al.*, 2004a
ABA responsiveness	GPA1	*Arabidopsis*	Negatively regulates inhibition of seed germination and root elongation by ABA and ABA-induced gene expression	Ullah *et al.*, 2002; Lapik and Kaufman, 2003; Pandey *et al.*, 2006
	AGB1	*Arabidopsis*	Negatively regulates inhibition of seed germination and root elongation by ABA and ABA-induced gene expression	Pandey and Assmann, 2004; Pandey *et al.*, 2006
	GCR1	*Arabidopsis*	Negatively regulates inhibition of seed germination, root elongation by ABA and ABA-induced gene expression	Chen *et al.*, 2004a; Pandey *et al.*, 2006
	AtRGS1	*Arabidopsis*	Up-regulates genes required for ABA biosynthesis Positively regulates ABA inhibition of root elongation and seed germination Overexpression of *AtRGS1* enhances drought tolerance	Chen *et al.*, 2003, 2006c
BR responsiveness	GPA1	*Arabidopsis*	Positively regulates BR-stimulated seed germination	Ullah *et al.*, 2002; Chen *et al.*, 2004a
	RGA1	Rice	Positively regulates inhibition of root elongation by BR, BR-induced lamina inclination and coleoptile elongation	Wang *et al.*, 2006b

	AGB1	*Arabidopsis*	Positively regulates BR-stimulated seed germination	Chen *et al.*, 2004a
	GCR1	*Arabidopsis*	Positively regulates BR-stimulated seed germination	Chen *et al.*, 2004a
Jasmonic acid responsiveness	AGB1	*Arabidopsis*	Promotes methyl jasmonate-induced expression of the plant defensin gene *PDF1.2* and inhibition of root elongation and seed germination by methyl jasmonate	Trusov *et al.*, 2006
	AGG1	*Arabidopsis*	Promotes inhibition of root elongation by MeJA	Trusov *et al.*, 2007
Light signaling	GPA1	*Arabidopsis*	Promotes blue light-mediated synthesis of phenylpyruvate and phenylalanine and blue light-induced gene expression	Warpeha *et al.*, 2006, 2007
	GCR1	*Arabidopsis*	Promotes blue light-induced gene expression	Warpeha *et al.*, 2007
Sugar sensing and signaling	GPA1	*Arabidopsis*	*gpa1* mutants are hypersensitive to high concentration of D-glucose in seed germination, early seedling development, and root growth	Ullah *et al.*, 2002; Huang *et al.*, 2006; Pandey *et al.*, 2006
	AGB1	*Arabidopsis*	Attenuates inhibitory effects of D-glucose on seed germination, early seedling development, and root growth	Ullah *et al.*, 2002; Huang *et al.*, 2006; Pandey *et al.*, 2006; Wang *et al.*, 2006a
	AGG1	*Arabidopsis*	*agg1* mutants are hypersensitive to high concentrations of both glucose and mannitol in seed germination	Trusov *et al.*, 2007
	AGG2	*Arabidopsis*	*agg2* mutants are hypersensitive to high concentrations of glucose but not mannitol in seed germination	Trusov *et al.*, 2007
	AtRGS1	*Arabidopsis*	*Atrgs1* mutants are insensitive to D-glucose and overexpression of *AtRGS1* confers hypersensitivity to D-glucose	Chen *et al.*, 2003; Chen and Jones, 2004; Chen *et al.*, 2006c
Responses to ozone	GPA1	*Arabidopsis*	Positively modulates response to both chronic ozone treatment (100 nmol mol^{-3} ozone for 12 days) and acute ozone treatment (500 or 700 nmol mol^{-3} ozone for 3 h)	Booker *et al.*, 2004; Joo *et al.*, 2005
	AGB1	*Arabidopsis*	Negatively modulates responses to acute ozone treatment (500 or 700 nmol mol^{-3} ozone for 3 h)	Joo *et al.*, 2005

(*continued*)

Table 2.1 *(continued)*

Physiological responses	G-protein element involved	Plant species	Regulation by G proteins	Reference(s)
Responses to fungi and fungal elicitors	GPA1	*Arabidopsis*	Negatively regulates defense response to necrotrophic fungus *Plectosphaerella cucumerina*	Llorente *et al.*, 2005
	RGA1	Rice	Positively regulates H_2O_2 production, *PR* gene expression, and OsMAPK6 protein accumulation induced by rice blast fungus and sphingolipid elicitor	Suharsono *et al.*, 2002; Lieberherr *et al.*, 2005
	AGB1	*Arabidopsis*	Positively regulates defense responses to necrotrophic fungi such as *Plectosphaerella cucumerina*, *A. brassicicola*, and *F. oxysporum*	Llorente *et al.*, 2005; Trusov *et al.*, 2006
	AGG1	*Arabidopsis*	Positively regulates defense responses to fungi *A. brassicicola* and *F. oxysporum*	Trusov *et al.*, 2007
UPR-associated cell death	AGB1	*Arabidopsis*	Modulates tunicamycin (a protein glycosylation inhibitor)-induced cell death	Wang *et al.*, 2007
Response to salinity	PsGα1	Pea	Overexpression of PsGα1 in tobacco confers salinity tolerance	Misra *et al.*, 2007
	PsGα2	Pea	Overexpression of PsGα2 in tobacco confers salinity tolerance	Misra *et al.*, 2007
Response to heat	PsGα1	Pea	Overexpression of PsGα1 in tobacco confers heat tolerance	Misra *et al.*, 2007
	PsGα2	Pea	Overexpression of PsGα2 in tobacco confers heat tolerance	Misra *et al.*, 2007
	PsGβ	Pea	Overexpression of PsGβ in tobacco confers heat tolerance	Misra *et al.*, 2007

Note: Information is derived from phenotypic analysis of loss-of-function mutants and/or transgenic overexpression lines, unless otherwise stated.

are rounder than those of wild-type plants. However, careful observation showed that the epidermal cell number decreases significantly in the mutant rosette leaves compared with the wild-type plants, while epidermal cell size is significantly larger than in wild type. Using cyc1At-CDB-GUS as a mitotic reporter, a weak and diffuse GUS staining was detected in the *gpa1* mutant leaves, while an intense and discrete GUS staining was observed in the wild-type plants, suggesting a reduction in cell division in the *gpa1* mutants. Increased cell division can be induced by ectopic expression of wild-type *GPA1* in *Arabidopsis*. Overexpression of GPA1 in synchronized tobacco BY-2 cells caused advance of the cell cycle from the G1 phase to the G2 phase and promoted formation of nascent cell plates. The reduced cell division in the *gpa1* mutant appears to be a result of an extended G1 phase during the cell cycle (Ullah *et al.*, 2001). These data demonstrate that GPA1 is a positive modulator of cell proliferation in *Arabidopsis*. Increased cell division is also observed in the primary root of the *Atrgs1* null mutants (Chen *et al.*, 2003, 2006a), while overexpression of *Arabidopsis* GCR1 in tobacco BY-2 cells confers increased incorporation of thymidine, an indication of increased DNA synthesis, and increased cell division (Colucci *et al.*, 2002; Apone *et al.*, 2003).

It was further illustrated that the activated form of GPA1 (GTP bound), the heterotrimer, and the Gβγ dimer each modulate cell production in the root (Chen *et al.*, 2006a). The inactive form of GPA1 (GDP bound), presumably in the form of the intact heterotrimer, acts as a negative modulator for cell production in the primary root (Chen *et al.*, 2006a). AGB1 also functions as a negative modulator of cell division but at a different point in the cell cycle (Ullah *et al.*, 2003; Chen *et al.*, 2006a). *agb1-2* null mutants have longer primary roots and more lateral roots compared with wild type (Chen *et al.*, 2006a). Molecular and genetic analyses indicate that AGB1 works together with the GPA1 to negatively regulate cell production in the primary root, whereas AGB1 functions downstream of GPA1 to negatively modulate lateral root formation (Chen *et al.*, 2006a).

In the shoot, similar to *gpa1* mutants, *agb1* null mutants display shorter hypocotyls, which are caused by a decrease in axial cell division (Ullah *et al.*, 2003). However, in contrast to the *gpa1* mutants, hypocotyls of the *agb1* mutants are wider than those of the wild-type plants, as a result of an increased circumferential cell division (Ullah *et al.*, 2003). Mutation in the *AGB1* gene also affects organ shape in leaf, flower, and fruit (Lease *et al.*, 2001; Ullah *et al.*, 2003).

2.3.2.2 Regulation of ion channels

Regulation of ion channels by heterotrimeric G-protein subunits is well established in mammalian cells (Brown and Birnbaumer, 1990). As in animal cells, heterotrimeric G proteins regulate ion fluxes across plant cell membranes. In an early study conducted in *Vicia faba* guard cells by patch clamping, application of GDPβS, a GDP analog that promotes the inactive state of G proteins, increased current through inwardly rectifying K^+ channels, while GTPγS, a nonhydrolyzable GTP analog that locks G proteins in the active state, decreased inward K^+ current (Fairley-Grenot and Assmann, 1991). Subsequent electrophysiological experiments additionally supported involvement of heterotrimeric G proteins in regulation of K^+ channels in higher plants (Li and Assmann, 1993; Wu and Assmann, 1994; Saalbach *et al.*, 1999).

In wild-type *Arabidopsis* plants, ABA inhibits inward K^+ channels in guard cells, an effect that contributes to inhibition of stomatal opening by ABA. However, loss of GPA1 abolishes the inhibition of the inward K^+ channels by ABA (Wang *et al.*, 2001). A similar impairment in the *gpa1* mutants is loss of inhibition of inward K^+ channels by S1P, a lipid metabolite that is both a second messenger for ABA in plants and a ligand for specific GPCRs in mammalian systems (Spiegel and Milstein, 2003; Coursol *et al.*, 2003). Because ABA activates sphingosine kinase in plants, leading to increased S1P production (Coursol *et al.*, 2003), these results imply a signal transduction chain extending from ABA through sphingosine kinase, S1P and then GPA1 to the downstream target consisting of inward K^+ channels. S1P and the related metabolite, phyto-S1P, also alter stomatal apertures in a manner similar to that of ABA, and the influence on stomatal apertures of these lipid metabolites is diminished in *gpa1* mutants (Ng *et al.*, 2001; Coursol *et al.*, 2003; Coursol *et al.*, 2005).

Besides K^+ channels, regulation of anion channels and Ca^{2+}-permeable channels is also mediated by heterotrimeric G proteins in plant cells. S1P, like ABA, activates slow anion channels in wild-type *Arabidopsis* guard cells, and the activation of slow anion channels by S1P is attenuated in the *gpa1* mutants (Wang *et al.*, 2001; Coursol *et al.*, 2003). In membrane patches from tomato protoplasts, race-specific fungal elicitors activate Ca^{2+}-permeable channels (Gelli *et al.*, 1997). This activation of the Ca^{2+} channels by the fungal elicitors is attenuated by application of the G-protein inhibitor GDPβS, whereas G-protein activators GTPγS and mastoparan mimic the stimulatory effect of fungal elicitors on the Ca^{2+} channels, implying possible involvement of heterotrimeric G proteins in activation of the Ca^{2+} channels by the fungal elicitors. Consistent with these results, application of recombinant TGA1 protein to tomato plasma membrane patches increases the open probability of the Ca^{2+} channels while channel conductance remains unchanged. A constitutively active form of TGA1 has a more significant stimulatory effect than the wild-type form (Aharon *et al.* 1998).

2.3.2.3 Plant hormone action

Early pharmacological experiments implicated heterotrimeric G proteins in plant hormone signaling (Ma, 1994; Hooley, 1998; Fujisawa *et al.*, 2001). Through phenotypic analysis of loss-of-function heterotrimeric G-protein mutants, there is extensive evidence that heterotrimeric G proteins play important roles in multiple phytohormone signaling cascades (Assmann, 2002; Jones and Assmann, 2004; Perfus-Barbeoch *et al.*, 2004).

Auxins. Early experiments using rice coleoptile membranes demonstrated that indole-3-acetic acid (IAA), an auxin, increased GTPγS binding and stimulated GTPase activity (Zaina *et al.*, 1990; Zaina *et al.*, 1991), suggesting a possible involvement of heterotrimeric G proteins in auxin signal transduction in rice. Auxin rapidly induces expression of GPA1 and inhibits expression of AGB1 via a G-protein-independent pathway, indicating crosstalk between different auxin signaling pathways (Ullah *et al.*, 2003). Multiple approaches showed that G proteins mediate auxin regulation of cell division and possibly also indirectly modulate auxin-regulated cell expansion. Overexpression of *Arabidopsis* GPA1 in synchronized tobacco BY-2 cells results in advancement of a greater percentage of cells to the G2 phase, and a similar effect was induced by applying the auxin analog 2,4-D (Ullah *et al.*, 2001). Based on the phenotypes of *agb1* mutant plants, AGB1 is a negative regulator for auxin-regulated cell division (Ullah *et al.*, 2003). More lateral roots are produced in the *agb1* null mutants than corresponding wild-type plants. Application of NAA, an auxin analog, to the *agb1* mutants elicits more lateral root primordia than in the wild-type plants. Similar effects of NAA on adventitious roots are also observed in hypocotyl explants of the *agb1* mutants.

Both *agg1* and *agg2* mutants showed increased sensitivity to induction of lateral roots by auxin; however, the hypersensitivity of *agg1* and *agg2* to auxin was attenuated or impaired by inhibition of acropetal or basipetal auxin transport respectively, suggesting a role of Gβγ1 in modulating induction of lateral roots by acropetally transported auxin and of Gβγ2 in modulating induction of lateral roots by basipetally transported auxin (Trusov *et al.*, 2007). In another study, overexpression of a truncated version of AGG1, which lacks the C-teminal CAAX isoprenylation motif and therefore has its membrane anchor disrupted (thus presumably disrupting membrane targeting of the βγ dimer), leads to increased lateral roots and primordia, a phenotype shared by *agb1* mutants (Chakravorty and Botella, 2007). Because the *agb1* and *gpa1* null mutants still retain significant responsiveness to auxin (Ullah *et al.*, 2003), heterotrimeric G proteins most likely participate in auxin signaling by modulating hormone sensitivity.

Gibberellins (GA). Genetic evidence from studies of rice and *Arabidopsis* G-protein mutants confirms involvement of heterotrimeric G proteins in gibberellin signaling. In rice, a gibberellin-insensitive dwarf mutant, *dwarf1* (*d1*), turned out to be caused by a mutation in a rice Gα subunit gene, *RGA1* (Ashikari *et al.*, 1999). Transgenic reduction of *RGA1* mRNA (Ishikawa *et al.*, 1995; Seo *et al.*, 1995) results in dwarfism in rice (Fujisawa *et al.*, 1999). Application of 10^{-7} M or lower concentrations of GA_3 to embryo-less half seeds results in an induction of α-amylase activity and *OsGAmyb* transcript in wild-type rice, but very low or no induction in the *d1* mutant (Ueguchi-Tanaka *et al.*, 2000). GA-induced internode elongation is also significantly slower in the *d1* mutant than in the wild type. However, high concentrations of GA_3 (10^{-4} M or higher) produce the same induction of α-amylase activity in both the *d*1 mutant and wild-type rice, suggesting that a Gα-independent GA-signaling pathway is present in rice at high GA concentration. The genetic evidence from the *d1* rice mutants confirms an implication from pharmacological experiments that heterotrimeric G proteins are involved in induction of α-amylase gene transcription in wild oat aleurone (Jones *et al.*, 1998).

In *Arabidopsis*, loss-of-function *gpa1* mutants are 100-fold less sensitive to GA_3 in seed germination than wild type, while ectopic overexpression of *GPA1* makes seed germination at least a million-fold more responsive to GA_3 than wild type (Ullah *et al.*, 2002). Yet, even for lines that ectopically overexpress GPA1, GA remains essential for seed germination, indicating that GPA1 modulates GA signaling, rather than serving as a direct transducer of this hormone (Ullah *et al.*, 2002). Similarly, loss-of-function *agb1* and *gcr1* mutants also show less sensitivity to GA_3 in seed germination (Chen *et al.*, 2004a). The effects of *gpa1*, *agb1*, and *gcr1* on GA hyposensitivity are additive or synergistic, as shown by analysis of their double and triple mutants, indicating that besides the G-protein-dependent pathway, GCR1 may also modulate GA signaling in seed germination via a G-protein-independent pathway (Chen *et al.*, 2004a).

Abscisic acid (ABA). GPA1 appears to be positively involved in mediating aspects of guard cell ABA response but is a direct or indirect negative regulator for ABA responses in other cell and tissue types, as schematized in Fig. 2.4. As discussed earlier, GPA1 is involved in inhibition of guard cell inward K^+ channels and slow anion channels by ABA and thus the inhibition of stomatal opening by ABA. The lipid metabolite, S1P, and its relative, phyto-S1P, also alter stomatal apertures in a manner similar to that of ABA, and the influence on stomatal apertures of these lipid metabolites is diminished in *gpa1* mutants (Ng *et al.*, 2001; Coursol *et al.*, 2003, 2005), suggesting that a sphingosine kinase functions upstream of GPA1 in guard cell ABA signaling.

GPA1 physically interacts with PLDα1 (Zhang *et al.*, 2004), a phospholipase that produces phosphatidic acid (PA). ABA activates PLD activity (Jacob *et al.*, 1999) and PA inhibits inward K^+ channels and stomatal opening in wild-type plants (Jacob *et al.*, 1999; Mishra *et al.*, 2006). PA inhibition of stomatal opening is abrogated in *gpa1* mutants, indicating that PA acts upstream of GPA1 to promote ABA-linked inhibition of stomatal opening. In response to ABA, *pldα1* single mutants, *gpa1* single mutants, and *pldα1 gpa1* double mutants all exhibit similar ABA hyposensitivity of stomatal opening, again consistent with the idea that PLDα1 acts upstream and/or at the level of GPA1. However, because PLDα1 also accelerates the GTPase activity of GPA1, a second effect of PLDα1 is as a negative regulator of GPA1: when wild-type PLDα1 is replaced by a mutant form of PLDα1 with a 10-fold lower affinity for GPA1, stomatal opening becomes hypersensitive to ABA (Mishra *et al.*, 2006). Thus, PLDα1 has both negative and positive roles in G-protein-mediated ABA inhibition of stomatal opening. PLDα1-produced PA binds to the ABI1 PP2C to signal ABA-promoted stomatal closure, whereas PLDα1 and PA interact with GPA1 to mediate ABA inhibition of stomatal opening.

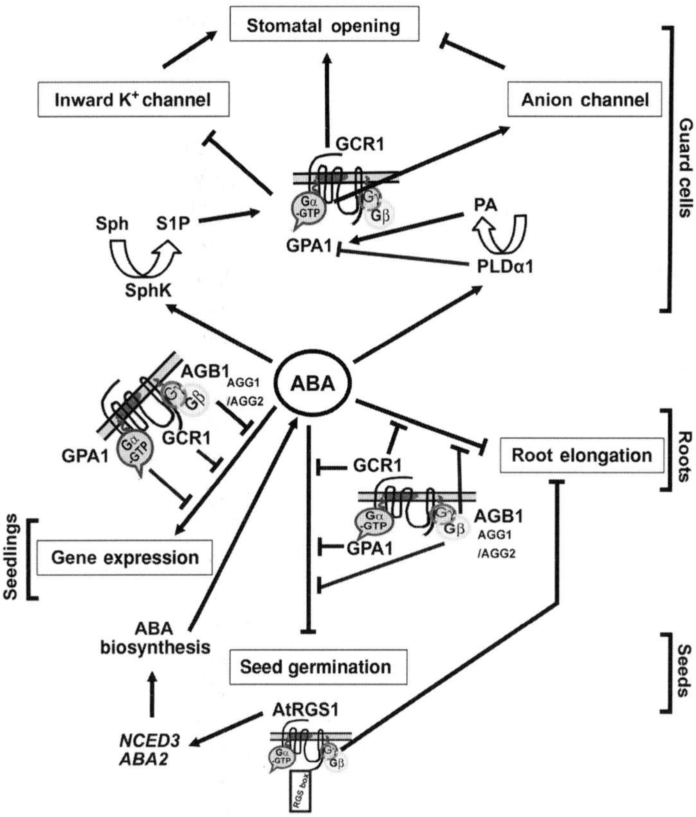

Figure 2.4 *Diverse roles of G-protein components in ABA signaling in Arabidopsis.* GPA1 acts in guard cells as a positive regulator of ABA inhibition of stomatal opening; however in other tissue types, GPA1 and/or AGB1 acts as a negative regulator of ABA action. In guard cells, GPA1, as a positive regulator, mediates both inhibition of inward K^+ channels and activation of anion channels by ABA, which are required for ABA inhibition of stomatal opening. ABA activates GPA1 through sphingosine-1-phosphate (S1P) and/or phosphatidic acid (PA), which are produced by two processes catalyzed respectively by sphingosine kinase (SphK) and phospholipase Dα1 (PLDα1). PLDα1 also has a negative effect on GPA1 via elevation of its GTPase activity; this figure is a simplification of a more complex model (see Zhang *et al.*, 2004; Mishra *et al.*, 2006). AGB1 negatively regulates ABA inhibition of both root elongation and seed germination, and GPA1 may play an indirect role in the two responses through modulating availability of AGB1. Both GPA1 and AGB1 negatively regulate ABA-induced gene expression in young seedlings. GCR1 negatively regulates all these physiological processes. AtRGS1 positively affects ABA action by up-regulating expression of genes required for ABA biosynthesis (such as *NCED3* and *ABA2*), thus contributing to ABA inhibition of both root elongation and seed germination. Arrows indicate positive regulation and blunted arrows indicate negative regulation.

While *gpa1* guard cells are hyposensitive in some ABA responses, seed germination of the *gpa1* null mutants is more sensitive to ABA (Ullah *et al.*, 2002; Lapik and Kaufman, 2003; Pandey *et al.*, 2006), and *gpa1* seedlings also show hypersensitivity to ABA inhibition of primary root growth, and to ABA induction of stress gene expression (Pandey *et al.*, 2006). In contrast, GCR1 consistently acts as a negative regulator of ABA sensitivity in all responses assayed: stomatal apertures, root elongation, and gene expression. A detailed comparison of germination, root growth, and gene expression responses in *gpa1*, *agb1*, and *gcr1* single mutants, as well as all double (*agb1 gpa1*, *gcr1 gpa1*, *agb1 gcr1*) and triple (*agb1 gcr1 gpa1*) mutant combinations, reveals ABA hypersensitivity to these three responses in all these genotypes (Pandey and Assmann, 2004; Pandey *et al.*, 2006).

For germination and root growth inhibition, single, double, and triple mutants containing an *agb1* allele show stronger hypersensitivity to ABA than do mutants containing a *gpa1* allele, suggesting that AGB1 and its downstream effectors play the predominant role in these ABA responses, and that GPA1 modulates this pathway through controlling the availability of free βγ subunits. However, such a clear dichotomy is not seen for gene regulation, suggesting that gene regulation probably occurs through both GPA1- and AGB1-based pathways.

Brassinosteroids (BR). The *gpa1*, *agb1*, and *gcr1* single mutants all show decreased sensitivity to BR in seed germination relative to wild type, suggesting that heterotrimeric G proteins participate in, or modulate, BR signaling (Ullah *et al.*, 2002; Chen *et al.*, 2004a). In root elongation inhibition assays and coleoptile elongation analysis, the loss-of-function mutant of rice G-protein α subunit (RGA), *d1*, also displayed reduced sensitivity to BR (Wang *et al.*, 2006b). When *gcr1* is combined with *gpa1* or *agb1* in double or triple mutant combinations, the hyposensitivity to BR increases (Chen *et al.*, 2004a). This result is unexpected if GCR1 is signaling solely via the heterotrimer, and therefore implies that, just as for GA signaling, GCR1 may modulate BR signaling during seed germination via both G-protein-dependent and G-protein-independent pathways (Chen *et al.*, 2004a).

Ethylene *gpa1* mutant seeds retain wild-type sensitivity to 1-aminocyclopropane-1-carboxylate (ACC), an immediate precursor for ethylene, in a seed germination assay (Ullah *et al.*, 2002), therefore it was proposed that GPA1 does not couple ethylene regulation of sugar inhibition of germination (Ullah *et al.*, 2002). However, there is pharmacological evidence that G proteins may have a role in ethylene-mediated leaf abscission, presumably through regulation of ACC synthase and ACC oxidase expression (Yuan *et al.*, 2005).

Jasmonic acid. *agb1* mutants have decreased sensitivity to a number of methyl jasmonate-induced responses including induction of the plant defensin gene

PDF1.2, inhibition of root elongation, and seed germination, whereas *gpa1* mutants display increased sensitivities (Trusov *et al.*, 2006). Therefore, it is proposed that jasmonic acid signaling is mainly influenced by AGB1 but not by GPA1, and that AGB1 acts as a direct or indirect enhancer of jasmonate signaling (Trusov *et al.*, 2006). Consistent with this hypothesis, impaired inhibition of root elongation by MeJA was also found in the *agg1* mutants, albeit not in the *agg2* mutant (Trusov *et al.*, 2007).

2.3.2.4 Sugar sensing and response

Arabidopsis G-protein α and β subunits mutants, *gpa1* and *agb1*, are hypersensitive to high concentrations of D-glucose in seed germination, early seedling development, and root growth (Ullah *et al.*, 2002; Huang *et al.*, 2006; Pandey *et al.*, 2006), indicating that G proteins are involved in sugar sensing. Consistent with this premise, overexpression of a constitutive active form of GPA1 ($GPA1^{QL}$) confers D-glucose hyposensitivity (Huang *et al.*, 2006). As would be predicted, null mutants of *AtRGS1*, a negative regulator of G-protein signaling, are D-glucose insensitive, and overexpression of *AtRGS1* confers hypersensitivity to high concentrations of D-glucose (Chen *et al.*, 2003, 2006c; Chen and Jones, 2004). Because sugar metabolism and phosphorylation by hexokinase (HXK) are not required for AtRGS1-mediated sugar signaling, it was proposed that AtRGS1 functions in a HXK-independent glucose signaling pathway (Chen and Jones 2004). Hypersensitivity to D-glucose was also observed in loss-of-function mutants of THF1, a GPA1-interacting protein (Huang *et al.*, 2006). Because THF1 is a plastid protein, and acts downstream of the plasma membrane-localized GPA1, these findings suggest that sugar signaling occurs between plastids and the plasma membrane.

G proteins also have a role in sugar transport. A Golgi-localized hexose transporter, suppressor of Gβ (SGB1), was identified from a genetic screen for AGB1 modifiers that could suppress the altered cell division in the hypocotyl and glucose hypersensitivity of the *agb1-2* mutant (Wang *et al.*, 2006a). *SGB1* has a similar tissue expression pattern as that of *AGB1*, and its expression increases in the presence of D-glucose or sucrose. Interestingly, in the absence of exogenous sugar, SGB1 traffics to small vesicular compartments suggestive of the trans-Golgi network and the addition of sugar collapses these vesicles to the Golgi proper. Loss-of-function mutants of *SGB1* phenocopy *agb1-2* mutants, whereas overexpression of *SGB1* suppresses the cell division and sugar hypersensitivity of *agb1-2* mutants. These findings provide genetic evidence that SGB1 acts together with AGB1 in regulating sugar transport. These findings establish SGB1 as the first potential effector protein for AGB1, although it is yet to be determined if SGB1 physically interacts with AGB1. This is the first report of a Golgi-localized transporter and the first example linking sugar transport to G-protein signaling.

One Gγ subunit mutant, *agg1*, is hypersensitive to both high concentration of glucose and mannitol in seed germination while the other Gγ mutant, *agg2*, is only hypersensitive to high concentration of glucose but not to mannitol. These sensitivities are dependent on light intensity (Trusov *et al.*, 2007), and provide further evidence that G-protein signaling is involved in sugar sensing and response.

2.3.2.5 Light

Heterotrimeric G proteins were suggested to modulate signal transduction by phytochrome and blue light photoreceptors, based on early experiments with GTP analogs and bacterial toxins (Warpeha *et al.*, 1991; Muschietti *et al.*, 1993; Neuhaus *et al.*, 1993; Raghuram *et al.*, 1999).

When grown in darkness, loss-of-function *gpa1* and *agb1* mutants, but not *gcr1* mutants, exhibit shorter hypocotyls, a partial de-etiolation phenotype (Ullah *et al.*, 2001, 2003; Jones *et al.*, 2003; Chen *et al.*, 2004a), which suggests a possible connection between heterotrimeric G proteins and photomorphogenesis. Okamoto *et al.* (2001) overexpressed both wild type (wGα) and constitutively active (cGα) forms of GPA1 in *Arabidopsis* under control of a DEX-inducible promoter and tested responses of one overexpression line for each construct to different wavelengths of light. Under continuous 10 μmol m^{-2} s^{-1}white light, induction of either wGα or cGα by DEX caused shortened hypocotyls compared with empty vector control lines, while no difference was observed when these seedlings were grown in darkness. It was suggested that overexpression of wGα and cGα led to hypersensitivity of hypocotyls to white light. However, in a study with loss-of-function *gpa1* and *agb1* single mutants as well as *gpa1 agb1* double mutants, all these mutants showed the same sensitivity to red and far-red light as the wild-type plants (Jones *et al.*, 2003). This result ruled out the possibility that heterotrimeric G proteins are directly involved in red and far-red light-mediated hypocotyl growth (Jones *et al.*, 2003). By contrast, as discussed above, PD1, a GPA1-interacting protein, has a specific role in blue light-mediated synthesis of phenylpyruvate and phenylalanine (Warpeha *et al.*, 2006). Because PD1 activity is doubled by active GPA1, it is likely that a GPA1-PD1 signaling complex regulates the blue light-mediated synthesis of phenylpyruvate, phenylalanine, and metabolites derived from phenylalanine. GCR1, GPA1, AtPirin1 and a nuclear factor Y (NF-Y) complex transcription factor participate in a signaling cascade that mediates both blue light and ABA responses in *Arabidopsis* (Warpeha et al., 2007).

2.3.2.6 Biotic and abiotic stress

As mentioned earlier, heterotrimeric G proteins may be involved in fungal elicitor-induced activation of Ca^{2+} channels in tomato (Gelli *et al.*, 1997). A study with the rice Gα defective mutant *d1* provided evidence that heterotrimeric G proteins play a part in the response to rice blast fungus *Magnaporthe grisea* (Suharsono *et al.*, 2002). An increase in Gα transcript is induced by the fungus in wild-type rice but not in the *d1* mutant. Application of sphingolipids, which are rice blast elicitors, also results in an increased Gα transcription in wild-type rice plants (Suharsono *et al.*, 2002).

Corresponding to a reduction in the hypersensitive response of the *d1* mutants to infection with an avirulent strain of rice blast (observed as a reduction in HR-induced cell death), induction of two rice *PR* genes known to be involved in pathogen responses is delayed in the *d1* mutants compared with the wild-type rice plants. Application of sphingolipid elicitors promotes release of reactive oxygen molecules (ROS) signaling elements in plant pathogen defense responses, and boosts the transcription of the PR gene, *PBZ1*, in wild type, but has no significant effect in the *d1* mutants (Suharsono *et al.*, 2002). Sphingolipid activation of a defense-related MAPK, OsMAPK6, is also reduced in *d1* plants. A small GTPase, OsRAC1, functions upstream of both OsMAPK6 and ROS production during the HR response of wild-type plants; a reduction in *OsRAC1* transcript levels occurs in *d1* plants, and this effect may thus account for the reduced activation of downstream defense molecules (Lieberherr *et al.*, 2005). Together, these data indicate that the Gα subunit is an important player in the hypersensitive response of rice to avirulent rice blast fungus (Assmann, 2005).

In *Arabidopsis*, *agb1* mutants show enhanced susceptibility to the necrotrophic fungus *Plectosphaerella cucumerina*, whereas *gpa1* mutants show enhanced resistance to this pathogen (Llorente *et al.*, 2005). Consistent with these observations, *agb1* and *agg1* (but not *agg2*) mutants are also impaired in defense response to the necrotrophic pathogens *Alternaria brassicicola* and *Fusarium oxysporum*, whereas *gpa1* mutants display increased resistance (Trusov *et al.*, 2006; Trusov *et al.*, 2007), suggestive of the involvement of GPA1 and Gβγ1 (but not Gβγ2) in responses to these fungus pathogens. In contrast, both *agb1* and *gpa1* mutants have wild-type response to inoculation with virulent (DC3000) and avirulent (JL1065) strains of *Pseudomonas syringae* (Trusov *et al.*, 2006). These findings may suggest that heterotrimeric G proteins have a regulatory role in plant–fungal interactions, whereas some plant–bacterial interactions may be independent of heterotrimeric G proteins.

Heterotrimeric G proteins also play roles in plant responses to abiotic stresses. Heterotrimeric G-protein involvement in signaling by the stress hormone, ABA, was discussed previously. Overexpressing the *Pisum sativum Gα1* and *Gα2* subunits in tobacco confers tolerance of the transgenic plants to both salinity and heat, while overexpressing *P. sativum* Gβ in tobacco only confers heat tolerance (Misra *et al.*, 2007). Booker *et al.* (2004) used *Arabidopsis gpa1*, *agb1*, *gcr1*, and *Atrgs1* single mutants plus *gpa1 agb1* double mutants to test responsiveness to ozone. Following a chronic ozone treatment, consisting of 12 days of 100 nmol mol^{-3} ozone treatment, two alleles of *gpa1* mutants and a *gpa1 agb1* double mutant did not exhibit the leaf curling shown by wild-type rosette leaves. This amount of ozone is physiologically relevant in that large cities in the US can experience this level during the summer months. However, after treatment with higher concentrations of ozone (175 and 250 nmol mol^{-3}), no significant differences were detected among these genotypes in terms of damage that ozone caused to rosette leaves. In another study by Joo *et al.* (2005), an acute ozone treatment consisting of 350 nmol mol^{-3} ozone for 6 h triggered transient increase of both *GPA1* and *AGB1* transcripts 1 h later. A difference between the two genes is that after ozone treatment, the *GPA1* gene had two expression peaks, while the *AGB1* gene had only one. After acute exposure to 500 or 700 nmol mol^{-3} ozone for 3 h, the *gpa1-4* mutant displayed fewer lesions and less damage than the wild-type plants, while the *agb1-2* mutant showed more severe phenotypes than the wild type, revealing roles for G proteins in acute oxidative stress. Joo *et al.* (2005) detected two oxidative bursts of H_2O_2 (one of the reactive oxygen species [ROS]) in response to ozone treatment in wild-type plants. The *agb1* mutant is defective in the first H_2O_2burst, while the *gpa1* mutant lacks both bursts.

AGB1 is also involved in regulation of unfolded protein response (UPR)-associated cell death (Wang *et al.*, 2007). Activation of the UPR by the antibiotic tunicamycin, a protein glycosylation inhibitor, prompts programmed cell death, and this cell death response is attenuated in the *agb1* mutant. Regulation of UPR-associated cell death by AGB1 is probably independent of a Gαβγ heterotrimer, since the *gpa1* mutant shows a wild-type-like response to tunicamycin-induced cell death.

2.3.3 Heterotrimeric G-protein signaling models in higher plants

As for mammalian heterotrimeric G-protein signaling, plant Gα subunits bind and hydrolyze GTP, and GTPase activity is regulated by an RGS protein. Also similar to mammalian systems, some plant G-protein-regulated pathways appear primarily dependent on Gα, and others on Gβ. For example, the shape of rosette leaves is affected similarly by loss of GPA1 or AGB1, and the double mutant exhibits a similar phenotype as the single mutants, consistent with a Gα-based signaling pathway in which loss of AGB1 disrupts GPA1 coupling with receptor or effector proteins. By contrast, other plant responses, e.g., lateral root formation, are much more severely affected by *AGB1* knockout than *GPA1* knockout, and the double *gpa1 agb1* mutants exhibit a similar phenotype as the *agb1* single mutants (Chen *et al.*, 2006a), consistent with a signaling pathway primarily dependent on Gβ in which AGB1 acts downstream of GPA1, and loss of GPA1 releases the sequestration of AGB1 by GPA1. The studies of overexpression of *GPA1* or *AGB1* in *agb1* or *gpa1* mutant backgrounds provide evidence that the heterotrimer functions as a negative modulator for cell production in the primary root (Chen *et al.*, 2006a).

One aspect where the classical model appears to diverge between metazoan systems and plants is in the absence of numerous GPCRs with conserved sequence similarity to mammalian GPCRs, and in the small number of Gα and Gβ subunits and RGS proteins. Considering the existence of about 50 putative 7TM domain proteins in *Arabidopsis* (Schwacke *et al.*, 2003; Moriyama *et al.*, 2006) and only two G-protein heterotrimer combinations in *Arabidopsis* and in rice, plants may take advantage of novel types of GPCRs to achieve specificity in their regulation. In addition, crosstalk in plant cell signaling (Genoud and Metraux, 1999; Gazzarrini and McCourt, 2003) makes it more complicated to identify signaling pathways that are directly regulated by heterotrimeric G proteins in plants. For example, it appears that neither the auxin-induced cell division negatively regulated by AGB1 nor the GA-induced seed germination positively regulated by GPA1 is a direct effect of AGB1 or GPA1, because loss of GPA1 or AGB1 confers altered sensitivity but not complete insensitivity.

On the other hand, there is the possibility that a noncanonical heterotrimeric G-protein signaling model may also operate in plant cells. As discussed previously, GCR1 may act independently of GPA1 and AGB1 in GA and BR-regulated seed germination (Chen *et al.*, 2004a). Whether the converse mechanism, namely, receptor-independent heterotrimeric G-protein signaling, also operates in plant systems awaits further investigation. Moreover, the unique structure of AtRGS1 suggests that plants may have a receptor GAP, which has not been described in other eukaryotes.

There is also the possibility that plants employ unique G-protein signaling mechanisms. The presence of unconventional Gα proteins in plants, such as extra-large GTP-binding proteins (XLGs) (Lee and Assmann, 1999; Assmann, 2002; Ding *et al.*, 2008), may highlight this possibility. No homologs to these proteins have been found in nonplant genomes to date. There are three XLG proteins encoded in the *Arabidopsis* genome and four XLG homologs in rice. Of these seven, *Arabidopsis* XLG1 has been studied in greatest detail. The predicted open reading frame (ORF) of the *XLG1* cDNA encodes a 99-kDa protein composed of 888 amino acids. The C-terminal region of the protein is 32% identical and 54% similar to GPA1. This C-terminal region also has ~26% identity and ~50% similarity with yeast and mouse Gα proteins. Besides the Gα region, XLG1 also has a 400+ amino acid N-terminal region with no significant homology to any proteins, other than to other XLG family members. The N-termini of the three *Arabidopsis* XLG proteins share conservation of a cysteine-rich region and a nuclear localization signal. Not all functional motifs are equally conserved within the Gα-like domain of the XLGs (Temple and Jones, 2007). The P-loop shows some variation from the canonical form in all seven of the XLGs, with AtXLG2 and two of the rice XLGs showing the most divergence. In the region corresponding to the DxxGQ motif, not all XLGs have the D, G, and Q residues that are conserved in canonical Gαs. The NKxD motif is invariant in all XLGs except for one rice XLG protein and AtXLG2. All three switches in the XLGs show significant modification from the canonical form. Switch I has a 15-residue insert in the middle, and shows little sequence conservation with the canonical form. Switch II of the XLGs has a single-residue insert in the middle of a highly conserved region and otherwise shows significant divergence in sequence. Similarly, switch III is quite divergent from the canonical switch III and has undergone both deletions and mutations depending on the particular XLG. Recombinant XLG1 protein expressed in *E. coli* binds GTP–γ–^{35}S and the binding of GTP–γ–^{35}S by XLG1 can be competed by cold GTP but much less so by ATP, indicating that XLG1 specifically binds GTP (Lee and Assmann, 1999). However, to date there is no evidence that XLG1 interacts with AGB1, so whether the XLG proteins participate in heterotrimeric G-protein signaling remains an open question.

With respect to plant-specific Gβ subunits, it was originally proposed that another seven-WD40 repeat protein, RACK1A, could also be a G-protein β subunit (Ishida *et al.*, 1993). *RACK1A* was originally cloned as an auxin-inducible gene (Ishida *et al.*, 1993), and the RACK1A protein has very similar domain structures to Gβ. However, data from molecular modeling reveal that *Arabidopsis* RACK1A protein lacks the N-terminal helix of Gβ that is critical for Gγ interaction (Chen *et al.*, 2006b). In addition, RACK1A does not interact with the sole *Arabidopsis* Gα in a yeast split-ubiquitin assay (Chen *et al.*, 2006b). Therefore, it is unlikely that RACK1A could function as a Gβ. Interestingly, in mammalian cells, RACK1 protein can physically bind the Gβγ dimer and Gαβγ trimer, and regulate specific functions of Gβγ, such as Gβγ-mediated activation of phospholipase Cβ2 and adenylyl cyclase (Chen *et al.*, 2004b). It has yet to be determined if RACK1A could also regulate Gβγ function in plants.

2.4 Conclusions and future directions

Although it has been established that heterotrimeric G proteins play regulatory roles in multiple developmental processes and hormone responses in plants, the upstream (presumably GPCR) and downstream (effector) components in G-protein signaling pathway remain largely elusive. GPCRs are proteins that typically have 7TM domains. So far, two such proteins, GCR1 and AtRGS1, have been shown to physically bind GPA1 (Chen *et al.*, 2003; Pandey and Assmann, 2004). However, no ligand has been identified for either GCR1 or AtRGS1. Sphingosine kinase and its product, S1P, appear to be upstream intermediaries in GPA1-based ABA response. There are several proteins that can act as direct downstream effectors for GPA1, including AtPirin1 (Lapik and Kaufman, 2003), PLDα1 (Zhao and Wang, 2004), PD1 (Warpeha *et al.*, 2006), and THF1 (Huang *et al.*, 2006), all of which have been shown to directly interact with GPA1. K^+, Ca^{2+}, and anion channels are also targets of GPA1 signaling. SGB1 is the only putative downstream effector for AGB1 identified at this time (Wang *et al.*, 2006a). The complete cascade in any given developmental process or hormonal modulation mediated by heterotrimeric G proteins remains unknown. The field is wide open for identification of the mechanisms by which G proteins regulate phenotypic plasticity (Assmann, 2002). Future studies are expected to reveal additional upstream and downstream components of heterotrimeric G-protein signal transduction pathways.

Acknowledgments

Research on heterotrimeric G proteins in the authors' laboratories was supported by NSF grant MCB-0209711 to SMA and AJ, by USDA grant 2006-35100-17254 to SMA, by NIH grant GM65989-01 and DOE grant DE-FG02-05er15671 to AJ, and by NSERC grant RGPIN311651-05 and CFI grant 10496 to J-GC.

References

Adjobo-Hermans, M.J., Goedhart, J. and Gadella, T.W., Jr. (2006) Plant G protein heterotrimers require dual lipidation motifs of Gα and Gγ and do not dissociate upon activation. *J Cell Sci*, **119**, 5087–5097.

Aharon, G.S., Gelli, A., Snedden, W.A. and Blumwald, E. (1998) Activation of a plant plasma membrane Ca^{2+} channel by TGα1, a heterotrimeric G protein α-subunit homologue. *FEBS Lett*, **424**, 17–21.

Anderson, D.J. and Botella J.R. (2007) Expression analysis and subcellular localization of the *Arabidopsis thaliana* G-protein β-subunit AGB1. *Plant Cell Rep*, **26**, 1469–1480.

Apone, F., Alyeshmerni, N., Wiens, K., Chalmers, D., Chrispeels, M.J. and Colucci, G. (2003) The G-protein-coupled receptor GCR1 regulates DNA synthesis through activation of phosphatidylinositol-specific phospholipase C. *Plant Physiol*, **133**, 571–579.

Ashikari, M., Wu, J., Yano, M., Sasaki, T. and Yoshimura, A. (1999) Rice gibberellin-insensitive dwarf mutant gene *Dwarf 1* encodes the α-subunit of GTP-binding protein. *Proc Natl Acad Sci USA*,**96**, 10284–10289.

Assmann, S.M. (2002) Heterotrimeric and unconventional GTP binding proteins in plant cell signaling. *Plant Cell*, **14** (Suppl), S355–S373.

Assmann, S.M. (2005) G protein regulation of disease resistance during infection of rice with rice blast fungus. *Sci STKE*, **310**, cm13.

Berman, D.M. and Gilman, A.G. (1998) Mammalian RGS proteins: Barbarians at the gate. *J Biol Chem*, **273**, 1269–1272.

Booker, F.L., Burkey, K.O., Overmyer, K. and Jones, A.M. (2004) Differential responses of G-protein *Arabidopsis thaliana* mutants to ozone. *New Phytol*, **162**, 633–641.

Bourne, H.R., Sanders, D.A. and McCormick, F. (1990) The GTPase superfamily: a conserved switch for diverse cell functions. *Nature*, **348**, 125–132.

Bourne, H.R., Sanders, D.A. and McCormick, F. (1991) The GTPase superfamily: conserved structure and molecular mechanism. *Nature*, **349**, 117–127.

Brito, M., Guzman, L., Romo, X., Soto, X., Hinrichs, M.V. and Olate, J. (2002) S111N mutation in the helical domain of human Gsα reduces its GDP/GTP exchange rate. *J Cell Biochem*, **85**, 615–620.

Brown, A.M. and Birnbaumer, L. (1990) Ionic channels and their regulation by G protein subunits. *Annu Rev Physiol*, **52**, 197–213.

Cao, X., Cismowski, M.J., Sato, M., Blumer, J.B. and Lanier, S.M. (2004) Identification and characterization of AGS4: a protein containing three G-protein regulatory motifs that regulate the activation state of $G_i\alpha$. *J Biol Chem*, **279**, 27567–27574.

Casey, P.J. (1994) Lipid modifications of G proteins. *Curr Opin Cell Biol*, **6**, 219–225.

Chakravorty, D. and Botella, J.R. (2007) Over-expression of a truncated *Arabidopsis thaliana* heterotrimeric G protein γ subunit results in a phenotype similar to α and β subunit knockouts. *Gene*, **393**, 163–170.

Chen, J.G., Gao, Y. and Jones, A.M. (2006a) Differential roles of *Arabidopsis* heterotrimeric G-protein subunits in modulating cell division in roots. *Plant Physiol*, **141**, 887–897.

Chen, J.G. and Jones, A.M. (2004) AtRGS1 function in *Arabidopsis thaliana*. *Methods Enzymol*, **389**, 338–350.

Chen, J.G., Pandey, S., Huang, J., Alonso, J.M., Ecker, J.R., Assmann, S.M. and Jones, A.M. (2004a) GCR1 can act independently of heterotrimeric G-protein in response to brassinosteroids and gibberellins in *Arabidopsis* seed germination. *Plant Physiol*, **135**, 907–915.

Chen, J.-G., Ullah, H., Temple, B. Liang, J., Alonso, J.M., Ecker, J. and Jones, A.M. (2006b) RACK1 mediates multiple hormone responsiveness and developmental processes in *Arabidopsis*. *J Exp Bot*, **57**, 2697–2708.

Chen, J.-G., Willard, F.S., Huang, J., Liang, J.,Chasse, S.A.,Jones, A.M. and Siderovski, D.P. (2003) A seven-transmembrane RGS protein that modulates plant cell proliferation. *Science*, **301**, 1728–1731.

Chen, S., Dell, E.J., Lin, F., Sai, J. and Hamm, H.E. (2004b) RACK1 regulates specific functions of Gβγ. *J Biol Chem*, **279**, 17861–17868.

Chen, Y., Ji, F., Xie, H., Liang, J. and Zhang, J. (2006c) The regulator of G-protein signaling proteins involved in sugar and abscisic acid signaling in *Arabidopsis* seed germination. *Plant Physiol*, **140**, 302–310.

Chen, Z., Hartmann, H.A., Wu, M.-J., Friedman, E.J., Chen, J.-G., Pulley, M., Schulze-Lefert, P., Panstruga, R. and Jones, A.M. (2006d) Expression analysis of the AtMLO gene family encoding plant-specific seven-transmembrane domain proteins. *Plant Mol Biol*, **60**, 583–597.

Cismowski, M.J., Takesono, A., Bernard, M.L., Duzic, E. and Lanier, S.M. (2001) Receptor-independent activators of heterotrimeric G-proteins. *Life Sci*, **68**, 2301–2308.

Cismowski, M.J., Takesono, A., Ma, C., Lizano, J.S., Xie, X., Fuernkranz, H., Lanier, S.M., Duzic, E. (1999) Genetic screens in yeast to identify mammalian nonreceptor modulators of G-protein signaling. *Nat Biotechnol*, **17**, 878–883.

Colucci, G., Apone, F., Alyeshmerni, N., Chalmers, D. and Chrispeels, M.J. (2002) GCR1, the putative *Arabidopsis* G-protein-coupled receptor gene is cell cycle-regulated, and its overexpression abolishes seed dormancy and shortens time to flowering. *Proc Natl Acad Sci USA*, **99**, 4736–4741.

Coursol, S., Fan, L.M., Le Stunff, H., Spiegel, S., Gilroy, S. and Assmann, S.M. (2003) Sphingolipid signalling in *Arabidopsis* guard cells involves heterotrimeric G-proteins. *Nature*, **423**, 651–654.

Coursol, S., Le Stunff, H., Lynch, D.V., Gilroy, S., Assmann, S.M. and Spiegel, S. (2005) *Arabidopsis* sphingosine kinase and the effects of phytosphingosine-1-phosphate on stomatal aperture. *Plant Physiol*, **137**, 724–737.

De Vries, L., Zheng, B., Fischer, T., Elenko, E. and Farquhar, M.G. (2000a) Activator of G protein signaling 3 is a guanine dissociation inhibitor for $G\alpha_i$ subunits. *Proc Natl Acad Sci USA*, **97**, 14364–14369.

De Vries, L., Zheng, B., Fischer, T., Elenko, E. and Farquhar, M.G. (2000b) The regulator of G-protein signaling family. *Annu Rev Pharmacol Toxicol*, **40**, 235–271.

Devoto, A., Hartmann, H.A., Piffanelli, P., Elliott, C., Simmons, C., Taramino, G., Goh, C.S., Cohen, F.E., Emerson, B.C., Schulze-Lefert, P. and Panstruga, R. (2003) Molecular phylogeny and evolution of the plant-specific seven-transmembrane MLO family. *J Mol Evol*, **56**, 77–88.

Devoto, A., Piffanelli, P., Nilsson, I., Wallin, E., Panstruga, R., von Heijne, G. and Schulze-Lefert, P. (1999) Topology, subcellular localization, and sequence diversity of the Mlo family in plants. *J Biol Chem*, **274**, 34993–35004.

Ding, L., Pandey, S. and Assmann, S.M. (2008) *Arabidopsis* extra-large G-proteins (XLGs) regulate root morphogenesis. *Plant J*, **53**, 248–263.

Echeverría, V., Hinrichs, M.V., Torrejón, M., Ropero, S., Martinez, J., Toro, M.J. and Olate, J. (2000) Mutagenesis in the switch IV of the helical domain of the human Gsα reduces its GDP/GTP exchange rate. *J Cell Biochem* **76**, 368–375.

Fairley-Grenot, K.A. and Assmann, S.M. (1991) Evidence for G-protein regulation of inward K^+ channel current in guard cells of fava bean. *Plant Cell*, **3**, 1037–1044.

Fredriksson, R. and Schioth, H.B. (2005) The repertoire of G-protein-coupled receptors in fully sequenced genomes. *Mol Pharmacol*, **67**, 1414–1425.

Fujisawa, Y., Kato, H. and Iwasaki, Y. (2001) Structure and function of heterotrimeric G-proteins in plants. *Plant Cell Physiol*, **42**, 789–794.

Fujisawa, Y., Kato, T., Ohki, S., Ishikawa, A., Kitano, H., Sasaki, T., Asahi, T. and Iwasaki, Y. (1999) Suppression of the heterotrimeric G-protein causes abnormal morphology, including dwarfism, in rice. *Proc Natl Acad Sci USA*, **96**, 7575–7580.

Gao, Y., Zeng, Q., Guo, J., Cheng, J., Ellis, B.E. and Chen, J.G. (2007) Genetic characterization reveals no role for the reported ABA receptor, GCR2, in ABA control of seed germination and early seedling development in *Arabidopsis*. *Plant J*, **52**, 1001–1113.

Gaudet, R., Bohm, A. and Sigler, P.B. (1996) Crystal structure at 2.4 Å resolution of the complex of transducin betagamma and its regulator, phosducin. *Cell*, **87**, 577–588.

Gazzarrini, S. and McCourt, P (2003) Cross-talk in plant hormone signalling: what *Arabidopsis* mutants are telling us. *Ann Bot (Lond)*, **91**, 605–612.

Gelli, A., Higgins, V.J. and Blumwald, E. (1997) Activation of plant plasma membrane Ca^{2+}-permeable channels by race-specific fungal elicitors. *Plant Physiol*, **113**, 269–279.

Genoud, T. and Metraux, J.P. (1999) Crosstalk in plant cell signaling: structure and function of the genetic network. *Trends Plant Sci*, **4**, 503–507.

Gurevich, V.V. and Gurevich, E.V. (2004) The molecular acrobatics of arrestin activation. *Trends Pharmacol Sci*, **25**, 105–111.

Gutkind, J.S. (1998) The pathways connecting G-protein-coupled receptors to the nucleus through divergent mitogen-activated protein kinase cascades. *J Biol Chem*, **273**, 1839–1842.

Hamm, H.E. (1998) The many faces of G-protein signaling. *J Biol Chem*, **273**, 669–672.

Hampoelz, B. and Knoblich, J.A. (2004) Heterotrimeric G-proteins: new tricks for an old dog. *Cell*, **119**, 453–456.

Hepler, J.R. and Gilman, A.G. (1992) G-proteins. *TrendsBiochem Sci*, **17**, 383–387.

Hooley, R. (1998) Plant hormone perception and action: a role for G-protein signal transduction? *Philos Trans R Soc Lond B Biol Sci*, **353**, 1425–1430.

Huang, H., Weiss, C.A. and Ma, H. (1994) Regulated expression of the *Arabidopsis* G-protein-subunit gene *GPA1*. *Int J Plant Sci*, **155**, 3–14.

Huang, J., Taylor, J.P., Chen, J.G., Wang, M., Uhrig, J.F., Nakagawa, T., Korth, K.L., Jones, A.M. (2006) The plastid protein THYLAKOID FORMATION1 and the plasma membrane G-protein GPA1 interact in a novel sugar-signaling mechanism in *Arabidopsis*. *Plant Cell*, **18**, 1226–1238.

Ishida, S., Takahashi, Y. and Nagata, T. (1993) Isolation of cDNA of an auxin-regulated gene encoding a G-protein β subunit-like protein from tobacco BY2-cells. *Proc Natl Acad Sci USA*, **90**, 11152–11156.

Ishikawa, A., Iwasaki, Y. and Asahi, T. (1996) Molecular cloning and characterization of a cDNA for the β subunit of a G-protein from rice. *Plant Cell Physiol*, **37**, 223–228.

Ishikawa, A., Tsubouchi, H., Iwasaki, Y. and Asahi, T. (1995) Molecular cloning and characterization of a cDNA for the α subunit of a G protein from rice. *Plant Cell Physiol*, **36**, 353–359.

Jacob, T., Ritchie, S., Assmann, S.M. and Gilroy, S. (1999) Abscisic acid signal transduction in guard cells is mediated by phospholipase D activity. *Proc Natl Acad Sci USA*, **96**, 12192–12197.

Johnston, C.A., Taylor, J.P., Gao, Y., Kimple, A.J., Chen, J.G., Siderovski, D.P., Jones, A.M. and Willard, F.S. (2007a) GTPase acceleration as the rate-limiting step in *Arabidopsis* G-protein coupled sugar sensing. *Proc Natl Acad Sci USA*, **104**, 17317–17322.

Johnston, C.A., Temple, B.R., Chen, J.G., Gao, Y., Moriyama, E.N., Jones, A.M., Siderovski, D.P. and Willard, F.S. (2007b) Comment on 'A G-protein-coupled receptor is a plasma membrane receptor for the plant hormone abscisic acid'. *Science*, **318**, 914c.

Jones, A.M. (2002) G-protein-coupled signaling in *Arabidopsis*. *Curr Opin Plant Biol*, **5**, 402–407.

Jones, A.M. and Assmann, S.M. (2004) Plants: the latest model system for G-protein research. *EMBO Rep*, **5**, 572–578.

Jones, A.M., Ecker, J.R. and Chen, J.G. (2003) A reevaluation of the role of the heterotrimeric G protein in coupling light responses in *Arabidopsis*. *Plant Physiol*, **131**, 1623–1627.

Jones, H.D., Smith, S.J., Desikan, R., Plakidou-Dymock, S., Lovegrove, A. and Hooley, R. (1998) Heterotrimeric G-proteins are implicated in gibberellin induction of α-amylase gene expression in wild oat aleurone. *Plant Cell*, **10**, 245–254.

Joo, J.H., Wang, S., Chen, J.G., Jones, A.M. and Fedoroff, N.V. (2005) Different signaling and cell death roles of heterotrimeric G-protein α and β subunits in the *Arabidopsis* oxidative stress response to ozone. *Plant Cell*, **17**, 957–970.

Josefsson, L.G. (1999) Evidence for kinship between diverse G-protein coupled receptors. *Gene*, **239**, 333–340.

Josefsson, L.G. and Rask, L. (1997) Cloning of putative G-protein–coupled receptor from *Arabidopsis thaliana*. *Eur J Biochem*, **249**, 415–420.

Kato, C., Mizutani, T., Tamaki, H., Kumagai, H., Kamiya, T., Hirobe, A., Fujisawa, Y., Kato, H. and Iwasaki, Y. (2004) Characterization of heterotrimeric G-protein complexes in rice plasma membrane. *Plant J*, **38**, 320–331.

Kim, M.C., Panstruga, R., Elliott, C., Müller, J., Devoto, A., Yoon, H.W., Park, H.C., Cho, M.J. and Schulze-Lefert, P. (2002) Calmodulin interacts with MLO protein to regulate defence against mildew in barley. *Nature*, **416**, 447–451.

Lanier, S.M. (2004) AGS proteins, GPR motifs and the signals processed by heterotrimeric G proteins. *Biol Cell*, **96**, 369–372.

Lapik, Y.R. and Kaufman, L.S. (2003) The *Arabidopsis* cupin domain protein AtPirin1 interacts with the G-protein α-subunit GPA1 and regulates seed germination and early seedling development. *Plant Cell,* **15**, 1578–1590.

Lease, K.A, Wen, J., Li, J., Doke, J.T., Liscum, E. and Walker, J.C. (2001) A mutant *Arabidopsis* heterotrimeric G-protein β subunit affects leaf, flower, and fruit development. *Plant Cell,* **13**, 2631–2641.

Lee, Y.R. and Assmann, S.M. (1999) *Arabidopsis thaliana* 'extra-large GTP-binding protein' (AtXLG1): a new class of G-protein. *Plant Mol Biol,* **40**, 55–64.

Li, W. and Assmann, S.M. (1993) Characterization of a G-protein-regulated outward K^+ current in mesophyll cells of *Vicia faba L. Proc Natl Acad Sci USA,* **90**, 262–266.

Lieberherr, D., Thao, N.P., Nakashima, A., Umemura, K., Kawasaki, T. and Shimamoto, K. (2005) A sphingolipid elicitor-inducible mitogen-activated protein kinase is regulated by the small GTPase OsRac1 and heterotrimeric G-protein in rice. *Plant Physiol,* **138**, 1644–1652.

Liu, W. and Northup, J.K. (1998) The helical domain of a G-protein α subunit is a regulator of its effector. *Proc Natl Acad Sci USA,* **95**, 12878–12883.

Liu, X., Yue, Y., Li, B., Nie, Y., Li, W., Wu, W.H. and Ma, L. (2007) A G-protein-coupled receptor is a plasma membrane receptor for the plant hormone abscisic acid. *Science,* **315**, 1712–1716.

Llorente, F., Alonso-Blanco, C., Sanchez-Rodriguez, C., Jorda, L. and Molina, A. (2005) ERECTA receptor-like kinase and heterotrimeric G-protein from *Arabidopsis* are required for resistance to the necrotrophic fungus *Plectosphaerella cucumerina. Plant J,* **43**, 165–180.

Luttrell, L.M. (2002) Big G, little G: G-proteins and actin cytoskeletal reorganization. *Mol Cell,* **9**, 1152–1154.

Ma, H. (1994) GTP-binding proteins in plants: new members of an old family. *Plant Mol Biol,* **26**, 1611–1636.

Ma, H., Yanofsky, M.F. and Huang, H. (1991) Isolation and sequence analysis of *TGA1* cDNAs encoding a tomato G-protein α subunit. *Gene,* **107**, 189–195.

Ma, H., Yanofsky, M.F. and Meyerowitz, E.M. (1990) Molecular cloning and characterization of GPA1, a G-protein α subunit gene from *Arabidopsis thaliana. Proc Natl Acad Sci USA,* **87**, 3821–3825.

Mason, M.G. and Botella, J.R. (2001) Isolation of a novel G-protein γ-subunit from *Arabidopsis thaliana* and its interaction with Gβ. *Biochim Biophys Acta,* **1520**, 147–153.

Mason, M.G. and Botella, J.R. (2000) Completing the heterotrimer: isolation and characterization of an *Arabidopsis thaliana* G-protein γ-subunit cDNA. *Proc Natl Acad Sci USA,* **97**, 14784–14788.

McCudden, C.R., Hains, M.D., Kimple, R.J., Siderovski, D.P. and Willard, F.S. (2005) G-protein signaling: back to the future. *Cell Mol Life Sci,* **62**, 551–577.

Milligan, G. and Grassie, M.A. (1997) How do G-proteins stay at the plasma membrane? *Essays Biochem,* **32**, 49–60.

Mishra, G., Zhang, W., Deng, F., Zhao, J. and Wang, X. (2006) A bifurcating pathway directs abscisic acid effects on stomatal closure and opening in *Arabidopsis. Science,* **312**, 264–266.

Misra, S., Wu, Y., Venkataraman, G., Sopory, S.K. and Tuteja, N. (2007) Heterotrimeric G-protein complex and G-protein-coupled receptor from a legume (*Pisum sativum*): role in salinity and heat stress and cross-talk with phospholipase C. *Plant J,* **51**, 656–669.

Moore, C.A.C., Milano, S.K. and Benovic, J.L. (2007) Regulation of receptor trafficking by GRKs and arrestins. *Ann Rev Physiol,* **69**, 451–482.

Moriyama, E.N., Strope, P.K., Opiyo, S.O., Chen, Z. and Jones, A.M. (2006) Mining the *Arabidopsis thaliana* genome for highly-divergent seven transmembrane receptors. *Genome Biology,* **7**, R96.

Muschietti, J.P., Martinetto, H.E., Coso, O.A., Farber, M.D., Torres, H.N. and Flawia, M.M. (1993) G-protein from *Medicago sativa*: functional association to photoreceptors. *Biochem J,* **291**, 383–388.

Natochin, M., Lester, B., Peterson, Y.K., Bernard, M.L., Lanier, S.M. and Artemyev, N.O. (2000) AGS3 inhibits GDP dissociation from Gα subunits of the G_i family and rhodopsin-dependent activation of transducin. *J Biol Chem,* **275**, 40981–40985.

Neer, E.J. (1995) Heterotrimeric G proteins: Organizers of transmembrane signals. *Cell,* **80**, 249–257.

Neuhaus, G., Bowler, C., Kern, R. and Chua, N.H. (1993) Calcium/calmodulin-dependent and -independent phytochrome signal transduction pathways. *Cell,* **73**, 937–952.

New, D.C. and Wong, J.T. (1998) The evidence for G-protein-coupled receptors and heterotrimeric G proteins in protozoa and ancestral metazoa. *Biol Signals Recept,* **7**, 98–108.

Ng, C.K., Carr, K., McAinsh, M.R., Powell, B. and Hetherington, A.M. (2001) Drought-induced guard cell signal transduction involves sphingosine-1-phosphate. *Nature,* **410**, 596–599.

Offermanns, S., Mancino, V., Revel, J.P. and Simon, M.I. (1997) Vascular system defects and impaired cell chemokinesis as a result of $G\alpha_{13}$ deficiency. *Science,* **275**, 533–536.

Okamoto, H., Matsui, M. and Deng, X.W. (2001) Overexpression of the heterotrimeric G-protein α-subunit enhances phytochrome-mediated inhibition of hypocotyl elongation in *Arabidopsis*. *Plant Cell,* **13**, 1639–1652.

Pandey, S. and Assmann, S.M. (2004) The *Arabidopsis* putative G-protein-coupled receptor GCR1 interacts with the G protein α subunit GPA1 and regulates abscisic acid signaling. *Plant Cell,* **16**, 1616–1632.

Pandey, S., Chen, J.G., Jones, A.M. and Assmann, S.M. (2006) G-protein complex mutants are hypersensitive to abscisic acid regulation of germination and postgermination development. *Plant Physiol,* **141**, 243–256.

Perfus-Barbeoch, L., Jones, A.M. and Assmann, S.M. (2004) Plant heterotrimeric G protein function: insights from *Arabidopsis* and rice mutants. *Curr Opin Plant Biol,* **7**, 719–731.

Raghuram, N., Chandok, M.R. and Sopory, S.K. (1999) Light regulation of nitrate reductase gene expression in maize involves a G-protein. *Mol Cell Biol Res Commun,* **2**, 86–90.

Restrepo, D., Teeter, J.H. and Schild, D. (1996) Second messenger signaling in olfactory transduction. *J Neurobiol,* **30**, 37–48.

Rhee, S.G. (2001) Regulation of phosphoinositide-specific phospholipase C. *Annu Rev Biochem,* **70**, 281–312.

Saalbach, G., Natura, G., Lein, W., Buschmann, P., Dahse, I., Rohrbeck, M. and Nagy F. (1999) The α-subunit of a heterotrimeric G-protein from tobacco, NtGPα1, functions in K^+ channel regulation in mesophyll cells. *J Exp Bot,* **50**, 53–61.

Sanada, K. and Tsai, L.H. (2005) G protein βγ subunits and AGS3 control spindle orientation and asymmetric cell fate of cerebral cortical progenitors. *Cell,* **122**, 119–131.

Schioth, H.B. and Fredriksson, R. (2005) The GRAFS classification system of G-protein coupled receptors in comparative perspective. *Gen Comp Endocrinol,* **142**, 94–101.

Schulz, R. (2001) The pharmacology of phosducin. *Pharmacol Res,* **43**, 1–10.

Schwacke, R., Schneider, A., van der Graaff, E., Fischer, K., Catoni, E., Desimone, M., Frommer, W.B., Flügge, U.I. and Kunze, R. (2003) ARAMEMNON, a novel database for *Arabidopsis* integral membrane proteins. *Plant Physiol,* **131**, 16–26.

Seo, H.S., Choi, C.H., Lee, S.Y., Cho, M.J. and Bahk, J.D. (1997) Biochemical characteristics of a rice (*Oryza sativa L.*, IR36) G-protein α-subunit expressed in *Escherichia coli*. *Biochem J,* **324**, 273–281.

Seo, H.S., Kim, H.Y., Jeong, J.Y., Lee, S.Y., Cho, M.J. and Bahk, J.D. (1995) Molecular cloning and characterization of *RGA1* encoding a G protein α subunit from rice (*Oryza sativa L.* IR-36). *Plant Mol Biol,* **27**, 1119–1131.

Spiegel, S. and Milstien, S. (2003) Sphingosine-1-phosphate: an enigmatic signalling lipid. *Nat Rev Mol Cell Biol,* **4**, 397–407.

Sprang, S.R. (1997) G-protein mechanisms: Insights from structural analysis. *Annu Rev Biochem,* **66**, 639–678.

Suharsono, U., Fujisawa, Y., Kawasaki, T., Iwasaki, Y., Satoh, H. and Shimamoto, K. (2002) The heterotrimeric G-protein α subunit acts upstream of the small GTPase Rac in disease resistance of rice. *Proc Natl Acad Sci USA,* **99**, 13307–13312.

Takesono, A., Cismowski, M.J., Ribas, C., Bernard, M., Chung, P., Hazard, S., III, Duzic, E. and Lanier, S.M. (1999) Receptor-independent activators of heterotrimeric G-protein signaling pathways. *J Biol Chem,* **274**, 33202–33205.

Tang, X. and Downes, C.P. (1997) Purification and characterization of Gβγ-responsive phosphoinositide 3-kinases from pig platelet cytosol. *J Biol Chem,* **272**, 14193–14199.

Temple, B.R.S. and Jones, A.M. (2007) The plant heterotrimeric G-protein complex. *Annu Rev Plant Biol,* **58**, 249–266.

Trusov, Y., Rookes, J.E., Chakravorty, D., Armour, D., Schenk, P.M. and Botella, J.R. (2006) Heterotrimeric G-proteins facilitate *Arabidopsis* resistance to necrotrophic pathogens and are involved in jasmonate signaling. *Plant Physiol,* **140**, 210–220.

Trusov, Y., Rookes, J.E., Tilbrook, K., Chakravorty, D., Mason, M.G., Anderson, D., Chen, J.G., Jones, A.M. and Botella, J.R. (2007) Heterotrimeric G-protein γ subunits provide functional selectivity in Gβγ dimer signaling in *Arabidopsis*. *Plant Cell,* **19**, 1235–1250.

Ueguchi-Tanaka, M., Fujisawa, Y., Kobayashi, M., Ashikari, M., Iwasaki, Y., Kitano, H. and Matsuoka, M. (2000) Rice dwarf mutant *d1*, which is defective in the α subunit of the heterotrimeric G-protein, affects gibberellin signal transduction. *Proc Natl Acad Sci USA,* **97**, 11638–11643.

Ullah, H., Chen, J.-G., Temple, B., Boyes, D.C., Alonso, J.M., Davis, K.R., Ecker, J.R. and Jones, A.M. (2003) The β-subunit of the *Arabidopsis* G protein negatively regulates auxin-induced cell division and affects multiple developmental processes. *Plant Cell,* **15**, 393–409.

Ullah, H., Chen, J.G., Wang, S. and Jones, A.M. (2002) Role of a heterotrimeric G-protein in regulation of *Arabidopsis* seed germination. *Plant Physiol,* **129**, 897–907.

Ullah, H., Chen, J.G., Young, J.C., Im, K.H., Sussman, M.R. and Jones, A.M. (2001) Modulation of cell proliferation by heterotrimeric G protein in *Arabidopsis*. *Science,* **92**, 2066–2069.

Wang, H.X., Perdue, T., Weerasinghe, R., Taylor, J.P., Cakmakci, N.G., Marzluff, W.F. and Jones, A.M. (2006a) A Golgi-localized hexose transporter is involved in heterotrimeric G protein-mediated early development in *Arabidopsis*. *Mol Biol Cell,* **17**, 4257–4269.

Wang, L., Xu, Y.-Y., Ma, Q.-B., Li, D., Xu, Z.-H. and Chong, K. (2006b) Heterotrimeric G-protein α subunit is involved in rice brassinosteroid response. *Cell Res,* **16**, 916–922.

Wang, S., Narendra, S. and Fedoroff, N. (2007) Heterotrimeric G-protein signaling in the *Arabidopsis* unfolded protein response. *Proc Natl Acad Sci USA,* **104**, 3817–3822.

Wang, X.Q., Ullah, H., Jones, A.M. and Assmann, S.M. (2001) G-protein regulation of ion channels and abscisic acid signaling in *Arabidopsis* guard cells. *Science,* **292**, 2070–2072.

Warpeha, K.M., Hamm, H.E., Rasenick, M.M. and Kaufman, L.S. (1991) A blue-light-activated GTP-binding protein in the plasma membranes of etiolated peas. *Proc Natl Acad Sci USA,* **88**, 8925–8929.

Warpeha, K.M., Lateef, S.S., Lapik, Y., Anderson, M., Lee, B.S. and Kaufman, L.S. (2006) G-protein-coupled receptor 1, G-protein Gα-subunit 1, and prephenate dehydratase 1 are required for blue light-induced production of phenylalanine in etiolated *Arabidopsis*. *Plant Physiol,* **140**, 844–855.

Warpeha, K.M., Upadhyay, S., Yeh, J., Adamiak, J., Hawkins, S.I., Lapik, Y.R., Anderson, M.B. and Kaufman, L.S. (2007) The GCR1, GPA1, PRN1, NF-Y signal chain mediates both blue light and abscisic acid responses in *Arabidopsis*. *Plant Physiol,* **143**, 1590–1600.

Weiss, C.A., Garnaat, C.W., Mukai, K., Hu, Y. and Ma, H. (1994) Isolation of cDNAs encoding guanine nucleotide-binding protein β-subunit homologues from maize (ZGB1) and *Arabidopsis* (AGB1). *Proc Natl Acad Sci USA,* **91**, 9554–9558.

Weiss, C.A., Huang, H. and Ma, H. (1993) Immunolocalization of the G protein α subunit encoded by the GPA1 gene in *Arabidopsis*. *Plant Cell,* **5**, 1513–1528.

Wickman, K. and Clapham, D.E. (1995) Ion channel regulation by G proteins. *Physiol Rev,* **75**, 865–885.

Willard, F.S., Kimple, R.J. and Siderovski, D.P. (2004) Return of the GDI: the GoLoco motif in cell division. *Annu Rev Biochem,* **73**, 925–951.

Wise, A., Thomas, P.G., Carr, T.H., Murphy, G.A. and Millner, P.A. (1997) Expression of the *Arabidopsis* G-protein GPα1: purification and characterisation of the recombinant protein. *Plant Mol Biol,* **33**, 723–728.

Wu, W.-H. and Assmann, S.M. (1994) A membrane-delimited pathway of G-protein regulation of the guard-cell inward K^+ channel. *Proc Natl Acad Sci USA,* **91**, 6310–6314.

Yuan, R., Wu, Z., Kostenyuk, I.A. and Burns, J.K. (2005) G-protein-coupled α_{2A}-adrenoreceptor agonists differentially alter citrus leaf and fruit abscission by affecting expression of ACC synthase and ACC oxidase. *J Exp Bot,* **56,** 1867–1875.

Zaina, S., Mapelli, S., Reggiani, R. and Bertani, A. (1991) Auxin and GTPase activity in membranes from aerobic and anaerobic rice coleoptile. *J Plant Physiol,* **138**, 760–762.

Zaina, S., Reggiani, R. and Bertani, A. (1990) Preliminary evidence for involvement of GTP-binding protein(s) in auxin signal transduction in rice (*Oryza sativa* L.) coleoptiles. *J Plant Physiol,* **136**, 653–658.

Zheng, Y. (2004) G protein control of microtubule assembly. *Annu Rev Cell Dev Biol,* **20**, 867–894.

Zeng, Q., Wang, X. and Running, M.P. (2007) Dual lipid modification of *Arabidopsis* Gγ subunits is required for efficient plasma membrane targeting. *Plant Physiol,* **143**, 1119–1131.

Zhang, W., Qin, C., Zhao, J. and Wang, X. (2004) Phospholipase Dα1-derived phosphatidic acid interacts with ABI1 phosphatase 2C and regulates abscisic acid signaling. *Proc Natl Acad Sci USA,* **101**, 9508–9513.

Zhao, J. and Wang, X. (2004) *Arabidopsis* phospholipase Dα1 interacts with the heterotrimeric G-protein α-subunit through a motif analogous to the DRY motif in G-protein-coupled receptors. *J Biol Chem,* **279**, 1794–1800.

Annual Plant Reviews (2008) **33**, 64–99
doi: 10.1111/b.9781405160025.2008.00003.x www.interscience.wiley.com

Chapter 3

ROP/RAC GTPases

Ying Fu[1,2], Tsutomu Kawasaki[3], Ko Shimamoto[3], and Zhenbiao Yang[2]

[1] *State Key Laboratory of Plant Physiology and Biochemistry, College of Biological Sciences, China Agricultural University, Beijing 100094, People's Republic of China*
[2] *Department of Botany and Plant Sciences, University of California, Riverside, CA 92521, USA*
[3] *Laboratory of Plant Molecular Genetics, Nara Institute of Science and Technology, Takayama 8916-5, Ikoma, Nara 630-0192, Japan*

Abstract: Rho-family small GTPases are monomeric guanine nucleotide-binding proteins that act as key molecular switches in the regulation of many important cellular processes. ROP (*Rho*-related GTPase from *p*lants)/RAC is a plant-specific subfamily of Rho GTPases that plays a versatile role in the regulation of plant growth, development, and responses to the environment. Plant ROP/RAC proteins share conserved structural features, cellular functions, and functional partners with their counterparts in fungi and animals. However, plant ROP/RAC GTPases contain unique structural motifs not present in their fungal and animal counterparts. Furthermore, recent studies have revealed several classes of plant-specific regulators and effector proteins for the ROP subfamily of Rho GTPases. Several ROP-dependent signaling networks have also emerged from the investigation of ROP-interacting proteins. This chapter summarizes our current knowledge of ROP/RAC signaling mechanisms and pathways/networks in the two model plant systems, *Arabidopsis* and rice.

Keywords: G proteins; small GTPases; cell polarity; cytoskeleton; hormones; defense

3.1 Introduction

Small guanine nucleotide-binding proteins (or GTPases), belonging to the Ras superfamily of monomeric G proteins, are structurally related to the heterotrimeric G-protein α subunit (Gα) (see Chapter 2). Small GTPase and Gα share a conserved Ras-like GTPase domain (G-domain), the minimal signaling unit that carries out the function of nucleotide binding and hydrolysis in most GTPases. Like heterotrimeric G protein, small GTPases cycle between the "OFF" (GDP-bound) and "ON" (GTP-bound) states. However, several aspects distinguish small GTPases from heterotrimeric G proteins. First, small GTPases do not have the independently folding α-helical domain, nor the Asp/Glu-rich loop found in Gα subunits. Second, in contrast to the heterotrimer as a functional structure in heterotrimeric G proteins, small GTPases are composed of a single protein as the basic signaling unit. As a result, they are also named monomeric G proteins. Third, instead of direct activation by transmembrane receptors as found for heterotrimeric G proteins, small GTPases are activated by intracellular signaling molecules, guanine nucleotide exchange factors (GEFs) that convert the GDP-bound inactive form into the GTP-bound active form. Fourth, the intrinsic GTPase activity of small GTPase is weak, underscoring the importance of GTPase-activating proteins (GAPs) for efficient GTP hydrolysis in small GTPase signaling. Fifth, Gα subunits usually remain associated with the plasma membrane (PM), but most small GTPases cycle between the PM and the cytosol, and only the PM-associated form can be activated by GEFs. Therefore, a cytosolic sequestering protein called guanine nucleotide dissociation inhibitor (GDI) plays a critical role in the regulation of small GTPase signaling. Last but not the least, the mechanism by which small GTPases regulate downstream effectors is distinct. Upon activation, small GTPases interact with multiple downstream effectors usually through its single effector domain. These unique structural and functional features underscore the importance of the diversity of their interacting partners in the signaling of small GTPases (Bourne *et al.*, 1991; Vetter and Wittinghofer, 2001; Yang, 2002; Wennerberg *et al.*, 2005; Yang and Fu, 2007).

The Ras superfamily is divided into five families (Ras, Rho, Rab, Arf, and Ran) with distinct structures and functions. Rho and Ras families can be directly involved in relaying extracellular signals, whereas Arf, Rab, and Ran proteins usually participate in the control of fundamental cellular processes common to all eukaryotic cells including vesicle trafficking and RNA and protein transport through the nuclear pore. Ras GTPase homologs are absent from plants, while Rho GTPases are found in all eukaryotic kingdoms. The Rho (*Ras homologous*) family was identified as a new member of the Ras superfamily in early 1990s. Rho GTPases are well known for their conserved function in the signaling pathways regulating the actin cytoskeleton. There are 22 members of Rho GTPases in mammals and 6 in the budding yeast (Hall, 1998; Wennerberg *et al.*, 2005; Brembu *et al.*, 2006; Brennwald and Rossi, 2007).

Plants possess a single subfamily of Rho GTPases, termed ROPs (*Rho*-related GTPases from *p*lants). Phylogenetic analysis suggests that ROPs evolved prior to the divergence of other three subfamilies of Rho GTPases from animals and fungi (Zheng and Yang, 2000; Vernoud *et al.*, 2003). At the primary amino acid sequence level, ROPs are most similar to RACs and thus have also been referred to as RACs (Yang, 2002; Christensen *et al.*, 2003; Vernoud *et al.*, 2003). This chapter aims to summarize a wealth of knowledge on ROP/RAC signaling with regards to the structure, function, and regulation of these GTPases.

3.2 Structural conservation and diversification

3.2.1 The primary sequence

Since the first plant homolog of Rho-related GTPases was discovered in the garden pea (Yang and Watson, 1993), other members of the ROP subfamily have been identified widely from different plant species, including the lower plant moss, gymnosperm, and angiosperm (Winge *et al.*, 2000; Brembu *et al.*, 2006). Genome sequencing reveals 11 *ROP* genes in *Arabidopsis* and 7 in rice, and at least 9 *ROP* genes have been found in maize (Li *et al.*, 1998; Winge *et al.*, 2000; Christensen *et al.*, 2003).

The comparison of the primary amino acid sequence between ROP proteins with other Ras-superfamily small GTPases has been summarized in previous reviews (Zheng and Yang, 2000; Yang, 2002; Berken, 2006). Generally, there are five highly conserved sequence motifs (G1–G5), which have also been referred to as guanine nucleotide-binding domain I–IV and effector binding domain E (Yang, 2002) (Fig. 3.1a, Color plate 5). The G1 loop (domain I), also called the P loop, is responsible for the binding of the α- and β-phosphate groups. The G3 loop (domain II) provides residues for binding Mg^{2+} and the γ-phosphate group. The G2 loop (E domain) is the site that effectors and GAPs bind. This domain also contains a conserved Thr residue responsible for Mg^{2+} binding. The G5 loop (domain IV) recognizes the guanine base, and the G4 loop (domain III) contains Lys and Asp residues interacting with the nucleotide. Certain specific residues in guanine nucleotide-binding domains (G1, G3, G4, and G5 or domain I–IV) are critical for the "ON/OFF" status of Rho GTPases.

Sequence comparison among Rho family members reveals that ROPs are clearly distinct from RAC, RHO, and CDC42 and belong to a unique subfamily of Rho GTPases. Most striking differences are several conserved amino acid substitutions found in the GTP-binding motifs (Zheng and Yang, 2000; Yang, 2002; Berken, 2006). Another significant difference is 2–4-amino acid deletion in the Rho insert region, a domain found only in the Rho GTPases of the Ras superfamily and proposed to function in effector interaction or effector activation (Bishop and Hall, 2000; Thapar *et al.*, 2002; Berken, 2006).

Within the plant kingdom, ROPs share over 70% amino acid similarity with each other and are divided into four phylogenetic groups I–IV (Zheng and Yang, 2000; Yang, 2002). The most variable region is located at the C-terminus. Those ROPs in group I (*Arabidopsis* ROP8), group III (ROP7, cotton GhRAC9, GhRAC13, rice OsROP5), and group IV (ROP1–ROP6, GhRAC1, maize ZmROPB and ZmROPD) contain a signature motif CAAL (C, cysteine; L, leucine; A, aliphatic amino acids) at the C-terminus. This motif is the target for geranylgeranyl transferase I (GGTase I). The prenylation of the cysteine residue in CAAL motif is involved in the anchoring of ROPs to cellular membranes. Group II ROPs (ROP9–ROP11, OsRAC1–OsRAC4,

ZmROP6–ZmROP8) contain diverse prenylation sites having either farnesylation motif CAAX (X, any amino acid except for L) or the geranylgeranylation motif CXX for GGTase II (Zheng and Yang, 2000; Yang, 2002). ROP11 is an exception, which lacks a prenylation motif. Additional one or two cysteine residues are found in the C-terminus of some proteins from group II and may be palmitoylated (Ivanchenko *et al.*, 2000). These post-translational modifications of the C-terminal region of ROPs are crucial for ROP membrane association, which is usually a prerequisite for the regulation of ROP GTPase activities (see below).

3.2.2 The three-dimensional structure

The crystal structure of Rho small GTPases and their regulators has been resolved for certain mammalian members (Vetter and Wittinghofer, 2001; Dvorsky and Ahmadian, 2004). Based on the crystal structure of the human RAC1, a structure of *Arabidopsis* ROP4 was predicted (Berken, 2006). The predicted structure features the basic RAS GTPase α/β fold, consisting of six stranded β-sheets ($\beta1$–$\beta6$) and five helices ($\alpha1$–$\alpha5$). Surprisingly, the resolution of the first crystal structure of a small GTPase in the plant kingdom (ROP9) reveals only four helices ($\alpha1$, $\alpha3$–$\alpha5$) surrounding the hydrophobic core of six β-sheets (Sormo *et al.*, 2006). The $\alpha2$ helix in the switch II region of human RHO, CDC42, and RAC GTPases is missing in ROP9. Probably due to a distinct serine residue (Ser68) in ROP9 instead of aspartic acid (Asp) in its human counterparts, the ROP9 switch II region cannot form a stable secondary structure. This flexibility together with the conserved SYR site in ROPs (instead of SYP site in other Rho GTPases) may facilitate ROP9's specific interaction with novel proteins in plants. However, this Ser residue is replaced by Asn residues in other ROPs, and thus it is possible that the structure of the switch II region for other ROPs varies from that of ROP9 and Rho GTPases from nonplant systems. Another remarkable difference between ROP9 and human Rho proteins is found in the Rho insertion region. ROP9 has a shortened Rho insertion region (4-amino acid deletion), although it still forms a smaller insert helix (α_i) (Sormo *et al.*, 2006). The Rho insertion region was suggested to be responsible for binding to certain effectors (Bishop and Hall, 2000; Thapar *et al.*, 2002; Berken, 2006); the unique feature of the primary and secondary structure in this region may imply that plants have specific signaling pathways not found in other systems.

3.3 Physiological functions and downstream signaling

Signaling small GTPases are important molecular switches that regulate a wide range of cellular processes and functions. In fungi and animal systems, each cellular process is often coordinately controlled by several distinct small GTPases. For example, cell polarity establishment in the budding yeast requires the coordination of Ras, CDC42, and RHO (Chant, 1999; Fu and Yang, 2001). However, plants do not possess Ras homologs and RAC, CDC42, and RHO subfamilies of Rho GTPases. ROP is the sole subfamily of signaling small GTPases in plants. The solitude of ROP in small GTPase signaling in plants could explain why ROPs have emerged as signal integrators and co-ordinators of a wide range of signaling pathways that regulate fundamental developmental processes and responses to hormones, pathogens, and abiotic stimuli.

3.3.1 ROPs and plant development

3.3.1.1 Cell polarity control and polarized cell growth

Cell polarity is critical for development in plants as in other organisms. For example, the formation of cell polarity is a prerequisite for asymmetric cell division that is a key mechanism for cell differentiation. In plants, polarized localization of the auxin efflux carriers PIN proteins to a particular side of the cell determines the direction of auxin flows and is required for the establishment of auxin concentration gradients. Cell polarity is also required for cell morphogenesis and is associated with the development of specialized cells, such as root hair initiation, pollen germination, and trichome morphogenesis. Because of lack of cell mobility and the presence of cell walls, cell morphogenesis in plants strictly relies on the expansion of the cell surface in specific sites of the cell, i.e., polarized growth. Polarized growth is also essential for the formation of specialized cells, such as pollen tubes, root hairs, and trichomes. Studies over the last decade have shown that ROPs play an important role in the control of cell polarity, polarized growth, and cell morphogenesis in various cell types in plants (Yang, 2002; Berken, 2006).

3.3.1.1.1 Pollen tubes. Immunolocalization studies indicate that ROPs are localized to the apical region of the PM in pea pollen tubes, providing the first hint that ROPs are involved in the control of cell polarity (Lin *et al.*, 1996). Pollen tubes expand by tip growth, an extreme form of polarized growth, in which vesicles are targeted to and fused with the apical region of the PM. Subsequently, it was shown that microinjected anti-ROP antibody inhibited pollen tube growth, suggesting that ROPs are required for polarized tip growth in pollen tubes (Lin and Yang, 1997). An essential role for ROPs in tip growth was further demonstrated using dominant negative (DN) forms of ROPs or loss-of-function approaches. Overexpression of DN-rop1 or -rop5/AtRAC2 blocked pollen tube elongation in transgenic *Arabidopsis* plants or transiently transfected tobacco pollen (Kost *et al.*, 1999; Li *et al.*, 1999; Fu *et al.*, 2001). In *Arabidopsis*, three closely related ROPs, ROP1, ROP3, and ROP5, are expressed in pollen (Li *et al.*, 1998). Expression of an antisense RNA for the pollen-specific ROP1 weakly inhibited pollen tube growth in *Arabidopsis* (Li *et al.*, 1999), but expression of a *ROP3 RNAi* construct that presumably silences all three ROPs completely blocked pollen tube growth (Lee and Yang, unpublished results). Taken together, ROP1, ROP3, and ROP5 are functionally redundant in their control of pollen tube tip growth.

Overexpression of wild-type ROP1 or ROP5 and a constitutively active form of these ROPs induced depolarization of cell growth in pollen tubes; that is, growth was not restricted to the tip any more (Kost *et al.*, 1999; Li *et al.*, 1999). These results suggest that a proper regulation of ROP activity is required for the control of cell polarity in pollen tubes. Using an active ROP1 reporter, it was shown that ROP1 activity is localized to the apical region of the pollen tube PM as an apical cap (Hwang *et al.*, 2005). Consistent with the requirement for ROP1 activity in the control of growth polarity, the ROP1 activity cap shows an interesting spatiotemporal dynamics. It oscillates along with pollen tube growth oscillation—an increase in the ROP1 activity is followed by an increase in the pollen tube growth rate. Disrupting the dynamics of the ROP1 activity cap causes growth depolarization as well as loss of growth oscillation (Hwang *et al.*, 2005). Growth depolarization induced by ROP1 overexpression is associated with the enlargement and stabilization of the apical cap of active ROP1. Further studies support the notion that the active ROP1 cap spatially determines the tip growth region and temporally drives tip growth (Hwang *et al.*, submitted).

3.3.1.1.2 Root hairs. Development of root hairs can be divided into several stages: selection of root hair sites, bulge/swelling formation, transition from diffuse growth to tip growth, and continuous tip growth to form a hair-like structure (Fu and Yang, 2001; Jones *et al.*, 2002). An anti-ROP4 antibody detected ROPs at the root hair formation site and at the tip of swelling bulge (Molendijk *et al.*, 2001). GFP-ROP2, when expressed in a transgenic line to a low level, was also localized to the future site of root hair formation before any detectable swelling, and is maintained at the tip of bulge or elongating root hairs during remaining developmental stages (Jones *et al.*, 2002). These localization patterns support a role for ROP GTPases in the regulation of cell polarity establishment and maintenance throughout root hair development. Overexpression of ROP2, a group IV ROP expressed in root hairs, produced extra root hairs initiated at abnormal sites as well as multiple tips during tip growth, and expression of constitutive active (CA)-rop2 induced depolarized growth in addition to additional and misplaced root hairs, further arguing a critical role for ROP2 in the regulation of cell polarity during root hair development (Jones *et al.*, 2002). On the contrary, overexpression of DN-rop2 inhibited both the initiation and maintenance of root hair tip growth (Jones *et al.*, 2002). Molendijk and coworkers also reported that expression of GFP-CA-rop4 and GFP-CA-rop6 led to depolarized growth in root hairs, but there is no direct evidence that ROP4 and ROP6 are expressed in root hairs (Molendijk *et al.*, 2001; Jones *et al.*, 2002).

Recently, Jones and coworkers identified six novel genes involved in root hair morphogenesis by comparing transcriptomes in the root hair differentiation zone between *Arabidopsis* wild-type and root hair defective mutant (*rhd-2*). These six genes (*MRH1–MRH6*) encode signaling proteins such as leucine-rich repeat (LRR) receptor-like kinases and glycosylphosphatidylinositol-anchored proteins. The latter hints a link between lipid rafts and the regulation of root hair development. Interestingly, ROP6 and perhaps other ROPs have been suggested to localize to lipid rafts (Sorek *et al.*, 2007). Furthermore, mutants for some of these *MRH* genes show phenotypic resemblance to the ROP2 overexpression phenotype (e.g., *mrh6* displays wider root hair initiate site and root hair base) (Jones *et al.*, 2002, 2006). Further studies to determine the relationship between these genes and ROPs may uncover new players in ROP signaling pathways underlying root hair development.

3.3.1.1.3 Pavement cells. Unlike root hairs and pollen tubes, most cells in plants expand and develop into specific shapes by a form of polar growth, termed polarized diffuse growth. Diffuse growth occurs throughout the entire cell surface, whereas the polarity is controlled by cortical microtubule (MT)-mediated microfibril arrangement (Kropf *et al.*, 1998). Diffuse growth is unique to plants. The involvement of ROP2 in the bulge formation during root hair development suggests a role for ROPs in the regulation of diffuse growth (Jones *et al.*, 2002). The demonstration of ROP2 function in the regulation of pavement cell morphogenesis suggests a wider role for ROPs in the regulation of diffuse growth (Fu *et al.*, 2002). *Arabidopsis* leave pavement cells exhibit jigsaw-puzzle appearance (with pronounced interlocking lobes and indentations) and provide a multicellular model system to study the regulation of polarized diffuse growth. *CA-rop2* expression triggered ectopic cell expansion that eliminates the indenting neck region, whereas *DN-rop2* expression caused an opposite effect, producing pavement cells with narrower neck regions and shorter lobes (Fu *et al.*, 2002). Similar to *DN-rop2* expression, *ROP2RNAi/rop4* double mutant dramatically restricted the expansion in the neck region and inhibited the elongation of the lobe (Fu *et al.*, 2002, 2005). Since *ROP2RNAi/rop4* has stronger effect compared to either *ROP2RNAi* or *rop4-1* single mutants, ROP4 is believed to function redundantly to ROP2. Transiently expressed GFP-ROP2 was localized preferentially to the initiation and growth site of lobes. *In vivo* FRET (fluorescence resonance energy transfer) analysis between CFP-RIC4 (a ROP downstream effector that preferentially interacts with active form ROP2) and YFP-ROP2 provided evidence that ROP2 is activated at the initiation site and the tip of lobes (Fu *et al.*, 2005). These results clearly indicate ROP2 directs the position of polarized cell expansion as found in the formation of root hair budges. It is interesting that the ROP2-mediated spatial regulation of diffuse growth in pavement cells involves two antagonistic signaling pathways: positive regulation of an F-actin promoting pathway required for lobe outgrowth, and negative regulation of an MT-promoting pathway that inhibits lobe outgrowth. The regulation of these cytoskeleton elements (F-actin and MTs) through ROP2 downstream targets RIC4 and RIC1 (see below) provides further evidence for the role of ROP2 in regulating polarized diffuse growth in pavement cells (Fu *et al.*, 2002, 2005).

3.3.1.2 ROP downstream signaling to the control of cell polarity

An interesting question regarding the ROP regulation of cell polarity described above is how ROPs control cell polarity in both tip growing and diffusely growing cells, given they involve very different mechanisms for polarized growth. Recent studies support the hypothesis that ROPs are able to coordinate a common pathway with various cell-type-specific downstream pathways, providing a unifying mechanism underlying polar cell growth in different systems (Fu *et al.*, 2001, 2002, 2005; Gu *et al.*, 2005; reviewed in Gu *et al.*, 2004; Hwang *et al.*, 2005; Berken, 2006; Uhrig and Hulskamp, 2006; Yang and Fu, 2007). In fact, such a unifying mechanism is extended to Rho GTPase regulation of cell polarity in all eukaryotic cells. It is well known that in yeast and animal cells Rho GTPase controls cell polarity in part through its regulation of the actin cytoskeleton. To form a specific shape, immobile plant cells rely on wall restriction and polarized cell expansion, whereas shape formation in animal cells depends on cytoskeleton-driven membrane protrusion. Given the dramatic differences in the mechanisms of cell shape formation between plant and animal cells, it is striking that there is a common fundamental mechanism across plant, fungal, and animal kingdoms. In animal cells, Rho-mediated actin seems to provide a localized pushing force for membrane protrusion. In yeast, actin is involved in endocytosis and exocytosis (Smythe and Ayscough, 2006; Toret and Drubin, 2006). In plant cells, the function of localized actin is not clear, but may be to target secretory vesicles to the site of growth, which either provide membrane and cell wall growth in the case of tip growth or secrete factors that locally modify the cell wall in the case of diffuse growth.

3.3.1.2.1 Conservation of Rho GTPase downstream pathways. In fungi and animals, RAC/CDC42 GTPases promote actin polymerization through WASPs (Wiskott–Aldrich syndrome proteins) or WAVE (Wiskott–Aldrich syndrome protein family verprolin homologous) complex-mediated regulation of the ARP2/3 actin–nucleation complex. The ARP2/3 and the WAVE complexes are conserved in plants. Mutants of WAVE and ARP2/3 complex subunits all display similar defects in cell morphology of several epidermal cells (Smith and Oppenheimer, 2005; Szymanski, 2005). Defects in pavement cell shapes in these mutants exhibit some similarities to, but also are distinct from, those caused by loss of ROP2 and ROP4 function (Li *et al.*, 2003; Basu *et al.*, 2004; Brembu *et al.*, 2004; El-Din El-Assal *et al.*, 2004; Zimmermann *et al.*, 2004; Djakovic *et al.*, 2006), raising the possibility that ROP2 might regulate the ARP2/3 complex as it is regulated by RAC/CDC42 in animals. In support of this possibility, Szymanski's group reported that ROP2 interacted with a WAVE complex subunit, PIR121/SRA1, in a yeast two-hybrid assay (Basu *et al.*, 2004). However, a recent report from Hulskamp'group did not detect this interaction in their yeast two-hybrid assays (Uhrig *et al.*, 2007). Further investigation is needed to clarify whether ROPs regulate the activity of the WAVE complex.

An apparently unique mechanism of ROP regulation of actin polymerization has been reported. The mediator between ROP and F-actin is a ROP effector, RIC4. RIC4 belongs to a plant-specific family of ROP effectors, termed RICs (*R*OP-*i*nteractive CRIB motif containing proteins) (Wu *et al.*, 2001). There are 11 RIC proteins in *Arabidopsis*. Overexpression of RIC4 promotes fine F-actin formation in both pollen tube and pavement cell systems, and suppression of *RIC4* mRNA levels in a *RIC4* knockdown mutant, *ric4-1*, decreased the accumulation of fine F-actin (Fu *et al.*, 2005; Gu *et al.*, 2005). *The ric4-1* mutant also exhibited shorter pollen tubes and pavement cells with narrower necks and shallower lobes, resembling the phenotype observed in loss-of-function *rop1* and *rop2* mutants, respectively (Fu *et al.*, 2005; Gu *et al.*, 2005). Physical interaction between RIC4 and ROP1 or between RIC4 and ROP2 has been demonstrated using FRET analysis (Fu *et al.*, 2005; Gu *et al.*, 2005). Thus, RIC4 is a common target of ROP1 and ROP2 in their regulation of the actin cytoskeleton in the pollen tube and pavement cell systems, respectively.

Another novel ROP/RAC effector, interactor of CA-ROPs 1 (ICR1), has been reported as a scaffold protein that may link ROPs with vesicle trafficking. ICR1 interacts with the exocyst vesicle tethering complex subunit SEC3, and preferentially interacts with GTP-bound ROP6 and ROP10 (Lavy *et al.*, 2007). Overexpression of GFP-ICR1 induced deformed pavement cell and swollen root hair similar to ROP2 gain-of-function mutants, whereas T-DNA insertion into *ICR1* caused inhibition of pavement cell expansion. In yeast, CDC42 recruits SEC3 to the site of growth, activating vesicle tethering. Mutations in exocyst subunits also cause defect in pollen tube and root hair growth (Zhang *et al.*, 2001; Cole *et al.*, 2005; Cole and Fowler, 2006). Thus, ICR1 could mediate vesicle tethering in plant cells, representing another potential common mechanism for Rho GTPase regulation of cell polarity and morphogenesis.

Recently, increasing evidence suggests that reactive oxygen species or ROS (H_2O_2) served as an important second messenger to regulate polar cell growth. A knockout mutation (*rhd2*) of *AtRBOHC* (encoding NADPH oxidase; see below) impairs ROS production in growing root hairs, which results in inhibition of formation of root hairs (Foreman *et al.*, 2003). A mutation of a RhoGDP dissociation inhibitor (GDI), an inhibitor of ROP GTPase, develops ectopic root hair formation accompanied with ROS production (Carol *et al.*, 2005) likely resulting from constitutive activation of ROP GTPase. The ectopic root hair formation and ROS production are suppressed by the *rhd2* mutation. These results strongly suggest that the ROS production in growing root hairs is controlled in the ROP–RBOH (*r*espiratory *b*urst *o*xidase *h*omolog) pathway. Jones and coworkers showed that overexpression of ROP2 and CA-rop2 stimulates formation of ROS (Jones *et al.*, 2007). On the contrary, ROS production was decreased in *DN-rop2* plants, relative to wild-type plants. Expression of

CA-rop2 in the RBOHC loss-of-function *rhd2-1* mutant led to impaired ROS formation and root hair growth, suggesting that RHD2 is required for this ROP2-dependent ROS production in regulating root hairs formation (Jones *et al.*, 2007). It is believed that ROS either activate Ca^{2+} channel or loosen cell wall for cell expansion (Foreman *et al.*, 2003; Mori and Schroeder, 2004; Uhrig and Hulskamp, 2006). It would be interesting to investigate whether ROP–RBOH–ROS pathway plays similar role in controlling tip growth in pollen tubes. However, RAC GTPase activates NADPH oxidase by directly interacting with its regulatory subunit p67 (Miyano and Sumimoto, 2007), again suggesting the conservation of this Rho-family GTPase downstream pathways across eukaryotic kingdoms.

3.3.1.2.2 Plant-specific downstream pathways. As discussed above, cell polarity regulation by cortical MTs and microfibril arrangement is unique to plants, suggesting the likelihood that plants use unique downstream signaling pathways. In pavement cells, cortical MTs restrict localized outgrowth through guiding the microfibril orientation. In addition to the regulation of RIC4 and ICR1 as discussed above, ROP2 modulates the organization of cortical MTs in its role in pavement cell morphogenesis. RIC1 is the downstream effector that relay signal from ROP to MTs (Fu *et al.*, 2005). RIC1 promotes the formation of well-ordered cortical MTs to restrict the expansion of the neck region. Active ROP2, which is localized to the tip of lobes, inhibits the association of RIC1 with MTs and thus restricts the RIC1 activity to the indenting neck region (Fig. 3.2). The ROP–RIC1 pathway is unique to plants and may specifically regulates the polarity of diffuse growth, which requires cortical MTs.

In tip-growing cells, other unique pathways seem to operate. Tip-focused calcium gradient is critical for tip growth of pollen tubes and root hairs. In pollen tubes, another member of the RIC family, RIC3, was suggested to stimulate Ca^{2+} influx and accumulation in the tip region, which promotes depolymerization of F-actin at the pollen tube tip (Gu *et al.*, 2005). It is interesting that RIC3 overexpression suppressed depolarization of pollen tubes caused by RIC4 overexpression, although RIC3 overexpression itself led to depolarized growth. Based on these observations, it was proposed that the RIC4–actin pathway and the RIC3–Ca^{2+} pathway check and balance to control actin dynamics, which is essential for polarized pollen tube tip growth (Fu *et al.*, 2001; Gu *et al.*, 2005; Hwang *et al.*, 2005). Although direct evidence is lacking, it would not be surprising if similar RIC4- and RIC3-dependent counteracting pathways also regulate tip growth in root hairs.

Phosphatidylinositol monophosphate kinase (PIPK) was also proposed as ROP downstream effector that is involved in actin and Ca^{2+} signaling (Kost *et al.*, 1999). ROPs were reported to associate with a PIPK activity, which synthesizes phosphatidylinositol 4,5-bisphosphate (PI4,5-P2). Using a GFP-tagged pleckstrin homology domain, which binds PIP4,5-P2 specifically, it was shown that PIP4,5-P2 is localized to the apical region of the pollen tube PM (Kost *et al.*, 1999), resembling the localization of active ROP1 (Hwang

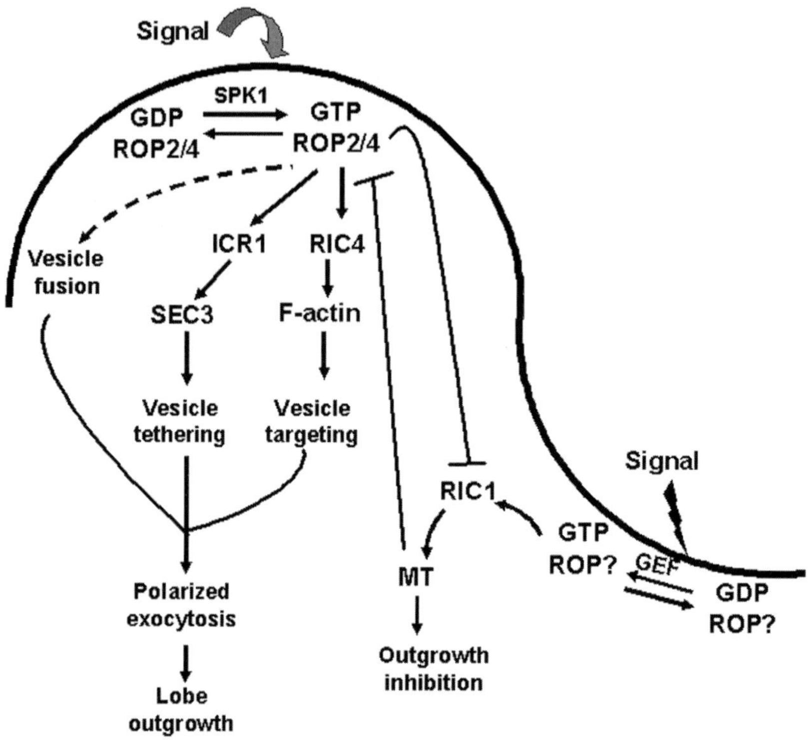

Figure 3.2 ROP2/ROP4 GTPases coordinate multiple downstream pathways in the control of pavement cell morphogenesis. In the outgrowing region, SPK1 (a putative RopGEF) is postulated to activate ROP2/4, because its knockout mutant shows a phenotype similar to that caused by the loss of ROP2/4 function. The RIC4–actin and ICR1–SEC3 pathways are proposed to regulate polarized exocytosis, but no direct evidence is available. The localized exocytosis together with ROP2/4 inhibition of the RIC1–MT pathway allows the localized expansion during the lobe formation. In the indenting region, RIC1 is postulated to be activated by a signaling pathway that may involve another ROP, and the activation of RIC1 promotes the ordering of cortical MTs.

et al., 2005). PI4,5-P2 may not only regulate actin organization by translocating actin-binding proteins, but may also affect Ca^{2+} level by serving as the substrate of phospholipase C to produce inositol 1,4,5-triphosphate (Kost *et al.*, 1999; Berken, 2006). Further investigation is needed to determine whether PIPK–PI4,5-P2indeed acts downstream of ROP1 in the complex signaling network that controls growth in pollen tubes.

From the above analysis, a conceptual framework has emerged that underpins ROP GTPase regulation of cell polarity and polar cell growth in different cell systems (see Fig. 3.2). A ROP GTPase is locally activated in the cell cortex and then coordinately regulates multiple pathways, which all contribute to the formation of cell polarity or cell shape. Some of these pathways (e.g., the RIC4 pathway) are conserved among different cell types, while others are cell-specific. This framework can explain why a simple ROP GTPase can orchestrate cell polarity and cell morphogenesis in different cell types that apparently involve distinct cellular mechanisms (e.g., tip growth versus diffuse growth). Interestingly, this conceptual framework can be extended to Rho GTPase signaling to similar processes across eukaryotic kingdoms, although the details within this framework vary among different kingdoms. For example, ROP regulates the organization of the actin cytoskeleton as its counterpart in other systems, but ROP does so through a plant-specific RIC4 downstream pathway.

3.3.1.3 ROP/RAC GTPases in the regulation of hormone responses

Phenotype characterization of transgenic lines expressing *CA-* or *DN-rop2* mutants suggests that ROP signaling may impact on several aspects of plant development, such as seed dormancy, lateral shoot initiation, and hypocotyl elongation, which are well known to be modulated by abscisic acid (ABA), auxin, and brassinosteroids (BRs) (Li *et al.*, 2001). Further investigation indeed links ROP signaling to the action of these hormones.

ABA is a phytohormone important for the establishment of seed dormancy, stomatal movement, and plant responses to abiotic stresses. ROP10 has been demonstrated to be a negative regulator of ABA signaling (Zheng *et al.*, 2002). Both *Arabidopsis rop10* knockout mutants and *DN-rop10* lines show increased ABA responses, including seed dormancy, stomata closure, and inhibition of root elongation. Conversely, *CA-rop10* lines display reduced ABA sensitivity. ABA regulation of ROP10 activity seems to be important in ABA responses, but ABA was also found to downregulate the transcription of ROP10 in root tips (Zheng *et al.*, 2002).

Several other ROPs have also been implicated in ABA responses. *CA-rop6* lines showed reduced stomata closure induced by ABA, whereas *DN-rop6* enhanced it (Lemichez *et al.*, 2001). Similar enhancement and reduction of seed germination responses to ABA were observed for *Arabidopsis* plants expressing *DN-rop2* and *CA-rop2*, respectively (Li *et al.*, 2001). Because the function of ROP2 and ROP6 in ABA signaling is supported only by gain-of-function analysis, the use of loss-of-function approaches will be necessary to determine whether ROP2, ROP6, and ROP10 govern distinct ABA signaling pathways. It is possible that those phenotypes caused by *CA- or DN-rop* mutants were the consequence of their interference with the function of ROP10.

Auxin is a key phytohormone essential for the regulation of many important cellular processes during plant growth and development, such as cell division, expansion, and differentiation, as well as plant responses to phototropism and gravitropism. A link between auxin responses and ROP GTPase signaling was revealed by Cheung's group (Tao *et al.*, 2002). They reported that expression of wild-type and CA form of a tobacco ROP, NtRAC1, stimulated expression of auxin-responsive genes, whereas overexpression of DN-NtRAC1, NtRAC1 RNAi, or a ROP negative regulator (GDI or GAP from *Arabidopsis*) suppressed auxin-induced gene expression. Moreover, exogenous auxin rapidly promoted the accumulation of GTP-bound NtRACs. These results suggest that ROPs transmit auxin signals to ROP downstream effectors and auxin responsive factors (Tao *et al.*, 2002).

CA-NtRAC1 showed higher affinity to the PM than wild-type or inactive NtRAC1, suggesting that membrane targeting of NtRAC1 may be important for NtRAC1 signaling. This also raises the possibility that ROPs may relay signal from a PM-localized auxin receptor. A recent report demonstrated that auxin is required to properly position ROP2 on the PM to establish planar polarity in *Arabidopsis* root hair development (Fischer *et al.*, 2006). Root hair usually initiate from the site that is adjacent to the basal end of trichoblasts. ROPs are localized to root hair initiation sites (Molendijk *et al.*, 2001; Jones *et al.*, 2002). Both *aux1* mutant (knocking out auxin influx carrier AUX1) and ethylene-insensitive mutant *ein2* display the apical shift of ROP localization, whereas overproduction of auxin and ethylene enhances the polar bias of ROP localization (Fischer *et al.*, 2006). Since ethylene signaling can act upstream of auxin biosynthesis in *Arabidopsis* roots (Stepanova *et al.*, 2005), ethylene may play an indirect role in the regulation of ROP positioning.

Polar auxin transport is critical for auxin function. Both auxin influx carrier AUX1 and the efflux carrier PINs exhibit distinct asymmetrical subcellular localization pattern at the PM, which correlates with the direction of auxin flow (Swarup *et al.*, 2004; Paponov *et al.*, 2005; Xu and Scheres, 2005). *CA-rop2* enhanced polar accumulation of PIN2 protein in the root elongation region and increased gravitropism, which is significantly affected by disruption of F-actin assembly. *DN-rop2* lines show delayed tropic responses, strongly supporting that ROP2 modulates the PIN2 location through regulating F-actins (Li *et al.*, 2005).

Besides ROP2 and ROP4, ROP11 is another candidate that is potentially involved in auxin signaling. CA-rop11 inhibits endocytosis and induces phenotypes indicative of altered auxin responses (Bloch *et al.*, 2005). Auxin has been shown to promote PM localization of PIN1 by inhibiting the endocytosis of PIN1 (Friml *et al.*, 2004). ROP11 could participate in auxin regulation of endocytosis of PINs and/or other cell membrane molecules important for auxin signaling.

Brassinosteroids (BRs) cross talk with auxin signaling. The transcription of *PIN* genes was differentially regulated by treatment with BR brassinolide or in mutants defective in BR biosynthesis (Li *et al.*, 2005). Brassinolide also promotes the accumulation of the PIN2 protein in root and triggers the expression and dispersed localization of ROP2 during tropic responses. Thus, ROP2 was suggested to mediate the BR effects on polar auxin transport and tropic responses (Li *et al.*, 2005). It would not be surprising if future studies link ROPs to the action of other phytohormones such as cytokinin and gibberellin acids.

3.3.1.4 Other developmental processes

Several studies have implicated ROPs in the regulation of broader developmental processes than those described above. *CA-rop2* and *DN-rop2* transgenic plants display pleiotropic phenotypes that could be partially explained by ROP regulation of auxin and ABA responses described above (Li *et al.*, 2001). However, certain *CA-rop2* and *DN-rop2* phenotypes, such as defect in late stages of embryo development and altered orientation of lateral organs, could not be easily explained by ROP involvement in hormone signaling (Li *et al.*, 2001). ROP7 may play a role in the regulation of xylem differentiation. The *ROP7* promoter-driven reporter genes (GUS or GFP) are specifically expressed in xylem cells and procambium of roots, hypocotyls, stems, and leaves (Brembu *et al.*, 2005; Yang and Fu, 2007). Using global comparative transcriptome analysis, Ko and coworkers investigated gene networks that regulate secondary xylem development in *Arabidopsis*. Five genes in the core xylem gene set encode essential components of ROP signaling cascades (Ko *et al.*, 2006). ROPs may influence meristem maintenance as well. The CLAVATA1 complex is the well-known complex that controls the maintenance of shoot apical meristems. This complex includes a ROP protein (Trotochaud *et al.*, 1999). The ROP downstream effector ICR1 was implicated in root meristem maintenance, since knockdown or silencing of ICR1 led to loss of root stem cell population (Lavy *et al.*, 2007).

3.3.2 Regulation of plant innate immunity

3.3.2.1 The regulation of ROS production

Animals and plants produce ROS, such as superoxide and hydrogen peroxide, which are able to directly kill microorganisms, and also function as second messengers to induce a series of immune responses. In animal phagocytes, the ROS production is mediated by the activation of PM-localized NADPH oxidase, a large complex composed of two membrane-bound proteins, gp91^{phox} and p22^{phox}, and the cytosolic proteins p67^{phox}, p47^{phox}, p40^{phox}, and small GTPase RAC (Babior, 2004). Genome sequences of plants such as rice and *Arabidopsis* reveal homologs of gp91^{phox} and RAC-related ROPs, but not other components of the phagocytic NADPH oxidase complex (Torres and Dangl, 2005). In plants, the gp91^{phox} homologs are termed *r*espiratory *b*urst *o*xidase *h*omolog (RBOH). Unlike gp91^{phox}, the RBOH proteins have an extended N-terminus, which contains two Ca^{2+}-binding EF-hand motifs (Torres and Dangl, 2005). Since plants possess only ROP/RAC small GTPase as cytosolic components of NADPH oxidase, an interesting question is whether ROP/RAC is able to regulate the RBOH activity in plants. In many plant species including rice, *Arabidopsis*, cotton, tobacco, and soybean, it was shown that activation of ROP GTPase stimulates ROS production in plant cells (Kawasaki *et al.*, 1999; Potikha *et al.*, 1999; Baxter-Burrell *et al.*, 2002; Jones *et al.*, 2007). In addition, maize ROP GTPase enhances ROS production in mammalian cells (Hassanain *et al.*, 2000), suggesting that ROP GTPase has an ability to activate NADPH oxidase as has been shown in the animal system.

Rice contains seven members of the ROP/RAC (OsRAC) family (Miki *et al.*, 2005). CA-OsRAC1 induces ROS production in rice cells, which is inhibited by

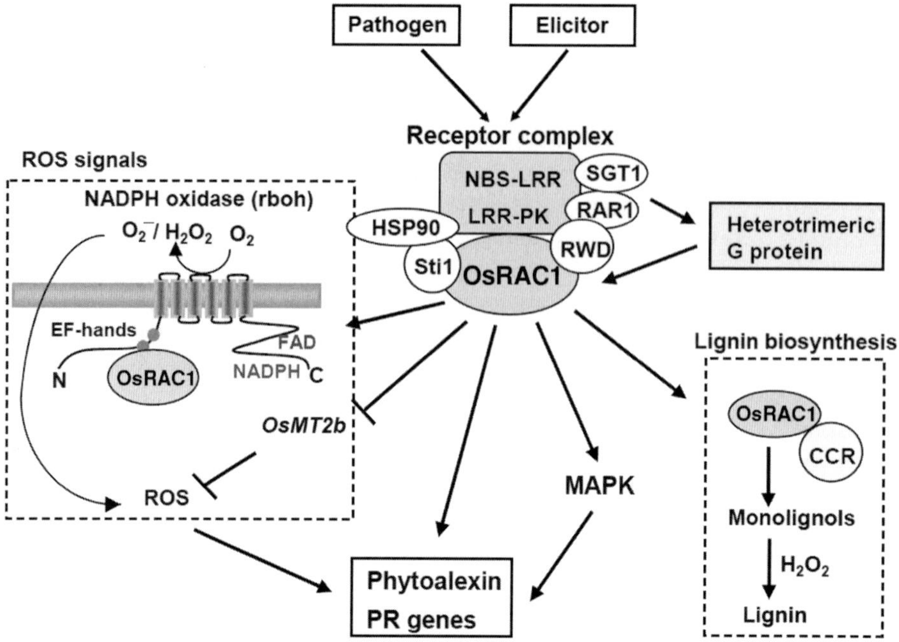

Figure 3.3 A model of OsRAC1-mediated innate immunity in rice. As described for the ROP2/4 regulation of cell morphogenesis in Figure 3.4 (Color plate 6), OsRAC1 regulation of defense responses also involves its coordination of multiple downstream pathways, including ROS accumulation, phytoalexin production, and lignin production. OsRAC1 appears to be regulated by heterotrimeric G proteins, but how pathogen signals and elicitors activate OsRAC1 remains unclear. It is possible that NBS-LRR and LRR-RLK R-gene products may form a complex with OsRAC1/RopGEFs in the transmission of these defense signals. OsRAC1 forms a complex with NBS-LRR-type R proteins and LRR-RLK; chaperones such as HSP90, RAR1, SGT1, Sti1; and putative scaffold protein RWD, which are suggested to be involved in recognition of pathogens and induction of immunity responses. OsRAC1 controls production of ROS by regulating PM-localized NADPH oxidase (RBOH) activity by direct interaction with N-terminal extension of RBOHs. At the same time, OsRAC1 also downregulates expression of a scavenger, metallothionein (*OsMT2b*), to inhibit ROS removal. The reduction of *OsMT2* contributes to maintain transient accumulation of ROS, which results in potentiation of ROS signals to effectively induce immune responses. OsRAC1 controls lignin synthesis through the coordinate regulation of both cinnamoyl-CoA reductase (CCR) and NADPH oxidase activities.

diphenylene iodonium, an inhibitor for flavin-containing phagocytic NADPH oxidase (Kawasaki *et al.*, 1999). ROS produced by OsRAC1-mediated signaling functions as second messengers to induce a series of immune responses (Fig. 3.3). Direct interactions between OsRAC1 and RBOHs have been observed in the yeast two-hybrid system and *in vitro* pull-down experiments. OsRAC directly interacts with the N-terminal regions of RBOHs including EF-hand motifs (Wong *et al.*, unpublished results). The interaction is

GTP-dependent, which is consistent with the observation that expression of CA-OsRAC1 induces ROS production in plant cells, whereas DN-OsRAC1 inhibits ROS production induced with two different fungal elicitors, sphingolipid and *N*-acetylchitooligosaccharide elicitors (Ono *et al.*, 2001; Suharsono *et al.*, 2002). An experiment with a FRET technique confirmed *in vivo* interaction of ROP GTPase and RBOH protein in plant cells (Wong *et al.*, unpublished results). In tobacco, NtRBOHD and NtRAC5 were shown to colocalize in the lipid raft in the PM upon the elicitation of cells with a fungal elicitor, cryptogein (Mongrand *et al.*, 2004). Thus, it is likely that ROP regulates the NADPH oxidase activity by direct interaction with N-terminal extension of RBOHs. The investigation on roles of the ROP–ROS pathway in regulating root hair polar growth also genetically identified the interaction between ROP and RBOH in *Arabidopsis* (see above).

Although OsRAC1 activates NADPH oxidase-dependent ROS production, OsRAC1 downregulates expression of a scavenger, metallothionein (*OsMT2b*), to inhibit ROS removal (Wong *et al.*, 2004). The reduction of *OsMT2* contributes to maintain transient accumulation of ROS, which results in potentiation of ROS signals to effectively induce immune responses in the infected sites of pathogens. In fact, numerous reports have described reduction of ROS scavengers during immune responses (Takahashi *et al.*, 1997; Chamnongpol *et al.*, 1998; Mittler *et al.*, 1998, 1999).

3.3.2.2 Regulation of cell wall biosynthesis

Deposition of lignin on the cell wall is one of the immune responses. Lignin, which is polymerized through peroxidase activity using hydrogen peroxide in the cell wall, presents an undegradable mechanical barrier to most pathogens (Moerschbacher *et al.*, 1990). Rice cinnamoyl-CoA reductase 1 (OsCCR1), a key enzyme for synthesis of monolignols in lignin biosynthesis, has been identified as an effector of OsRAC1 (Kawasaki *et al.*, 2006). OsCCR1 is specifically expressed during defense responses, indicating that OsCCR1 contributes to lignin biosynthesis in the immune response, but not in development. The interaction of OsCCR1 with OsRAC1 drastically activates OsCCR1 activity *in vitro*, and transgenic cell cultures expressing *CA-OsRAC1* accumulates lignin through enhanced CCR activity (Kawasaki *et al.*, 2006). As mentioned above, OsRAC1 also stimulates NADPH oxidase-dependent ROS production that is required for the polymerization of monolignol on the cell wall. These data suggest that OsRAC1 controls lignin synthesis through the coordinate regulation of both NADPH oxidase and OsCCR1 activities during immune responses in rice (Yang and Fu, 2007). In another case, ROP protein was shown to interact with UDP-glucose transferase involved in callose synthesis at the forming cell plate during cytokinesis (Hong *et al.*, 2001). Although callose is produced at the infection sites of pathogens (Kim *et al.*, 2005), whether ROP GTPase is involved in callose formation during immune responses remain to be investigated. Thus, the ROP proteins play important roles in regulating cell wall biogenesis.

3.3.2.3 ROP/RAC GTPase is a key regulator for innate immunity

Immune responses mediated by the ROP proteins have been studied in rice. ROP GTPase controls ROS production through the regulation of the RBOH activity. Involvement of RBOH in innate immunity has been well investigated in many plant systems including *Arabidopsis* and tobacco (Torres *et al.*, 2002; Yoshioka *et al.*, 2003; Torres and Dangl, 2005). OsRAC1-mediated ROS production is accompanied with cell death in rice suspension cells and plants (Kawasaki *et al.*, 1999). The cell death is often found in disease resistance (R)-gene-mediated defense response, which is one of the characteristics of hypersensitive response (HR). Transgenic plants expressing *CA-OsRAC1* highly accumulate a well-characterized rice phytoalexin, momilactone A, and upregulate expression of *ent*-copalyl diphosphate synthase (D9/OsCPS2/OsCYC2) gene that is known to be responsible for the biosynthesis of rice phytoalexins including momilactone A (Ono *et al.*, 2001; Otomo *et al.*, 2004). A series of immune responses induced by CA-OsRAC1 leads to enhanced disease resistance to virulent races of *Magnaporthe grisea* and *Xanthomonas oryzae* (Ono *et al.*, 2001). In addition, DN-OsRAC1 suppresses R-gene-dependent defense responses including hypersensitive cell death. Suppression of OsRAC1 by RNAi also inhibits HR induced by an avirulent race of *M. grisea* (Chen *et al.*, unpublished results). The finding suggests that OsRAC1 plays important roles in both R-gene-mediated resistance and basal resistance.

In tobacco, expression of DN-OsRAC1 delays lesion formation in *N*gene-mediated resistance to tobacco mosaic virus, which is correlated with reduced level of ROS production, altered PR gene expression, and reduction of salicylic acid accumulation (Moeder *et al.*, 2005). In addition, DN-OsRAC1 also suppresses *Pto* gene-mediated hypersensitive responses and nonhost resistance to *Pseudomonas syringae* pv. *maculicola* (Moeder *et al.*, 2005). However, systemic acquired resistance is not affected in the DN-OsRAC1 tobacco plants. Thus, the effects of ROP GTPase may change depending on kinds of defense responses and plant species. Transgenic tobacco plants carrying an antisense construct of a *Medicago sativa* RAC cDNA inhibit necrotic lesion that is induced by infiltration with a yeast elicitor (Schiene *et al.*, 2000). Taken together, it is most likely that *ROP* genes have general roles in disease resistance in plants.

3.3.2.4 Cytoskeleton remodeling by ROP/RAC GTPase during immune responses

The actin cytoskeleton plays a crucial role in innate immune response at the early stage of fungal penetration (Kobayashi *et al.*, 1997). Actin microfilaments are polarized toward the penetration sites of the biotrophic powdery mildew fungus *Blumeria graminis* f.sp. *hordei* (*Bgh*) in barley mutant lacking the susceptibility factor mildew resistance locus (MLO) (Opalski *et al.*, 2005). The polarization of actin microfilaments is closely associated with successful prevention of penetration of *Bgh*, in which actin focusing is induced in resistant response, but not in susceptible response. An active mutant of barley, HvRACB, partially inhibits actin reorganization, which results in enhanced susceptibility against *Bgh* (Schultheiss *et al.*, 2005). Since HvRACB belongs to group II/type I subfamily of ROP proteins that function in the induction of subcellular Ca^{2+} gradients and polar actin patterns in development (see above), CA-HvRACB may induce diffuse instead of focal Ca^{2+} influx required for actin reorganization. In addition, CA-HvRACB does not affect HR and *mlo*-mediated disease resistance (Schultheiss *et al.*, 2005), suggesting that HvRACB is not simple negative regulator for defense response. Suppression of HvRACB by RNAi leads to partial activation of resistance to *Bgh*. This may result from that HvRACB regulates the polar membrane growth process that is involved in PM invagination in haustorium establishment of *Bgh* (Schultheiss *et al.*, 2002, 2003).

3.3.2.5 Roles of ROP GTPases as negative regulators in innate immunity

Several members of ROP proteins have been shown to be negative regulators of immune response. Rice OsRACB/OsRAC5 is downregulated in response to infection of *M. grisea*. Overexpression of OsRACB/OsRAC5 enhances susceptibility to a compatible race of *M. grisea* (Jung *et al.*, 2006). Similarly, silencing of *OsRAC4* induces constitutive expression of several PR genes, which results in enhanced resistance to virulent *M. grisea* (unpublished results).

However, how OsRACB/OsRAC5 and OsRAC4 negatively regulate immune responses remains to be identified. NtRAC5 is downregulated during elicitation with a fungal elicitor, cryptogein (Morel *et al.*, 2004), and overexpression of NtRAC5 suppresses ROS production. Interestingly, overexpression of NtRAC5 reduces the mRNA and protein levels of NtRBOHD, suggesting that NtRAC5 negatively regulates NtRBOHD at both transcriptional and translational levels (Morel *et al.*, 2004). Thus, ROP GTPase regulates immune responses in positive and negative manners.

3.3.2.6 Regulation of mitogen-activated protein kinase signaling by ROP/RAC GTPases

In plants, mitogen-activated protein kinase (MAPK) has been found to be involved in biotic and abiotic signal transductions (Ichimura *et al.*, 2000; Asai *et al.*, 2002; Ichimura *et al.*, 2002) (see Chapter 4). Rice MAPK6 (OsMAPK6) is known to be activated at the post-translational levels during immune responses as found in well-characterized tobacco MAPKs, salicylic acid-induced protein kinase and the wound-induced protein kinase (Lieberherr *et al.*, 2005). Silencing of the OsMAPK6 activity by RNAi causes alternation of mRNA levels of phenylalanine ammonia lyase and two other *MAPK* genes, *OsMAPK5a* and *BWMK1*, that are known to be associated with biotic and abiotic stresses (He *et al.*, 1999; Xiong and Yang, 2003). Suppression of *OsRAC1* mRNA level by RNAi and the *d 1* mutation lacking Gα destabilizes the OsMAPK6 protein in rice cells (Suharsono *et al.*, 2002). As described below, Gα and OsRAC1 likely function in the same signal transduction pathway, suggesting that the pathway regulates stability of OsMAPK6. Interestingly, the OsMAPK6 protein was co-immunoprecipitated with CA-OsRAC1, but not DN-OsRAC1. Activation of OsMAPK6 is suppressed in *OsRAC1 RNAi* and *DN-OsRAC1* cells. Thus, it seems that OsRAC1 regulates the OsMAPK6 activity in a protein complex. In *Arabidopsis*, an MAPK cascade associated with the immune response has been well characterized (Asai *et al.*, 2002; He *et al.*, 2006). Therefore, examining whether the ROP pathway is also connected with the MAPK cascade in *Arabidopsis* is an interesting issue.

3.3.2.7 Proteomic dissection of ROP GTPase signaling in defense

The OsRAC1-dependent signaling pathway has been analyzed using a proteomics approach (Fujiwara *et al.*, 2006). Proteins whose expression levels are altered by CA-OsRAC1, DN-OsRAC1, and/or sphingolipid elicitor were identified by the use of two-dimensional gel electrophoresis and mass spectrometry. Many proteins that are known to be involved in immune responses were changed at the protein levels by OsRAC1 and sphingolipid elicitors. Interestingly, among 100 proteins upregulated by sphingolipid elicitors, 87 were also induced by expression of CA-OsRAC1, indicating that OsRAC1 is a pivotal component in the elicitor-dependent defense signaling.

Identification of immune components in a complex containing OsRAC1 would give a new insight into the role of OsRAC1 in disease resistance. Many components of OsRAC1 complex were identified by affinity chromatography using protein extracts purified from rice cells treated with the sphingolipid elicitor (Nakashima *et al.*, unpublished results). The components include NBS-LRR-type disease resistance proteins, chaperones such as HSP90, and scaffold proteins, which are suggested to be involved in immune responses of other species. Thus, it is likely that OsRAC1 forms a large complex with important immune components.

3.3.3 Regulation of responses to abiotic stress

ROS production is often associated with plant responses to abiotic stresses in addition to biotic stress discussed above (see Chapter 7). However, the mechanism by which plants use ROS for tolerance against abiotic stress seems to differ from OsRAC1-mediated ROS production in defense responses. The latter is accompanied with oxidative-burst-induced localized cell death. The role of ROP regulation of ROS production in abiotic stress has been elegantly studied for plant tolerance to hypoxia (Baxter-Burrell *et al.*, 2002). In this system, oxygen deprivation triggers a rise in H_2O_2 levels, which is required to induce the expression of genes that are crucial for stress responses (Baxter-Burrell *et al.*, 2002). It was shown that *Arabidopsis* seedlings install an elegant regulatory system to prevent the accumulation of high toxic concentrations of H_2O_2 by tuning the activity of ROPs. As in the defense system, oxygen deprivation activates ROP2, which then promotes the production of H_2O_2. H_2O_2 induces the expression of alcohol dehydrogenase, which is important for tolerance to oxygen deprivation. Interestingly, H_2O_2 also activates the expression of RopGAP4, which is a negative regulator of ROPs that terminates

ROP activation (Baxter-Burrell *et al.*, 2002). The regulation forms a negative feedback loop to shut down ROP signaling, avoiding the accumulation of toxic levels of H_2O_2 (Baxter-Burrell *et al.*, 2002). This would be similar to the mechanism for ROP–ROS-dependent regulation of polar growth (see above), where the production of H_2O_2 has to reach a threshold for inducing sufficient downstream factors, but the accumulation of H_2O_2 should be tightly restricted to avoid toxic effects.

3.4 Mechanisms for the regulation of the ROP GTPase "ON/OFF" status

3.4.1 Regulation of GTP–GDP cycling

As for other small GTPases, ROP GTPases cycle between the GTP-bound active form (or "ON" state) and the GDP-bound inactive form (or "OFF" state). The switching between these two forms is precisely regulated by three major classes of regulatory proteins: GDIs, GAPs, and GEFs.

GEFs turn on small GTPases by stimulating the exchange of GDP with GTP. To date, at least three families of RhoGEFs have been identified, including the Db1 family, the Dock180 family, and the RopGEF family. The biggest RhoGEF family belongs to Db1 homology proteins; members of this family share a conserved Db1 homology (DH) domain and a pleckstrin homology (PH) domain. The DH domain is believed to associate with the switch region of GTPases, responsible for GTPase remodeling and the exchange of GDP with GTP. The function of PH domains in different GEFs varies. Some PH domains bind to the phospholipids to facilitate the PM targeting of GEFs. Other PH domains affect the nucleotide exchange by interacting with GTPases directly or facilitating the conformational change of the GEFs. Still some other PH domains actually inhibit GEFs activity by interacting with the DH domain (Schmidt and Hall, 2002; Rossman *et al.*, 2003, 2005; Lu and Ravichandran, 2006). However, no homolog of DH–PH containing RhoGEF has been discovered in plants.

The Dock180 family is an unconventional RhoGEF family. Members in this family do not contain the typical DH domain. Instead, they have a so-called Dock homology region 2 (DHR2), which is conserved in fungi, animals, and plants. This domain specifically binds to the nucleotide-free state of RHO GTPases and is crucial for the GEF activity (Lu and Ravichandran, 2006). Dock180 proteins usually function as a RhoGEF together with a cofactor named ELMO. In *Arabidopsis*, a single gene called *SPIKE1* (*SPK1*) was found to encode a Dock180-like protein (with a single DHR2 domain). Knocking out *SPK1* leads to altered pavement cell shape (Qiu *et al.*, 2002), which mimics the defect induced by transgenic *ROP* mutants (Fu *et al.*, 2002). It was reported in a conference abstract that *SPK1* has GEF activity toward ROPs (reviewed in

Berken, 2006). However, nothing is known about the regulation of *SPK1*, although genes encoding ELMO-like proteins are also present in the *Arabidopsis* genome.

Plants possess a novel family of GEFs that activate ROPs (Berken *et al.*, 2005; Gu *et al.*, 2006). Members of this family are characterized by a highly conserved domain annotated as DUF315 (domain of unknown function 315). Berken and colleagues named this domain PRONE (*p*lant-specific *ROP* *n*ucleotide *e*xchanger) since this domain was shown to stimulate nucleotide dissociation from ROPs (Fig. 3.1d, Color plate 5) (Berken *et al.*, 2005; Gu *et al.*, 2006). X-ray crystallographic analysis of the PRONE domain from ROPGEF8 demonstrates that the dimerization of PRONE is essential for GEF function (Thomas *et al.*, 2007). Like GEFs for other small G proteins, PRONE/DUF315 stimulates nucleotide exchange by forming a transient ternary complex with two molecules of GDP-bound ROP. Binding with GEF results in conformation change of ROPs that reduce the nucleotide affinity and facilitates the release of GDP. Dissociation of GDP is accompanied by spontaneous loading of GTP (and thus activation of ROP) since the cellular concentration of GTP is much higher than that of GDP (Wittinghofer, 1998; Cherfils and Chardin, 1999; Thomas *et al.*, 2007).

Evidence suggests that RopGEFs activate ROP1 in the control of polarized pollen tube growth. In a yeast two-hybrid screen for proteins that interact with the intracellular kinase domain of LePRK1 (a tomato receptor-like kinase expressed in pollen), Kaothien *et al.* identified a protein named KPP (Kaothien *et al.*, 2005), which turns out to be a homolog of RopGEFs (Berken *et al.*, 2005; Gu *et al.*, 2006). KPP overexpression induced depolarization of pollen tube growth as does ROP1 overexpression (Li *et al.*, 1999; Kaothien *et al.*, 2005). Among 14 members of the *Arabidopsis* RopGEF family, at least 5 are expressed in pollen (Gu *et al.*, 2006). GFP tagging suggests that all these 5 RopGEFs are localized to the apical region of the pollen tube PM corresponding to the site of ROP1 activation (Hwang *et al.*, 2005; Gu *et al.*, 2006). Interestingly, overexpression of RopGEF1 but not other RopGEFs induced severe growth depolarization in pollen tubes as did KPP, suggesting that RopGEF1 may be an ortholog of KPP. RopGEF1 overexpression phenotype was suppressed by DN-rop1. These results support a role for RopGEF1 in the activation of ROP1 in the control of pollen tube growth. However, RopGEF1 knockout mutants are not affected in pollen tube growth (Gu and Yang, unpublished results), suggesting that other RopGEFs are functionally redundant to RopGEF1. The differences in the overexpression phenotypes between RopGEF1 and others suggest that RopGEF1 and other RopGEFs may be differentially regulated.

Outside the PRONE domain, RopGEFs have variable N- and C-termini, which may play important roles in the regulation of RopGEFs. Gu and her coworkers reported that the interaction between the C-terminus and the catalytic domain within the molecule of RopGEF1 blocks its GEF activity and proposed an autoinhibitory mechanism for the regulation of *Arabidopsis* RopGEF1

(Gu *et al.*, 2006). Presumably, the autoinhibition could be released by an upstream activator of ROP1 signaling. It would be interesting to see whether receptor-like kinases (RLKs), such as LePRK1 and LePRK2 (Kaothien *et al.*, 2005), could be such upstream activators.

GDIs affect the subcellular localization and activation status of Rho-family GTPases by forming complexes with lipid-modified Rho GTPases. These high-affinity complexes prevent bound nucleotide (especially GDP) from being dissociated from the GTPases. Furthermore, GDIs extract Rho proteins from cellular membranes and sequester them in the cytosol (DerMardirossian and Bokoch, 2005). GDIs could also inhibit the interaction of Rho GTPases with GAPs and the GTP hydrolysis, making GDI a tricky player regulating activity of Rho GTPases. GDI homologs have been identified from *Arabidopsis* (three homologs) and tobacco (one homolog) (Bischoff *et al.*, 2000; Kieffer *et al.*, 2000), which are highly similar to mammalian RhoGDIs. There are four highly conserved regions among all GDIs (Fig. 3.1b, Color plate 5) (Bischoff *et al.*, 2000). RopGDIs play important roles in controlling polarized growth of pollen tubes and root hairs. Negative regulation of ROP signaling by *Arabidopsis* RopGDI1 restricts ROP1 to the apex of the pollen tube PM to maintain tip growth (Fu *et al.*, 2001; Yang, 2002; Hwang *et al.*, submitted). *Arabidopsis* RopGDI1 suppressed growth depolarization caused by ROP1 overexpression in pollen tubes (Fu *et al.*, 2001). A tobacco GDI (NtRhoGDI2) was found to preferentially interact with prenylated NtRAC5. A point mutation in NtRAC5 that abolishes its interaction with NtRhoGDI2 caused mislocalization of NtRAC5 to the flank of the pollen tube (Klahre *et al.*, 2006). These clearly indicate that GDI is important for polarized ROP localization in specific membrane domains.

Root hair is another model system for the study of spatially restricted cell growth. Normally each trichoblast has one root hair growth site where ROP2 is distributed. However, the disruption of RopGDI1 function in *supercentipede1* (*scn1*) mutant plants results in multiple root hairs on a single trichoblast due to ectopic accumulation of ROP2 at the cell surface (Carol *et al.*, 2005). It is expected that RopGDIs play a similar role in pavement cell system, since the formation of the jigsaw appearance in pavement cells is dependent on proper localization of ROP2 as well (Fu *et al.*, 2002, 2005).

GAPs interact with the active form of Rho GTPases and promote their GTP hydrolysis to recycle them back to the GDP-bound inactive status. The conserved arginine finger (an arginine residue in the catalytically active GAP domain of the protein) inserts into the GTPase active site to stabilize the transition state, whereby the intrinsic GTPase activity is promoted. Plants have a unique GAP family termed RopGAP (with six members in *Arabidopsis*) (Borg *et al.*, 1999; Wu *et al.*, 2000). RopGAPs contain a GAP catalytic domain that shares the greatest similarity with animal CDC42 GAPs. RopGAPs are unique in that they contain a CDC42/RAC-*i*nteractive *b*inding (CRIB) motif, which is absent from animal and fungal RhoGAPs, but is usually found in CDC42/RAC effectors. It was demonstrated that *Arabidopsis* RopGAPs promoted GTP hydrolysis of ROPs but not that of CDC42. The CRIB motif is required for RopGAP activity as well as its binding to ROPs. This motif may participate in the stabilization of the transitional state of ROPs during GTP hydrolysis (Wu *et al.*, 2000). The GAP domain and the CRIB motif are conserved in all RopGAPs from both dicot and monocot species. Their N- and C-terminal regions are highly variable, providing a potential functional specificity for different RopGAPs.

In vivo functions for most members of the RopGAP family are poorly understood. RopGAP4 takes part in ROP-regulated ROS production upon abiotic stress in *Arabidopsis* (see above) (Baxter-Burrell *et al.*, 2002). Overexpression of *Arabidopsis* RopGAP1 suppressed depolarized pollen tube growth induced by ROP1 overexpression, suggesting that RopGAP1 may act as a negative regulator of ROP1 (Fu *et al.*, 2001; Hwang *et al.*, unpublished results). Similarly, NtRopGAP1 suppressed NtRAC5 overexpression phenotype (Klahre and Kost, 2006). NtRhoGAP1 was found to associate with the region flanking the pollen tube PM apex where NtRAC5 is presumably activated. It was proposed that NtRopGAP1 acts in the flanking region to restrict NtRAC5 signaling to the PM apex (Klahre and Kost, 2006). However, loss of function approaches are required to test the role of these pollen-expressed RopGAPs in pollen tube growth.

3.4.2 Regulation of subcellular localization

Proper subcellular localization of signaling proteins is very important for their function. RAS-family GTPases are usually associated with the PM or an endomembrane compartment such as endoplasmic reticulum, Golgi apparatus, or vesicles (Hancock *et al.*, 1991; Choy *et al.*, 1999; Ivanchenko *et al.*, 2000). Rho GTPases are dynamically distributed in the cytosol and membrane systems, implying that shuffling between membranes and the cytosol may regulate their activities. Small GTPases are translated by free ribosomes and do not contain transmembrane domains. Their association with cellular membranes depends on their modification by lipid molecules and/or their binding to membrane-localized proteins. Their dissociation from membranes is controlled by RhoGDIs, which remove them from cellular membranes and sequestering them in the cytosol (see above). The information for membrane targeting lies in the C-terminal C(XX) motif that is usually subject to lipid modification and its proximal hypervariable region (HVR) (Hancock *et al.*, 1991; Choy *et al.*, 1999; Ivanchenko *et al.*, 2000; Heo *et al.*, 2006).

Like other Rho GTPases, ROP association with the PM requires prenylation and/or possibly acylation. Typical of Rho-family GTPases, most ROPs contain the C-terminal CAAL motif that is the target of protein geranylgeranyl transferase (see above) (Ivanchenko *et al.*, 2000; Zheng and Yang, 2000; Yang, 2002). The importance of geranylgeranylation for ROP membrane targeting has been investigated. Pea ROP1 and cotton GhRAC13 were reported to be geranylgeranylated *in vitro* (Lin *et al.*, 1996; Trainin *et al.*, 1996). ROP1 and ROP5, which contain the conical CAAL geranylgeranylation motif, were found in both the cytosol and the apical region of the pollen tube PM. C-to-S mutations in the CAAL motif, which eliminate its lipid modification capacity, abolish the PM localization and biological function of these ROPs in the regulation of pollen tube growth (Kost *et al.*, 1999; Li *et al.*, 1999). Similar mutations in rice OsRACB and barley HvRACB eliminate their PM targeting and function as regulators of disease resistance (Schultheiss *et al.*, 2003; Jung *et al.*, 2006). These results suggest that lipid modification (most likely through geranylgeranylation) of ROPs is involved in the membrane targeting and thereby the activity regulation of ROP GTPases. However, it remains to be determined whether ROPs are geranylgeranylated *in vivo*. Furthermore, evidence suggests that ROPs containing the CAAL motif could be modified by other lipid moieties (see below).

ROPs belonging to group II do not have the conventional CAAL motif. *Arabidopsis* ROP10 contains a CGKN motif, which can only be weakly farnesylated *in vitro* (Lavy *et al.*, 2002; Zheng *et al.*, 2002). Knocking out ERA1, the only β-subunit of farnesyltransferase in *Arabidopsis*, cannot remove ROP10 completely from the PM. These results indicate a farnesylation-independent lipid modification is required for ROP10 PM localization. ROP10 can be palmitoylated *in vitro* and palmitoylation inhibitor, 2-bromopalmitate, caused a localization shift of ROP10 from the PM to the cytosol, suggesting that the PM localization of ROP10 is dependent on palmitoylation (Lavy *et al.*, 2002; Zheng *et al.*, 2002). Palmitoylation inhibitor was shown to remove ROP9 and ROP11 (in the same group II as ROP10) from the PM as well, suggesting that they could also be subject to palmitoylation (Lavy *et al.*, 2002; Zheng *et al.*, 2002). Two maize group II ROPs, ZmROP6 and ZmROP7, contain the C-terminal CAA motif, and are exclusively localized to the PM (Ivanchenko *et al.*, 2000). Interestingly, the PM targeting of these two ROPs is not dependent on prenylation of the cysteine residue in the CAA motif, but requires a couple of internal cysteines that are absent from the ROP HVR regions of other ROPs. Therefore, the targeting of ZmROP6 and ZmROP7 appears not to be mediated by prenylation, but by other types of lipid modification such as palmitoylation (Ivanchenko *et al.*, 2000). However, further investigation will be needed to determine whether or not ZmROP6 and ZmROP7 are palmitoylated.

In addition to association of ROPs with the PM, ROPs have been shown to localize to other endomembrane systems as well as to a specific domain of the PM, suggesting that more precise regulation of subcellular localization operates to allow specific targeting of ROP GTPases. Different ROPs have been found to be associated with specific subcellular compartments including perinuclear organelle (ROP4), the tonoplast of developing vacuoles (pea ROPs), the whole PM, and PM domains. Interestingly, although all PM associated, ROP9/10, ZmROP6/7, HvRACB and OsRACB occupy the entire PM, whereas ROP1, ROP2, and ROP5 are localized to specific PM domains, and the localization pattern of each ROP is closely related to its function (Kost *et al.*, 1999; Li *et al.*, 1999; Bischoff *et al.*, 2000; Lin *et al.*, 2001; Fu *et al.*, 2002; Jones *et al.*, 2002; Schultheiss *et al.*, 2003; Jung *et al.*, 2006). Information for specific ROP targeting appears to be provided by residues within the HVR that is proximal to the site of lipid modification (Bischoff *et al.*, 2000).

It was recently reported that an activation-dependent acylation of ROP6 may be involved in the localization of ROPs to specific membrane domains. It was shown that wild-type ROP6 is only prenylated, mostly by geranylgeranylation. Upon activation, the cysteine 156 within the HVR region is transiently S-acylated (probably palmitoylated) and the acylated form accumulated in detergent-resistant membrane (DRM), which is often associated with lipid rafts (Sorek *et al.*, 2007). Evidence suggests that since DRM (or lipid rafts) has function in plant cell polarity regulation (Schrick *et al.*, 2000; Fischer *et al.*, 2004), ROP6 partitioning in DRM could play an important role in plant cell polarity establishment. Since cysteine 156 is conserved in many ROPs, activation-dependent S-acylation and DRM targeting may be a common mechanism for regulating the localization and function of ROPs.

As discussed above, the localization to a specific PM domain is regulated by cell polarity signaling and plays an important role in the formation of cell polarity. For example, an auxin gradient regulates the localization of ROP2 to the site of root hair formation in trichoblasts, where it initiates root hair formation (Fischer *et al.*, 2006). How polarity signals control the subcellular localization of ROPs is an interesting topic of future studies.

3.5 Potential upstream regulators of ROP signaling

3.5.1 Receptor-like kinases

How are upstream extracellular signals (such as hormonal, developmental, and pathogen-induced signals) relayed to ROPs remains a mystery. One class of candidate upstream regulators is the superfamily of receptor-like serine/threonine kinases (RLKs) (see Chapter 1). ROPs were reported to be part of an active CLAVATA1 complex that controls the maintenance of shoot apical meristem (Trotochaud *et al.*, 1999). Interestingly, the tomato RopGEF homolog KPP has been shown to interact with the tomato RLKs, LePRK1 and LePRK2, both *in vitro* and *in vivo* (Kaothien *et al.*, 2005). Two important questions in the RopGEF–RLK interaction need to be addressed: (1) How widespread is this interaction? (2) What is the functional significance of this interaction in ROP signaling? KPP has been shown to be phosphorylated (Kaothien *et al.*, 2005). It is possible that phosphorylation of RopGEFs by RLKs may regulate RopGEF activity.

There are hints that this interaction could be widespread. RLK (RPK1) is involved in ABA perception/signaling (Osakabe *et al.*, 2005), and ROPs have been shown to be a PM regulator of ABA signaling (Zheng *et al.*, 2002). During defense responses, plants recognize pathogen-derived signals as pathogen-associated molecular patterns (PAMPs) (Zipfel and Felix, 2005; Jones and Dangl, 2006). So far, two plant PAMPs receptors, FLS2 and ERF, have been identified to recognize bacterial components, fragellin and EF-Tu (elongation factor Tu), respectively, by direct interaction (Gomez-Gomez and Boller, 2000; Bauer *et al.*, 2001; Zipfel *et al.*, 2006). Both proteins are receptor-type protein kinases with LRR. A rice bacterial blight resistance gene, *XA21*, encodes a similar LRR-protein kinase, which may recognize type I secreted protein from *X. oryzae* (Lee *et al.*, 2006). However, how these receptors transmit the PAMPs signals to downstream immune components is unknown. Interestingly, the PRONE-type RopGEFs are known to interact with LRR-protein kinases (Trotochaud *et al.*, 1999; Kaothien *et al.*, 2005); it is possible that these RopGEFs and ROPs transmit the PAMPs signals that are recognized by LRR-protein kinase receptors. Since a MAPK cascade is known to involve a signal transduction mediated by LRR-protein kinase including FLS2, an interaction between MAPK and ROP signaling would be possible.

3.5.2 Heterotrimeric G proteins

Heterotrimeric G proteins, composed of three subunits, α, β, and γ, are involved in many signal transduction pathways. G_{12} is one of four families of $G\alpha$ subunits. It has been well known that $G\alpha_{12}$ regulates cellular responses and function through Rho GTPases in animals. G_{12} family G-protein activates Rho by binding to RhoGEF and balancing unphosphorylated and phosphorylated forms of RhoGEF. $G\alpha_{12}$ also binds RasGAP through the latter's PH–BM module and stimulates its GTPase activity (Hepler and Gilman, 1992; Kurose, 2003). In plants, a number of pharmacological studies have indicated involvement of heterotrimeric G protein in plant disease resistance responses (Assmann, 1996; Jones *et al.*, 1998; see Chapter 2). Rice contains a single-copy $G\alpha$ gene. Therefore, a knockout mutant of rice $G\alpha$ is predicted to lose all the functions of heterotrimeric G protein. Rice $G\alpha$ mutants (*dwarf1:d1*) exhibit dwarf phenotype, suggesting the role of $G\alpha$ in gibberellin signal transduction (Ashikari *et al.*, 1999; Fujisawa *et al.*, 1999). Expression of the $G\alpha$ gene is induced by an avirulent race of *M. grisea*, but not by a virulent race (Suharsono *et al.*, 2002). The *d1* mutants lacking the functional $G\alpha$ gene have defects in expression of PR genes, ROS production, and resistance response to the avirulent race (Suharsono *et al.*, 2002), indicating that $G\alpha$ plays essential roles in innate immunity. Furthermore, expression of the CA-OsRAC1 restores ROS production, PR gene expression, and resistance to the avirulent race in the *d1* mutants, suggesting that OsRAC1 is downstream of $G\alpha$ in R-gene-mediated signal transduction pathways (Suharsono *et al.*, 2002). Involvement of heterotrimeric G protein in innate immunity has also been found in *Arabidopsis*

(Llorente *et al.*, 2005; Trusov *et al.*, 2006). However, how heterotrimeric G proteins are connected to ROP/RAC remains unknown. Furthermore, RLKs and heterotrimeric G proteins may not represent all upstream regulatory mechanisms for ROP activation. Identification of more interactors of RopGEF, RopGAP, and RopGDI will be important to fully reveal how the ROP signaling pathways are regulated in plants.

3.6 Future perspectives

It is clear that ROP GTPases play a crucial role in plant intracellular signaling. The function of ROPs has been linked to the regulation of many important processes during plant growth, development, and responses to the environment. Recent studies have uncovered a large number of ROP interacting partners, including conserved and novel regulators and effectors as well as a number of ROP-dependent signaling pathways/networks (Fig. 3.4, Color plate 6). However, the picture is far from being complete in our understanding of ROP signaling mechanisms and ROP-dependent ROP signaling pathways/networks. It is anticipated that many functions for ROP signaling have yet to be discovered, and the difficulty in elucidating their functions lies in the complex function of a given ROP GTPase, which could be functionally redundant with other ROPs as well as controlling multiple signaling networks or processes. New functional partners of ROPs will continue to be identified, and their functional analysis may enhance our understanding of ROP functions. Little is known about the initial signals that control ROP signaling pathways/networks, although two phytohormones, auxin and ABA, have been implicated. A major challenge in the future is to identify these signals and to determine how they are linked to ROP regulators such as RopGEFs and RopGAPs.

Acknowledgments

The work of Yang and Fu is supported by grants from US Department of Energy, National Institute of Health, and National Science Foundation. The work of Shimamoto and Kasawaki is supported by Grants-in-Aid from the Ministry of Agriculture, Forestry and Fisheries of Japan (Rice Genome Project), and the Japan Society for Promotion of Science.

References

Asai, T., Tena, G., Plotnikova, J., Willmann, M.R., Chiu, W.L., Gomez-Gomez, L., Boller, T., Ausubel, F.M. and Sheen, J. (2002) MAP kinase signalling cascade in *Arabidopsis* innate immunity. *Nature*, **415**, 977–983.

Ashikari, M., Wu, J., Yano, M., Sasaki, T. and Yoshimura, A. (1999) Rice gibberellin-insensitive dwarf mutant gene *Dwarf 1* encodes the alpha-subunit of GTP-binding protein. *Proc Natl Acad Sci USA*, **96**, 10284–10289.

Assmann, S.M. (1996) Guard cell G proteins. *Trends Plant Sci*, **1**, 73–74.

Babior, B.M. (2004) NADPH oxidase. *Curr Opin Immunol*, **16**, 42–47.

Basu, D., El-Assal Sel, D., Le, J., Mallery, E.L. and Szymanski, D.B. (2004) Interchangeable functions of *Arabidopsis* PIROGI and the human WAVE complex subunit SRA1 during leaf epidermal development. *Development*, **131**, 4345–4355.

Bauer, Z., Gomez-Gomez, L., Boller, T. and Felix, G. (2001) Sensitivity of different ecotypes and mutants of *Arabidopsis thaliana* toward the bacterial elicitor flagellin correlates with the presence of receptor-binding sites. *J Biol Chem*, **276**, 45669–45676.

Baxter-Burrell, A., Yang, Z., Springer, P.S. and Bailey-Serres, J. (2002) RopGAP4-dependent Rop GTPase rheostat control of *Arabidopsis* oxygen deprivation tolerance. *Science*, **296**, 2026–2028.

Berken, A. (2006) ROPs in the spotlight of plant signal transduction. *Cell Mol Life Sci*, **63**, 2446–2459.

Berken, A., Thomas, C. and Wittinghofer, A. (2005) A new family of RhoGEFs activates the Rop molecular switch in plants. *Nature*, **436**, 1176–1180.

Bischoff, F., Vahlkamp, L., Molendijk, A. and Palme, K. (2000) Localization of AtROP4 and AtROP6 and interaction with the guanine nucleotide dissociation inhibitor AtRhoGDI1 from *Arabidopsis*. *Plant Mol Biol*, **42**, 515–530.

Bishop, A.L. and Hall, A. (2000) Rho GTPases and their effector proteins. *Biochem J*, **348** (Pt 2), 241–255.

Bloch, D., Lavy, M., Efrat, Y., Efroni, I., Bracha-Drori, K., Abu-Abied, M., Sadot, E. and Yalovsky, S. (2005) Ectopic expression of an activated RAC in *Arabidopsis* disrupts membrane cycling. *Mol Biol Cell*, **16**, 1913–1927.

Borg, S., Podenphant, L., Jensen, T.J. and Poulsen, C. (1999) Plant cell growth and differentiation may involve GAP regulation of Rac activity. *FEBS Lett*, **453**, 341–345.

Bourne, H.R., Sanders, D.A. and Mccormick, F. (1991) The GTPase superfamily: conserved structure and molecular mechanism. *Nature*, **349**, 117–127.

Brembu, T., Winge, P. and Bones, A.M. (2005) The small GTPase AtRAC2/ROP7 is specifically expressed during late stages of xylem differentiation in *Arabidopsis*. *J Exp Bot*, **56**, 2465–2476.

Brembu, T., Winge, P., Bones, A.M. and Yang, Z. (2006) A RHOse by any other name: a comparative analysis of animal and plant Rho GTPases. *Cell Res*, **16**, 435–445.

Brembu, T., Winge, P., Seem, M. and Bones, A.M. (2004) NAPP and PIRP encode subunits of a putative wave regulatory protein complex involved in plant cell morphogenesis. *Plant Cell*, **16**, 2335–2349.

Brennwald, P. and Rossi, G. (2007) Spatial regulation of exocytosis and cell polarity: yeast as a model for animal cells. *FEBS Lett*, **581**, 2119–2124.

Carol, R.J., Takeda, S., Linstead, P., Durrant, M.C., Kakesova, H., Derbyshire, P., Drea, S., Zarsky, V. and Dolan, L. (2005) A RhoGDP dissociation inhibitor spatially regulates growth in root hair cells. *Nature*, **438**, 1013–1016.

Chamnongpol, S., Willekens, H., Moeder, W., Langebartels, C., Sandermann, H., Jr., Van Montagu, M., Inze, D. and Van Camp, W. (1998) Defense activation and enhanced pathogen tolerance induced by H_2O_2 in transgenic tobacco. *Proc Natl Acad Sci USA*, **95**, 5818–5823.

Chant, J. (1999) Cell polarity in yeast. *Annu Rev Cell Dev Biol*, **15**, 365–391.

Cherfils, J. and Chardin, P. (1999) GEFs: structural basis for their activation of small GTP-binding proteins. *Trends Biochem Sci*, **24**, 306–311.
Choy, E., Chiu, V.K., Silletti, J., Feoktistov, M., Morimoto, T., Michaelson, D., Ivanov, I.E. and Philips, M.R. (1999) Endomembrane trafficking of ras: the CAAX motif targets proteins to the ER and Golgi. *Cell*, **98**, 69–80.
Christensen, T.M., Vejlupkova, Z., Sharma, Y.K., Arthur, K.M., Spatafora, J.W., Albright, C.A., Meeley, R.B., Duvick, J.P., Quatrano, R.S. and Fowler, J.E. (2003) Conserved subgroups and developmental regulation in the monocot rop gene family. *Plant Physiol*, **133**, 1791–1808.
Cole, R.A. and Fowler, J.E. (2006) Polarized growth: maintaining focus on the tip. *Curr Opin Plant Biol*, **9**, 579–588.
Cole, R.A., Synek, L., Zarsky, V. and Fowler, J.E. (2005) SEC8, a subunit of the putative *Arabidopsis* exocyst complex, facilitates pollen germination and competitive pollen tube growth. *Plant Physiol*, **138**, 2005–2018.
Dermardirossian, C. and Bokoch, G.M. (2005) GDIs: central regulatory molecules in Rho GTPase activation. *Trends Cell Biol*, **15**, 356–363.
Djakovic, S., Dyachok, J., Burke, M., Frank, M.J. and Smith, L.G. (2006) BRICK1/HSPC300 functions with SCAR and the ARP2/3 complex to regulate epidermal cell shape in *Arabidopsis*. *Development*, **133**, 1091–1100.
Dvorsky, R. and Ahmadian, M.R. (2004) Always look on the bright site of Rho: structural implications for a conserved intermolecular interface. *EMBO Rep*, **5**, 1130–1136.
El-Din El-Assal, S., Le, J., Basu, D., Mallery, E.L. and Szymanski, D.B. (2004) DISTORTED2 encodes an ARPC2 subunit of the putative *Arabidopsis* ARP2/3 complex. *Plant J*, **38**, 526–538.
Fischer, U., Ikeda, Y., Ljung, K., Serralbo, O., Singh, M., Heidstra, R., Palme, K., Scheres, B. and Grebe, M. (2006) Vectorial information for *Arabidopsis* planar polarity is mediated by combined AUX1, EIN2, and GNOM activity. *Curr Biol*, **16**, 2143–2149.
Fischer, U., Men, S. and Grebe, M. (2004) Lipid function in plant cell polarity. *Curr Opin Plant Biol*, **7**, 670–676.
Foreman, J., Demidchik, V., Bothwell, J.H., Mylona, P., Miedema, H., Torres, M.A., Linstead, P., Costa, S., Brownlee, C., Jones, J.D., Davies, J.M. and Dolan, L. (2003) Reactive oxygen species produced by NADPH oxidase regulate plant cell growth. *Nature*, **422**, 442–446.
Friml, J., Yang, X., Michniewicz, M., Weijers, D., Quint, A., Tietz, O., Benjamins, R., Ouwerkerk, P.B., Ljung, K., Sandberg, G., Hooykaas, P.J., Palme, K. and Offringa, R. (2004) A PINOID-dependent binary switch in apical-basal PIN polar targeting directs auxin efflux. *Science*, **306**, 862–865.
Fu, Y., Gu, Y., Zheng, Z., Wasteneys, G. and Yang, Z. (2005) *Arabidopsis* interdigitating cell growth requires two antagonistic pathways with opposing action on cell morphogenesis. *Cell*, **120**, 687–700.
Fu, Y., Li, H. and Yang, Z. (2002) The ROP2 GTPase controls the formation of cortical fine F-actin and the early phase of directional cell expansion during *Arabidopsis* organogenesis. *Plant Cell*, **14**, 777–794.
Fu, Y., Wu, G. and Yang, Z. (2001) Rop GTPase-dependent dynamics of tip-localized F-actin controls tip growth in pollen tubes. *J Cell Biol*, **152**, 1019–1032.
Fu, Y. and Yang, Z. (2001) Rop GTPase: a master switch of cell polarity development in plants. *Trends Plant Sci*, **6**, 545–547.

Fujisawa, Y., Kato, T., Ohki, S., Ishikawa, A., Kitano, H., Sasaki, T., Asahi, T. and Iwasaki, Y. (1999) Suppression of the heterotrimeric G protein causes abnormal morphology, including dwarfism, in rice. *Proc Natl Acad Sci USA*, **96**, 7575–7580.

Fujiwara, M., Umemura, K., Kawasaki, T. and Shimamoto, K. (2006) Proteomics of Rac GTPase signaling reveals its predominant role in elicitor-induced defense response of cultured rice cells. *Plant Physiol*, **140**, 734–745.

Gomez-Gomez, L. and Boller, T. (2000) FLS2: an LRR receptor-like kinase involved in the perception of the bacterial elicitor flagellin in *Arabidopsis*. *Mol Cell*, **5**, 1003–1011.

Gu, Y., Fu, Y., Dowd, P., Li, S., Vernoud, V., Gilroy, S. and Yang, Z. (2005) A Rho family GTPase controls actin dynamics and tip growth via two counteracting downstream pathways in pollen tubes. *J Cell Biol*, **169**, 127–138.

Gu, Y., Li, S., Lord, E.M. and Yang, Z. (2006) Members of a novel class of *Arabidopsis* Rho guanine nucleotide exchange factors control Rho GTPase-dependent polar growth. *Plant Cell*, **18**, 366–381.

Gu, Y., Wang, Z. and Yang, Z. (2004) ROP/RAC GTPase: an old new master regulator for plant signaling. *Curr Opin Plant Biol*, **7**, 527–536.

Hall, A. (1998) Rho GTPases and the actin cytoskeleton. *Science*, **279**, 509–514.

Hancock, J.F., Cadwallader, K., Paterson, H. and Marshall, C.J. (1991) A CAAX or a CAAL motif and a second signal are sufficient for plasma membrane targeting of ras proteins. *EMBO J*, **10**, 4033–4039.

Hassanain, H.H., Sharma, Y.K., Moldovan, L., Khramtsov, V., Berliner, L.J., Duvick, J.P. and Goldschmidt-Clermont, P.J. (2000) Plant rac proteins induce superoxide production in mammalian cells. *Biochem Biophys Res Commun*, **272**, 783–788.

He, C., Fong, S.H., Yang, D. and Wang, G.L. (1999) BWMK1, a novel MAP kinase induced by fungal infection and mechanical wounding in rice. *Mol Plant Microbe Interact*, **12**, 1064–1073.

He, P., Shan, L., Lin, N.C., Martin, G.B., Kemmerling, B., Nurnberger, T. and Sheen, J. (2006) Specific bacterial suppressors of MAMP signaling upstream of MAPKKK in *Arabidopsis* innate immunity. *Cell*, **125**, 563–575.

Heo, W.D., Inoue, T., Park, W.S., Kim, M.L., Park, B.O., Wandless, T.J. and Meyer, T. (2006) PI(3,4,5)P3 and PI(4,5)P2 lipids target proteins with polybasic clusters to the plasma membrane. *Science*, **314**, 1458–1461.

Hepler, J.R. and Gilman, A.G. (1992) G proteins. *Trends Biochem Sci*, **17**, 383–387.

Hong, Z., Zhang, Z., Olson, J.M. and Verma, D.P. (2001) A novel UDP-glucose transferase is part of the callose synthase complex and interacts with phragmoplastin at the forming cell plate. *Plant Cell*, **13**, 769–779.

Hwang, J.U., Gu, Y., Lee, Y.J. and Yang, Z. (2005) Oscillatory ROP GTPase activation leads the oscillatory polarized growth of pollen tubes. *Mol Biol Cell*, **16**, 5385–5399.

Ichimura, K., Mizoguchi, T., Yoshida, R., Yuasa, T. and Shinozaki, K. (2000) Various abiotic stresses rapidly activate *Arabidopsis* MAP kinases ATMPK4 and ATMPK6. *Plant J*, **24**, 655–665.

Ichimura, K., Shinozaki, K., Tena, G., Sheen, J., Henri, Y., Champion, A., Kreis, M., Zhang, S.Q., Hirt, H., Wilson, C., Heberle-Bors, E., Ellis, B.E., Morris, P.C., Innes, R.W., Ecker, J.R., Scheel, D., Klessig, D.F., Machida, Y., Mundy, J., Ohashi, Y. and Walker, J.C. (2002) Mitogen-activated protein kinase cascades in plants: a new nomenclature. *Trends Plant Sci*, **7**, 301–308.

Ivanchenko, M., Vejlupkova, Z., Quatrano, R.S. and Fowler, J.E. (2000) Maize ROP7 GTPase contains a unique, CaaX box-independent plasma membrane targeting signal. *Plant J*, **24**, 79–90.

Jones, H.D., Smith, S.J., Desikan, R., Plakidou-Dymock, S., Lovegrove, A. and Hooley, R. (1998) Heterotrimeric G proteins are implicated in gibberellin induction of a-amylase gene expression in wild oat aleurone. *Plant Cell*, **10**, 245–254.

Jones, J.D. and Dangl, J.L. (2006) The plant immune system. *Nature*, **444**, 323–329.

Jones, M.A., Raymond, M.J. and Smirnoff, N. (2006) Analysis of the root-hair morphogenesis transcriptome reveals the molecular identity of six genes with roles in root-hair development in *Arabidopsis*. *Plant J*, **45**, 83–100.

Jones, M.A., Raymond, M.J., Yang, Z. and Smirnoff, N. (2007) NADPH oxidase-dependent reactive oxygen species formation required for root hair growth depends on ROP GTPase. *J Exp Bot*, **58**, 1261–1270.

Jones, M.A., Shen, J.J., Fu, Y., Li, H., Yang, Z. and Grierson, C.S. (2002) The *Arabidopsis* Rop2 GTPase is a positive regulator of both root hair initiation and tip growth. *Plant Cell*, **14**, 763–776.

Jung, Y.H., Agrawal, G.K., Rakwal, R., Kim, J.A., Lee, M.O., Choi, P.G., Kim, Y.J., Kim, M.J., Shibato, J., Kim, S.H., Iwahashi, H. and Jwa, N.S. (2006) Functional characterization of OsRacB GTPase—a potentially negative regulator of basal disease resistance in rice. *Plant Physiol Biochem*, **44**, 68–77.

Kaothien, P., Ok, S.H., Shuai, B., Wengier, D., Cotter, R., Kelley, D., Kiriakopolos, S., Muschietti, J. and Mccormick, S. (2005) Kinase partner protein interacts with the LePRK1 and LePRK2 receptor kinases and plays a role in polarized pollen tube growth. *Plant J*, **42**, 492–503.

Kawasaki, T., Henmi, K., Ono, E., Hatakeyama, S., Iwano, M., Satoh, H. and Shimamoto, K. (1999) The small GTP-binding protein rac is a regulator of cell death in plants. *Proc Natl Acad Sci USA*, **96**, 10922–10926.

Kawasaki, T., Koita, H., Nakatsubo, T., Hasegawa, K., Wakabayashi, K., Takahashi, H., Umemura, K., Umezawa, T. and Shimamoto, K. (2006) Cinnamoyl-CoA reductase, a key enzyme in lignin biosynthesis, is an effector of small GTPase Rac in defense signaling in rice. *Proc Natl Acad Sci USA*, **103**, 230–235.

Kieffer, F., Elmayan, T., Rubier, S., Simon-Plas, F., Dagher, M.C. and Blein, J.P. (2000) Cloning of Rac and Rho-GDI from tobacco using an heterologous two-hybrid screen. *Biochimie*, **82**, 1099–1105.

Kim, M.G., Da Cunha, L., Mcfall, A.J., Belkhadir, Y., Debroy, S., Dangl, J.L. and Mackey, D. (2005) Two *Pseudomonas syringae* type III effectors inhibit RIN4-regulated basal defense in *Arabidopsis*. *Cell*, **121**, 749–759.

Klahre, U., Becker, C., Schmitt, A.C. and Kost, B. (2006) Nt-RhoGDI2 regulates Rac/Rop signaling and polar cell growth in tobacco pollen tubes. *Plant J*, **46**, 1018–1031.

Klahre, U. and Kost, B. (2006) Tobacco RhoGTPase ACTIVATING PROTEIN1 spatially restricts signaling of RAC/Rop to the apex of pollen tubes. *Plant Cell*, **18**, 3033–3046.

Ko, J.H., Beers, E.P. and Han, K.H. (2006) Global comparative transcriptome analysis identifies gene network regulating secondary xylem development in *Arabidopsis thaliana*. *Mol Genet Genomics*, **276**, 517–531.

Kobayashi, Y., Kobayashi, H., Funaki, Y., Fujimoto, S., Takemoto, T. and Kunoh, H. (1997) Dynamic reorganization of microfilaments and microtubules is necessary for the expression of non-host resistance in barley coleoptile cells. *Plant J*, **11**, 525–537.

Kost, B., Lemichez, E., Spielhofer, P., Hong, Y., Tolias, K., Carpenter, C. and Chua, N.H. (1999) Rac homologues and compartmentalized phosphatidylinositol 4, 5-bisphosphate act in a common pathway to regulate polar pollen tube growth. *J Cell Biol*, **145**, 317–330.

Kropf, D.L., Bisgrove, S.R. and Hable, W.E. (1998) Cytoskeletal control of polar growth in plant cells. *Curr Opin Cell Biol*, **10**, 117–122.
Kurose, H. (2003) Galpha12 and Galpha13 as key regulatory mediator in signal transduction. *Life Sci*, **74**, 155–161.
Lavy, M., Bloch, D., Hazak, O., Gutman, I., Poraty, L., Sorek, N., Sternberg, H. and Yalovsky, S. (2007) A novel ROP/RAC effector links cell polarity, root-meristem maintenance, and vesicle trafficking. *Curr Biol*, **17**, 947–952.
Lavy, M., Bracha-Drori, K., Sternberg, H. and Yalovsky, S. (2002) A cell-specific, prenylation-independent mechanism regulates targeting of type II RACs. *Plant Cell*, **14**, 2431–2450.
Lee, S.W., Han, S.W., Bartley, L.E. and Ronald, P.C. (2006) From the Academy: Colloquium review. Unique characteristics of *Xanthomonas oryzae* pv. *oryzae* AvrXa21 and implications for plant innate immunity. *Proc Natl Acad Sci USA*, **103**, 18395–18400.
Lemichez, E., Wu, Y., Sanchez, J.P., Mettouchi, A., Mathur, J. and Chua, N.H. (2001) Inactivation of AtRac1 by abscisic acid is essential for stomatal closure. *Genes Dev*, **15**, 1808–1816.
Li, H., Lin, Y., Heath, R.M., Zhu, M.X. and Yang, Z. (1999) Control of pollen tube tip growth by a Rop GTPase-dependent pathway that leads to tip-localized calcium influx. *Plant Cell*, **11**, 1731–1742.
Li, H., Shen, J.J., Zheng, Z.L., Lin, Y. and Yang, Z. (2001) The Rop GTPase switch controls multiple developmental processes in *Arabidopsis*. *Plant Physiol*, **126**, 670–684.
Li, H., Wu, G., Ware, D., Davis, K.R. and Yang, Z. (1998) *Arabidopsis* Rho-related GTPases: differential gene expression in pollen and polar localization in fission yeast. *Plant Physiol*, **118**, 407–417.
Li, L., Xu, J., Xu, Z.H. and Xue, H.W. (2005) Brassinosteroids stimulate plant tropisms through modulation of polar auxin transport in *Brassica* and *Arabidopsis*. *Plant Cell*, **17**, 2738–2753.
Li, S., Blanchoin, L., Yang, Z. and Lord, E.M. (2003) The putative *Arabidopsis* arp2/3 complex controls leaf cell morphogenesis. *Plant Physiol*, **132**, 2034–2044.
Lieberherr, D., Thao, N.P., Nakashima, A., Umemura, K., Kawasaki, T. and Shimamoto, K. (2005) A sphingolipid elicitor-inducible mitogen-activated protein kinase is regulated by the small GTPase OsRac1 and heterotrimeric G-protein in rice 1[w]. *Plant Physiol*, **138**, 1644–1652.
Lin, Y., Seals, D.F., Randall, S.K. and Yang, Z. (2001) Dynamic localization of rop GTPases to the tonoplast during vacuole development. *Plant Physiol*, **125**, 241–251.
Lin, Y., Wang, Y., Zhu, J.K. and Yang, Z. (1996) Localization of a Rho GTPase implies a role in tip growth and movement of the generative cell in pollen tubes. *Plant Cell*, **8**, 293–303.
Lin, Y. and Yang, Z. (1997) Inhibition of pollen tube elongation by microinjected anti-Rop1Ps antibodies suggests a crucial role for Rho-type GTPases in the control of tip growth. *Plant Cell*, **9**, 1647–1659.
Llorente, F., Alonso-Blanco, C., Sanchez-Rodriguez, C., Jorda, L. and Molina, A. (2005) ERECTA receptor-like kinase and heterotrimeric G protein from *Arabidopsis* are required for resistance to the necrotrophic fungus *Plectosphaerella cucumerina*. *Plant J*, **43**, 165–180.
Lu, M. and Ravichandran, K.S. (2006) Dock180-ELMO cooperation in Rac activation. *Methods Enzymol*, **406**, 388–402.
Miki, D., Itoh, R. and Shimamoto, K. (2005) RNA silencing of single and multiple members in a gene family of rice. *Plant Physiol*, **138**, 1903–1913.

Mittler, R., Feng, X. and Cohen, M. (1998) Post-transcriptional suppression of cytosolic ascorbate peroxidase expression during pathogen-induced programmed cell death in tobacco. *Plant Cell*, **10**, 461–473.

Mittler, R., Herr, E.H., Orvar, B.L., Van Camp, W., Willekens, H., Inze, D. and Ellis, B.E. (1999) Transgenic tobacco plants with reduced capability to detoxify reactive oxygen intermediates are hyperresponsive to pathogen infection. *Proc Natl Acad Sci USA*, **96**, 14165–14170.

Miyano, K. and Sumimoto, H. (2007) Role of the small GTPase Rac in p22(phox)-dependent NADPH oxidases. *Biochimie*, **89**, 1133–1144.

Moeder, W., Yoshioka, K. and Klessig, D.F. (2005) Involvement of the small GTPase Rac in the defense responses of tobacco to pathogens. *Mol Plant Microbe Interact*, **18**, 116–124.

Moerschbacher, B.M., Noll, U., Gorrichon, L. and Reisener, H.J. (1990) Specific inhibition of lignification breaks hypersensitive resistance of wheat to stem rust. *Plant Physiol*, **93**, 465–470.

Molendijk, A.J., Bischoff, F., Rajendrakumar, C.S., Friml, J., Braun, M., Gilroy, S. and Palme, K. (2001) *Arabidopsis thaliana* Rop GTPases are localized to tips of root hairs and control polar growth. *EMBO J*, **20**, 2779–2788.

Mongrand, S., Morel, J., Laroche, J., Claverol, S., Carde, J.P., Hartmann, M.A., Bonneu, M., Simon-Plas, F., Lessire, R. and Bessoule, J.J. (2004) Lipid rafts in higher plant cells: purification and characterization of Triton X-100-insoluble microdomains from tobacco plasma membrane. *J Biol Chem*, **279**, 36277–36286.

Morel, J., Fromentin, J., Blein, J.P., Simon-Plas, F. and Elmayan, T. (2004) Rac regulation of NtrbohD, the oxidase responsible for the oxidative burst in elicited tobacco cell. *Plant J*, **37**, 282–293.

Mori, I.C. and Schroeder, J.I. (2004) Reactive oxygen species activation of plant Ca^{2+} channels: a signaling mechanism in polar growth, hormone transduction, stress signaling, and hypothetically mechanotransduction. *Plant Physiol*, **135**, 702–708.

Ono, E., Wong, H.L., Kawasaki, T., Hasegawa, M., Kodama, O. and Shimamoto, K. (2001) Essential role of the small GTPase Rac in disease resistance of rice. *Proc Natl Acad Sci USA*, **98**, 759–764.

Opalski, K.S., Schultheiss, H., Kogel, K.H. and Huckelhoven, R. (2005) The receptor-like MLO protein and the RAC/ROP family G-protein RACB modulate actin reorganization in barley attacked by the biotrophic powdery mildew fungus *Blumeria graminis* f.sp. *hordei*. *Plant J*, **41**, 291–303.

Osakabe, Y., Maruyama, K., Seki, M., Satou, M., Shinozaki, K. and Yamaguchi-Shinozaki, K. (2005) Leucine-rich repeat receptor-like kinase1 is a key membrane-bound regulator of abscisic acid early signaling in *Arabidopsis*. *Plant Cell*, **17**, 1105–1119.

Otomo, K., Kenmoku, H., Oikawa, H., Konig, W.A., Toshima, H., Mitsuhashi, W., Yamane, H., Sassa, T. and Toyomasu, T. (2004) Biological functions of *ent*- and *syn*-copalyl diphosphate synthases in rice: key enzymes for the branch point of gibberellin and phytoalexin biosynthesis. *Plant J*, **39**, 886–893.

Paponov, I.A., Teale, W.D., Trebar, M., Blilou, I. and Palme, K. (2005) The PIN auxin efflux facilitators: evolutionary and functional perspectives. *Trends Plant Sci*, **10**, 170–177.

Potikha, T.S., Collins, C.C., Johnson, D.I., Delmer, D.P. and Levine, A. (1999) The involvement of hydrogen peroxide in the differentiation of secondary walls in cotton fibers. *Plant Physiol*, **119**, 849–858.

Qiu, J.L., Jilk, R., Marks, M.D. and Szymanski, D.B. (2002) The *Arabidopsis SPIKE1* gene is required for normal cell shape control and tissue development. *Plant Cell*, **14**, 101–118.

Rossman, K.L., Cheng, L., Mahon, G.M., Rojas, R.J., Snyder, J.T., Whitehead, I.P. and Sondek, J. (2003) Multifunctional roles for the PH domain of Dbs in regulating Rho GTPase activation. *J Biol Chem*, **278**, 18393–18400.

Rossman, K.L., Der, C.J. and Sondek, J. (2005) GEF means go: turning on RHO GTPases with guanine nucleotide-exchange factors. *Nat Rev Mol Cell Biol*, **6**, 167–180.

Schiene, K., Puhler, A. and Niehaus, K. (2000) Transgenic tobacco plants that express an antisense construct derived from a *Medicago sativa* cDNA encoding a Rac-related small GTP-binding protein fail to develop necrotic lesions upon elicitor infiltration. *Mol Gen Genet*, **263**, 761–770.

Schmidt, A. and Hall, A. (2002) Guanine nucleotide exchange factors for Rho GTPases: turning on the switch. *Genes Dev*, **16**, 1587–1609.

Schrick, K., Mayer, U., Horrichs, A., Kuhnt, C., Bellini, C., Dangl, J., Schmidt, J. and Jurgens, G. (2000) FACKEL is a sterol C-14 reductase required for organized cell division and expansion in *Arabidopsis* embryogenesis. *Genes Dev*, **14**, 1471–1484.

Schultheiss, H., Dechert, C., Kogel, K.H. and Huckelhoven, R. (2002) A small GTP-binding host protein is required for entry of powdery mildew fungus into epidermal cells of barley. *Plant Physiol*, **128**, 1447–1454.

Schultheiss, H., Dechert, C., Kogel, K.H. and Huckelhoven, R. (2003) Functional analysis of barley RAC/ROP G-protein family members in susceptibility to the powdery mildew fungus. *Plant J*, **36**, 589–601.

Schultheiss, H., Hensel, G., Imani, J., Broeders, S., Sonnewald, U., Kogel, K.H., Kumlehn, J. and Huckelhoven, R. (2005) Ectopic expression of constitutively activated RACB in barley enhances susceptibility to powdery mildew and abiotic stress. *Plant Physiol*, **139**, 353–362.

Smith, L.G. and Oppenheimer, D.G. (2005) Spatial control of cell expansion by the plant cytoskeleton. *Annu Rev Cell Dev Biol*, **21**, 271–295.

Smythe, E. and Ayscough, K.R. (2006) Actin regulation in endocytosis. *J Cell Sci*, **119**, 4589–4598.

Sorek, N., Poraty, L., Sternberg, H., Bar, E., Lewinsohn, E. and Yalovsky, S. (2007) Activation status-coupled transient S acylation determines membrane partitioning of a plant Rho-related GTPase. *Mol Cell Biol*, **27**, 2144–2154.

Sormo, C.G., Leiros, I., Brembu, T., Winge, P., Os, V. and Bones, A.M. (2006) The crystal structure of *Arabidopsis thaliana* RAC7/ROP9: the first RAS superfamily GTPase from the plant kingdom. *Phytochemistry*, **67**, 2332–2340.

Stepanova, A.N., Hoyt, J.M., Hamilton, A.A. and Alonso, J.M. (2005) A link between ethylene and auxin uncovered by the characterization of two root-specific ethylene-insensitive mutants in *Arabidopsis*. *Plant Cell*, **17**, 2230–2242.

Suharsono, U., Fujisawa, Y., Kawasaki, T., Iwasaki, Y., Satoh, H. and Shimamoto, K. (2002) The heterotrimeric G protein alpha subunit acts upstream of the small GTPase Rac in disease resistance of rice. *Proc Natl Acad Sci USA*, **99**, 13307–13312.

Swarup, R., Kargul, J., Marchant, A., Zadik, D., Rahman, A., Mills, R., Yemm, A., May, S., Williams, L., Millner, P., Tsurumi, S., Moore, I., Napier, R., Kerr, I.D. and Bennett, M.J. (2004) Structure–function analysis of the presumptive *Arabidopsis* auxin permease AUX1. *Plant Cell*, **16**, 3069–3083.

Szymanski, D.B. (2005) Breaking the WAVE complex: the point of *Arabidopsis* trichomes. *Curr Opin Plant Biol*, **8**, 103–112.

Takahashi, H., Chen, Z., Du, H., Liu, Y. and Klessig, D.F. (1997) Development of necrosis and activation of disease resistance in transgenic tobacco plants with severely reduced catalase levels. *Plant J*, **11**, 993–1005.

Tao, L.Z., Cheung, A.Y. and Wu, H.M. (2002) Plant Rac-like GTPases are activated by auxin and mediate auxin-responsive gene expression. *Plant Cell*, **14**, 2745–2760.

Thapar, R., Karnoub, A.E. and Campbell, S.L. (2002) Structural and biophysical insights into the role of the insert region in Rac1 function. *Biochemistry*, **41**, 3875–3883.

Thomas, C., Fricke, I., Scrima, A., Berken, A. and Wittinghofer, A. (2007) Structural evidence for a common intermediate in small G protein-GEF reactions. *Mol Cell*, **25**, 141–149.

Toret, C.P. and Drubin, D.G. (2006) The budding yeast endocytic pathway. *J Cell Sci*, **119**, 4585–4587.

Torres, M.A. and Dangl, J.L. (2005) Functions of the respiratory burst oxidase in biotic interactions, abiotic stress and development. *Curr Opin Plant Biol*, **8**, 397–403.

Torres, M.A., Dangl, J.L. and Jones, J.D. (2002) *Arabidopsis* gp91phox homologues AtrbohD and AtrbohF are required for accumulation of reactive oxygen intermediates in the plant defense response. *Proc Natl Acad Sci USA*, **99**, 517–522.

Trainin, T., Shmuel, M. and Delmer, D.P. (1996) *In vitro* prenylation of the small GTPase Rac13 of cotton. *Plant Physiol*, **112**, 1491–1497.

Trotochaud, A.E., Hao, T., Wu, G., Yang, Z. and Clark, S.E. (1999) The CLAVATA1 receptor-like kinase requires CLAVATA3 for its assembly into a signaling complex that includes KAPP and a Rho-related protein. *Plant Cell*, **11**, 393–406.

Trusov, Y., Rookes, J.E., Chakravorty, D., Armour, D., Schenk, P.M. and Botella, J.R. (2006) Heterotrimeric G proteins facilitate *Arabidopsis* resistance to necrotrophic pathogens and are involved in jasmonate signaling. *Plant Physiol*, **140**, 210–220.

Uhrig, J.F. and Hulskamp, M. (2006) Plant GTPases: regulation of morphogenesis by ROPs and ROS. *Curr Biol*, **16**, R211–R213.

Uhrig, J.F., Mutondo, M., Zimmermann, I., Deeks, M.J., Machesky, L.M., Thomas, P., Uhrig, S., Rambke, C., Hussey, P.J. and Hulskamp, M. (2007) The role of *Arabidopsis* SCAR genes in ARP2–ARP3-dependent cell morphogenesis. *Development*, **134**, 967–977.

Vernoud, V., Horton, A.C., Yang, Z. and Nielsen, E. (2003) Analysis of the small GTPase gene superfamily of *Arabidopsis*. *Plant Physiol*, **131**, 1191–1208.

Vetter, I.R. and Wittinghofer, A. (2001) The guanine nucleotide-binding switch in three dimensions. *Science*, **294**, 1299–1304.

Wennerberg, K., Rossman, K.L. and Der, C.J. (2005) The Ras superfamily at a glance. *J Cell Sci*, **118**, 843–846.

Winge, P., Brembu, T., Kristensen, R. and Bones, A.M. (2000) Genetic structure and evolution of RAC-GTPases in *Arabidopsis thaliana*. *Genetics*, **156**, 1959–1971.

Wittinghofer, F. (1998) Ras signalling: caught in the act of the switch-on. *Nature*, **394**, 317, 319–320.

Wong, H.L., Sakamoto, T., Kawasaki, T., Umemura, K. and Shimamoto, K. (2004) Down-regulation of metallothionein, a reactive oxygen scavenger, by the small GTPase OsRac1 in rice. *Plant Physiol*, **135**, 1447–1456.

Wu, G., Gu, Y., Li, S. and Yang, Z. (2001) A genome-wide analysis of *Arabidopsis* Rop-interactive CRIB motif-containing proteins that act as Rop GTPase targets. *Plant Cell*, **13**, 2841–2856.

Wu, G., Li, H. and Yang, Z. (2000) *Arabidopsis* RopGAPs are a novel family of rho GTPase-activating proteins that require the Cdc42/Rac-interactive binding motif for rop-specific GTPase stimulation. *Plant Physiol*, **124**, 1625–1636.

Xiong, L. and Yang, Y. (2003) Disease resistance and abiotic stress tolerance in rice are inversely modulated by an abscisic acid-inducible mitogen-activated protein kinase. *Plant Cell*, **15**, 745–759.

Xu, J. and Scheres, B. (2005) Cell polarity: ROPing the ends together. *Curr Opin Plant Biol*, **8**, 613–618.

Yang, Z. (2002) Small GTPases: versatile signaling switches in plants. *Plant Cell*, **14** (Suppl), S375–S388.

Yang, Z. and Fu, Y. (2007) ROP/RAC GTPase signaling. *Curr Opin Plant Biol*, **10**, 490–494.

Yang, Z. and Watson, J.C. (1993) Molecular cloning and characterization of rho, a ras-related small GTP-binding protein from the garden pea. *Proc Natl Acad Sci USA*, **90**, 8732–8736.

Yoshioka, H., Numata, N., Nakajima, K., Katou, S., Kawakita, K., Rowland, O., Jones, J.D. and Doke, N. (2003) *Nicotiana benthamiana* gp91phox homologs NbrbohA and NbrbohB participate in H_2O_2 accumulation and resistance to *Phytophthora infestans*. *Plant Cell*, **15**, 706–718.

Zhang, X., Bi, E., Novick, P., Du, L., Kozminski, K.G., Lipschutz, J.H. and Guo, W. (2001) Cdc42 interacts with the exocyst and regulates polarized secretion. *J Biol Chem*, **276**, 46745–46750.

Zheng, Z.L., Nafisi, M., Tam, A., Li, H., Crowell, D.N., Chary, S.N., Schroeder, J.I., Shen, J. and Yang, Z. (2002) Plasma membrane-associated ROP10 small GTPase is a specific negative regulator of abscisic acid responses in *Arabidopsis*. *Plant Cell*, **14**, 2787–2797.

Zheng, Z.L. and Yang, Z. (2000) The Rop GTPase: an emerging signaling switch in plants. *Plant Mol Biol*, **44**, 1–9.

Zimmermann, I., Saedler, R., Mutondo, M. and Hulskamp, M. (2004) The *Arabidopsis* GNARLED gene encodes the NAP125 homolog and controls several actin-based cell shape changes. *Mol Genet Genomics*, **272**, 290–296.

Zipfel, C. and Felix, G. (2005) Plants and animals: a different taste for microbes? *Curr Opin Plant Biol*, **8**, 353–360.

Zipfel, C., Kunze, G., Chinchilla, D., Caniard, A., Jones, J.D., Boller, T. and Felix, G. (2006) Perception of the bacterial PAMP EF-Tu by the receptor EFR restricts *Agrobacterium*-mediated transformation. *Cell*, **125**, 749–760.

Annual Plant Reviews (2008) **33**, 100–136
doi: 10.1111/b.9781405160025.2008.00004.x www.interscience.wiley.com

Chapter 4

MITOGEN-ACTIVATED PROTEIN KINASE CASCADES IN PLANT SIGNALING

Shuqun Zhang

Department of Biochemistry, 371G Life Sciences Center, University of Missouri-Columbia, Columbia, MO 65211, USA

Abstract: Plants have an expanded family of mitogen-activated protein kinases (MAPKs) in comparison with yeast and animals. Analysis of the completely sequenced *Arabidopsis*, poplar, and rice genomes revealed 20, 21, and 15 members, respectively. These MAPKs have the potential to form numerous functional cascades with their upstream MAPK kinases (MAPKKs or MEKs) and MAPKK kinases (MAPKKKs or MEKKs). Biochemical and genetic analyses demonstrated that plant MAPK cascades play important roles in signaling plant growth, development, and response to environmental cues. It is also becoming clear that different functional cascades frequently share common components, suggesting complicated signaling networks centered on these MAPKs. How specificity is maintained when distinct functional pathways share common components is central to our understanding of plant intracellular signaling. Future research aimed at identifying upstream receptors/sensors and downstream MAPK substrates will reveal the molecular mechanisms underlying MAPK functions and shed light on how signaling specificity is maintained.

Keywords: mitogen-activated protein kinase (MAPK); cellular signaling; plant hormones; growth and development; stress response; disease resistance

4.1 Mitogen-activated protein kinase cascades are evolutionarily conserved signaling modules in eukaryotic cells

Mitogen-activated protein kinase (MAPK) cascades are highly conserved signaling modules in eukaryotes. They function downstream of sensors/receptors and convert signals generated at the sensors/receptors into cellular responses, typically through altering the expression of specific sets of genes (Widmann *et al.*, 1999; Davis, 2000; Chang and Karin, 2001; Hazzalin and Mahadevan, 2002; Schwartz and Madhani, 2004). The core of an MAPK cascade consists of three interconnected kinases. MAPK, the last kinase in the cascade, is activated by dual phosphorylation of the threonine (Thr) and tyrosine (Tyr) residues in a tripeptide motif (Thr–Xaa–Tyr) located in the activation loop (T-loop) between subdomains VII and VIII of the kinase catalytic domain. This phosphorylation is mediated by an MAPK kinase (MAPKK or MEK), which is activated, in turn, by an MAPKK kinase (MAPKKK or MEKK) through phosphorylation. There are multiple members in each of the three tiers of kinases, which contributes to the specificity of the transmitted signal. In *Arabidopsis*, there are 20 MAPKs, 10 MAPKs, and about 60 MAPKKKs (MAPK Group, 2002). Other plant species have similar numbers (Hamel *et al.*, 2006). When added together, they form the second-largest group of kinases in *Arabidopsis*.

4.2 History of plant MAPK research

Plant MAPK research began with the molecular cloning of kinases with high homology to mammalian/yeast MAPKs in 1993 independently by several groups (Duerr *et al.*, 1993; Jonak *et al.*, 1993; Mizoguchi *et al.*, 1993; Stafstrom *et al.*, 1993). At about the same time, several MAPKKK genes were also identified in plants (Banno *et al.*, 1993). Earlier molecular and biochemical characterizations linked plant MAPKs to various processes including cell division, hormonal response, and stress response (Banno *et al.*, 1993; Jonak *et al.*, 1993; Mizoguchi *et al.*, 1994; Seo *et al.*, 1995). However, because of the lack of member-specific antibodies, there was a disconnection between the molecular data and the identity of the kinase activity detected using biochemical assays.

The first attempt of using a member-specific antibody to link a specific MAPK gene to a kinase activity was made in alfalfa by the Heribert Hirt group (Jonak *et al.*, 1996). This research concluded that MMK4 (*Medicago* MAP kinase 4; later renamed as SAMK) is specifically activated by cold and drought stresses based on an antibody raised against a peptide corresponding to the C-terminus of MMK4. However, this antibody is likely to recognize MMK1 (later renamed as SIMK, salt stress-induced MAP kinase) because of the highly conserved C-termini of MMK4 and MMK1. MMK1 is the first alfalfa MAPK identified (Jonak *et al.*, 1993). Unfortunately, it was not included in the study to determine the specificity of the antibody (Jonak *et al.*, 1996).

In search for kinases induced by salicylic acid (SA) in tobacco, Daniel Klessig's laboratory purified and cloned *s*alicylic-acid-*i*nduced *p*rotein *k*inase (SIPK), which encodes a tobacco MAPK sharing high homology with alfalfa MMK1 (Zhang and Klessig, 1997). SIPK was later shown to be activated by a variety of stress/defense stimuli independent of SA (Zhang *et al.*, 1998; Zhang and Klessig, 1998a,b; Romeis *et al.*, 1999; Hoyos and Zhang, 2000; Mikolajczyk *et al.*, 2000; Samuel and Ellis, 2002). As a result, SIPK becomes the acronym of *s*tress-*i*nduced *p*rotein *k*inase. In-gel assay revealed a second smaller MAPK that was activated by tobacco mosaic virus (TMV) infection and elicitin treatment. This MAPK was determined to be WIPK (wound-induced protein kinase) using a member-specific antibody (Seo *et al.*, 1995; Zhang *et al.*, 1998; Zhang and Klessig, 1998b).

Similar to animal and yeast systems, plant MAPKs are encoded by multigene families. A total of seven MAPKs were identified in *Arabidopsis*, a record held by Kazuo Shinozaki's group in the pregenome era (Mizoguchi *et al.*, 1993). With the completion of the *Arabidopsis* genome sequencing project, it finally came to the age that we could catalog all the MAPKs, MAPKKs, and, to certain extent, MAPKKKs in a plant species (MAPK Group, 2002). This information is vital to the understanding of MAPK functions in a number of ways.

Forward genetic studies identified several genes in MAPK cascades, including CTR1, MPK4, EDR1, YDA, and BUD1/MKK7 (Kieber *et al.*, 1993; Petersen *et al.*, 2000; Frye *et al.*, 2001; Bergmann *et al.*, 2004; Lukowitz *et al.*, 2004; Dai *et al.*, 2006). The available T-DNA insertional mutants and genomic information permitted the use of a reverse genetic approach, which can overcome the functional redundancy issue. This approach begins to reveal novel functions of plant MAPKs (Krysan *et al.*, 2002; Chaiwongsar *et al.*, 2006; Wang *et al.*, 2007).

4.3 Plant MAPK cascades

After the identification of MAPKs, search for their upstream MAPKKs and MAPKKKs began. The criteria for placing an MAPKK or MAPKKK upstream of a specific MAPK or MAPKK, which can be coined from the definition of an MAPK cascade, are (1) the upstream kinase can phosphorylate the downstream substrate MAPK, or MAPKK; (2) the phosphorylation by upstream kinase leads to the activation of downstream kinase; and (3) in vivo data to show that all three kinases in the cascade function in the same biological process.

It is not difficult to determine whether a gene encodes an MAPKK because its sequences are highly conserved in all eukaryotes. However, it is not trivial to place an MAPKK upstream of a specific MAPK or MAPKs based on in vivo evidence. It is even more difficult to place an MAPKKK upstream of a specific MAPKK for several reasons. Firstly, it is not easy to assay the activity of MAPKKKs, which are large protein kinases with C-terminal and/or N-terminal extensions outside of the kinase domains. These extra sequences are involved in regulating the activity and specificity of the kinase domain of the MAPKKKs. As a result, the truncated constitutively active version with only the kinase domain may be unable to maintain the specificity in activating the downstream MAPKK(s). Secondly, the mechanisms underlying the activation of MAPKKKs after the sensing of stimuli are largely unknown. Early evidence supporting the placement of an MAPKK or MAPKKK upstream of an MAPK or MAPKK, respectively, included (1) direct interaction based on yeast two-hybrid assay, (2) complementation of yeast mutants, and (3) coregulation at transcriptional level (Mizoguchi *et al.*, 1996; Ichimura *et al.*, 1998; Mizoguchi *et al.*, 1998).

4.3.1 Constitutively active MAPKK and MAPKKK mutants

In yeast and animals, MAPKKs are activated through the phosphorylation of two Ser/Thr (S/T) residues in a conserved S–TxxxS–T motif by MAPKKKs (Widmann *et al.*, 1999). When these two S/T residues are replaced with Glu (E) or Asp (D), the mutant MAPKK becomes constitutively active (Mansour *et al.*, 1994; Wurgler-Murphy *et al.*, 1997). It turns out that plant MAPKKs have a similar activation motif, with the exception that the two S/T residues are separated by five amino acids rather than the three in animals/yeast (Ichimura *et al.*, 1998; Yang *et al.*, 2001). Mutation of the Thr227 and Ser233 residues in the activation loop of tobacco NtMEK2 to Asp (D) residues gave rise to a constitutively active form of NtMEK2. Based on both the in vitro and in vivo evidence, it was concluded that NtMEK2 is upstream of both SIPK and WIPK (Yang *et al.*, 2001). Recently, it was found that a third tobacco MAPK, Ntf4, is also downstream of NtMEK2 (Ren *et al.*, 2006). Ntf4 shares high homology with SIPK, and might have originated from an ancient gene duplication event in the ancestral plant of Solanaceae family.

The placement of an MAPKKK in a specific MAPK cascade reported is mostly based on protoplast gain-of-function analysis and/or genetic evidence (Asai *et al.*, 2002; Takahashi *et al.*, 2004; Ichimura *et al.*, 2006; Nakagami *et al.*, 2006; Suarez-Rodriguez *et al.*, 2007; Wang *et al.*, 2007). A complete *Arabidopsis* MAPK cascade (MEKK1–MKK4/MKK5–MPK3/MPK6) was established using a protoplast-based system (Asai *et al.*, 2002). This cascade is activated in *Arabidopsis* treated with flg22, a peptide elicitor derived from bacterial flagellin. However, recent genetic results from several laboratories showed that MEKK1 functions upstream of MPK4 in the flg22 signaling pathway (Ichimura *et al.*, 2006; Nakagami *et al.*, 2006; Suarez-Rodriguez *et al.*, 2007). In *mekk1* mutants, activation of MPK3/MPK6 after flg22 treatment is not altered. It remains to be determined whether the full activation of MPK3 and MPK6 in *mekk1* plants after flg22 treatment is a result of redundant MAPKKKs or MEKK1 is not the endogenous MAPKKK upstream of MPK3 and MPK6. Three other complete MAPK cascades have been established, including tobacco NPK1–NQK1–NRK1 in cytokinesis, tobacco MAPKKKα–MEK2–SIPK in avrPto-induced cell death, and *Arabidopsis* YDA–MKK4/MKK5–MPK3/MPK6 in stomatal development and patterning based on both biochemical and genetic evidence (del Pozo *et al.*, 2004; Takahashi *et al.*, 2004; Wang *et al.*, 2007).

4.3.2 Are plant Raf-like kinases bona fide MAPKKKs?

MAPKs and MAPKKs are highly conserved across species and between different members within a species. However, MAPKKKs are highly divergent. In *Arabidopsis*, more than 60 kinases were predicted to encode MAPKKKs (MAPK Group, 2002; Champion *et al.*, 2004). Among them, 12 belong to the MEKK subfamily and the remaining 50 or so belong to Raf-like kinase subfamily. There are ample data, both biochemical and genetic, to implicate the MEKK-like members in plant MAPK cascades. However, no evidence is available to support any of the Raf-like members yet. Based on the definition of MAPKKK, a kinase has to function upstream of an MAPKK–MAPK module before it qualifies as an MAPKKK. At present, it is unknown whether the Raf-like plant kinases are bona fide MAPKKKs.

The founding members of the plant Raf-like kinase subgroup include CTR1 and EDR1, two kinases identified in genetic screens (Kieber *et al.*, 1993; Frye *et al.*, 2001). CTR1 is a negative regulator of ethylene signaling, and EDR1 is a negative regulator of plant disease resistance. So far, there is no biochemical evidence to support that they function upstream of an MAPKK–MAPK module. CTR1 and EDR1 were classified in the B3 subgroup of plant Raf-like kinases (MAPK Group, 2002), with a C-terminal kinase domain and a putative regulatory N-terminal region. More detailed phylogenetic analysis leads to the conclusion that all plant Raf-like kinases are quite divergent from animal MAPKKKs. They are slightly more closely related to the kinase in the mixed lineage kinase (MLK) subfamily than to the Raf kinase (Tang and Innes, 2002). In mammals, kinases from several kinase subfamilies including MEKK, Raf, and MLK can all function as MAPKKKs (Widmann *et al.*, 1999). However, not all MLK members function as MAPKKKs. Based on the fact that EDR1 shares no detectable similarity to any animal kinases outside the kinase domain, it is concluded that the EDR1/CTR1 kinase subfamily is unique to flowering plants and it might be more appropriate not to infer them as MAPKKKs functionally until biochemical evidence is obtained (Tang and Innes, 2002).

4.4 Negative regulation of plant MAPK cascades

The physiological outcomes of the activation of an MAPK cascade are dependent on the duration of its activation. How long an MAPK stays active can be affected by either how long the upstream kinase remains active and/or whether an MAPK phosphatase is involved. Compared with the activation of MAPKs, their inactivation processes are less understood. There is almost no information about how plant MAPKKs and MAPKKKs are inactivated, which will certainly affect MAPK activities. In plants, both dual-specificity Tyr protein phosphatases and S/T protein phosphatases have been implicated in the inactivation of MAPKs. Alfalfa MP2C, a PP2C-type protein phosphatase, was shown to be a specific negative regulator of SIMK (also known as MMK1) (Meskiene *et al.*, 1998, 2003). Inactivation of SIMK by MP2C is through Thr dephosphorylation of the pTEpY motif. Expression of *MP2C* is wound inducible, which correlates with the timing of SIMK inactivation, suggesting that MP2C might be responsible for the transient nature of SIMK activation by wounding stress.

Arabidopsis DsPTP1, the first dual-specificity protein Tyr phsophatase identified in plants, was demonstrated to be able to dephosphorylate and inactivate MPK4 (Gupta *et al.*, 1998). A genetic screen for mutants that are hypersensitive to genotoxic stresses (UV-C and methyl methanesulfonate) revealed the involvement of AtMKP1 in regulating the activity of MAPKs (Ulm *et al.*, 2002). AtMKP1, which specifically interacts with MPK3, MPK6, and MPK4, may regulate plant stress responses through the negative regulation of these MAPKs. Homologs of AtMPK1 were also identified in tobacco and rice (Katou *et al.*, 2005a; Katou *et al.*, 2007). NtMPK1 negatively regulates the activity of SIPK, which is dependent on its interaction with SIPK via the common docking domain of SIPK.

In addition to DsPTP1 and AtMPK1, another dual-specificity protein Tyr phosphatases, IBR5, was shown to modulate auxin and abscisic acid responsiveness in *Arabidopsis* (Monroe-Augustus *et al.*, 2003). *ibr5* was isolated as an *Arabidopsis* indole-3-butyric acid–response mutant, and it is less responsive to other auxins and auxin transport inhibitors. In addition, it is less responsive to abscisic acid (ABA), another phytohormone. At this stage, it is unknown whether IBR5 confers these biological functions through inactivating plant MAPKs.

4.5 Important tools/techniques in MAPK research

Despite their late discovery and highly complicated regulation, MAPKs have become one of the best-studied kinase subfamilies in plants. This can be mostly attributed to the covalent modification of MAPKs, which makes it possible to detect the in vivo activation of MAPKs. This information can then be used to link them to specific biological functions.

4.5.1 In-gel kinase activity assay

MAPKs are relatively small kinases and can be renatured after sodium dodecyl sulfate–polyacrylamide gel electrophoresis (SDS–PAGE), which permits the use of in-gel kinase activity assay to determine their activities (Zhang and Klessig, 1997). Making this assay a more important tool is that MAPKs are activated by phosphorylation, a covalent modification that can be preserved throughout the assay procedure. As a result, the activity detected in vitro using the in-gel kinase assay reflects their in vivo activation. For these reasons, in-gel kinase assay has become a very important assay to study the regulation/activation of MAPKs and to link a specific MAPK to a biological function.

In-gel activity assays can reveal the size of the kinase and relative activity of a kinase. However, not all MAPKs phosphorylate myelin basic protein (MBP), the artificial substrate used in the assay, at equal efficiency. As a result, no conclusion can be drawn about the relative activities between different MAPKs. The other caution is that not all MAPKs can use MBP as a substrate. As a result, this assay is not universal for all the MAPKs.

4.5.2 Immune-complex kinase assay using member-specific antibodies

In-gel kinase assay is a powerful method in determining the activity of MAPKs. In addition, the separation of kinases based on their sizes in the in-gel kinase assay allows the detection of multiple MAPKs with different sizes, and the identification of kinases to a certain extent. However, assigning the identity to a kinase based solely on size is not reliable, because many different MAPKs have similar size. As a result, we cannot use this method to positively link the kinase activity to a specific encoding gene.

To make up this shortcoming, researchers often use so-called immune-complex kinase assay, which combines immunoprecipitation with a kinase activity assay to detect the activity of a specific kinase. Member-specific antibodies generated against peptides corresponding to the unique regions of MAPKs are used in the assay, which allows one to assign the activity to a specific MAPK. A number of member-specific antibodies have been raised against MAPKs from different plant species (Jonak *et al.*, 1996; Zhang *et al.*, 1998; Zhang and Klessig, 1998a,b; Seo *et al.*, 1999; Cardinale *et al.*, 2000; Ichimura *et al.*, 2000; Samuel and Ellis, 2002; Gomi *et al.*, 2005). *Arabidopsis* MAPK antibodies are also commercially available.

Great care should be paid to the specificity of the immunoprecipitation, which is determined by the antibody and the buffer conditions (mainly the concentration of salt and detergents added). In addition, the nonspecific covalent cross-linking of kinases to the agarose beads, protein A, or IgG may contribute to the nonspecific pull-down of other kinases. This problem is more pronounced when one works with protein extracts rich in phenonic compounds, such as tobacco and stress-treated plants. Because of the high sensitivity of the in-solution kinase assay, even the smallest amount of kinases will phosphorylate enough MBP to give a band on the autoradiogram. Frequently, the nonspecifically immunoprecipitated kinase activity follows the change of the major kinase in the extracts.

There are two remedies for this problem. First is to perform a competition experiment using the peptide to which the antibody was raised. Only the specific binding can be competed off the immune complex. In contrast, nonspecific binding or the cross-linking of other kinases to the beads, protein A, or IgG will not be affected. We have used this competition assay to confirm the identities of the MAPKs (Zhang *et al.*, 1998; Zhang and Klessig, 1998a,b). Alternatively, in-gel kinase assay, rather than in-solution kinase assay, can be performed after immunoprecipitation. Nonspecifically cross-linked kinase(s) may remain covalently bound to agarose beads and be removed. In addition, the nonspecifically bound kinases may have a different size as that of the subject kinase. Furthermore, the nonspecifically bound kinases may not be able to renature after SDS–PAGE and, therefore, are eliminated from the assay (Ren *et al.*, 2006).

4.5.3 Commercial phospho-specific antibodies

Plant MAPKs have either TEY or TDY activation motif, the first of which is identical to that of the ERK subfamily of mammalian MAPKs. A number of companies offer anti-active ERK antibodies, which are specific for the dually phosphorylated TEY motif (pTEpY). They have been used successfully to detect the activation of MAPKs in plants by immunoblot analysis (Samuel and Ellis, 2002; Yang *et al.*, 2002). This detection method is similar to the in-gel kinase assay, which detects the relative activation of an MPAK and its size but is unable to reveal the identity of the MAPK. The advantage of this detection method is that no radioactive material is needed for the assay. The disadvantage is that it is not as reliable and its sensitivity is lower than that of the in-gel kinase assay.

4.5.4 MAPK-specific inhibitors

Specific inhibitors, which were identified in drug screenings, are widely used in studying animal MAPK signaling pathways (reviewed in Davies *et al.*, 2000). The most frequently used are PD98059 and U0126. They inhibit the activation of ERKs by preventing the activation of MKK1 via slightly different mechanisms. They are effective at micro- or submicromolar concentrations. At high concentrations, both have additional kinase targets (Davies *et al.*, 2000). In plants, there are mixed reports about the effectiveness of these inhibitors (Zhang *et al.*, 2000; Desikan *et al.*, 2001; Lee *et al.*, 2001). In our hands, neither U0126 nor PD98059 was able to inhibit the activation of SIPK/WIPK in cell suspension at nonlethal concentrations. It was also reported that these two inhibitors potentiate the activation of an auxin-induced MAPK (Mockaitis and Howell, 2000). As a result, these inhibitors should be used with caution.

4.5.5 Genetic gain- and loss-of-function mutant/transgenic approaches

Almost always, a stimulus will activate multiple signaling pathways in a cell. To isolate and study the responses downstream of a specific MAPK cascade, researchers frequently use the gain-of-function constitutively active mutants. There was no success story in generating an active MAPK mutant yet. In contrast, both active MAPKK and MAPKKK mutants have been reported (Yang *et al.*, 2001; Asai *et al.*, 2002; Takahashi *et al.*, 2007a).

Forward genetic screens identified only a few components in the MAPK cascade, including MPK4, YDA, and BUD1/MKK7 (Petersen *et al.*, 2000; Bergmann *et al.*, 2004; Lukowitz *et al.*, 2004; Dai *et al.*, 2006). Part of the reason is the gene redundancy in the MAPK signaling pathway. With the availability of various collections of insertional mutants and the use of reverse genetic approaches, novel functions have been assigned to plant MAPK cascades (Krysan *et al.*, 2002; Chaiwongsar *et al.*, 2006; Wang *et al.*, 2007).

4.6 Biological functions of MAPK cascades in plants

In spite of the highly conserved organization, plant MAPK cascades have evolved to carry out functions that are unique to plants, such as phragmoplast formation in cytokinesis, stomatal development and patterning, embryo development, pollen development, regulation of plant hormone biosynthesis, and plant hormone signaling. A subset of plant MAPKs including *Arabidopsis* MPK3, MPK6, and MPK4, as well as their orthologs in other plant species, is also involved in signaling plant stress/defense responses, which is functionally analogous to mammalian SAPK/JNK and p38 MAPKs (Widmann *et al.*, 1999; Davis, 2000). However, plant-stress-responsive MAPKs have the TEY activation motif and were likely to evolve from the ancient MAPK that also gave rise to mammalian ERK subfamily of MAPKs (Caffrey *et al.*, 1999). As a result, functions of MAPKs in stress/defense responses evolved independently in plants and animals, implying divergent mechanisms of their functions. Consistent with this, plant-stress-responsive MAPKs phosphorylate plant-specific substrates (Liu and Zhang, 2004; Andreasson *et al.*, 2005).

4.6.1 MAPK cascades in plant stress/defense signaling

Based on gain- and loss-of-function analyses, stress-responsive MAPK cascades represented by *Arabidopsis* MPK3, MPK6, and MPK4 have been shown to play important roles in defense gene activation, reactive oxygen species (ROS) induction, stress-hormone biosynthesis, stress-hormone signaling, cell death, and disease resistance (reviewed in Mizoguchi *et al.*, 1997; Tena *et al.*, 2001; Zhang and Klessig, 2001; Nakagami *et al.*, 2005; Pedley and Martin, 2005). Functions of plant MAPK cascades in stress/defense signaling are summarized in Fig. 4.1.

4.6.1.1 MAPK cascade in plant innate immunity and disease resistance

The active defense of plants against invading pathogens often includes the generation of ROS, the activation of a complex array of defense genes, the production of antimicrobial phytoalexins, and rapid cell death known as hypersensitive response (HR) (Dangl and Jones, 2001; Dixon, 2001; Martin *et al.*,

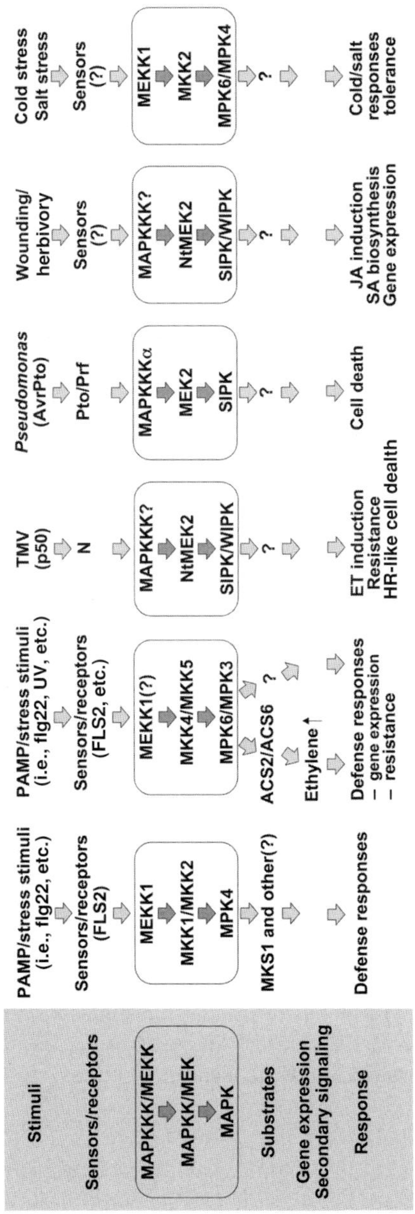

Figure 4.1 Roles of plant MAPK cascades in response to biotic and abiotic stresses. A typical linear signaling pathway from the sensing of a stimulus to the cellular response is depicted in the shaded area on the left.

2003; Greenberg and Yao, 2004; Torres and Dangl, 2005). In addition to these local responses, the uninfected portions of the plant usually develop systemic acquired resistance (SAR), which is manifested as enhanced resistance to a subsequent challenge by pathogens (Durrant and Dong, 2004). These defense responses are regulated by a complex signaling network initiated by pathogen recognition, which is mediated either by gene-for-gene interactions between plant resistance (R) genes and pathogen avirulence (Avr) genes or by the binding of non-host-specific pathogen-associated molecular patterns (PAMPs) to their receptors (Staskawicz *et al.*, 1995; Dangl and Jones, 2001; Martin *et al.*, 2003; Boller, 2005).

Biochemical studies from a number of laboratories demonstrated that the tobacco MAPKs, SIPK, and WIPK, and their orthologs in other plant species, including MPK6 and MPK3 in *Arabidopsis*, SIMK and SAMK in alfalfa, LeMPK1/2 (LeSIPK) and LeMPK3 (LeWIPK) in tomato, and PcMPK6 and PcMPK3 in parsley, are activated in plants treated with PAMPs or after pathogen infection (Stratmann and Ryan, 1997; Zhang *et al.*, 1998; Zhang and Klessig, 1998b; Romeis *et al.*, 1999; Cardinale *et al.*, 2000; Kovtun *et al.*, 2000; Nühse *et al.*, 2000; Zhang *et al.*, 2000; Desikan *et al.*, 2001; Lee *et al.*, 2001; Yuasa *et al.*, 2001; Asai *et al.*, 2002; Link *et al.*, 2002; Ekengren *et al.*, 2003; Holley *et al.*, 2003; Kroj *et al.*, 2003; del Pozo *et al.*, 2004; Lee *et al.*, 2004; Pedley and Martin, 2004). In tobacco, SIPK and WIPK share a common upstream MAPKK, NtMEK2 (Yang *et al.*, 2001). There are two NtMEK2 orthologs in *Arabidopsis*, MKK4 and MKK5, which are likely to have arisen from a relatively recent gene duplication (Asai *et al.*, 2002; MAPK Group, 2002; Ren *et al.*, 2002; Hamel *et al.*, 2006). Recently, additional MAPKKs including tomato LeMKK4 and tobacco NbMKK1, which are more closely related to *Arabidopsis* MKK7 and MKK9, were shown to activate the same MAPKs and induce HR-like cell death (Pedley and Martin, 2004; Takahashi *et al.*, 2007b). The MAPKKKs upstream of NtMEK2/MKK4/MKK5 include MEKK1 and MAPKKKα (Asai *et al.*, 2002; del Pozo *et al.*, 2004).

Loss- and gain-of-function studies provided genetic evidence, supporting a positive role of this MAPK cascade in signaling plant disease resistance, although the underlying mechanism(s) remains to be determined (Yang *et al.*, 2001; Asai *et al.*, 2002; Ren *et al.*, 2002; Jin *et al.*, 2003; Kroj *et al.*, 2003; Sharma *et al.*, 2003; Yoshioka *et al.*, 2003; del Pozo *et al.*, 2004; Menke *et al.*, 2004). Activation of MPK3/MPK6 by PAMPs or Avr factors occurs within 1–5 min, representing one of the earliest detectable defense responses. The rapid activation of these two MAPKs potentially allows them to regulate a variety of other early, intermediate, and late defense responses. The identification of the first plant MAPK substrate revealed that MPK3/MPK6 regulates ethylene production by stabilizing a subset of 1-aminocyclopropane-1-carboxylic acid synthase (ACS) isoforms after direct phosphorylation (Kim *et al.*, 2003; Liu and Zhang, 2004). Ethylene induction is associated with plant defense responses, and positively regulates defense gene expression. However, its function in plant disease resistance is complex. It seems that ethylene can modulate both resistance and susceptibility responses of plants (Wang *et al.*, 2002; Broekaert *et al.*, 2006). In addition to regulating ethylene biosynthesis, this MAPK pathway is also involved in regulating defense gene activation (a subset is likely to be downstream of ethylene), ROS generation, and HR-like cell death (Yang *et al.*, 2001; Zhang and Liu, 2001; Ren *et al.*, 2002; Kroj *et al.*, 2003; Yoshioka *et al.*, 2003; Kim and Zhang, 2004). These findings indicate that the MPK3/MPK6 cascade regulates multiple defense responses, either in parallel or sequentially, and the compromised resistance in the loss-of-function plants could be due to defects in multiple defense responses.

4.6.1.2 MPK4 negatively regulates SAR

Besides MPK3 and MPK6, a number of biotic and abiotic stresses also activate MPK4 (Ichimura *et al.*, 2000; Droillard *et al.*, 2004; Teige *et al.*, 2004; Ichimura *et al.*, 2006; Nakagami *et al.*, 2006; Suarez-Rodriguez *et al.*, 2007). Transposon inactivation of *Arabidopsis* MPK4 results in constitutive SAR, including elevated SA levels, increased resistance to virulent pathogens, and constitutive pathogenesis-related gene expression. As a result, it was concluded that *Arabidopsis* MPK4 functions as a negative regulator of SAR (Petersen *et al.*, 2000). Analysis of *mpk4/NahG* and *mpk4/npr1* double mutants indicated that SAR in *mpk4* is dependent on elevated SA levels but is independent of *NPR1*. Interestingly, jasmonic acid (JA)-induced *PDF1.2* and *THI2.1* gene expression was blocked in *mpk4*, suggesting that MPK4 is a positive regulator of the JA response. Recently, it was shown that *mpk4* mutants are also defective in defense gene induction in response to ethylene (Brodersen *et al.*, 2006).

MKS1, which was identified as MPK4-interacting protein in a yeast two-hybrid screen, was shown to be an MPK4 substrate (Andreasson *et al.*, 2005). MKS1 has potential MAPK phosphorylation sites. Biochemical analysis showed that active MPK4 immunoprecipitated from *Arabidopsis* seedlings can phosphorylate recombinant MKS1 in vitro. Overexpression of *MKS1* results in the dwarf phenotype similar to *mpk4* null mutants, indicating the functional link between these two proteins. Like the *mpk4* mutant, pathogenesis-related proteins that are normally induced in SAR are upregulated in *MKS1* overexpression transgenic plants and the transgenic plants are more resistant to pathogen attack. Furthermore, the *mpk4* mutant can be partially rescued by reducing *MKS1* expression, indicating that MPK4 negatively regulates MKS1 activity after phosphorylation.

4.6.1.3 An MAPK cascade in plant freezing and salt tolerance

MPK3 and MPK6, as well as their orthologs in other plant species, are activated in plants subjected to a diverse array of abiotic stresses, including salinity, osmotic stress, cold, drought, wounding, and even touching (reviewed in Mizoguchi *et al.*, 1997; Tena *et al.*, 2001; Zhang and Klessig, 2001; Nakagami *et al.*, 2005; Pedley and Martin, 2005). Ethylene biosynthesis, which is induced by all these stresses, is likely one of the outputs of MPK3/MPK6 activation (Kim *et al.*, 2003; Liu and Zhang, 2004). MPK4 is activated by a similar set of abiotic stresses, including cold, low humidity, hyperosmolarity, touch, and wounding (Ichimura *et al.*, 2000). Recently, it was reported that the loss-of-function mutant *mpk4* is more tolerant to hyperosmotic stress, although no activation of MPK4 was detected in plants subjected to hyperosmotic stress (Droillard *et al.*, 2004). It is therefore concluded that MPK4 is a negative regulator of hyperosmotic stress responses. MKK1 (AtMEK1) may function upstream of MPK4 in the stress-response pathway (Matsuoka *et al.*, 2002). MKK1 protein immunoprecipitated from *Arabidopsis* seedlings subjected to drought, high salt, wounding, and cold stresses showed higher activity toward recombinant kinase-inactive MPK4.

MKK2, a close homolog of MKK1, was reported to be upstream of both MPK4 and MPK6 (Teige *et al.*, 2004). In *Arabidopsis* protoplasts, MKK2 was specifically activation by cold and salt stress and by the stress-induced MEKK1. Plants overexpressing MKK2 exhibited constitutive MPK4 and MPK6 activity, constitutively upregulated expression of stress-induced marker genes, and increased freezing and salt tolerance. In contrast, *mkk2* mutant plants were impaired in MPK4 and MPK6 activation and were hypersensitive to salt and cold stress. At this stage, the function of MKK1 in the cold and salt stress response is unknown. Biochemical evidence revealed that MKK1 is activated by these two stresses and it functions upstream of MPK4 (Matsuoka *et al.*, 2002). On the other hand, loss-of-function MKK2 is sufficient to give a mutant phenotype, which suggests that MKK1 is not sufficient to complement the loss of MKK2 in the process (Teige *et al.*, 2004). In this study, it was also shown that MKK2 mediates the activation of MPK4/MPK6 by gain-of-function MEKK1, indicating that MEKK1 could be an upstream MAPKKK of MKK2 in the cold and salt stress response.

4.6.1.4 MAPK cascade and ROS, two interconnected signaling pathways

Several reports relate ROS to the activation of stress-responsive MAPKs in plants under stresses (Romeis *et al.*, 1999; Kovtun *et al.*, 2000; Yuasa *et al.*, 2001; Apel and Hirt, 2004; Rentel *et al.*, 2004). Ozone (O_3) treatment rapidly activates SIPK and WIPK in tobacco (Samuel and Ellis, 2002). High concentrations of H_2O_2, when exogenously added, activate SIPK/WIPK or their orthologs in other plants. However, whether ROS generation after pathogen infection or elicitor treatment is required for the activation of MAPKs by these stimuli remains controversial. It was shown that Avr9-induced SIPK/WIPK activation is not dependent on a burst of ROS from membrane-associated NADPH oxidases (Romeis *et al.*, 1999). Consistent with this report, we found that the rapid oxidative burst induced by elicitin is not required for SIPK activation by elicitin either. In addition, elicitin-induced activation of SIPK/Ntf4/WIPK, which occurs within 1 min after elicitin treatment, precedes the oxidative burst, which is only detectable 10 min after elicitin treatment (Y. Liu and S. Zhang, unpublished data). These results suggest that the MAPK activation after the perception of pathogens/elicitors is independent of ROS burst. They are two parallel downstream events after the sensing step, and the ROS burst may then positively feed into MAPK activation.

It was also reported that NADPH oxidase is a downstream target of SIPK/WIPK in the induction of H_2O_2 generation. Activation of SIPK/WIPK by the active MEK2^{DD} induced *NbrbohB* expression (Yoshioka *et al.*, 2003). Using the luminal-based chemiluminescence method, we failed to detect a rapid H_2O_2 burst in *NtMEK2^{DD}* plants after dexamethasone (DEX) treatment (Yang *et al.*, 2001; Ren *et al.*, 2002). Similar experiments using conditional gain-of-function *Arabidopsis* seedlings yielded negative results as well (Y. Liu and S. Zhang, unpublished data), suggesting that SIPK/Ntf4/WIPK activation is not involved in the early ROS burst from NADPH oxidase in pathogen-infected plants. Alternatively, SIPK/Ntf4/WIPK activation is not sufficient to induce ROS generation from NADPH oxidases and the activation of additional pathways is needed. In either scenario, the transcriptional activation of *NbrbohB* expression by an MAPK cascade (Yoshioka *et al.*, 2003) is likely to be too slow to account for the rapid ROS burst. Recently, we found that chloroplast-generated ROS are involved in HR-like cell death after SIPK/Ntf4/WIPK activation (Liu *et al.*, 2007).

4.6.2 MAPK cascades in phytohormone biosynthesis and signaling

Plant hormones are critical to plant growth, development, and response to the environment. Recent molecular genetic analyses greatly enriched our knowledge of the perception of hormones and downstream signaling pathways that lead to the physiological responses. Components of plant MAPK cascades have been implicated in both the regulation of hormone biosynthesis and the signaling events downstream of hormone sensing. Functions of plant MAPK cascades in plant hormone signaling are summarized in Fig. 4.2.

4.6.2.1 MAPK cascades in regulating ethylene biosynthesis and ethylene signaling

Ethylene, a gaseous plant hormone, plays important roles in regulating plant growth, development, and response to biotic/abiotic stress stimuli (Abeles *et al.*, 1992; Wang *et al.*, 2002; Broekaert *et al.*, 2006). Ethylene is sensed by a group of five transmembrane histidine kinase receptors, ETR1, ETR2, ERS1, ERS2, and EIN4, which initiate ethylene signaling pathways and activate ethylene responses (reviewed in Schaller and Kieber, 2002; Wang *et al.*, 2002; Chang and Bleecker, 2004).

4.6.2.1.1 A putative MAPKKK in ethylene signaling. CTR1, whose mutation results in a constitutive ethylene response, functions as a negative regulator downstream of the ethylene receptors. The C-terminal kinase domain of CTR1 bears similarity to the kinase domain of animal Raf MAPKKK (Kieber *et al.*, 1993). As a result, it was proposed that an MAPK cascade is involved in

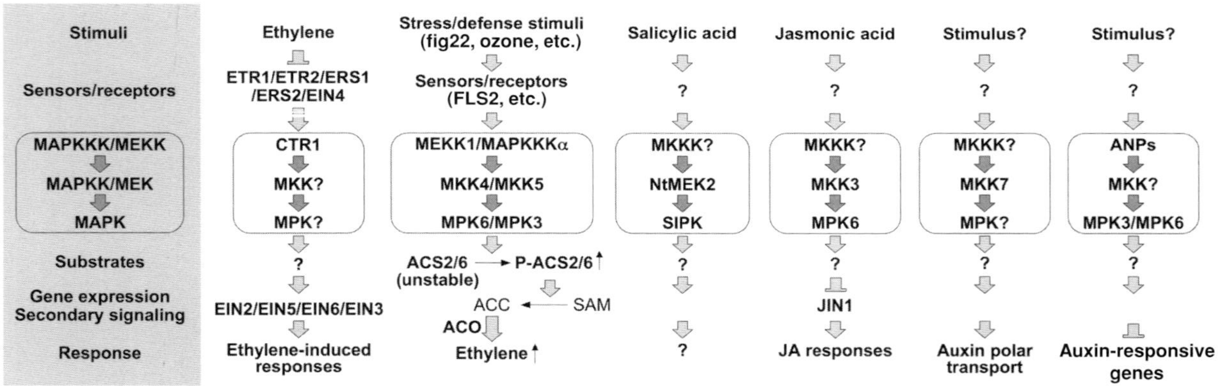

Figure 4.2 Involvement of plant MAPK cascades in regulating plant hormone biosynthesis and plant hormone signaling. A typical linear signaling pathway from the sensing of a stimulus to the cellular response is depicted in the shaded area on the left.

ethylene signaling. However, an intense searches for its downstream MAPKK and MAPK have since then not been able to generate indisputable evidence that CTR1 functions as an MAPKKK in an MAPK cascade. It was reported that alfalfa MAPKs, SIMK and MMK3, and their upstream MAPKK, SIMKK, are downstream of CTR1 in an MAPK cascade (Ouaked *et al.*, 2003). However, the critical experiment described in the report is seriously flawed. In order to fit with the ethylene signaling pathway based on genetic evidence, the CTR1 MAPKKK has to negatively regulate the MAPKK (SIMKK) (Ouaked *et al.*, 2003). No data were provided to support such unprecedented MAPKKK–MAPKK relationship. The *Arabidopsis* orthologs of alfalfa SIMK and MMK3 are MPK6 and MPK13, respectively. It was found that double-mutant seedlings of *mpk6* and *mpk13* have no ethylene phenotype (Ecker, 2004). As a result, it remains an open question whether an MAPK cascade is involved in ethylene signaling downstream of ethylene receptors.

4.6.2.1.2 Stress-responsive MAPK cascade in regulating ethylene biosynthesis. Ethylene-regulated processes begin with the induction of its biosynthesis, which is under tight regulation (Zarembinski and Theologis, 1994; Kende, 2001; Wang *et al.*, 2002; Chae and Kieber, 2005). Plants produce dramatically higher levels of ethylene during seed germination, leaf/flower senescence, fruit ripening, and exposure to stress stimuli (Abeles *et al.*, 1992; Wang *et al.*, 2002; Broekaert *et al.*, 2006; van Loon *et al.*, 2006). There are only two enzymes that are specific for the ethylene biosynthetic pathway, the conversion of *S*-adenosyl-L-methionine (SAM) to 1-aminocyclopropane-1-carboxylic acid (ACC), which is catalyzed by ACC synthase (ACS), and the oxidative cleavage of ACC to form ethylene, which is catalyzed by ACC oxidase (ACO) (Abeles *et al.*, 1992). Both enzymes are encoded by small gene families. In general, ACS activity is very low in tissues that do not produce significant amount of ethylene. In contrast, ACO activity is constitutively present in most vegetative tissues. On stimulation, ACS activity is rapidly induced. Therefore, ACS is the rate-limiting enzyme and is the major regulatory step in the induction of ethylene (Wang *et al.*, 2002; Chae and Kieber, 2005). Based on studies using general kinase and phosphatase inhibitors, protein phosphorylation and dephosphorylation were implicated in the regulation of ACS activity and ethylene induction in plants under stress (Spanu *et al.*, 1994). However, the specific kinase involved and the underlying mechanism are unknown.

Biochemical and genetic analyses using *Arabidopsis* mutants/transgenics functionally placed the stress-responsive MPK6 cascade in the signaling pathway regulating ethylene biosynthesis in plants under stress (Liu and Zhang, 2004). This research began with the finding that the activation of SIPK and WIPK by the constitutively active NtMEK2^{DD} resulted in a dramatic increase in ethylene production in the conditional gain-of-function *GVG-NtMEK2*^{DD} transgenic tobacco (Kim *et al.*, 2003). The increase in ethylene after the activation of SIPK/WIPK coincides with a dramatic increase in ACS activity. It was later found that ACS2 and ACS6, and likely ACS1, which forms a subgroup of the *Arabidopsis* ACS family, are substrates of MPK6. Phosphorylation of ACS2 and ACS6 by MPK6 leads to the accumulation of ACS protein, the rate-limiting enzyme of ethylene biosynthesis, thus elevating cellular ACS activity and ethylene production. These conclusions are based on the following evidence: (1) stress-induced activation of MPK6/SIPK precedes and correlates with the stress induction of ethylene; (2) activation of MPK6/SIPK and MPK3/WIPK in the conditional gain-of-function *Nt-MEK2*^{DD} transgenic plants leads to ethylene production; (3) full induction of ethylene by NtMEK2^{DD} and flg22 elicitor requires a functional *MPK6* gene; (4) the enhanced ethylene production after MPK6 activation is associated with the increase in ACS6 activity; (5) ACS6, ACS2, and ACS1, which form a unique clade in the ACS family, contain conserved MAPK phosphorylation sites; (6) recombinant ACS6 and ACS2 can be phosphorylated by active MPK6 very efficiently in vitro; (7) ACS6 is phosphorylated in vivo after MPK6 activation by NtMEK2^{DD} and flg22 elicitor; (8) MPK6 phosphorylation sites in ACS6 are essential for ACS6 accumulation after MPK6 activation in vivo; (9) ACS6^{DDD}, a mutant that mimics phospho-ACS6, accumulates in the transgenic plants in the absence of stress stimuli, and the transgenic plants constitutively overproduce ethylene and show ethylene-induced phenotypes; (10) genetic analysis of *acs6* and *acs2* mutants placed them in the MAPK-activation-induced ethylene production; and (11) ethylene signaling pathway is required for phenotypes in the gain-of-function *ACS6*^{DDD} and *ACS2*^{DDD} transgenic plants (Kim *et al.*, 2003; Liu and Zhang, 2004) (Y. Liu and S. Zhang, unpublished data).

ACS is the first plant MAPK substrate identified, which not only uncovers a signaling pathway that modulates the biosynthesis of ethylene, an important plant hormone, but also reveals one mechanism by which stress-responsive MAPK regulates plant stress response.

4.6.2.2 Stress-responsive MAPKs in SA and JA biosynthesis and signaling pathways

In addition to ethylene, there are two additional plant hormones/regulators, SA and JA, that are involved in regulating plant stress/defense response (Creelman and Mullet, 1997; Durrant and Dong, 2004). The levels of SA and JA are responsive to stress/defense stimuli. MAPKs have been implicated in both SA and JA signaling pathways and the induction of their biosynthesis, although the underlying mechanisms are still unknown.

4.6.2.2.1 Tobacco SIPK. One of the early pieces of evidence implicating the involvement of plant MAPKs in stress signaling is the purification and identification of SIPK, or salicylic-acid-induced protein kinase (Zhang and Klessig, 1997). SIPK was later found to be activated by a variety of stress stimuli independent of SA, based on two facts: (1) stresses including wounding, which do not induce SA, activate SIPK, and (2) SIPK activation by stress/PAMPs precedes the induction of SA (Zhang *et al.*, 1998; Zhang and Klessig, 1998a,b). As a result, SIPK became the acronym of *s*tress-*i*nduced *p*rotein *k*inase, which reflects its nature better. At this stage, the function of SIPK in SA signaling pathway is unknown.

Based on research in Brian Ellis's laboratory, SIPK may function as a negative feedback regulator of SA biosynthesis. SA, JA, and ethylene are involved in ozone-induced responses, including cell death (Rao *et al.*, 2002). Ozone induces rapid activation of SIPK (Samuel and Ellis, 2002). Interestingly, overexpressing SIPK in tobacco plants potentiates ozone-induced ethylene production but suppresses ozone-induced SA accumulation (Samuel *et al.*, 2005). As detailed below, SIPK is also implicated in JA signaling and the induction of JA biosynthesis, which may reflect the complex role of this MAPK cascade in plant stress/defense signaling. It is also likely that some of the reported findings are secondary and/or nonspecific effects of the mutants/transgenics. At this stage, we need to make efforts to identify additional SIPK/WIPK substrates, which will reveal the underlying molecular mechanisms of MAPK action and allow the establishment of a direct link between these events.

4.6.2.2.2 Stress-responsive MAPKs in JA biosynthesis. JA and other oxylipins play important roles in plant stress/defense and development (Creelman and Mullet, 1997). JA is highly induced by wounding or insect chewing. In tobacco, wounding causes rapid activation of SIPK and Ntf4, two highly homologous MAPKs (Seo *et al.*, 1995; Zhang and Klessig, 1998a; Seo *et al.*, 1999; Ren *et al.*, 2006). In plants with elevated levels of WIPK protein either due to preexposure to stress, which induces WIPK, or due to ectopic overexpression, activation of WIPK is also detected (Liu *et al.*, 2003). It was concluded that activation of WIPK is required for the production of wound-induced JA (Seo *et al.*, 1995). Later, it was shown that overexpression of WIPK is sufficient to induce JA (Seo *et al.*, 1999). However, in the conditional gain-of-function *NtMEK2*^{DD} plants, activation of SIPK and WIPK is not sufficient to induce JA accumulation (Kim *et al.*, 2003). As a result, the exact role of WIPK in JA induction remains unresolved. It could also be possible that these two MAPKs are required, but not sufficient, to induce JA biosynthesis. Recently, it was reported that wound-induced JA production was reduced in *WIPK-*, *SIPK-*, or *WIPK/SIPK* RNAi gene silencing plants. In contrast, both SA and transcripts of SA-responsive genes accumulated abnormally in these plants (Seo *et al.*, 2007). Based on these findings, it is concluded that WIPK and SIPK play an important role in JA production in response to wounding and that they function cooperatively to control SA biosynthesis.

4.6.2.2.3 An MAPKK–MAPK module in JA signaling. Making things even more complex, MPK6 (the *Arabidopsis* SIPK ortholog) was recently shown to be an important part of the JA signal transduction pathway in *Arabidopsis* (Takahashi *et al.*, 2007a). MKK3 is the upstream kinase of MPK6 in this pathway. ATMYC2/JASMONATE-INSENSITIVE1 (JIN1) is a positive regulator of JA-inducible gene expression and essential for JA-dependent developmental processes in *Arabidopsis* (Lorenzo *et al.*, 2004). Both positive and negative regulation by JA may be used to fine-tune ATMYC2/JIN1 expression to control JA signaling. It was shown that the MKK3–MPK6 module is involved in JA-dependent negative regulation of ATMYC2/JIN1 expression based on microarray analysis. Moreover, JA-regulated root growth inhibition is affected by mutations in the MKK3–MPK6 cascade, implicating them in the JA signaling pathway. A model was proposed to explain how MPK6 can convert three distinct signals, including JA, pathogen, and cold and salt stress, into three different sets of responses in *Arabidopsis* (Takahashi *et al.*, 2007a).

4.6.2.2.4 MPK4 negatively regulates SA biosynthesis and positively regulates JA signaling. In addition to MPK3 and MPK6, the other stress-responsive MAPK, MPK4, was also implicated in the JA signaling pathway. Mutant *mpk4* plants exhibit constitutive activation of SA-dependent defenses, but fail to induce JA defense marker genes in response to JA (Petersen *et al.*, 2000). Expressing *NahG* (a bacterial salicylate hydroxylase that converts SA into catechol) in *mpk4* plants partially rescue the *mpk4* mutant phenotypes, indicating that *mpk4* is upstream of SA accumulation in the SAR signaling. However, removing SA in *NahG/mpk4* double mutant was unable to reverse the suppression of JA-inducible gene expression (Petersen *et al.*, 2000). This result indicates that MPK4 positively regulates JA-inducible responses independently of its negative regulatory role in SA-mediated SAR.

With the recent report that *mpk4* mutant plants are also defective in defense gene induction in response to ethylene (Brodersen *et al.*, 2006), the picture, again, becomes very complicated, with all three stress/defense-related hormones (SA, JA, and ethylene) implicated as the downstream targets of MPK4. As detailed below, MPK3, MPK6, and MPK4 also play important roles in plant growth and development (Petersen *et al.*, 2000; Wang *et al.*, 2007). As a result, it is important to sort out primary defense responses from the secondary and, possibly, nonspecific defense activation due to growth/development defects in the loss-of-function mutants. Only the primary specific responses regulated by these MAPKs will eventually be supported by the information from their MAPK substrates.

4.6.2.3 MAPK cascade in auxin signaling

Auxin is essential to plant growth and development. Recently, it was shown that auxin also plays important roles in plant defense (Navarro *et al.*, 2006). The mechanism of auxin action has been intensely investigated for many decades. Recently, TIR1, an F-box protein, was shown to be an auxin receptor. Direct binding of auxin by SCF^{TIR1} triggers SCF^{TIR1} to interact with transcriptional regulators Aux/IAA and target them to degradation via the ubiquitin–proteasome pathway (Dharmasiri *et al.*, 2005; Kepinski and Leyser, 2005). The degradation of Aux/IAA releases auxin response factor, allowing auxin-responsive transcription activation. Although it could be possible that the TIR1 receptor accounts for a large set of the auxin-mediated responses, additional signaling mechanisms may exist in plants (Badescu and Napier, 2006).

4.6.2.3.1 Tobacco NPK1 and Arabidopsis *ANPs in auxin signaling.* Several components of MAPK cascades were implicated in auxin signaling and transport. Transient overexpression of ΔNPK1, the kinase domain of NPK1, was shown to suppress the induction of an early auxin-response gene in a protoplast system (Kovtun *et al.*, 1998). Expression of the kinase domain of ANP1, one of the *Arabidopsis* orthologs of tobacco NPK1, gave similar results. Based on these findings, it was concluded that NPK1 is a negative regulator of auxin signaling (Kovtun *et al.*, 1998). Later, it was found that constitutively active ΔANP1 activates MPK3 and MPK6, which mimics the effect of H_2O_2 treatment and induces specific stress-responsive genes, but blocks the action of auxin. Therefore, it was proposed that oxidative stress suppresses auxin-inducible genes through the activation of ANP1 MAPK cascade (Kovtun *et al.*, 2000).

Evidence supporting the role of tobacco NPK1 and *Arabidopsis* ANP1 in auxin signaling is based on gain-of-function analysis, using the truncated MAPKKK without the regulatory domain (Kovtun *et al.*, 1998; Kovtun *et al.*, 2000). Interestingly, the same MAPKKK was demonstrated to be involved in cytokinesis (discussed in more detail later). The later function of NPK1 and ANP1 is supported by loss-of-function genetic evidence from several groups (Nishihama *et al.*, 2001; Jin *et al.*, 2002; Krysan *et al.*, 2002). *Arabidopsis* has two additional homologs of NPK1 besides ANP1. They are ANP2 and ANP3. Single-mutant plants of all three genes displayed no obvious abnormal phenotypes, while *anp2/anp3* double-mutant combinations displayed defects in cell division and growth. The triple-mutant combination was not transmitted through either male or female gametes. Plants with *anp2/anp3* genotype display no change in auxin sensitivity. GeneChip analysis failed to reveal change in auxin-regulated gene expression in the *anp2/anp3* double mutant either (Krysan *et al.*, 2002). Further experiments are needed to address these apparently contradictory results.

Rapid MAPK activation was detected in *Arabidopsis* roots after auxin treatment (Mockaitis and Howell, 2000). More importantly, the MAPK activation was abolished in the auxin-resistant mutant *axr4*, linking MAPK activation to a known auxin signaling component. However, a prior report showed that the auxin-induced MAPK activation is a result of cytoplastic acidification, a nonspecific effect of auxin treatment (Tena and Renaudin, 1998). The identity of the MAPK induced by auxin is still unknown. A more definitive answer to this question will have to come from functional analysis after the identity of the MAPK is determined.

4.6.2.3.2 Arabidopsis *MKK7 in polar auxin transport.* Polar auxin transport (PAT) is critical to the auxin function. Recently, MKK7 was found to play a role in PAT. In a T-DNA insertional mutant screening, Dai *et al.* identified a bushy and dwarf1 (*bud1*) mutant (Dai *et al.*, 2003). Molecular genetic analysis indicated that the *bud1* phenotype is a result of increased expression of MKK7 (Dai *et al.*, 2006). Compared with the wild type, the *bud1* plants develop significantly fewer lateral roots, simpler venation patterns, and a quicker and greater curvature in the gravitropism assay. In addition, the *bud1* plants have shorter hypocotyls at high temperature (29°C) under light. These phenotypes are consistent with a defective auxin action in the *bud1* mutant at the levels of auxin synthesis, transport, or signaling. Direct IAA transport assays showed that the increased expression of *MKK7* in *bud1* or the repressed expression in *MKK7* antisense transgenic plants causes deficiency or enhancement in auxin transport, indicating that MKK7 negatively regulates PAT. Further double-mutant analysis of *bud1/axr3* indicates that reduced PAT can be partially compensated by elevated auxin sensitivity in the auxin-hypersensitive mutant *axr3*. In contrast, a double mutant of *bud1* and *doc-1*, an auxin transport mutant, exacerbates the PAT defect, reinforcing the idea that *bud1* is a negative regulator of PAT (Dai *et al.*, 2006).

Surprisingly, in the initial report of *bud1* mutant from the same group, it is concluded that the *bud1* phenotype is a result of an altered auxin metabolism because the auxin sensitivity and transport assay indicates that these two processes are normal in *bud1* (Dai *et al.*, 2003).

4.6.2.4 MAPKs in ABA signaling

ABA mediates plant responses to environmental stress, particularly to water status. In addition, it plays important roles in seed dormancy and germination (Finkelstein and Rock, 2002). ABA treatment activates an MBP kinase in *Pisum sativum* epidermal peels, which correlates with stomatal responses to ABA (Burnett *et al.*, 2000). This unidentified kinase has the characteristics of an MAPK. It requires Tyr phosphorylation for activity and is Tyr phosphorylated on activation. Inhibition of its activation by PD98059, a specific inhibitor of MAPKK, and thus MAPK activation, correlated with PD98059 inhibition of ABA-induced stomatal closure and dehydrin gene expression, suggesting that this MAPK is involved in the ABA signaling in pea epidermal peels. The identification of this MAPK activity will allow the functional analysis of its role in ABA signaling in future.

MAPK was also implicated in the ABA-induced antioxidant defense in maize leaves (Zhang *et al.*, 2006). Treatments with ABA or H_2O_2 induced the activation of a 46-kDa MAPK, which correlates with the induction of antioxidant gene expression and the total activities of the antioxidant enzymes. Pretreatment with MAPKK inhibitors and ROS inhibitors or scavengers abolished the gene activation. Pretreatment with MAPKK inhibitors also inhibited the ABA-induced H_2O_2 production. These results suggest that the activation of this MAPK is involved in the ABA-induced antioxidant defense. It was hypothesized that ABA-induced H_2O_2 production activates MAPK, which in turn induces the expression and the activities of antioxidant enzymes.

During seed germination, the embryo emerges from dormancy as the ABA concentration declines. Exposure to exogenous ABA can reverse the germination process. *Arabidopsis* MPK3 was found to be an important component in signaling this process (Lu *et al.*, 2002). Transgenic plants overexpressing MPK3 are hypersensitive to ABA, whereas inhibition of MAPK signaling decreases the sensitivity to postgermination growth arrest by ABA.

4.6.3 MAPK cascades in plant growth and development

Normal plant growth and development requires coordinated activities of different cells/tissues/organs, illustrating the importance of cell–cell communication. In eukaryotes, cell–cell communication often involves cell-surface receptors that are coupled to signaling molecules such as GTPases and protein kinases, which ultimately influence gene expression programs and growth/development. Plant MAPK cascades play important roles in regulating cytokinesis, cell differentiation, and reproduction. Functions of plant MAPK cascades in signaling growth and development are summarized in Fig. 4.3.

4.6.3.1 Role of NPK1–NQK1–NRK1 MAPK cascade in cytokinesis

Plant cytokinesis is unique with the involvement of new cell-plate formation. New cell plate is generated from the phragmoplast, a membranous structure located between two daughter nuclei that contains a network of microtubules and proteins related to vesicle trafficking. The Golgi-originated vesicles peripherally fuse to the cell plate and eventually reach the parental plasma membrane and cytokinesis is accomplished (Jurgens, 2005). A complete MAPK cascade, NPK1–NQK1–NRK1, as well as its direct downstream MAPK substrate MAP65-1 and upstream regulator NACK1/NACK2 have been shown

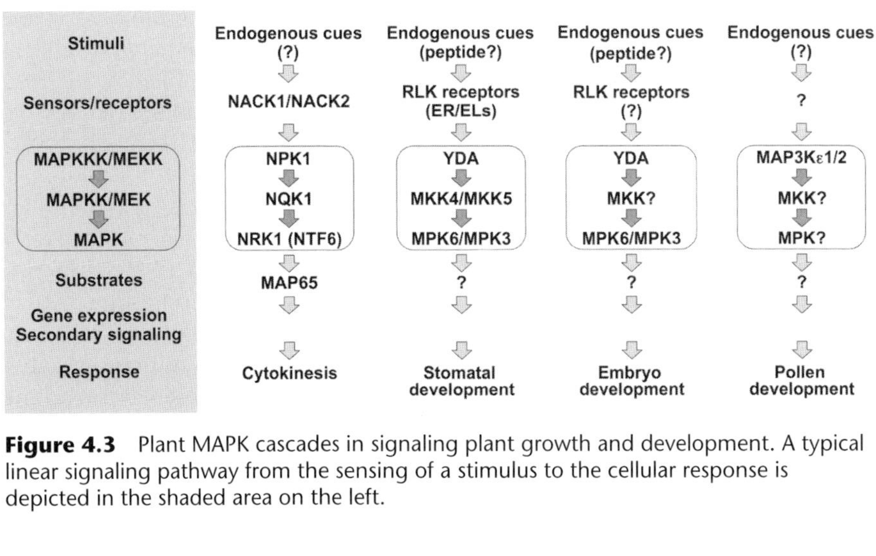

Figure 4.3 Plant MAPK cascades in signaling plant growth and development. A typical linear signaling pathway from the sensing of a stimulus to the cellular response is depicted in the shaded area on the left.

to play essential roles in this process. For more detailed discussions relating to this topic, see Chapter 12

4.6.3.2 MAPK in plant embryogenesis

Embryogenesis begins with the elongation and then first cell division of the zygote. This first cell division in plants is asymmetric, which results in a small apical cell with embryonic fate and a larger basal cell with mostly extraembryonic fate. The apical cell undergoes two rounds of longitudinal divisions followed by one round of transverse division and differentiates into the proembryo. In contrast, the basal cell undergoes a series of transverse divisions and gives rise to a file of cells that form the suspensor (Berleth and Chatfield, 2002). The first report implicating an MAPK cascade in *Arabidopsis* embryogenesis is the identification of YODA, an MAPKKK (Lukowitz *et al.*, 2004). In *yda* loss-of-function mutants, the zygote fails to elongate and the cells of the basal lineage are eventually incorporated into the embryo instead of differentiating to the extraembryonic suspensor. In the gain-of-function transgenic plants that express a constitutively active YODA mutant with N-terminal deletion (ΔN-YDA), apical cell identity is suppressed, which causes exaggerated growth of the suspensor. As a result, it is concluded that YODA functions as a molecular switch to promote the extraembryonic fate of the basal cell. The phenotypes of both the loss-of-function *yda* and the gain-of-function ΔN-YDA are likely a result of the disruption of normal asymmetric cell division. As discussed later, YODA is also involved in asymmetric cell division during stomatal development.

MPK3 and MPK6 are likely to function as a pair of redundant MAPKs downstream of YODA based on a similar mutant phenotype. MPK3 and MPK6 play overlapping functions in signaling plant stress/defense responses, which are the focus of a number of studies. In an attempt to generate double *mpk3* and *mpk6* mutant for loss-of-function analysis, we found that the *mpk3/mpk6* mutant is embryo lethal. Microscopic examination revealed abnormality at a very early stage after the first cell division of zygotes. Instead of an asymmetric cell division to form an apical cell and a basal cell, the products of the first cell division of *mpk3/mpk6* zygotes were about equal size, which developed into a group of disorganized cells, leading to aborted seeds (Wang *et al.*, 2007). Additional biochemical and/or genetic analyses are needed to formerly establish the link between YODA and MPK3/MPK6. The potential MAPKK(s) in the MAPK cascade remains to be identified.

4.6.3.3 YDA–MKK4/MKK5–MPK3/MPK6 in stomatal development and patterning

YODA also functions in stomatal development and patterning (Bergmann *et al.*, 2004). Stomata are specialized epidermal structures formed by two guard cells surrounding a pore, through which plants absorb CO_2 from and release O_2 to their environment. Asymmetric cell divisions precede stomatal cell fate specification in *Arabidopsis*. Perturbation of the frequency of asymmetric cell division, the orientation of the asymmetric division plane, and the polarity of the progenies of the asymmetric division ultimately disrupt stomatal development and patterning (Nadeau and Sack, 2002). In *Arabidopsis*, stomatal patterning follows the one-cell spacing rule; i.e., no two stomata are directly adjacent to each other and there is at least one pavement cell in between. In the *yda* mutant, disrupted asymmetric division of stomatal precursor cells results in the clustering of stomata. In contrast, stomatal development is suppressed in the gain-of-function ΔN-YDA transgenic plants (Bergmann *et al.*, 2004).

In *Arabidopsis*, several stomatal patterning mutants in the signaling pathway(s) have been characterized. Mutations in genes encoding TOO MANY MOUTHs (TMM, a leucine-rich repeat receptor protein), STOMATA DENSITY AND DISTRIBUTION 1 (SDD1, a subtilisin-like serine protease), and ER/ERL1/ERL2 (ERECTA and ERECTA-like, leucine-rich repeat receptor kinases) disrupt stomatal patterning and result in clustered stomata (Nadeau and Sack, 2002; Shpak *et al.*, 2005). It is proposed that unknown ligands processed by SDD1 bind to the TMM/ERs receptors in the target cells, which triggers the activation of YODA MAPK cascade and inhibits the development of stomata.

In our attempt to generate double mutant of *mpk3* and *mpk6* for functional analysis, we found that these two stress-responsive MAPKs are also involved in regulating stomatal development and patterning (Wang *et al.*, 2007). Double mutant of *mpk3/mpk6* is embryo lethal. To conditionally rescue the embryo-lethal double mutant, we transformed a DEX-inducible promoter *MPK6* (*GVG-MPK6*) into $mpk3^{-/-}/mpk6^{+/-}$ plants. Homozygous *GVG-MPK6* plants with $mpk3^{-/-}/mpk6^{+/-}$ genotype were sprayed with DEX during the flowering stage. In the next generation, about a quarter of the progeny were double mutants. The growth and development of the double mutant are arrested at the seedling stage. Microscopy revealed that the epidermis of the rescued *mpk3/mpk6* seedlings has clustered stomata. We further establish that MKK4/MKK5 are upstream of MPK3/MPK6 with YDA as the MAPKKK in the cascade. The establishment of a complete MAPK signaling cascade as a key regulator of stomatal development and patterning advances our understanding of the intercellular signaling events that control stomatal development (Bergmann *et al.*, 2004; Wang *et al.*, 2007).

4.6.3.4 MAPKKKs in *Arabidopsis* pollen development

Using reverse genetic analysis, the Krysan laboratory identified a pair of closely related MAPKKKs, MAP3Kε1 and MAP3Kε2, that function in *Arabidopsis* pollen development (Chaiwongsar *et al.*, 2006). Single-mutant plants have no apparent phenotype, whereas the double-mutant combination caused pollen lethality. In contrast, transmission of the double-mutant combination through the female gametophyte was not affected. Tetrad analysis performed using the *Arabidopsis* quartet mutation demonstrated that the pollen grains with $map3k\varepsilon1^{-}/map3k\varepsilon2^{-}$ genotype are lethal. Transmission electron microscopy revealed that double-mutant pollen grains develop plasma membrane irregularities following pollen mitosis I. Analysis of the subcellular localization of a yellow fluorescent protein (YFP)–MAP3Kε1 fusion protein using confocal microscopy and biochemical fractionation indicated that MAP3Kε1 protein is predominantly localized to the plasma membrane. Based on these results, it is concluded that MAP3Kε1 and MAP3Kε2 are required for the normal functioning of the plasma membrane in developing *Arabidopsis* pollen.

4.7 Signaling specificity of plant MAPK cascades

Although plant MAPKs are implicated in a variety of biological functions, from cell division, growth, and development to response to endogenous signals (e.g., plant hormones) and exogenous stimuli (e.g., biotic and abiotic stresses), the underlying molecular mechanisms are largely unknown. As discussed above, components in different MAPK cascades/pathways are frequently shared. How specificity is maintained when distinct functional pathways share common components is a central issue in cell biology that is highly relevant to our understanding of intracellular signaling.

In yeast and animal systems, several different mechanisms that can program distinct outcomes have been reported, including (1) different organ/tissue/cell types expressing distinct receptors and/or MAPK substrates, (2) quantitative differences in signaling strength or kinetics resulting in distinct outcomes, (3) different combinations of signaling pathways being activated by specific ligand–receptor pairs, (4) spatial restriction of signaling by pathway-specific scaffold proteins or formation of complexes, and (5) cross-pathway suppression of downstream components (Widmann *et al.*, 1999; Chang and Karin, 2001; Vaudry *et al.*, 2002; Schwartz and Madhani, 2004; Kolch, 2005; Remenyi *et al.*, 2005; Dard and Peter, 2006; Mor and Philips, 2006). However, little is known about how specificity is conferred in plant MAPK signaling pathways.

Spatiotemporal separation of the pathways via developmental-stage-dependent and/or tissue-specific expression is a common approach to achieve signaling specificity. Based on promoter–reporter fusion transgenics, most components in MAPK cascades including MPK3, MPK6, MPK4, MKK4, MKK5, and YDA are universally expressed with certain cells/tissues having higher levels of expression (D. Ren, H. Wang and S. Zhang, unpublished data) (Petersen *et al.*, 2000; Bergmann *et al.*, 2004). However, the expression of sensors/receptors could be limited to certain cells. So do the MAPK substrates. Cell/tissue-specific expression of the input molecules (receptors/sensors) and output molecules (substrates) can define a specific function of the MAPK cascade. As a result, to understand the signaling specificity of an MAPK cascade, it is essential to identify the receptor(s)/sensor(s) that activate the cascade in response to a stimulus and the in vivo substrate(s) of the MAPK.

Different substrates are likely to be involved in the diverse functions of MAPKs revealed by biochemical and genetic analyses. Recently, several MAPK substrates were reported, including ACS, MAP65, and MKS1 (Liu and Zhang, 2004; Andreasson *et al.*, 2005; Sasabe *et al.*, 2006). Additional proteins including tobacco NtWIP, NtWRKY1, and NbPPS3 were identified as potential MAPK substrates (Katou *et al.*, 2005b; Menke *et al.*, 2005; Yap *et al.*, 2005). It is predicted that each kinase has on average 20–40 in vivo substrates (Johnson and Hunter, 2005). A number of approaches including classical biochemical purification, yeast two-hybrid interaction screening, high-throughput protein array, and phosphoproteomics will lead to the identification of new MAPK substrates. About 48 potential substrates for MPK3 and MPK6 were recently identified, using a protein-array-based phosphorylation assay (Feilner *et al.*, 2005). Phosphoproteomic approach has the potential to identify new MAPK substrates as well (Peck, 2006). The identification of new MAPK substrates will reveal how MAPKs carry out their diverse functions in plants.

Signaling specificity can be maintained by complex formation with or without the involvement of a scaffold protein (Schwartz and Madhani, 2004; Kolch, 2005; Dard and Peter, 2006). Phosphorylation is most efficient when the enzyme and substrate interact physically. Previously, we showed that NtMEK2 interacts with SIPK and WIPK by co-immunoprecipitation (Jin *et al.*, 2003). A number of other MAPK components were also shown to interact with each other (Ichimura *et al.*, 1998; Mizoguchi *et al.*, 1998; Teige *et al.*, 2004). At this stage, there is no evidence indicating the involvement of plant scaffold proteins in MAPK signaling pathway yet.

The kinetics of MAPK activation and substrate specificity will also affect the signaling output. For instance, SIPK/WIPK activation can be either transient or long lasting depending on the stimuli. Long-lasting activation of these two MAPKs is associated with HR-like cell death in cultured cells treated with elicitin or plants infected with TMV (Zhang *et al.*, 1998; Zhang and Klessig, 1998a,b). Pharmacological studies linked the long-lasting activation of SIPK with cell death in tobacco cells (Suzuki *et al.*, 1999; Zhang *et al.*, 2000). In the conditional gain-of-function *NtMEK2*^{DD} transgenic system, induction of NtMEK2^{DD} expression results in long-lasting activation of downstream SIPK/WIPK, resulting in HR-like cell death, which mimics the long-lasting activation of SIPK/WIPK induced by pathogen infection (Yang *et al.*, 2001; Jin *et al.*, 2003). In mammalian cells, the kinetics of MAPK activation has been shown to influence the fate of cells under stress. Transient activation of SAPK/JNK and p38 induces various defense responses and allows the cells to adapt to adverse environments; in contrast, persistent activation of these two MAPKs leads to apoptosis (Widmann *et al.*, 1999).

Although MPK3/MPK6/MPK4 all phosphorylate MBP, an artificial substrate, it is likely that they have distinct substrates in vivo. Even when they all phosphorylate a common substrate, the specificity could still be maintained, depending on the kinetic properties of the reaction. An order difference in

K_m and/or V_{max} will allow the phosphorylation of a specific substrate by the dominant kinase, allowing the signaling of a particular response.

Although it has not been demonstrated in plants yet, cross-pathway suppression of downstream components could be another avenue to achieve signaling specificity. In yeast, eight proteins are shared between two MAPK cascades that signal two distinct developmental programs: (1) the mating pheromone response, and (2) the switch to filamentous growth. Pheromone signaling activates the mating pathway MAPK Fus3. However, due to the sharing of upstream kinases, a fraction of the filamentation pathway MAPK Kss1 is also activated. The effects of cross talk are suppressed because Fus3 induces the phosphorylation and destruction of Ted, the transcription factor that is activated by the filamentation pathway. The reduced half-life of Tec1 prevents pheromone-activated Kss1 from inducing filamentation-specific transcription via Tec1 (Bao *et al.*, 2004; Schwartz and Madhani, 2004). With the identification of additional MAPK substrates, we can begin to address the cross talk or cross suppression at this level.

4.8 Conclusion remarks

In the past decade, great progress has been made in our understanding of the biological functions of plant MAPK cascades. Seemingly conflicting results were reported, which can be partially attributed to the complexity of MAPK signaling pathways/networks. It is also possible that some of the phenotypes/responses are secondary and/or nonspecific responses of the transgenic/mutant plants. As a general comment, gain- and loss-of-function data have to be interpreted carefully because of the multifunctional nature of the MAPK cascades. In addition, stress signaling pathways are double-edged swords. Activation of an MAPK cascade that functions as a positive regulator will induce stress/defense response. Loss of function of such a positive regulator may also result in the activation of defense responses as a secondary effect because the plants may not be able to adapt, which causes general stress to the mutant plants. As a result, a conditional system, in which the long-term nonspecific effect can be avoided, will be the ideal system to elucidate the function of these pathways. The gap can also be filled by the identification of specific MAPK substrates.

Searching for MAPK substrates is likely to become the focal point in the next phase of MAPK research. The specific function of an MAPK cascade needs to be eventually backed up by the identification of specific substrates, which will reveal the molecular mechanisms underlying a specific MAPK function. It is also critical for us to identify the specific receptors/sensors that function upstream of an MAPK cascade and understand how the recognition of stimuli/ligands activates the MAPK cascade. After the identification of each individual component in the MAPK signaling networks, in vivo functional analyses will allow us to piece together the linear functional pathways and eventually the signaling networks illustrating the roles of MAPK cascades in plant growth, development, and stress/defense responses. By superimposing the spatiotemporal information of these components and their activities onto the networks, we will be able to understand how signaling specificity is maintained and the cross talk between different functional pathways.

Acknowledgments

I apologize for not being able to cite all related references because of space limitations. I thank Clayton Larue for comments. The work in the Zhang laboratory is supported by grants from the National Science Foundation.

References

Abeles, F.B., Morgan, P.W. and Saltveit, M.E.J. (1992) *Ethylene in Plant Biology*. Academic Press, San Diego, CA.

Andreasson, E., Jenkins, T., Brodersen, P., Thorgrimsen, S., Petersen, N.H.T., Zhu, S., Qiu, J.-L., Micheelsen, P., Rocher, A., Petersen, M., Newman, M.-A., Nielsen, H.B., Hirt, H., Somssich, I., Mattsson, O. and Mundy, J. (2005) The MAP kinase substrate MKS1 is a regulator of plant defense responses. *EMBO J*, **24**, 2579–2589.

Apel, K. and Hirt, H. (2004) Reactive oxygen species: metabolism, oxidative stress, and signal transduction. *Annu Rev Plant Biol*, **55**, 373–399.

Asai, T., Tena, G., Plotnikova, J., Willmann, M.R., Chiu, W.-L., Gomez-Gomez, L., Boller, T., Ausubel, F.M. and Sheen, J. (2002) MAP kinase signalling cascade in *Arabidopsis* innate immunity. *Nature*, **415**, 977–983.

Badescu, G.O. and Napier, R.M. (2006) Receptors for auxin: will it all end in TIRs? *Trends Plant Sci*, **11**, 217–223.

Banno, H., Hirano, K., Nakamura, T., Irie, K., Nomoto, S., Matsumoto, K. and Machida, Y. (1993) NPK1, a tobacco gene that encodes a protein with a domain homologous to yeast BCK1, STE11, and Byr2 protein kinases. *Mol Cell Biol*, **13**, 4745–4752.

Bao, M.Z., Schwartz, M.A., Cantin, G.T., Yates, J.R., III, and Madhani, H.D. (2004) Pheromone-dependent destruction of the Tec1 transcription factor is required for MAP kinase signaling specificity in yeast. *Cell*, **119**, 991–1000.

Bergmann, D.C., Lukowitz, W. and Somerville, C.R. (2004) Stomatal development and pattern controlled by a MAPKK kinase. *Science*, **304**, 1494–1497.

Berleth, T. and Chatfield, S. (2002) Embryogenesis: pattern formation from a single cell. In: *The Arabidopsis Book* (eds C.R. Somerville and E.M. Meyerowitz), pp. 1–22. American Society of Plant Biologists, Rockville, MD; http://www.aspb.org/publications/arabidopsis/.

Boller, T. (2005) Peptide signalling in plant development and self/non-self perception. *Cur Opin Cell Biol*, **17**, 116–122.

Brodersen, P., Petersen, M., Bjorn Nielsen, H., Zhu, S., Newman, M.-A., Shokat, K.M., Rietz, S., Parker, J. and Mundy, J. (2006) *Arabidopsis* MAP kinase 4 regulates salicylic acid- and jasmonic acid/ethylene-dependent responses via EDS1 and PAD4. *Plant J*, **47**, 532–546.

Broekaert, W.F., Delaure, S.L., De Bolle, M.F.C. and Cammue, B.P.A. (2006) The role of ethylene in host-pathogen interactions. *Annu Rev Phytopathol*, **44**, 393–416.

Burnett, E.C., Desikan, R., Moser, R.C. and Neill, S.J. (2000) ABA activation of an MBP kinase in *Pisum sativum* epidermal peels correlates with stomatal responses to ABA. *J Exp Bot*, **51**, 197–205.

Caffrey, D.R., O'Neill, L.A.J. and Shields, D.C. (1999) The evolution of the MAP kinase pathways: coduplication of interacting proteins leads to new signaling cascades. *J Mol Evol*, **49**, 567–582.

Cardinale, F., Jonak, C., Ligterink, W., Niehaus, K., Boller, T. and Hirt, H. (2000) Differential activation of four specific MAPK pathways by distinct elicitors. *J Biol Chem*, **275**, 36734–36740.

Chae, H.S. and Kieber, J.J. (2005) Eto Brute? Role of ACS turnover in regulating ethylene biosynthesis. *Trends Plant Sci*, **10**, 291–296.

Chaiwongsar, S., Otegui, M.S., Jester, P.J., Monson, S.S. and Krysan, P.J. (2006) The protein kinase genes MAP3Kε1 and MAP3Kε2 are required for pollen viability in *Arabidopsis thaliana*. *Plant J*, **48**, 193–205.

Champion, A., Picaud, A. and Henry, Y. (2004) Reassessing the MAP3K and MAP4K relationships. *Trends Plant Sci*, **9**, 123–129.

Chang, C. and Bleecker, A.B. (2004) Ethylene biology: more than a gas. *Plant Physiol*, **136**, 2895–2899.

Chang, L. and Karin, M. (2001) Mammalian MAP kinase signaling cascades. *Nature*, **410**, 37–40.

Creelman, R.A. and Mullet, J.E. (1997) Biosynthesis and action of jasmonates in plants. *Annu Rev Plant Biol*, **48**, 355–381.

Dai, Y., Fu, Z.M. and Li, J.Y. (2003) Isolation and characterization of an *Arabidopsis* bushy and dwarf mutant. *Acta Botanica Sinica*, **45**, 621–625.

Dai, Y., Wang, H., Li, B., Huang, J., Liu, X., Zhou, Y., Mou, Z. and Li, J. (2006) Increased expression of MAP KINASE KINASE7 causes deficiency in polar auxin transport and leads to plant architectural abnormality in *Arabidopsis*. *Plant Cell*, **18**, 308–320.

Dangl, J.L. and Jones, J.D.G. (2001) Plant pathogens and integrated defense responses to infection. *Nature*, **411**, 826–833.

Dard, N. and Peter, M. (2006) Scaffold proteins in MAP kinase signaling: more than simple passive activating platforms. *Bioessays*, **28**, 146–156.

Davies, S.P., Reddy, H., Caivano, M. and Cohen, P. (2000) Specificity and mechanism of action of some commonly used protein kinase inhibitors. *Biochem J*, **351**, 95–105.

Davis, R. (2000) Signal transduction by the JNK group of MAP kinases. *Cell*, **103**, 239–252.

del Pozo, O., Pedley, K.F. and Martin, G.B. (2004) MAPKKKα is a positive regulator of cell death associated with both plant immunity and disease. *EMBO J*, **23**, 3072–3082.

Desikan, R., Hancock, J.T., Ichimura, K., Shinozaki, K. and Neill, S.J. (2001) Harpin induced activation of the *Arabidopsis* mitogen-activated protein kinases AtMPK4 and AtMPK6. *Plant Physiol*, **126**, 1579–1587.

Dharmasiri, N., Dharmasiri, S. and Estelle, M. (2005) The F-box protein TIR1 is an auxin receptor. *Nature*, **435**, 441–445.

Dixon, R.A. (2001) Natural products and plant disease resistance. *Nature*, **411**, 843–847.

Droillard, M.-J., Boudsocq, M., Barbier-Brygoo, H. and Lauriere, C. (2004). Involvement of MPK4 in osmotic stress response pathways in cell suspensions and plantlets of *Arabidopsis thaliana*: activation by hypoosmolarity and negative role in hyperosmolarity tolerance. *FEBS Lett*, **574**, 42–48.

Duerr, B., Gawienowski, M., Ropp, T. and Jacobs, T. (1993) MsERK1: a mitogen-activated protein kinase from a flowering plant. *Plant Cell*, **5**, 87–96.

Durrant, W.E. and Dong, X. (2004) Systemic acquired resistance. *Annu Rev Phytopathol*, **42**, 185–209.

Ecker, J.R. (2004) Reentry of the ethylene MPK6 module. *Plant Cell*, **16**, 3169–3173.

Ekengren, S.K., Liu, Y., Schiff, M., Dinesh-Kumar, S.P. and Martin, G.B. (2003) Two MAPK cascades, NPR1, and TGA transcription factors play a role in Pto-mediated disease resistance in tomato. *Plant J*, **36**, 905–917.

Feilner, T., Hultschig, C., Lee, J., Meyer, S., Immink, R.G.H., Koenig, A., Possling, A., Seitz, H., Beveridge, A., Scheel, D., Cahill, D.J., Lehrach, H., Kreutzberger, J. and Kersten, B. (2005) High throughput identification of potential *Arabidopsis* mitogen-activated protein kinases substrates. *Mol Cell Proteomics*, **4**, 1558–1568.

Finkelstein, R.R. and Rock, C.D. (2002) Abscisic acid biosynthesis and response. In: *The Arabidopsis Book* (eds C.R. Somerville and E.M. Meyerowitz), pp. 1–48. American Society of Plant Biologists, Rockville, MD; http://www.aspb.org/publications/arabidopsis/.

Frye, C.A., Tang, D. and Innes, R.W. (2001) Negative regulation of defense responses in plants by a conserved MAPKK kinase. *Proc Natl Acad Sci USA*, **98**, 373–378.

Gomi, K., Ogawa, D., Katou, S., Kamada, H., Nakajima, N., Saji, H., Soyano, T., Sasabe, M., Machida, Y., Mitsuhara, I., Ohashi, Y. and Seo, S. (2005) A mitogen-activated protein kinase NtMPK4 activated by SIPKK is required for jasmonic acid signaling and involved in ozone tolerance via stomatal movement in tobacco. *Plant Cell Physiol*, **46**, 1902–1914.

Greenberg, J.T. and Yao, N. (2004) The role and regulation of programmed cell death in plant-pathogen interactions. *Cell Microbiol*, **6**, 201–211.

Gupta, R., Huang, Y., Kieber, J. and Luan, S. (1998) Identification of a dual-specificity protein phosphatase that inactivates a MAP kinase from *Arabidopsis*. *Plant J*, **16**, 581–589.

Hamel, L.-P., Nicole, M.-C., Sritubtim, S., Morency, M.-J., Ellis, M., Ehlting, J., Beaudoin, N., Barbazuk, B., Klessig, D., Lee, J., Martin, G., Mundy, J., Ohashi, Y., Scheel, D., Sheen, J., Xing, T., Zhang, S., Seguin, A. and Ellis, B.E. (2006) Ancient signals: comparative genomics of plant MAPK and MAPKK gene families. *Trends Plant Sci*, **11**, 192–198.

Hazzalin, C.A. and Mahadevan, L.C. (2002) MAPK-regulated transcription: a continuously variable gene switch? *Nat Rev Mol Cell Biol*, **3**, 30–40.

Holley, S.R., Yalamanchili, R.D., Moura, D.S., Ryan, C.A. and Stratmann, J.W. (2003) Convergence of signaling pathways induced by systemin, oligosaccharide elicitors, and ultraviolet-B radiation at the level of mitogen-activated protein kinases in *Lycopersicon peruvianum* suspension-cultured cells. *Plant Physiol*, **132**, 1728–1738.

Hoyos, M.E. and Zhang, S. (2000) Calcium-independent activation of salicylic acid-induced protein kinase and a 40-kilodalton protein kinase by hyperosmotic stress. *Plant Physiol*, **122**, 1355–1363.

Ichimura, K., Casais, C., Peck, S.C., Shinozaki, K. and Shirasu, K. (2006) MEKK1 is required for MPK4 activation and regulates tissue-specific and temperature-dependent cell death in *Arabidopsis*. *J Biol Chem*, **281**, 36969–36976.

Ichimura, K., Mizoguchi, T., Irie, K., Morris, P., Giraudat, J., Matsumoto, K. and Shinozaki, K. (1998) Isolation of AtMEKK1 (a MAP kinase kinase kinase)-interacting proteins and analysis of a MAP kinase cascade in *Arabidopsis*. *Biochem Biophys Res Comm*, **253**, 532–543.

Ichimura, K., Mizoguchi, T., Yoshida, R., Yuasa, T. and Shinozaki, K. (2000) Various abiotic stresses rapidly activate *Arabidopsis* MAP kinases ATPMK4 and ATMPK6. *Plant J*, **24**, 655–665.

Jin, H., Axtell, M.J., Dahlbeck, D., Ekwenna, O., Zhang, S., Staskawicz, B. and Baker, B. (2002) NPK1, an MEKK1-like mitogen-activated protein kinase kinase kinase, regulates innate immunity and development in plants. *Dev Cell*, **3**, 291–297.

Jin, H., Liu, Y., Yang, K.-Y., Kim, C.Y., Baker, B. and Zhang, S. (2003) Function of a mitogen-activated protein kinase pathway in *N*-gene mediated resistance in tobacco. *Plant J*, **33**, 719–731.

Johnson, S.A. and Hunter, T. (2005) Kinomics: methods for deciphering the kinome. *Nat Methods*, **2**, 17–25.

Jonak, C., Kiegeri, S., Ligterink, W., Barker, P.J., Huskisson, N.S. and Hirt, H. (1996) Stress signaling in plants: a mitogen-activated protein kinase pathway is activated by cold and drought. *Proc Natl Acad Sci USA*, **93**, 11274–11279.

Jonak, C., Páy, A., Bögre, L., Hirt, H. and Heberle-Bors, E. (1993) The plant homologue of MAP kinase is expressed in a cell cycle-dependent and organ-specific manner. *Plant J*, **3**, 611–617.

Jurgens, G. (2005) Cytokinesis in higher plants. *Annu Rev Plant Biol*, **56**, 281–299.

Katou, S., Karita, E., Yamakawa, H., Seo, S., Mitsuhara, I., Kuchitsu, K. and Ohashi, Y. (2005a) Catalytic activation of the plant MAPK phosphatase NtMKP1 by its physiological substrate salicylic acid-induced protein kinase but not by calmodulins. *J Biol Chem*, **280**, 39569–39581.

Katou, S., Kuroda, K., Seo, S., Yanagawa, Y., Tsuge, T., Yamazaki, M., Miyao, A., Hirochika, H. and Ohashi, Y. (2007) A calmodulin-binding mitogen-activated protein kinase phosphatase is induced by wounding and regulates the activities of stress-related mitogen-activated protein kinases in rice. *Plant Cell Physiol*, **48**, 332–344.

Katou, S., Yoshioka, H., Kawakita, K., Rowland, O., Jones, J.D.G., Mori, H. and Doke, N. (2005b) Involvement of PPS3 phosphorylated by elicitor-responsive mitogen-activated protein kinases in the regulation of plant cell death. *Plant Physiol*, **139**, 1914–1926.

Kende, H. (2001) Hormone response mutants: a plethora of surprises. *Plant Physiol*, **125**, 81–84.

Kepinski, S. and Leyser, O. (2005) The *Arabidopsis* F-box protein TIR1 is an auxin receptor. *Nature*, **435**, 446–451.

Kieber, J.J., Rothenberg, M., Roman, G., Feldmann, K.E. and Ecker, J. (1993) CTR1, a negative regulator of the ethylene response pathway in *Arabidopsis*, encodes a member of the Raf family of protein kinases. *Cell*, **72**, 427–441.

Kim, C.Y., Liu, Y., Thorne, E.T., Yang, H., Fukushig, H., Gassmann, W., Hildebrand, D., Sharp, R.E. and Zhang, S. (2003) Activation of a stress-responsive mitogen-activated protein kinase cascade induces the biosynthesis of ethylene in plants. *Plant Cell*, **15**, 2707–2718.

Kim, C.Y. and Zhang, S. (2004) Activation of a mitogen-activated protein kinase cascade induces WRKY family of transcription factors and defense genes in tobacco. *Plant J*, **38**, 142–151.

Kolch, W. (2005) Coordinating ERK/MAPK signalling through scaffolds and inhibitors. *Nat Rev Mol Cell Biol*, **6**, 827–837.

Kovtun, Y., Chiu, W.L. and Sheen, J. (1998) Suppression of auxin signal transduction by a MAPK cascade in higher plants. *Nature*, **395**, 716–720.

Kovtun, Y., Chiu, W.-L., Tena, G. and Sheen, J. (2000) Functional analysis of oxidative stress-activated mitogen-activated protein kinase cascade in plants. *Proc Natl Acad Sci USA*, **97**, 2940–2945.

Kroj, T., Rudd, J.J., Nürnberger, T., Gäbler, Y., Lee, J. and Scheel, D. (2003) Mitogen-activated protein kinases play an essential role in oxidative burst-independent expression of pathogenesis-related genes in parsley. *J Biol Chem*, **278**, 2256–2264.

Krysan, P.J., Jester, P.J., Gottwald, J.R. and Sussman, M.R. (2002) An *Arabidopsis* mitogen-activated protein kinase kinase kinase gene family encodes essential positive regulators of cytokinesis. *Plant Cell*, **14**, 1109–1120.

Lee, J., Klessig, D.F. and Nürnberger, T. (2001) A harpin binding site in tobacco plasma membranes mediates activation of the pathogenesis-related gene *HIN1* independent of extracellular calcium but dependent on mitogen-activated protein kinase activity. *Plant Cell*, **13**, 1079–1093.

Lee, J., Rudd, J.J., Macioszek, V.K. and Scheel, D. (2004) Dynamic changes in the localization of MAPK cascade components controlling pathogenesis-related (PR) gene expression during innate immunity in parsley. *J Biol Chem*, **279**, 22440–22448.

Link, V.L., Hofmann, M.G., Sinha, A.K., Ehness, R., Strnad, M. and Roitsch, T. (2002) Biochemical evidence for the activation of distinct subsets of mitogen-activated protein kinases by voltage and defense-related stimuli. *Plant Physiol*, **128**, 271–281.

Liu, Y., Jin, H., Yang, K.-Y., Kim, C.Y., Baker, B. and Zhang, S. (2003) Interaction between two mitogen-activated protein kinases during tobacco defense signaling. *Plant J*, **34**, 149–160.

Liu, Y., Ren, D., Pike, S., Pallardy, S., Gassmann, W. and Zhang, S. (2007) Chloroplast-generated reactive oxygen species are involved in hypersensitive response-like cell death mediated by a mitogen-activated protein kinase cascade. *Plant J*, **51**, 941–954.

Liu, Y. and Zhang, S. (2004). Phosphorylation of 1-aminocyclopropane-1-carboxylic acid synthase by MPK6, a stress-responsive mitogen-activated protein kinase, induces ethylene biosynthesis in *Arabidopsis*. *Plant Cell*, **16**, 3386–3399.

Lorenzo, O., Chico, J.M., Sanchez-Serrano, J.J. and Solano, R. (2004) *JASMONATE-INSENSITIVE1* encodes a MYC transcription factor essential to discriminate between different jasmonate-regulated defense responses in *Arabidopsis*. *Plant Cell*, **16**, 1938–1950.

Lu, C., Han, M.H., Guevara-Garcia, A. and Fedoroff, N.V. (2002) Mitogen-activated protein kinase signaling in postgermination arrest of development by abscisic acid. *Proc Natl Acad Sci USA*, **99**, 15812–15817.

Lukowitz, W., Roeder, A., Parmenter, D. and Somerville, C. (2004) A MAPKK kinase gene regulates extra-embryonic cell fate in *Arabidopsis*. *Cell*, **116**, 109–119.

Mansour, S.J., Matten, W.T., Hermann, A.S., Candia, J.M., Rong, S., Fukasawa, K., VandeWounde, G.F. and Ahn, N.G. (1994) Transformation of mammalian cells by constitutively active MAP kinase kinase. *Science*, **265**, 966–970.

MAPK Group (2002) Mitogen-activated protein kinase cascades in plants: a new nomenclature. *Trends Plant Sci*, **7**, 301–308.

Martin, G.B., Bogdanove, A.J. and Sessa, G. (2003) Understanding the functions of plant disease resistance proteins. *Annu Rev Plant Biol*, **54**, 23–61.

Matsuoka, D., Nanmori, T., Sato, K.-i., Fukami, Y., Kikkawa, U. and Yasuda, T. (2002) Activation of AtMEK1, an *Arabidopsis* mitogen-activated protein kinase kinase, in vitro and in vivo: analysis of active mutants expressed in *E. coli* and generation of the active form in stress response in seedlings. *Plant J*, **29**, 637–647.

Menke, F.L.H., Kang, H.-G., Chen, Z., Park, J.M., Kumar, D. and Klessig, D.F. (2005) Tobacco transcription factor WRKY1 is phosphorylated by the MAP kinase SIPK and mediates HR-like cell death in tobacco. *Mol Plant Microbe Interact*, **18**, 1027–1034.

Menke, F.L.H., van Pelt, J.A., Pieterse, C.M.J. and Klessig, D.F. (2004) Silencing of the mitogen-activated protein kinase MPK6 compromises disease resistance in *Arabidopsis*. *Plant Cell*, **16**, 897–907.

Meskiene, I., Baudouin, E., Schweighofer, A., Liwosz, A., Jonak, C., Rodriguez, P.L., Jelinek, H. and Hirt, H. (2003) Stress-induced protein phosphatase 2C is a negative regulator of a mitogen-activated protein kinase. *J Biol Chem*, **278**, 18945–18952.

Meskiene, I., Bögre, L., Glaser, W., Balog, J., Brandstötter, M., Zwerger, K., Ammerer, G. and Hirt, H. (1998) MP2C, a plant protein phosphatase 2C, functions as a negative regulator of mitogen-activated protein kinase pathways in yeast and plants. *Proc Natl Acad Sci USA*, **95**, 1938–1943.

Mikolajczyk, M., Awotunde, O.S., Muszynska, G., Klessig, D.F. and Dobrowolska, G. (2000) Osmotic stress induces rapid activation of a salicylic acid-induced protein kinase and a homolog of protein kinase ASK1 in tobacco cells. *Plant cell*, **12**, 165–178.

Mizoguchi, T., Gotoh, Y., Nishida, E., Yamaguchi-Shinozaki, K., Hayashida, N., Iwasaki, T., Kamada, H. and Shinozaki, K. (1994) Characterization of two cDNAs that encode MAP kinase homologues in *Arabidopsis thaliana* and analysis of the possible role of auxin in activating such kinase activities in cultured cells. *Plant J*, **5**, 111–122.

Mizoguchi, T., Hayashida, N., Yamaguchi-Shinozaki, K., Kamada, H. and Shinozaki, K. (1993) ATMPKs: a gene family of plant MAP kinases in *Arabidopsis thaliana*. *FEBS Lett*, **336**, 440–444.

Mizoguchi, T., Ichimura, K., Irie, K., Morris, P., Giraudat, J., Matsumoto, K. and Shinozaki, K. (1998) Identification of a possible MAP kinase cascade in *Arabidopsis thaliana* based on pairwise yeast two-hybrid analysis and functional complementation tests of yeast mutants. *FEBS Lett*, **437**, 56–60.

Mizoguchi, T., Ichimura, K. and Shinozaki, K. (1997) Environmental stress response in plants: the role of mitogen-activated protein kinases. *Trends Biotechnol*, **15**, 15–19.

Mizoguchi, T., Irie, K., Hirayama, T., Hayashida, N., Yamaguchi-Shinozaki, K., Matsumota, K. and Shinozaki, K. (1996) A gene encoding a mitogen-activated protein kinase kinase kinase is induced simultaneously with genes for a mitogen-activated protein kinase and an S6 ribosomal protein kinase by touch, cold, and water stress in *Arabidopsis thaliana*. *Proc Natl Acad Sci USA*, **93**, 765–769.

Mockaitis, K. and Howell, S.H. (2000) Auxin induced mitogenic activated protein kinase (MAPK) activation in roots of *Arabidopsis* seedlings. *Plant J*, **24**, 785–796.

Monroe-Augustus, M., Zolman, B.K. and Bartel, B. (2003) IBR5, a dual-specificity phosphatase-like protein modulating auxin and abscisic acid responsiveness in *Arabidopsis*. *Plant Cell*, **15**, 2979–2991.

Mor, A. and Philips, M.R. (2006) Compartmentalized RAS/MAPK signaling. *Annu Rev Immunol*, **24**, 771–800.

Nadeau, J.A. and Sack, F.D. (2002) Stomatal development in *Arabidopsis*. In: *The Arabidopsis Book* (eds C.R. Somerville and E.M. Meyerowitz), pp. 1–28. American Society of Plant Biologists, Rockville, MD; http://www.aspb.org/publications/arabidopsis/.

Nakagami, H., Pitzschke, A. and Hirt, H. (2005) Emerging MAP kinase pathways in plant stress signalling. *Trends Plant Sci*, **10**, 339–346.
Nakagami, H., Soukupova, H., Schikora, A., Zarsky, V. and Hirt, H. (2006) A mitogen-activated protein kinase kinase kinase mediates reactive oxygen species homeostasis in *Arabidopsis*. *J Biol Chem*, **281**, 38697–38704.
Navarro, L., Dunoyer, P., Jay, F., Arnold, B., Dharmasiri, N., Estelle, M., Voinnet, O. and Jones, J.D.G. (2006) A plant miRNA contributes to antibacterial resistance by repressing auxin signaling. *Science*, **312**, 436–439.
Nishihama, R., Ishikawa, M., Araki, S., Soyano, T., Asada, T. and Machida, Y. (2001) The NPK1 mitogen-activated protein kinase kinase kinase is a regulator of cell-plate formation in plant cytokinesis. *Genes Dev*, **15**, 352–363.
Nühse, T., Peck, S.C., Hirt, H. and Boller, T. (2000) Microbial elicitors induce activation and dual phosphorylation of the *Arabidopsis thaliana* MAPK6. *J Biol Chem*, **275**, 7521–7526.
Ouaked, F., Rozhon, W., Lecourieux, D. and Hirt, H. (2003) A MAPK pathway mediates ethylene signaling in plants. *EMBO J*, **22**, 1282–1288.
Peck, S.C. (2006) Analysis of protein phosphorylation: methods and strategies for studying kinases and substrates. *Plant J*, **45**, 512–522.
Pedley, K.F. and Martin, G.B. (2004) Identification of MAPKs and their possible MAPK kinase activators involved in the Pto-mediated defense response of tomato. *J Biol Chem*, **279**, 49229–49235.
Pedley, K.F. and Martin, G.B. (2005) Role of mitogen-activated protein kinases in plant immunity. *Curr Opin Plant Biol*, **8**, 541–547.
Petersen, M., Brodersen, P., Naested, H., Andreasson, E., Lindhart, U., Johansen, B., Nielsen, H.B., Lacy, M., Austin, M.J., Parker, J.E., Sharma, S.B., Klessig, D.F., Martienssen, R., Mattsson, O., Jensen, A.B. and Mundy, J. (2000) *Arabidopsis* MAP kinase 4 negatively regulates systemic acquired resistance. *Cell*, **103**, 1111–1120.
Rao, M.V., Lee, H.-I. and Davis, K.R. (2002) Ozone-induced ethylene production is dependent on salicylic acid, and both salicylic acid and ethylene act in concert to regulate ozone-induced cell death. *Plant J*, **32**, 447–456.
Remenyi, A., Good, M.C., Bhattacharyya, R.P. and Lim, W.A. (2005) The role of docking interactions in mediating signaling input, output, and discrimination in the yeast MAPK network. *Mol Cell*, **20**, 951–962.
Ren, D., Yang, H. and Zhang, S. (2002) Cell death mediated by mitogen-activated protein kinase pathway is associated with the generation of hydrogen peroxide in *Arabidopsis*. *J Biol Chem*, **277**, 559–565.
Ren, D., Yang, K.-Y., Li, G.-J., Liu, Y. and Zhang, S. (2006) Activation of Ntf4, a tobacco mitogen-activated protein kinase, during plant defense response and its involvement in hypersensitive response-like cell death. *Plant Physiol*, **141**, 1482–1493.
Rentel, M.C., Lecourieux, D., Ouaked, F., Usher, S.L., Petersen, L., Okamoto, H., Knight, H., Peck, S.C., Grierson, C.S., Hirt, H. and Knight, M.R. (2004) OXI1 kinase is necessary for oxidative burst-mediated signalling in *Arabidopsis*. *Nature*, **427**, 858–861.
Romeis, T., Piedras, P., Zhang, S., Klessig, D.F., Hirt, H. and Jones, J. (1999) Rapid Avr9- and Cf-9-dependent activation of MAP kinases in tobacco cell cultures and leaves: convergence of resistance gene, elicitor, wound and salicylate responses. *Plant Cell*, **11**, 273–287.
Samuel, M.A. and Ellis, B.E. (2002) Double jeopardy: both overexpression and suppression of a redox-activated plant mitogen-activated protein kinase render tobacco plants ozone sensitive. *Plant Cell*, **14**, 2059–2069.

Samuel, M.A., Walia, A., Mansfield, S.D. and Ellis, B.E. (2005) Overexpression of SIPK in tobacco enhances ozone-induced ethylene formation and blocks ozone-induced SA accumulation. *J Exp Bot*, **56**, 2195–2201.

Sasabe, M., Soyano, T., Takahashi, Y., Sonobe, S., Igarashi, H., Itoh, T.J., Hidaka, M. and Machida, Y. (2006) Phosphorylation of NtMAP65-1 by a MAP kinase down-regulates its activity of microtubule bundling and stimulates progression of cytokinesis of tobacco cells. *Genes Dev*, **20**, 1004–1014.

Schaller, G.E. and Kieber, J.J. (2002) Ethylene. In: *The Arabidopsis Book* (eds C.R. Somerville and E.M. Meyerowitz), pp. 1–19. American Society of Plant Biologists, Rockville, MD; http://www.aspb.org.publications/arabidopsis/.

Schwartz, M.A. and Madhani, H.D. (2004) Principle of MAP kinase signaling specificity in *Saccharomyces cerevisiae*. *Annu Rev Genet*, **38**, 725–748.

Seo, S., Katou, S., Seto, H., Gomi, K. and Ohashi, Y. (2007) The mitogen-activated protein kinases WIPK and SIPK regulate the levels of jasmonic and salicylic acids in wounded tobacco plants. *Plant J*, **49**, 899–909.

Seo, S., Okamoto, M., Seto, H., Ishizuka, K., Sano, H. and Ohashi, Y. (1995) Tobacco MAP kinase: a possible mediator in wound signal transduction pathways. *Science*, **270**, 1988–1992.

Seo, S., Sano, H. and Ohashi, Y. (1999) Jasmonate-based wound signal transduction requires activation of WIPK, a tobacco mitogen-activated protein kinase. *Plant Cell*, **11**, 289–298.

Sharma, P.C., Ito, A., Shimizu, T., Terauchi, R., Kamoun, S. and Saitoh, H. (2003) Virus-induced silencing of WIPK and SIPK genes reduces resistance to a bacterial pathogen, but has no effect on the INF1-induced hypersensitive response (HR) in *Nicotiana benthamiana*. *Mol Genet Genomics*, **269**, 583–591.

Shpak, E.D., McAbee, J.M., Pillitteri, L.J. and Torii, K.U. (2005) Stomatal patterning and differentiation by synergistic interactions of receptor kinases. *Science*, **309**, 290–293.

Spanu, P., Grosskopf, D.G., Felix, G. and Boller, T. (1994) The apparent turnover of 1-aminocyclopropane-1-carboxylate synthase in tomato cells is regulated by protein phosphorylation and dephosphorylation. *Plant Physiol*, **106**, 529–535.

Stafstrom, J.P., Altschuler, M. and Anderson, D.H. (1993) Molecular cloning and expression of a MAP kinase homologue from pea. *Plant Mol Biol*, **22**, 83–90.

Staskawicz, B.J., Ausubel, F.M., Baker, B.J., Ellis, J.G. and Jones, J.D.G. (1995) Molecular genetics of plant disease resistance. *Science*, **268**, 661–667.

Stratmann, J.W. and Ryan, C.A. (1997) Myelin basic protein kinase activity in tomato leaves is induced systemically by wounding and increases in response to systemin and oligosaccharide elicitors. *Proc Natl Acad Sci USA*, **94**, 11085–11089.

Suarez-Rodriguez, M.C., Adams-Phillips, L., Liu, Y., Wang, H., Su, S.-H., Jester, P.J., Zhang, S., Bent, A.F. and Krysan, P.J. (2007) MEKK1 is required for flg22-induced MPK4 activation in *Arabidopsis* plants. *Plant Physiol*, **143**, 661–669.

Suzuki, K., Yano, A. and Shinshi, H. (1999) Slow and prolonged activation of the p47 protein kinase during hypersensitive cell death in a culture of tobacco cells. *Plant Physiol*, **119**, 1465–1472.

Takahashi, F., Yoshida, R., Ichimura, K., Mizoguchi, T., Seo, S., Yonezawa, M., Maruyama, K., Yamaguchi-Shinozaki, K. and Shinozaki, K. (2007a) The mitogen-activated protein kinase cascade MKK3-MPK6 is an important part of the jasmonate signal transduction pathway in *Arabidopsis*. *Plant Cell*, **19**, 805–818.

Takahashi, Y., Nasir, K.H.B., Ito, A., Kanzaki, H., Matsumura, H., Saitoh, H., Fujisawa, S., Kamoun, S. and Terauchi, R. (2007b) A high-throughput screen of cell-death-inducing factors in *Nicotiana benthamiana* identifies a novel MAPKK that mediates INF1-induced cell death signaling and non-host resistance to *Pseudomonas cichorii*. *Plant J*, **49**, 1030–1040.

Takahashi, Y., Soyano, T., Sasabe, M. and Machida, Y. (2004) A MAP kinase cascade that controls plant cytokinesis. *J Biochem (Tokyo)*, **136**, 127–132.

Tang, D. and Innes, R.W. (2002) Overexpression of a kinase-deficient form of the EDR1 gene enhances powdery mildew resistance and ethylene-induced senescence in *Arabidopsis*. *Plant J*, **32**, 975–983.

Teige, M., Scheikl, E., Eulgem, T., Doczi, R., Ichimura, K., Shinozaki, K., Dangl, J.L. and Hirt, H. (2004) The MKK2 pathway mediates cold and salt stress signaling in *Arabidopsis*. *Mol Cell*, **15**, 141–152.

Tena, G., Asai, T., Chiu, W.-L. and Sheen, J. (2001) Plant mitogen-activated protein kinase signaling cascades. *Cur Opin Plant Biol*, **4**, 392–400.

Tena, G. and Renaudin, J.-P. (1998) Cytosolic acidification but not auxin at physiological concentration is an activator of MAP kinases in tobacco cells. *Plant J*, **16**, 173–182.

Torres, M.A. and Dangl, J.L. (2005) Functions of the respiratory burst oxidase in biotic interactions, abiotic stress and development. *Cur Opin Plant Biol*, **8**, 397–403.

Ulm, R., Ichimura, K., Mizoguchi, T., Peck, S.C., Zhu, T., Wang, X., Shinozaki, K. and Paszkowski, J. (2002) Distinct regulation of salinity and genotoxic stress responses by *Arabidopsis* MAP kinase phosphatase 1. *EMBO J*, **21**, 6483–6493.

van Loon, L.C., Geraats, B.P.J. and Linthorst, H.J.M. (2006) Ethylene as a modulator of disease resistance in plants. *Trends Plant Sci*, **11**, 184–191.

Vaudry, D., Stork, P.J.S., Lazarovici, P. and Eiden, L.E. (2002) Signaling pathways for PC12 cell differentiation: making the right connections. *Science*, **296**, 1648–1649.

Wang, H., Ngwenyama, N., Liu, Y., Walker, J.C. and Zhang, S. (2007) Stomatal development and patterning are regulated by environmentally responsive MAP kinases in *Arabidopsis*. *Plant Cell*, **19**, 63–73.

Wang, K.L.-C., Li, H. and Ecker, J.R. (2002) Ethylene biosynthesis and signaling networks. *Plant Cell*, **14**, S131–S151.

Widmann, C., Gibson, S., Jarpe, M.B. and Johnson, G.L. (1999) Mitogen-activated protein kinase: conservation of a three-kinase module from yeast to human. *Physiol Rev*, **79**, 143–180.

Wurgler-Murphy, S.M., Maeda, T., Witten, E.A. and Saito, H. (1997) Regulation of the *Saccharomyces cerevisiae* HOG1 mitogen-activated protein kinase by the PTP2 and PTP3 protein tyrosine phosphatases. *Mol Cell Biol*, **17**, 1289–1297.

Yang, K.-Y., Blee, K.A., Zhang, S. and Anderson, A.J. (2002) Oxycom^{TM} treatment suppresses *Pseudomonas syringae* infection and activates a mitogen-activated protein kinase pathway. *Physiol Mol Plant Pathol*, **61**, 249–256.

Yang, K.-Y., Liu, Y. and Zhang, S. (2001) Activation of a mitogen-activated protein kinase pathway is involved in disease resistance in tobacco. *Proc Natl Acad Sci USA*, **98**, 741–746.

Yap, Y.-K., Kodama, Y., Waller, F., Chung, K.M., Ueda, H., Nakamura, K., Oldsen, M., Yoda, H., Yamaguchi, Y. and Sano, H. (2005) Activation of a novel transcription factor through phosphorylation by WIPK, a wound-induced mitogen-activated protein kinase in tobacco plants. *Plant Physiol*, **139**, 127–137.

Yoshioka, H., Numata, N., Nakajima, K., Katou, S., Kawakita, K., Rowland, O., Jones, J.D.G. and Doke, N. (2003) *Nicotiana benthamiana* gp91phox homologs NbrbohA and NbrbohB participate in H_2O_2 accumulation and resistance to *Phytophthora infestans*. *Plant Cell*, **15**, 706–718.

Yuasa, T., Ichimura, K., Mizoguchi, T. and Shinozaki, K. (2001) Oxidative stress activates ATMPK6, an *Arabidopsis* homologue of MAP kinase. *Plant Cell Physiol*, **42**, 1012–1016.

Zarembinski, T.I. and Theologis, A. (1994) Ethylene biosynthesis and action: a case of conservation. *Plant Mol Biol*, **26**, 1579–1597.

Zhang, A., Jiang, M., Zhang, J., Tan, M. and Hu, X. (2006) Mitogen-activated protein kinase is involved in abscisic acid-induced antioxidant defense and acts downstream of reactive oxygen species production in leaves of maize plants. *Plant Physiol*, **141**, 475–487.

Zhang, S., Du, H. and Klessig, D.F. (1998) Activation of tobacco SIP kinase by both a cell wall-derived carbohydrate elicitor and purified proteinaceous elicitins from *Phytophthora* spp. *Plant Cell*, **10**, 435–449.

Zhang, S. and Klessig, D.F. (1997) Salicylic acid activates a 48 kD MAP kinase in tobacco. *Plant Cell*, **9**, 809–824.

Zhang, S. and Klessig, D.F. (1998a) The tobacco wounding-activated MAP kinase is encoded by *SIPK*. *Proc Natl Acad Sci USA*, **95**, 7225–7230.

Zhang, S. and Klessig, D.F. (1998b) *N* resistance gene-mediated de novo synthesis and activation of a tobacco MAP kinase by TMV infection. *Proc Natl Acad Sci USA*, **95**, 7433–7438.

Zhang, S. and Klessig, D.F. (2001) MAPK cascades in plant defense signaling. *Trends Plant Sci*, **6**, 520–527.

Zhang, S. and Liu, Y. (2001) Activation of salicylic acid-induced protein kinase, a mitogen-activated protein kinase, induces multiple defense responses in tobacco. *Plant Cell*, **13**, 1877–1889.

Zhang, S., Liu, Y. and Klessig, D.F. (2000) Multiple levels of tobacco WIPK activation during the induction of cell death by fungal elicitins. *Plant J*, **23**, 339–347.

Annual Plant Reviews (2008) **33**, 137–162
doi: 10.1111/b.9781405160025.2008.00005.x

www.interscience.wiley.com

Annual Plant Reviews

Chapter 5

CALCIUM SIGNALS AND THEIR REGULATION

Zhen-Ming Pei[1] and Simon Gilroy[2]

[1] *Department of Biology, Duke University, Box 90338, Durham, NC 27708, USA*
[2] *Botany Department, University of Wisconsin, Birge Hall, 430 Lincoln Drive, Madison, WI 53706, USA*

Abstract: Calcium is recognized as a ubiquitous cellular regulator and changes in cytosolic Ca^{2+} levels are known to be involved in plant processes as diverse as response to cold shock, hormone action, and touch signaling. The specificity of subsequent cellular response appears to be encoded, at least in part, in the temporal and spatial dynamics of the Ca^{2+} change and the spectrum of Ca^{2+}-responsive proteins expressed in the cell. In this chapter we review the evidence for the informational content of the Ca^{2+} signal in plant cells and discuss our current knowledge of the channel and pump systems that shape these Ca^{2+} changes. We will also use the Ca^{2+}-dependent Nod-factor signaling system that underlies rhizobium:plant interactions to highlight our current understanding of how the Ca^{2+} change may be transduced to the appropriate cellular response. In addition we use the example of Ca^{2+} uptake from soil and its subsequent translocation through the plant to show how, in addition to its role in cell signaling, this ion can act as a long-range messenger, integrating the cellular activity of the stomatal guard cell with water and nutrient uptake activities in the root. The theme that emerges from these examples is that although Ca^{2+} is involved in regulating a remarkably diverse array of plant processes, the more we understand about the systems that generate and respond to these Ca^{2+} changes, the more we realize just how sophisticated the molecular networks that encode and decode them turn out to be.

Keywords: ACA; calcium; CAS; CAX; ECA; guard cell; Nod factor; signal transduction

5.1 Introduction

Changes in cytosolic Ca^{2+} have been linked to the regulation of diverse plant responses ranging from hormone perception to the direction of pollen tube growth. Indeed, Ca^{2+} signals have been implicated in such a wide range of plant processes that the question has been raised as to whether such a ubiquitous cellular regulator could actually encode any useful information for the plant. This has led to the proposition that Ca^{2+} changes may simply act as an "on switch," nonspecifically activating the cell. In this scenario, the specific information about the stimulus is carried by other cellular regulators and Ca^{2+} merely acts to facilitate the response. However, recent work has begun to establish the molecular identity of the receptors that trigger Ca^{2+} changes, the pumps and channels that effect the change and the Ca^{2+}-dependent response systems that decode the alteration in Ca^{2+} to a cellular response. This work suggests that there is indeed information conveyed in the spatial and temporal character of the Ca^{2+} change in some systems, i.e., in the Ca^{2+} "signature" of the stimulus. Such information encoding would provide the cell with a complex Ca^{2+} signaling network with which to integrate many stimuli to appropriate cellular response. In this chapter, we will therefore outline some of the arguments for and against stimulus-specific Ca^{2+}-dependent signaling. We will also describe some of the newest insights into the molecular machinery that may decode the changes in Ca^{2+} and how these insights have come together to help us understand processes as diverse as how stomatal guard cells integrate transpiration and Ca^{2+} uptake, to how symbiotic signaling aids in the establishment of the rhizobium:plant symbiosis. Such studies are revealing the elegance with which information can be conveyed to the cell through its perhaps most ubiquitous second messenger, the Ca^{2+} ion.

5.2 Ca^{2+} as a second messenger in plants: of signatures and switches

In animal cells it is well established that the spatial and temporal kinetics of changes in cytosolic Ca^{2+} carry information that differentially controls cellular response (Iino, 2007). For example, in B lymphocytes the transcriptional regulators NF$_{\kappa}$-B and c-Jun N-terminal kinase (JNK) are selectively activated by a large transient increase in Ca^{2+}, whereas, in the same cells, upregulation of the NFAT transcription factor requires a sustained, low-level increase in the same ion (Dolmetsch *et al.*, 1997). Similarly, oscillations in Ca^{2+} have been shown to decrease the threshold for such transcriptional activation providing a further functional role for such Ca^{2+} "spiking" (Dolmetsch *et al.*, 1998). In addition, the locale of ion influx into the cell can also be critical for the Ca^{2+}-dependent triggering of only a specific subset of transcription factors from the range of Ca^{2+}-sensitive transcriptional machinery in the cell (West *et al.*, 2001; Kornhauser *et al.*, 2002). In plants, the precise role for stimulus-specific Ca^{2+} signatures is less clear, in large part due to a lack of equivalent in-depth analysis to that conducted on animal systems. While it is well established that Ca^{2+} is acting as a key regulator of plant cell function, several observations argue that this ion may play the role of a generic facilitator for cell function, with the specificity of response being encoded in parallel by some other signaling molecule (Scrase-Field and Knight, 2003; Plieth, 2005). For example, plants show a suite of transcriptional changes to cold stress, a stimulus which also triggers a Ca^{2+} transient thought to be part of the cold signaling system (e.g., Knight *et al.*, 1996). However, the response to cold stress is actually to the rate of chilling rather than the absolute temperature attained. As chilling rate falls, the size of the related Ca^{2+} transient declines (Plieth *et al.*, 1999), yet the cold-induced transcriptional profile remains the same (Knight *et al.*, 1996; Tahtiharju *et al.*, 1997). Thus, response does not appear tightly coupled to a precise pattern of the Ca^{2+} change in this case. Similarly, Ca^{2+}-dependent activation of several genes such as *KIN1/2* and *LTI78* occurs in response to both low temperature and osmotic stress despite these two stimuli showing very different Ca^{2+} signatures (Knight *et al.*, 1997; Tahtiharju *et al.*, 1997; Boyce *et al.*, 2003). Such observations have been used to argue that Ca^{2+} may be acting more like a switch where the precise signature of the change is not as important as the increase exceeding an activation threshold. The complex spatial and temporal kinetics seen in plant Ca^{2+} changes would then likely reflect the inherent oscillatory character of the Ca^{2+} regulatory network, with subtle mismatches in the response times and kinetics of channels and pumps causing oscillatory changes in Ca^{2+} that do not inherently carry information but are just the noise in the biochemistry of the Ca^{2+} homeostasis system.

However, in perhaps the most intensively studied Ca^{2+}-dependent signaling system in plants, the stomatal guard cell, data strongly support the animal paradigm that in addition to its role as a biochemical switch, Ca^{2+} changes can carry information about the stimulus that elicited them. Thus, abscisic acid (ABA), external Ca^{2+} (see below) and H_2O_2 all elicit Ca^{2+} transients in guard cells that have stimulus-specific signatures (McAinsh *et al.*, 1990, 1992, 1995; Gilroy *et al.*, 1991; Pei *et al.*, 2000; Han *et al.*, 2003). Application of molecules such as inositol-1,4,5-trisphophate, cADP-ribose, and sphingosoine-1-phosphate that are thought to trigger Ca^{2+} release from internal stores in response to these stimuli are all also capable of eliciting Ca^{2+} release and the subsequent biological response (stomatal closure) on their own, i.e., in the absence of the initial signal that is thought to lead to their production (Gilroy *et al.*, 1990; Leckie *et al.*, 1998; Ng *et al.*, 2001). Further, by clamping cytosolic Ca^{2+} levels, Allen and colleagues (2001) were able to show that in *Arabidopsis*, short-term stomatal closure is triggered by elevated Ca^{2+} in a switch-like fashion, whereas long-term steady-state closure requires Ca^{2+} oscillations within a defined range of frequency, duration, and amplitude. In the *det3* and *gca2* mutants the ABA- and H_2O_2-related Ca^{2+} responses are disrupted and stomatal closure is inhibited, whereas artificially imposing Ca^{2+} oscillations that mimic the naturally occurring ones elicits normal stomatal closure (Allen *et al.*, 2000, 2001). Thus, for guard cells, both the switch and the Ca^{2+} signature hypothesis have strong experimental backing. Defining the role of Ca^{2+} as switch versus subtle information carrier in other systems must await an equivalent in-depth analysis, although as data accrue on Ca^{2+} spiking phenomena in root hairs responding to rhizobial Nod factors, a similar picture of likely information encoding is also beginning to emerge (see below).

It is also important to remember that how stimuli are presented to the plant may well have profound effects on the ensuing Ca^{2+} signaling. Thus, modeling and empirical verification in animal cells has shown that when stimuli are given as periodic changes that more closely mimic how the signal would normally be seen by the cell in nature, they elicit more robust Ca^{2+} oscillations than when applied continuously, as occurs in many experimental protocols (Prank *et al.*, 2005). It may well be that experimental design and the influence of the experimental environment can alter or obscure the fine detail of Ca^{2+} changes that are of importance to the cell, a theme we will expand upon later in this chapter.

In tip-growing cells it is clear that the spatial and temporal changes in Ca^{2+} are critical for encoding response, i.e., localizing growth. Thus in root hairs, pollen tubes, algal rhizoids, and even fungal hyphae, a tip-focused Ca^{2+} gradient has been characterized that is critical for the restriction of growth to the apex of the elongating cell (reviewed in Bibikova and Gilroy, 2000). The Ca^{2+} gradient is localized to the apical 5–10 μm of these cells where it is thought to promote the targeted secretion of new wall materials and insertion of new membrane required to support tip growth. In addition to facilitating membrane fusion, the elevated Ca^{2+} is thought to integrate the activity of a host of proteins that maintain the secretory activity at the tip including modulators of cytoskeletal dynamics such as ADF and villin, the annexins, Ca^{2+}-dependent protein kinases, and phospholipid modulating enzymes (reviewed in Malho *et al.*, 2006). In pollen tubes, the gradient has been characterized as showing complex oscillatory behavior that correlates with the periodic nature of growth, although in this case the maximum of the Ca^{2+} gradient actually lags maximal growth by a few seconds (Holdaway-Clarke *et al.*, 1997; Messerli *et al.*, 1999). This observation has led to the suggestion that stretch-activated channels at the apex of the tube may be responsible for generating the Ca^{2+} influx that generates the gradient (Dutta and Robinson, 2004). However, the system regulating this influx appears more complex than a simple mechanical feedback oscillator regulating channel activity as there is evidence for roles of many other factors such as the rop family of monomeric G-proteins (Li *et al.*, 1999), rop interacting proteins (the RICs; Gu *et al.*, 2005), and reactive oxygen species generated through the NADPH oxidase system (Foreman *et al.*, 2003) in regulating the gradient formation. Importantly, the tip-focused gradient appears functionally important for driving growth. Thus, its dissipation through manipulations such as application of Ca^{2+} channel blockers or Ca^{2+} buffers leads to immediate growth arrest (e.g., Speksnijder *et al* 1989; Taylor *et al* 1996; Wymer *et al* 1997; De-Rejuiter *et al* 1998), whereas experimentally redirecting the gradient changes the growth direction to the site of the new gradient (Malho and Trewavas, 1996; Bibikova *et al.*, 1997). Thus, it seems clear for these tip-growing cells that the cytosolic Ca^{2+} gradient carries important information as to the direction of growth.

5.3 Ca^{2+} channels and pumps

Having established that signaling-related changes in Ca^{2+} do regulate plant cell activities, whether acting as a switch or carrying significant stimulus-related information for the cell, the next obvious question is how do stimuli trigger these Ca^{2+} changes in the first place. In recent years, there have been key advances in our understanding of how this is brought about by the major players in this signal generation system, the Ca^{2+} permeable channels.

Electrophysiology has identified a large range of conductances (putative channels) that are permeable to Ca^{2+} in the plasma membrane, endoplasmic reticulum, and the chloroplast, vacuolar and nuclear membranes of plant cells (reviewed in White, 2000). In addition, using correlation of expression pattern, empirically determined electrophysiological profiles and sequence homologies to known Ca^{2+} conductances from other kingdoms, a series of candidate plant Ca^{2+} channels have been deduced such as the annexin gene family (putative hyperpolarization-activated Ca^{2+} channel) and AtSKOR (putative depolarization-activated outward-rectifying Ca^{2+}-permeable channel; White *et al.*, 2002). Even so, we still have remarkably few channels that have been defined at the molecular level as supporting signaling-related changes in Ca^{2+} flux. This remains in large part due to difficulties in heterologously expressing putative Ca^{2+} channels to robustly assign function. However, plants with mutations in the CNGC (cyclic nucleotide-gated channels) show loss of Ca^{2+} conductances (Ali *et al.*, 2007) and these genes appear to encode Ca^{2+} channels when assayed via expression in yeast (Ali *et al.*, 2006). Also in the case of the GLR plant glutamate receptor/channel homologs, knockout mutants have a loss of the corresponding cytosolic Ca^{2+} signal elicited in response to glutamate (Qi *et al.*, 2006). Thus the CNGCs and GLRs are extremely strong candidates for plasma membrane localized stimulus responsive Ca^{2+} influx channels. Although the putative channel LCT1 has been expressed in yeast and determined to conduct Ca^{2+} (Clemens *et al.*, 1998), its subcellular locale in the plant and potential role in Ca^{2+} signaling/transport systems remains to be uncovered.

At the vacuole, two Ca^{2+} release channels have been defined through electrophysiology (VVCa, vacuolar voltage-gated Ca^{2+} channel and SV, the slow vacuolar channel), although both these activities may actually reflect characterization of the same SV channel (Pottosin and Schönknecht, 2007). Using a combination of proteomics (Carter *et al.*, 2004; Szponarski *et al.*, 2005) and analysis of knockout mutants (Peiter *et al.*, 2005), the SV channel has been defined as the product of AtTPC1 in *Arabidopsis*. Reports visualizing TPC1 localization at the plasma membrane most likely reflect miss-targeting of the GFP-tagged protein as this does appear to be a vacuolar channel. The SV channel is known to show Ca^{2+} activation suggesting that it could act as a Ca^{2+}-induced Ca^{2+} release channel (CICR channel) that would amplify a small initial cytosolic Ca^{2+} signal via triggering release from the vacuole. Inositol-1,4,5-trisphosphate- and cyclic ADP-ribose-gated Ca^{2+} conductances have also been observed from the vacuole suggesting a role for vacuolar Ca^{2+} release in these second messenger-mediated signaling systems. However, the lack of homologs of the respective animal channels in plant genomes has severely limited progress in identifying the genes for these putative plant ligand-gated vacuolar channels.

Channel activity is clearly the key to generating Ca^{2+} entry into the cell to trigger downstream response. However, removal and sequestration of the elevated Ca^{2+} is equally critical to establishing the spatial and temporal character of the Ca^{2+} signal, i.e., to the features that likely carry stimulus-specific information. To this end, plant cells employ myriad Ca^{2+}-transporting pumps, ranging from ATPases to cotransporters, that regulate the extent of the Ca^{2+} change. Just as for the channels, our increasing appreciation for how these pumps operate and are regulated is beginning to reveal how cells can impose complex regulatory patterns on the Ca^{2+} messenger.

At the plasma and internal membranes P-type ATPases of the ACA and ECA classes (P2A, P2B and possibly P5 classes, Baxter *et al.* 2003) are involved in removing Ca^{2+} from the cytosol. These Ca^{2+} transporting ATPases form a large family with 3 P2A and 11 P2B members in the rice genome (4 and 10 respectively in *Arabidopsis*). These pumps reside in many different membranes. For example, an ER location for AtECA1 (P2A) and AtACA4 (P2B) has been confirmed (Harper *et al.*, 1998; Hong *et al.*, 1999), whereas AtACA4 and 11 (P2B) have been localized to the vacuole (Geisler *et al.*, 2000; Lee *et al.*, 2007) and AtACA8 (P2B) to the plasma membrane (Bonza *et al.*, 2000). The P2B pumps are regulated by calmodulin and an N-terminal autoinhibitory domain (Carafoli and Brini, 2000; Geisler *et al.*, 2000; Sze *et al.*, 2000). In mammals, the P2A enzymes are regulated by the phospholamban peptides. However, it is unclear if the phospholamban regulatory system exists in plants and the conserved phospholamban binding site is missing in the plant P2A proteins.

The molecular identification of these transporters has allowed the testing of their functional significance through a combination of expression and mutant analysis. Thus, plants defective in AtECA1 show hypersensitivity to $Mn^{2+}$and grow poorly on low Ca^{2+} media, suggesting a role in Ca^{2+} homeostasis and heavy metal transport (Wu *et al.*, 2002). Similarly, mutants in AtACA9 have been shown to affect pollen tube function, suggesting a possible role in shaping the tip-focused Ca^{2+} gradient that drives pollen tube growth as described above. At the vacuole, Ca^{2+} uptake mediated by the high affinity P-type ATPases (Sze *et al.*, 2000) is supplemented by relatively low affinity but high capacity H^+/Ca^{2+} cotransporters of the CAX family (Shigaki and Hirschi, 2006). Manipulation of CAX expression leads to enhanced sensitivity to stresses such as altered ion levels in the growth media, an effect likely directly related to the role of these transporters in Ca^{2+} homeostasis (e.g., Shigaki *et al.*, 2002).

The combination of these diverse channel and pump systems coupled to less well characterized contributions from mitochondria and plastids and even the nucleus (Xiong *et al.*, 2006) provides the plant with the extremely flexible system required to generate and shape Ca^{2+} signals in response to the wide range of stimuli the plant must sense. Clearly such a complex system to encode information into the Ca^{2+} change requires an equally sophisticated system to subsequently decode the information to specific downstream cellular responses. Recent work from both animal and plant systems is now revealing that the subcellular localization and Ca^{2+}-dependent biochemical properties of many proteins may hardwire the cell to be able to respond to subtle changes in the dynamics of a Ca^{2+} change with highly specific biochemical outputs. For example, approximately 1% of the proteins in the *Arabidopsis* genome contain the EF-hand Ca^{2+}-binding motif (Day *et al.*, 2002), and these proteins range from protein kinases and the NADPH oxidases responsible for ROS production, to a huge number of proteins of unknown function. It is impossible to cover all the Ca^{2+} response systems that have been defined in plants and so in the next section we will describe some insights into the diversification of the plant Ca^{2+} response system drawn from the huge number of calmodulin (CaM) and CaM-like proteins (CMLs) that may shed some light onto how Ca^{2+} signals are processed. The readers are referred to Chapter 6 for a thorough discussion of Ca^{2+} sensing systems in plants.

5.4 Decoding the Ca^{2+} signal

Calmodulin is perhaps the most ubiquitous Ca^{2+}-responsive regulator in eukaryotic cells. Binding of Ca^{2+} induces a conformational change in the protein that then modulates the activity of other target proteins. *Arabidopsis* has 7 CaM genes (encoding 4 different protein isoforms) and 50 calmodulin-like (CML) genes. Ca^{2+}-binding occurs in the EF-hand motifs within these proteins and the large divergence in these domains amongst these proteins suggests a wide range of sensitivities to Ca^{2+} changes. Indeed, the CMLs show a large degree of differential expression at the levels of cell type, developmental stage and stimulus response (McCormack *et al.*, 2005). This diversity in the CML protein structure, Ca^{2+}-binding site, and expression patterns suggests that the CMLs along with the CaMs may help decode responses to a wide range of cellular stimuli. For example, despite the very high sequence similarity between the four *Arabidopsis* CaM isoforms, they interact with NAD kinase, cyclic nucleotide-gated ion channels, and a kinsein-like protein with different efficiencies (Liao *et al.*, 1996; Reddy *et al.*, 1999; Kohler and Neuhaus, 2000), implying differential regulatory capacities for each of these substrates. There are also a range of CMLs seen in other plants that have no clear homologs in *Arabidopsis*, with for example hexaploid wheat having at least 13 CaMs arranged in four distinct subfamilies (Yang *et al.*, 1996), reinforcing the idea that plants have invested heavily in CaM/CML diversity to expand the flexibility of the biochemical systems responding to Ca^{2+} signals.

CaM action has been implicated in an enormous range of plant responses, largely from inferences from the effects of CaM antagonists (Zielinski, 1998). However, there are many molecular studies confirming the central role of these proteins. For example, overexpression of the soybean CaMs ScaM-4 and ScaM-5 can trigger a stress response that is normally seen to be Ca^{2+} dependent (Heo *et al.*, 1999). This observation implies either that these Ca^{2+}-responsive elements are limiting for subsequent response, or that increasing their level has raised the Ca^{2+} sensitivity of the system to respond to resting Ca^{2+} levels.

Microdomains of activated calmodulin have also been reported that do not simply reflect sites of elevated Ca^{2+} levels (e.g., Torok *et al.*, 1998). In addition, the activity of CaM may be regulated by the proteins with which it interacts (Zielinski, 1998), suggesting that the cellular microenvironment is likely to alter CaM responsiveness to Ca^{2+}. Coupling these effects to locally increased levels of CaM enhancing the activity of low affinity CaM response elements may allow the cell to fine-tune its Ca^{2+} response system with high spatial resolution.

In animal cells, CaM has also emerged as an important element in the channeling of responses dependent on the nature of the origin of the Ca^{2+} change. Thus in neurons, Ca^{2+} influx through L-type channels triggers phosphorylation and activation of the CREB transcription factor. Ca^{2+} entry through non-L-type channels causes a Ca^{2+} increase of similar magnitude but fails to elicit the same signaling cascade leading to CREB activation (Dolmetsch *et al.*, 2001; West *et al.*, 2001). The spatial restriction in signaling appears to arise from a CaM molecule bound via an "IQ" motif on the inner surface of the L-type channel pore. Specific activation of this CaM triggers the downstream events, effectively limiting signaling to the Ca^{2+} flowing through these specific channels. We must await similar in-depth analysis of plant response elements to see whether similar spatial control of information flow is operating in plant Ca^{2+}-dependent signaling pathways.

5.5 Ca^{2+} and Nod-factor signaling: a role for kinases in decoding the Ca^{2+} signal?

Protein kinases present another ubiquitous Ca^{2+} response component where we are beginning to understand how biochemistry can interpret the complexities of the Ca^{2+} signal (Harper *et al.*, 2004). For example, in animal cells, CaM-activated kinases have been shown to be capable of decoding the frequency information in Ca^{2+} spiking phenomena. Thus, CaM kinase II has been shown to be activated by specific frequencies of Ca^{2+} spiking, with different subunit isoforms showing different frequency responses (De Koninck and Schulman, 1998). In plants a similar mechanism may well be hinted at in the role of DMI3, a Ca^{2+}- and calmodulin-dependent protein kinase (CCaMK). This kinase appears essential in responding to the Ca^{2+} spiking elicited by Nod-factor action during the establishment of the rhizobial nitrogen-fixing symbiosis as well as in mychoorrhizal interactions (Ane *et al.*, 2004; Levy *et al.*, 2004; Mitra *et al.*, 2004a,b; Gleason *et al.*, 2006; Tirichine *et al.*, 2006). Indeed, the Ca^{2+} signaling system behind Nod-factor signaling serves as an excellent example of how we are beginning to place Ca^{2+} changes within a firm molecular framework of signal generators and response components.

Thus, Ca^{2+} plays an essential role in the communication between legumes and nitrogen-fixing bacteria (rhizobia) in forming a symbiosis (nodule; Long, 1996). This symbiotic relationship starts with interchanging signals. Legume plants secrete flavonoids from the root to the soil, which trigger soil-born rhizobia to produce host-specific lipochitooligosaccharides, Nod factors. Nod factors then evoke a series of signaling events in plant roots, leading to rhizobia infection and nodule development (Riely *et al.*, 2004; Geurts *et al.*, 2005). The earliest detectable events in Nod-factor perception are depolarization of the plasma membrane potential, ion fluxes, and $[Ca^{2+}]$ oscillations (spiking; Oldroyd and Downie, 2004). Through genetic analysis of legume mutants defective in the Nod-factor signaling pathway, several key components have been identified (Stacey *et al.*, 2006), which include putative cell surface receptors for Nod factor, putative ion channels possibly conducting K^+ and/or Ca^{2+} ions, cytosolic Ca^{2+} sensors, and transcription factors that decode $[Ca^{2+}]$ changes. Although the exact biochemical mechanisms underlying the regulation and interaction of these components have not yet been elucidated, it seems that the Nod-factor signaling pathway represents the best molecularly defined Ca^{2+}-mediated pathway in plants, in which an external signal is converted into cytosolic $[Ca^{2+}]$ changes and transduced further to transcriptional regulation (Fig. 5.1; Riely *et al.*, 2004; Geurts *et al.*, 2005; Oldroyd and Downie, 2006).

Nod factor induces a transient depolarization of the plasma membrane electrical potential in root hairs within 1 min (Ehrhardt *et al.*, 1992; Radutoiu *et al.*, 2003). This depolarization results from Ca^{2+} influx and Cl^- efflux across the plasma membrane. The Ca^{2+} influx is required for the Cl^- efflux as well as the membrane depolarization (Felle *et al.*, 1999). The plasma membrane potential is likely repolarized by an initial K^+ efflux and additional long-lasting ion fluxes which have not yet been well characterized (Felle *et al.*, 1996, 1998).

Several methods have been used to detect Nod-factor-induced $[Ca^{2+}]$ changes. The most widely used method is microinjection of the Ca^{2+}-sensitive fluorescent dyes Oregon green-BAPTA (Ehrhardt *et al.*, 1996; Wais *et al.*, 2000; Shaw and Long, 2003; Miwa *et al.*, 2006a) and Fura-2 (Cardenas *et al.*, 1999; Gehring *et al.*, 1997). Although direct loading of these dyes into plant cells is very difficult, acid loading of Indo-1 has been used to monitor Nod-factor-evoked $[Ca^{2+}]$ changes (de Ruijter *et al.*, 1998). Recently, the GFP-based Ca^{2+} indicator, cameleon, has been employed to monitor $[Ca^{2+}]_i$ changes in various cell types in response to Nod factor (Miwa *et al.*, 2006b). Consistent with Nod-factor-evoked membrane depolarization, the initial increase in $[Ca^{2+}]$ is biphasic, with a rapid initial increase followed by a long-lasting decline of the $[Ca^{2+}]$ level (Ehrhardt *et al.*, 1996; Wais *et al.*, 2000; Shaw and Long, 2003; Miwa *et al.*, 2006a). Based on the pharmacological studies, the initial $[Ca^{2+}]$

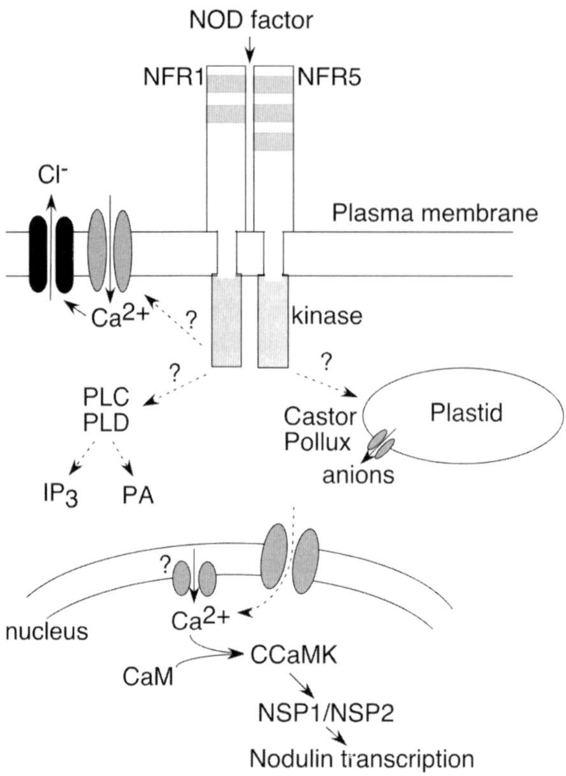

Figure 5.1 Simplified model of Nod-factor signaling. The NFR1 and 5 receptor kinases bind to the Nod factor, eliciting phosphorylation of intracellular messengers such as phospholipases that then transduce the signal to the nucleus, with the involvement of the Castor/Pollux putative plastid anion channel. In a nucleoporin-dependent fashion, this relay would then trigger nuclear Ca^{2+} release, activation of CCaMK and activation of transcription via the GRAS transcriptional regulators NSP1 and 2. Dashed lines represent activities yet to be defined. CaM, calmodulin; CCaMK, Ca^{2+} and Calmodulin binding protein kinase; IP_3, inositol-1,4,5-trisphosphate; PA, phosphatidic acid.

increase may result from Ca^{2+} influx across the plasma membrane, while the gradual $[Ca^{2+}]$ decline likely is due to transport of Ca^{2+} out of the cytosol via Ca^{2+} exchangers or pumps. Further studies have confirmed that the Ca^{2+} influx may begin at the cell periphery and move inward toward the nucleus (Shaw and Long, 2003).

Following the initial rapid $[Ca^{2+}]$ increase, oscillations in $[Ca^{2+}]$, known as Ca^{2+} spiking, occur 10 to 15 min after Nod-factor addition (Ehrhardt *et al.*, 1996; Shaw and Long, 2003; Miwa *et al.*, 2006a). Repetitive Ca^{2+} spikes originate in the nuclear area of the cell and propagate as a wave tip-ward (Shaw and Long, 2003), which is opposite to the initial Ca^{2+} influx. Several lines of evidence suggest that the initial Ca^{2+} increase and Ca^{2+} spiking are not correlated. The concentrations of Nod factor required for Ca^{2+} flux and spiking differ greatly, i.e., the concentration required for Ca^{2+} spiking is two to three orders of magnitude lower than that for Ca^{2+} flux (Shaw and Long, 2003). Ca^{2+} spiking can be induced by Nod-factor-like molecules without the induction of a Ca^{2+} flux (Walker *et al.*, 2000; Shaw and Long, 2003). These observations indicate that the ligand binding affinity and ligand specificity for the initial Ca^{2+} increase and the subsequent Ca^{2+} spiking differ significantly. For the initial Ca^{2+} increase, the receptor for Nod factor is very selective, and its binding affinity is low. For Ca^{2+} spiking, the Nod-factor receptor is less selective, and its apparent binding affinity is very high.

Several key molecular components responsible for Nod-factor-induced $[Ca^{2+}]$ changes have been identified by analyzing symbiosis-defective legume mutants (Riely *et al.*, 2004; Stacey *et al.*, 2006). Based on their effects on Nod-factor-induced $[Ca^{2+}]$ changes, these mutants can be classified into three groups: mutants lacking both the rapid Ca^{2+} flux and Ca^{2+} spiking, mutants defective only in Ca^{2+} spiking, and mutants that do not affect $[Ca^{2+}]$ changes. Mutations in lysine motif (LysM)-receptor-like kinases (RLKs), which have been proposed to be receptors for Nod factor, are defective in both the Ca^{2+} flux and Ca^{2+} spiking. These RLKs include *L. japonicus* NFR1 and NFR5 (Madsen *et al.*, 2003; Radutoiu *et al.*, 2003) and *M. truncatula* LYK3/4 and NFP (Limpens *et al.*, 2003; Arrighi *et al.*, 2006). It has been widely assumed that these LysM RLKs are localized to the plasma membrane and perceive Nod factor in the extracellular space (Oldroyd and Downie, 2004). Nevertheless, the subcellular location of these RLKs and the nature of Nod-factor binding to the LysM motif remain to be determined.

Mutations in putative ion channels with a conserved RCK domain, leucine rich repeat receptor-like kinases, and nucleoporins disrupt the Ca^{2+} spiking but not the Ca^{2+} flux. The putative ion channels include *M. truncatula* DMI1 (Ane *et al.*, 2004a) and *L. japonicus* CASTOR and POLLUX (Imaizumi-Anraku *et al.*, 2005). CASTOR and POLLUX are shown to be localized to the plastid, while DMI1 targets to the nuclear envelope (Riely *et al.*, 2007), possibly to the inner envelope. The nuclear envelope localization of DMI1 fits well with its proposed function for Ca^{2+} spiking that originates in the vicinity of the nucleus. The LRR RLKs include DMI2/SYMRK/NORK (Endre *et al.*, 2002; Stracke *et al.*, 2002). Several key issues remain to be resolved for these LRR RLK receptors, such as the subcellular localization, the nature of ligands and substrates, and their relationship with DMI1 ion channels. The nucleoporins include NUP133 (Kanamori *et al.*, 2006) and NUP85 (Saito *et al.*, 2007), and the subcellular localization of these putative nucleoporins has been determined to be in the nuclear envelope as expected.

As noted above, the chimeric Ca^{2+} calmodulin-dependent protein kinases (CCaMKs) have been proposed to be cytosolic Ca^{2+} sensors that perceive $[Ca^{2+}]$ changes resulting from Ca^{2+} flux and Ca^{2+} spiking (Levy *et al.*, 2004; Mitra *et al.*, 2004a,b). Consistent with a downstream role, mutations in these CCaMKs, DMI3 and SYM9, do not affect Nod-factor-evoked $[Ca^{2+}]$ changes at all (Mitra *et al.*, 2004a). CCaMKs have an autoinhibitory domain that negatively regulates kinase activity (Patil *et al.*, 1995; Harper and Harmon, 2005). It has been shown that the specific removal of the autoinhibition domain of DMI3 leads to the constitutive activation of the nodulation signaling pathway (Gleason *et al.*, 2006; Tirichine *et al.*, 2006).

Recent evidence from the Nod-factor system has also revealed that plants can count the number of Ca^{2+} transients and utilize this information as a threshold to eliciting response (Miwa *et al.*, 2006b). In *Medicago truncatula* it was observed that Nod-factor induction of ENOD11 transcription was dependent on accumulating 36 Ca^{2+} transients, irrespective of their frequency. Thus, in the case of nodulation evidence is accumulating for cellular mechanisms able to decode the information in the nuclear Ca^{2+} spiking.

It seems highly likely that with increases in $[Ca^{2+}]$, Ca^{2+} binds to CaM, which in turn interacts with DMI3, releases its autoinhibition, and switches the DMI3 kinase from an inactive state to an active state. Finally, DMI3 activates GRAS family transcription factors, NSP1 and NSP2 (Kalo *et al.*, 2005; Smit *et al.*, 2005), which mediate Nod-factor-induced transcriptional responses. Therefore, the Nod-factor signaling cascade would be as follows: cell surface LysM RLK receptor for Nod factor LYK3/4 → LRR receptor DMI2 → putative ion channel DMI1 → cytosolic Ca^{2+} sensor CCaMK DMI3 → GRAS family transcription factor NSP1/2 → nodulation (Fig. 5.1).

Although the major signaling framework for Nod factor is well established, the exact molecular mechanisms by which these key nodes are biochemically connected are still poorly understood (Fig. 5.1). To investigate the initial perception of Nod factor, the direct binding between Nod factor and the putative Nod-factor receptor LysM RLK should be analyzed, and the substrate of the receptor needs to be identified. Because the mutations in the receptors *nfr1*, *nfr5*, and *nfp* disrupt both Ca^{2+} flux and Ca^{2+} spiking and because no mutants have been isolated that are compromised in Ca^{2+} flux but not in spiking, it is likely that Ca^{2+} flux and Ca^{2+} spiking are also closely related. It is possible that Ca^{2+} flux could lead to Ca^{2+} spiking despite the fact that Ca^{2+} flux is not a prerequisite for Ca^{2+} spiking, as Ca^{2+} spiking can be induced independently of Ca^{2+} flux. Given that DMI1 ion channels are localized to the nuclear envelope, an additional Ca^{2+}-permeable channel in the plasma membrane is required for the rapid Ca^{2+} influx in response to Nod factor. Moreover, identification of the ligands for LRR RLKs, such as DMI2, would greatly improve our understanding of the interaction between Ca^{2+} flux and Ca^{2+} spiking.

With respect to the missing link in the Nod-factor signaling pathway, perception of Nod factor occurs at the plasma membrane, while the Ca^{2+} spiking is associated with the nucleus (Oldroyd and Downie, 2006). Clearly, a secondary messenger is needed to transduce the Nod-factor signal from the plasma membrane to the nucleus to activate Ca^{2+}-permeable channels that are located on the nuclear envelope, such as DMI1. Pharmacological studies have shown that Nod factor activates both phospholipase C and phospholipase D (den Hartog *et al.*, 2001) and inhibitors of both these enzymes block Nod-factor signaling (Engstrom *et al.*, 2002; Charron *et al.*, 2004). It would be interesting to test whether IP_3 activates DMI1 and mobilizes the Ca^{2+} stores within the nuclear envelope as seen in animal cells (Charron *et al.*, 2004; Gomes *et al.*, 2006). It might be also possible to monitor IP_3 changes in response to Nod factor using single-cell IP_3 imaging (Hirose *et al.*, 1999; Tang *et al.*, 2007). Finally, to understand the machinery responsible for Ca^{2+} spiking, the ion channel activity and ion selectivity of DMI1 need to be determined in vivo and in vitro.

The signaling pathway for rhizobial interactions shows how our understanding of plant Ca^{2+}-dependent signaling networks at the cellular level is advancing rapidly with the advent of molecular identification of these Ca^{2+}-responsive signaling components. However, cellular Ca^{2+} signaling must be integrated into the response status of the plant as a whole, allowing cellular activities to be coordinated across tissues and organs. In the next section we will look at how Ca^{2+} signaling can be entrained by features of whole-plant physiology such as Ca^{2+} uptake and transport and by the environment the plant is experiencing.

5.6 Ca^{2+} uptake and transport

Calcium is the most abundant divalent cation in higher plants and its essential nature as a plant nutrient has been recognized for over 100 years (Epstein, 1972; Kirkby and Pilbeam, 1984; Bush, 1995; Hepler, 2005). In roots, Ca^{2+} moves rapidly through the cortical extracellular space (apoplast) by diffusion and together with water enters the xylem in the root apical region (Clarkson, 1984; Hepler, 2005). In the xylem, Ca^{2+} is transported by mass flow and chromatographic movement along the Ca^{2+}-exchange sites in the cell wall (Clarkson, 1984). The xylem branches as it ascends the stem and permeates leaves and organs, and the growing tissues provide a "sink" for Ca^{2+} in the xylem. Thus, Ca^{2+} is transported from roots to shoots mainly through the apoplastic xylem. In contrast, the symplast is not an effective pathway for Ca^{2+} uptake and transport because the cytosolic free Ca^{2+} concentration is extremely low compared to extracellular free Ca^{2+} concentration (Sanders *et al.*, 2002; White and Broadley, 2003).

In plant tissues, a high proportion of the total Ca^{2+} is often located in the cell wall and at the exterior surface of the plasma membrane (Clarkson, 1984; Hepler, 2005). Thus, together with boron, Ca^{2+} is termed as an apoplastic element among 16 essential elements. In the cell wall, Ca^{2+} ions bind to pectins, which are polymers of galacturonic acid molecules and are very hydrophilic and soluble. Ca^{2+} then forms salt bridges with pectins to become pectic gels that are insoluble. The pectic gels occur primarily in a specialized region in the wall, called the middle lamella, which is shared by neighboring cells and cements them firmly together. Pectic gels are also found in the primary cell wall. Precise information on the concentration range of free extracellular Ca^{2+} is lacking (Sattelmacher, 2001), and data taken from the literature vary from

10 μM to 1000 μM (Sattelmacher, 2001; Roelfsema and Hedrich, 2002). Apart from apoplastic Ca^{2+} stores, Ca^{2+} ions are also stored in the vacuole, endoplasmic reticulum (ER), mitochondria, and possibly Golgi apparatus (Sze *et al.*, 2000; White and Broadley, 2003). The concentrations of free Ca^{2+} in the vacuole and ER are in the mM range.

The distribution of Ca^{2+} in the plant is affected by the rate of water transportation and evaporation (Clarkson, 1984; White and Broadley, 2003; Hepler, 2005). Calcium moves in relatively large amounts to highly transpiring old leaves, but much less to weakly transpiring young leaves. Overall, the Ca^{2+} content is higher in old leaves than in young leaves. In addition, Ca^{2+} cannot be remobilized from old leaves, i.e., Ca^{2+} is an immobile element. Consequently, Ca^{2+} deficiency symptoms often occur in the young and fast growing tissues. From the standpoint of Ca^{2+} supply, plants must downregulate transpiration when the Ca^{2+} unloading rate is high, and vice versa. To do so, plants need a system monitoring apoplastic Ca^{2+} as well as transpiration-facilitated Ca^{2+} unloading.

5.7 Sensing extracellular Ca^{2+}

Over 95% of transpirational water loss occurs from stomatal pores formed by pairs of guard cells on the leaf surface (Assmann, 1993; Schroeder *et al.*, 2001; Hetherington and Woodward, 2003). It has long been known that Ca^{2+} induces stomatal closure in vitro (MacRobbie, 1992). Extracellular Ca^{2+}-induced stomatal closure may represent a major mechanism by which plants regulate Ca^{2+} uptake and distribution. In stomatal guard cells, extracellular Ca^{2+} induces cytosolic $[Ca^{2+}]$ increases and oscillations that mediate stomatal closure (McAinsh *et al.*, 1995a). It has been shown that these extracellular Ca^{2+}-induced, cytosolic $[Ca^{2+}]$ increases are due to an influx of Ca^{2+} through Ca^{2+} channels (MacRobbie, 1992; McAinsh *et al.*, 1995a). Interestingly, through molecular identification of an extracellular Ca^{2+}-sensing receptor (CAS) from *Arabidopsis*, recent studies have shown that extracellular Ca^{2+} actually induces Ca^{2+} release from internal stores (Han *et al.*, 2003a).

CAS was isolated from a screen of an *Arabidopsis* cDNA library in mammalian cells using a Ca^{2+} imaging-based assay (Han *et al.*, 2003a). CAS localizes to the plasma membrane, exhibits low-affinity/high-capacity Ca^{2+}-binding, and mediates the extracellular Ca^{2+}-induced intracellular $[Ca^{2+}]$ increase in stomatal guard cells. Suppression of CAS activity disrupts these $[Ca^{2+}]$ increases as well as extracellular Ca^{2+}-induced stomatal closing. CAS is unique to plants and may represent the only plant cell surface receptor cloned so far that has been demonstrated to convert an external signal into $[Ca^{2+}]$ changes (Hetherington and Brownlee, 2004b; Reddy and Reddy, 2004).

Furthermore, it has been proposed that CAS is a cell surface receptor in the IP_3 signaling pathway, which mediates Ca^{2+} release from internal stores via a similar mechanism to that seen in animals (Berridge *et al.*, 2003; Han *et al.*, 2003; Tang *et al.*, 2007). In animal cells, external signals are perceived by cell surface receptors, which activate phospholipase C (PLC), increasing IP_3 concentration. IP_3 activates IP_3 receptors (IP_3R) in the ER, resulting in Ca^{2+} release. The Ca^{2+} ions are reabsorbed by the ER, leading to [Ca^{2+}] oscillations (Berridge *et al.*, 2003). In plants, while cell surface receptors and the IP_3R are unknown (Meijer and Munnik, 2003), several lines of evidence suggest that CAS might serve as a receptor triggering Ca^{2+} release (Tang *et al.*, 2007). First, PLC inhibitors block extracellular Ca^{2+}-induced intracellular[Ca^{2+}] increases in guard cells and mesophyll cells. Second, extracellular Ca^{2+} induces IP_3 generation in the leaves and PLC inhibitors eliminate the IP_3 production. Finally, both biochemical and single-cell imaging analyses of IP_3 have shown that CAS is required for extracellular Ca^{2+}-induced IP_3 production. Nevertheless, it is not known how CAS activates PLC in vivo.

5.8 Ca^{2+}, light, and circadian [Ca^{2+}] oscillations

The complex effects of the environment and status of the plant on Ca^{2+} signaling are well demonstrated by the effects of the circadian clock on Ca^{2+} responsiveness. Over 10 years ago, Johnson and colleagues (1995) reported that cytosolic [Ca^{2+}] displays circadian oscillations at the whole-plant level (Fig. 5.2a). The central issues remain, however, as to the origin as well as the physiological function of these circadian [Ca^{2+}] oscillations (Dodd *et al.*, 2005). It has been shown that the circadian [Ca^{2+}] oscillations occur in the cytoplasm but not in the nucleus, in contrast to Nod-factor-induced [Ca^{2+}]$_i$ spiking, and their phase varies between cell types (Wood *et al.*, 2001). Although it is not clear how different cells generate a specific phase for [Ca^{2+}] oscillations, overall the oscillations are controlled by photoperiod and light intensity (Love *et al.*, 2004).

A recent study has shown that diurnal (and possibly circadian) [Ca^{2+}] oscillations are controlled by soil Ca^{2+} levels and CAS activity (Fig. 5.2b; Tang *et al.*, 2007). Based on this study, the levels of soil Ca^{2+} and stomatal conductance act together to govern the oscillating amplitude, phase, and period of extracellular free Ca^{2+} concentration, which are then perceived by CAS and a yet to be identified Ca^{2+} influx channel, such as the guard cell ICa (Pei *et al.*, 2000a), and converted into [Ca^{2+}] oscillations. In other words, the oscillating amplitudes of extracellular [Ca^{2+}] and intracellular [Ca^{2+}] are controlled by soil Ca^{2+} levels and transpiration rates (Figs. 5.2c–5.2d), while their phases and periods are likely determined by stomatal conductance oscillations, which in turn are dictated by photoperiod and the central clock (Webb, 2003). The extracellular [Ca^{2+}] is determined by two opposite processes: supplying fresh Ca^{2+} to the apoplast primarily through transpiration and sequestrating it into both external and internal stores. For the sequestration, fresh Ca^{2+} ions bind to newly synthesized cell wall components, such as pectic acids, form Ca^{2+} oxalate, or move into internal stores (White and Broadley, 2003; Franceschi and Nakata, 2005; Hepler, 2005). Because most sequestrated Ca^{2+} ions become immobile, continuous Ca^{2+} supplies are needed and are

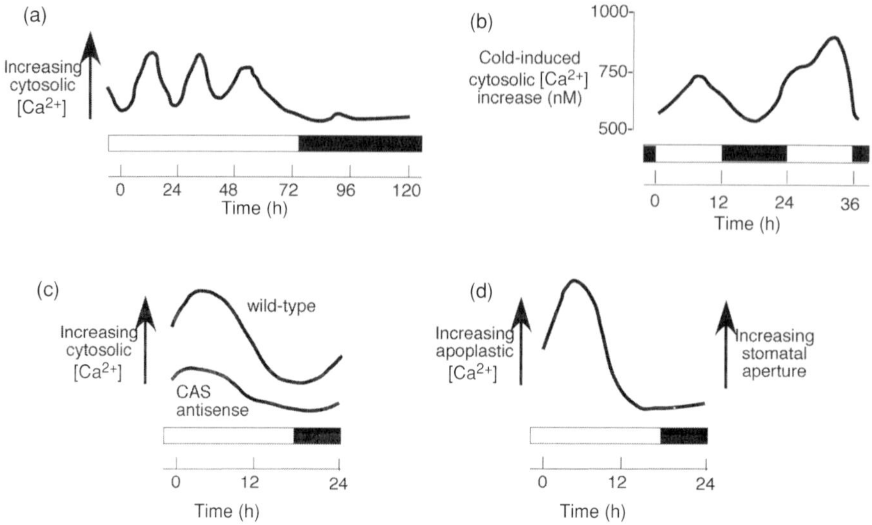

Figure 5.2 Diurnal rhythms in cytosolic Ca^{2+} and Ca^{2+} response. (a) Diurnal rhythm in cytosolic Ca^{2+} in the leaves of *Nicotiama tabacum* (Johnson *et al.*, 1995). (b) Changes in the extent of the Ca^{2+} transient in guard cells induced by cold shock depends upon when in the diurnal cycle the stimulus is applied (Dodd *et al.*, 2006). (c) Diurnal changes in cytosolic Ca^{2+} are depressed in plants expressing antisense to CAS (Tang *et al.*, 2007). (d) Diurnal changes in leaf extracellular Ca^{2+} are mirrored by changes in stomatal conductance (Tang *et al.*, 2007). All Ca^{2+} measurements were made using plants transformed with the Ca^{2+}-sensitive luminescent reporter aequorin. Black bars indicate darkness and white bars, periods of illumination. Data redrawn from Johnson *et al.* (1998), Dodd *et al.* (2006) and Tang *et al.* (2007).

likely to be the regulated step. Taken together, it seems that transpiration-mediated soil Ca^{2+} uptake and transport might provide a robust regulatory machinery that synchronizes the resting intracellular $[Ca^{2+}]$ throughout the plant in response to day and night changes.

Based on this model, it would be interesting to test how the central clock regulates cytosolic $[Ca^{2+}]$ oscillations, and whether manipulation of stomatal conductance oscillations affects these oscillations. For instance, CCA1/LHY and TOC1 are the well-characterized central oscillators, and the oscillating phase and period are altered in *cca1/lhy* and *toc1* mutants (Harmer *et al.*, 2001). Monitoring stomatal conductance and $[Ca^{2+}]$ oscillations simultaneously in these mutants could reveal their exact relationship. Similarly, analyzing stomatal conductance, Ca^{2+} uptake rate, and $[Ca^{2+}]$ oscillations in the extreme mutants, *spch* and *mute*, in which stomatal developmental is completely blocked (MacAlister *et al.*, 2007; Pillitteri *et al.*, 2007), could provide insight into the contribution of transpiration to circadian $[Ca^{2+}]$ oscillations.

The exact function of these circadian $[Ca^{2+}]$ oscillations remains to be determined (Dodd *et al.*, 2005). Several studies have suggested that $[Ca^{2+}]$ may act directly as a second messenger for light, and indirectly in the form of circadian [Ca^{2+}] oscillations as a key component in photoperiod signaling. Both red and blue light trigger cytosolic [Ca^{2+}] increases, and pharmacologically altering intracellular [Ca^{2+}] affects phytochrome-mediated gene expression (Shacklock *et al.*, 1992; Neuhaus *et al.*, 1993; Baum *et al.*, 1999; Lin and Shalitin, 2003). It has also been shown that Ca^{2+} is involved in the photoperiod floral induction of *Pharbitis nil* (Friedman *et al.*, 1989; Takeno, 1993; Szmidt-Jaworska *et al.*, 2006). Pharmacological agents that increase intracellular [Ca^{2+}] promote flowering, while agents that reduce [Ca^{2+}] delay flowering. Finally, changes in [Ca^{2+}] can shift the phase of circadian rhythms of several physiological processes (Kreps and Kay, 1997). Thus, it appears that cytosolic [Ca^{2+}] oscillations are likely related to outputs of the endogenous clock and photoperiod (Johnson, 2001; Dodd *et al.*, 2005), although their precise mechanisms are poorly understood. It is reasonable to presume that circadian [Ca^{2+}] oscillations may function in a subset of the processes that are controlled by the clock and photoperiod, including hypocotyl growth, cotyledon movements, leaf movements, stomatal movements, and the floral transition (Webb, 2003). In addition, a recent study has shown that low temperature-induced [Ca^{2+}] transients are significantly higher during the mid-photoperiod than at the beginning or end (Dodd *et al.*, 2006), which suggests that the clock may regulate cold perception and that the clock-associated cold perception may be mediated by Ca^{2+} oscillations. Together, these studies point to the control of [Ca^{2+}] as an important mechanism by which light regulates many physiological and developmental responses in plants.

5.9 Conclusions and perspectives

Calcium represents a ubiquitous regulator of plant activities. The cytoplasmic level has been shown to change in response to myriad stimuli ranging from hormones to cold shock, and fungal pathogens to mechanical stress (Hetherington and Brownlee, 2004a). In order to participate in signaling such varied stimuli it is likely that the spatial and temporal dynamics of the Ca^{2+} signal combine with cell-type specific expression and subcellular patterning of the Ca^{2+}-dependent components of the cell to yield the appropriate response. The 7 CaMs, approximately 50 CaM-like proteins (McCormack *et al.*, 2005), and the 43 calmodulin-like domain containing protein kinases (CDPKs) (Harper *et al.*, 2004) in *Arabidopsis* also hint at the expansion of the Ca^{2+}-responsive machinery in plants to support this diverse role in signaling. In contrast, humans have three CaM genes that encode an identical protein and *Saccaromyces cerevisae* has but a single CaM gene.

The sedentary lifestyle of plants has meant they have had to become exquisitely sensitive to the environment. Part of the sensing of the diverse range of environmental stimuli is likely encoded in Ca^{2+} signaling but it is also important to recognize that these stimuli can also affect the Ca^{2+}

signals themselves. For example, the environmental history of the plant has been shown to switch guard cell signaling between Ca^{2+}-dependent and -independent pathways (Allan *et al.*, 1994), and as noted above the magnitude of cold-induced Ca^{2+} transients is highly dependent on the time of day (Dodd *et al.*, 2006). The challenge for the future will undoubtedly be to place our increasingly detailed knowledge of the molecular machinery of Ca^{2+} signaling in the context of these environmentally modulated signaling networks. Plasticity is a fundamental feature of plant development and it seems likely that it will appear at all levels from the ultimate growth response down to the Ca^{2+}-dependent signaling networks participating in its regulation.

Acknowledgments

The authors thank Dr Sarah Swanson for critical reading of the manuscript. This work was supported by grants from the National Science Foundation.

References

Ali, R., Ma, W., Lemtiri-Chlieh, F., Tsaltas D., Leng Q., von Bodman S. and Berkowitz, G.A. (2007) Death don't have no mercy and neither does calcium: *Arabidopsis* cyclic NUCLEOTIDE GATED CHANNEL 2 and innate immunity. *Plant Cell*, **19**, 1081–1095.

Ali, R., Zielinski, R.E. and Berkowitz, G.A. (2006) Expression of plant cyclic nucleotide-gated cation channels in yeast. *J Exp Bot*, **57**, 125–138.

Allan, A.C., Fricker, M.D., Ward, J.L., Beale, M.H. and Trewavas, A.J. (1994) Two transduction pathways mediate rapid effects of abscisic acid in *Commelina* guard cells. *Plant Cell*, **6**, 1319–1328.

Allen, G.J., Chu, S.P., Schumacher, K., Shimazaki, C.T., Vafeados, D., Kemper, A., Hawke, S.D., Tallman, G., Tsien, R.Y., Harper, J.F., Chory, J. and Schroeder, J.I. (2000) Alteration of stimulus-specific guard cell calcium oscillations and stomatal closing in *Arabidopsis det3* mutant. *Science*, **289**, 2338–2342.

Ane, J.M., Kiss, G.B., Riely, B.K., Penmetsa, R.V., Oldroyd, G.E., Ayax, C., Lévy, J., Debellé, F., Baek, J.M., Kalo, P., Rosenberg, C., Roe, B.A., Long, S.R., Dénarié, J., and Cook, D.R. (2004) *Medicago truncatula* DMI1 required for bacterial and fungal symbioses in legumes. *Science*, **303**, 1364–1367.

Arrighi, J.F., Barre, A., Ben Amor, B. Bersoult, A., Soriano, L.C., Mirabella, R., de Carvalho-Niebel, F., Journet, E.P., Ghérardi, M., Huguet, T., Geurts, R., Dénarié, J., Rougé, P. and Gough, C. (2006) The *Medicago truncatula* lysine motif-receptor-like kinase gene family includes NFP and new nodule-expressed genes. *Plant Physiol*, **142**, 265–279.

Assmann, S.M. (1993) Signal transduction in guard cells. *Annu Rev Cell Biol*, **9**, 345–375.

Baum, G., Long, J.C., Jenkins, G.I. and Trewavas, A.J. (1999) Stimulation of the blue light phototropic receptor NPH1 causes a transient increase in cytosolic Ca^{2+}. *Proc Natl Acad Sci USA*, **96**, 13554–13559.

Baxter, I., Tchieu, J., Sussman, M.R. Boutry, M., Palmgren, M.G., Gribskov, M., Harper, J.F. and Axelsen, K.B. (2003) Genomic comparison of P-type ATPase ion pumps in *Arabidopsis* and rice. *Plant Physiol*, **132**, 618–628.

Berridge, M.J., Bootman, M.D. and Roderick, H.L. (2003) Calcium signalling: dynamics, homeostasis and remodelling. *Nat Rev Mol Cell Biol*, **4**, 517–529.

Bibikova, T.N. and Gilroy, S. (2000) Calcium in root hair growth and development. In: *The Cellular and Molecular Biology of Root Hairs* (eds R. Ridge and A.M. Emons), pp. 141–163. Springer, Berlin.

Bibikova, T.N., Zhigilei, A. and Gilroy, S. (1997) Root hair growth in *Arabidopsis thaliana* is directed by calcium and an endogenous polarity. *Planta*, **203**, 495–505.

Bonza, M.C., Morandini, P., Luoni, L. Geisler, M., Palmgren, M.G. and De Michelis, M.I. (2000) At-ACA8 encodes a plasma membrane-localized calcium-ATPase of *Arabidopsis* with a calmodulin-binding domain at the N terminus. *Plant Physiol*, **123**, 1495–1506.

Boyce, J.M., Knight, H., Deyholos, M. Openshaw, M.R., Galbraith, D.W., Warren, G. and Knight, M.R. (2003) The *sfr6* mutant of *Arabidopsis* is defective in transcriptional activation via CBF/DREB1 and DREB2 and shows sensitivity to osmotic stress. *Plant J*, **34**, 395–406.

Bush, D.S. (1995) Calcium regulation in plant cells and its role in signaling. *Annu Rev Plant Physiol Plant Mol Biol*, **46**, 95–122.

Carafoli, E. and Brini, M. (2000) Calcium pumps: structural basis for and mechanism of calcium transmembrane transport. *Curr Opin Chem Biol*, **4**, 152–161.

Cardenas, L., Feijo, J.A., Kunkel, J.G., Sanchez, F., Holdaway-Clarke, T., Hepler, P.K. and Quinto, C. (1999) Rhizobium nod factors induce increases in intracellular free calcium and extracellular calcium influxes in bean root hairs. *Plant J*, **19,** 347–352.

Carter, C., Pan, S., Zouhar, J., Avila, E.L., Girke, T. and Raikhel, N.V. (2004) The vegetative vacuole proteome of *Arabidopsis thaliana* reveals predicted and unexpected proteins. *Plant Cell*, **16**, 3285–3303.

Charron, D., Pingret, J.L., Chabaud, M., Journet, E.P. and Barker, D.G. (2004) Pharmacological evidence that multiple phospholipid signaling pathways link rhizobium nodulation factor perception in *Medicago truncatula* root hairs to intracellular responses, including Ca^{2+} spiking and specific ENOD gene expression. *Plant Physiol*, **136**, 3582–3593.

Clarkson, D.T. (1984) Calcium transport between tissues and its distribution in the plant. *Plant Cell Environ*, **7**, 449–456.

Clemens, S., Antosiewicz, D.M., Ward, J.M., Schachtman, D.P. and Schroeder, J.I. (1998) The plant cDNA LCT1 mediates the uptake of calcium and cadmium in yeast. *Proc Natl Acad Sci USA*, **95**, 12043–12048.

Day, I.S., Reddy, V.S., Shad Ali, G. and Reddy, A.S. (2002) Analysis of EF-hand-containing proteins in *Arabidopsis*. *Genome Biol*, **3**, RESEARCH0056.

De Koninck, P. and Schulman, H. (1998) Sensitivity of CaM kinase II to the frequency of Ca^{2+} oscillations. *Science*, **279**, 227–230.

de Ruijter, N.C.A., Rook, M.B., Bisseling, T. and Emons, A.M.C. (1998) Lipochito-oligosaccharides re-initiate root hair tip growth in *Vicia sativa* with high calcium and spectrin-like antigen at the tip. *Plant J*, **13**, 341–350.

den Hartog, M., Musgrave, A. and Munnik, T. (2001) Nod factor-induced phosphatidic acid and diacylglycerol pyrophosphate formation: a role for phospholipase C and D in root hair deformation. *Plant J*, **25**, 55–65.

Dodd, A.N., Jakobsen, M.K., Baker, A.J., Telzerow, A., Hou, S.W., Laplaze, L., Barrot, L., Poethig, R.S., Haseloff, J. and Webb, A.A. (2006) Time of day modulates low-temperature Ca^{2+} signals in *Arabidopsis*. *Plant J*, **48**, 962–973.

Dodd, A.N., Love, J. and Webb, A.A. (2005) The plant clock shows its metal: circadian regulation of cytosolic free Ca^{2+}. *Trends Plant Sci*, **10**, 15–21.

Dolmetsch, R.E., Lewis, R.S., Goodnow, C.C. and Healy, J.I. (1997) Differential activation of transcription factors induced by Ca^{2+} response amplitude and duration. *Nature*, **386**, 855–858.
Dolmetsch, R.E., Pajvani, U., Fife, K., Spotts, J.M. and Greenberg, M.E. (2001) Signaling to the nucleus by an L-type calcium channel-calmodulin complex through the MAP kinase pathway. *Science*, **294**, 333–339.
Dolmetsch, R.E., Xu, K. and Lewis, R.S. (1998) Calcium oscillations increase the efficiency and specificity of gene expression. *Nature*, **392**, 933–936.
Dutta, R. and Robinson, K.R. (2004) Identification and characterization of stretch-activated ion channels in pollen protoplasts. *Plant Physiol*, **135**, 1398–1406.
Ehrhardt, D.W., Atkinson, E.M. and Long, S.R. (1992). Depolarization of alfalfa root hair membrane-potential by *Rhizobium meliloti* Nod factors. *Science*, **256**, 998–1000.
Ehrhardt, D.W., Wais, R. and Long, S.R. (1996). Calcium spiking in plant root hairs responding to Rhizobium nodulation signals. *Cell*, **85**, 673–681.
Endre, G., Kereszt, A., Kevei, Z., Mihacea, S., Kalo, P. and Kiss, G.B. (2002). A receptor kinase gene regulating symbiotic nodule development. *Nature*, **417**, 962–966.
Engstrom, E.M., Ehrhardt, D.W., Mitra, R.M. and Long, S.R. (2002). Pharmacological analysis of nod factor-induced calcium spiking in *Medicago truncatula*. Evidence for the requirement of type IIA calcium pumps and phosphoinositide signaling. *Plant Physiol*, **128**, 1390–1401.
Epstein, E. (1972) *Mineral Nutrition of Plants: Principles and Perspectives*. Wiley, New York.
Felle, H.H., Kondorosi, E., Kondorosi, A. and Schultze, M. (1996) Rapid alkalinization in alfalfa root hairs in response to rhizobial lipochitooligosaccharide signals. *Plant J*, **10**, 295–301.
Felle, H.H., Kondorosi, E., Kondorosi, A. and Schultze, M. (1998) The role of ion fluxes in Nod factor signalling in *Medicago sativa*. *Plant J*, **13**, 455–463.
Felle, H.H., Kondorosi, E., Kondorosi, A. and Schultze, M. (1999) Elevation of the cytosolic free [Ca^{2+}] is indispensable for the transduction of the nod factor signal in alfalfa. *Plant Physiol*, **121**, 273–279.
Foreman, J., Demidchik, V., Bothwell, J.H.F. Mylona, P., Miedema, H., Torres, M.A., Linstead, P., Costa, S., Brownlee, C., Jones, J.D., Davies, J.M. and Dolan, L. (2003) Reactive oxygen species produced by NADPH oxidase regulate plant cell growth. *Nature*, **422**, 442–446.
Franceschi, V.R. and Nakata, P.A. (2005) Calcium oxalate in plants: formation and function. *Annu Rev Plant Biol*, **56**, 41–71.
Friedman, H., Goldschmidt, E.E. and Halevy, A.H. (1989) Involvement of calcium in the photoperiodic flower induction process of *Pharbitis nil*. *Plant Physiol*, **89**, 530–534.
Gehring, C.A., Irving, R.I., Kabbara, A.A., Parish, R.W., Boukli, N.M. and Broughton, W.J. (1997) Rapid, plateau-like increases in intracellular free calcium are associated with Nod-factor induced root hair deformation. *Mol Plant Microbe Interact*, **10**, 791–802.
Geisler, M., Frangne, N., Gomes, E., Martinoia, E. and Palmgren, M.G. (2000) The ACA4 gene of *Arabidopsis* encodes a vacuolar membrane calcium pump that improves salt tolerance in yeast. *Plant Physiol*, **124**, 1814–1827.
Geurts, R., Fedorova, E. and Bisseling, T. (2005) Nod factor signaling genes and their function in the early stages of Rhizobium infection. *Curr Opinplant Biol*, **8**, 346–352.
Gilroy, S., Fricker, M.D., Read, N.D. and Trewavas, A.J. (1991) Role of calcium in signal transduction of Commelina guard cells. *Plant Cell*, **3**, 333–344.

Gilroy, S., Read, N.D. and Trewavas, A.J. (1990) Elevation of cytoplasmic calcium by caged calcium or caged inositol triphosphate initiates stomatal closure. *Nature*, **346**, 769–771.

Gleason, C., Chaudhuri, S., Yang, T.B., Munoz, A., Poovaiah, B.W. and Oldroyd, G.E.D. (2006) Nodulation independent of rhizobia induced by a calcium-activated kinase lacking autoinhibition. *Nature*, **441**, 1149–1152.

Gomes, D.A., Leite, M.F., Bennett, A.M. and Nathanson, M.H. (2006) Calcium signaling in the nucleus. *Can J Physiol Pharmacol*, **84**, 325–332.

Gu, Y., Fu, Y., Dowd, P. Li, S., Vernoud, V., Gilroy, S. and Yang, Z. (2005) A Rho family GTPase controls actin dynamics and tip growth via two counteracting downstream pathways in pollen tubes. *J Cell Biol*, **169**, 127–138.

Han, S., Tang, R., Anderson, L.K., Woerner, T.E. and Pei, Z.-M. (2003) A cell surface receptor mediates extracellular Ca^{2+} sensing in guard cells. *Nature*, **425**, 196–200.

Harmer, S.L., Panda, S. and Kay, S.A. (2001) Molecular bases of circadian rhythms. *Annu Rev Cell Dev Biol*, **17**, 215–253.

Harper, J.F., Breton, G. and Harmon, A. (2004) Decoding Ca^{2+} signals through plant protein kinases. *Annu Rev Plant Biol*, **55**, 263–288.

Harper, J.F. and Harmon, A. (2005) Plants, symbiosis and parasites: a calcium signalling connection. *Nat Rev Mol Cell Biol*, **6**, 555–566.

Harper, J.F., Hong, B., Hwang, I., Guo, H.Q., Stoddard, R., Huang, J.F., Palmgren, M.G. and Sze, H. (1998) A novel calmodulin-regulated Ca^{2+}-ATPase (ACA2) from *Arabidopsis* with an N-terminal autoinhibitory domain. *J Biol Chem*, **273**, 1099–1106.

Heo, W.D., Lee, S.H., Kim, M.C., Kim, J.C., Chung, W.S., Chun, H.J., Lee, K.J., Park, C.Y., Park, H.C., Choi, J.Y. and Cho, M.J. (1999) Involvement of specific calmodulin isoforms in salicylic acid-independent activation of plant disease resistance responses. *Proc Natl Acad Sci USA*, **96**, 766–771.

Hepler, P.K. (2005) Calcium: A central regulator of plant growth and development. *Plant Cell*, **17**, 2142–2155.

Hetherington, A.M. and Brownlee, C. (2004) The generation of Ca^{2+} signals in plants. *Annu Rev Plant Biol*, **55**, 401–427.

Hetherington, A.M. and Woodward, F.I. (2003) The role of stomata in sensing and driving environmental change. *Nature*, **424**, 901–908.

Hirose, K., Kadowaki, S., Tanabe, M., Takeshima, H. and Iino, M. (1999) Spatiotemporal dynamics of inositol 1,4,5-trisphosphate that underlies complex Ca^{2+} mobilization patterns. *Science*, **284**, 1527–1530.

Holdaway-Clarke, T.L., Feijo, J.A., Hackett, G.R., Kunkel, J.G. and Hepler, P.K. (1997) Pollen tube growth and the intracellular cytosolic calcium gradient oscillate in phase while extracellular calcium influx is delayed. *Plant Cell*, **9**, 1999–2010.

Hong, B., Ichida, A., Wang, Y., Gens, J.S., Pickard, B.G. and Harper, J.F. (1999) Identification of a calmodulin-regulated Ca^{2+}-ATPase in the endoplasmic reticulum. *Plant Physiol*, **119**, 1165–1176.

Iino, M. (2007) Regulation of cell functions by Ca^{2+} oscillation. *Adv Expl Med Biol*, **592**, 305–312.

Imaizumi-Anraku, H., Takeda, N., Charpentier, M. Perry, J., Miwa, H., Umehara, Y., Kouchi, H., Murakami, Y., Mulder, L., Vickers, K., Pike, J., Downie, J.A., Wang, T., Sato, S., Asamizu, E., Tabata, S., Yoshikawa, M., Murooka, Y., Wu, G.J., Kawaguchi, M., Kawasaki, S., Parniske, M. and Hayashi, M. (2005) Plastid proteins crucial for symbiotic fungal and bacterial entry into plant roots. *Nature*, **433**, 527–531.

Johnson, C.H. (2001). Endogenous timekeepers in photosynthetic organisms. *Annu Rev Plant Physiol*, **63**, 695–728.

Johnson, C.H., Knight, M.R., Kondo, T., Masson, P., Sedbrook, J., Haley, A. and Trewavas, A. (1995) Circadian oscillations of cytosolic and chloroplastic free calcium in plants. *Science*, **269**, 1863–1865.

Kalo, P., Gleason, C., Edwards, A., Marsh, J., Mitra, R.M., Hirsch, S., Jakab, J., Sims, S., Long, S.R., Rogers, J., Kiss, G.B., Downie, J.A. and Oldroyd, G.E. (2005) Nodulation signaling in legumes requires NSP2, a member of the GRAS family of transcriptional regulators. *Science*, **308**, 1786–1789.

Kanamori, N., Madsen, L.H., Radutoiu, S. Frantescu, M., Quistgaard, E.M., Miwa, H., Downie, J.A., James, E.K., Felle, H.H., Haaning, L.L., Jensen, T.H., Sato, S., Nakamura, Y., Tabata, S., Sandal, N. and Stougaard, J. (2006) A nucleoporin is required for induction of Ca^{2+} spiking in legume nodule development and essential for rhizobial and fungal symbiosis. *Proc Natl Acad Sci USA*, **103**, 359–364.

Kirkby, E.A. and Pilbeam, D.J. (1984) Calcium as a plant nutrient. *Plant Cell Environ*, **7**, 397–405.

Knight, H., Trewavas, A.J. and Knight, M.R. (1996) Cold calcium signaling in *Arabidopsis* involves two cellular pools and a change in calcium signature after acclimation. *Plant Cell*, **8**, 489–503.

Knight, H., Trewavas, A.J. and Knight, M.R. (1997) Calcium signalling in *Arabidopsis thaliana* responding to drought and salinity. *Plant J*, **12**, 1067–1078.

Kohler, C. and Neuhaus, G. (2000) Characterisation of calmodulin binding to cyclic nucleotide-gated ion channels from *Arabidopsis thaliana*. *FEBS Lett*, **471**, 133–136.

Kornhauser, J.M., Cowan, C.W., Shaywitz, A.J., Dolmetsch, R.E., Griffith, E.C., Hu, L.S., Haddad, C, Xia, Z and Greenberg, M.E. (2002) CREB transcriptional activity in neurons is regulated by multiple, calcium-specific phosphorylation events. *Neuron*, **34**, 221–233.

Kreps, J.A. and Kay, S.A. (1997) Coordination of plant metabolism and development by the circadian clock. *Plant Cell*, **9**, 1235–1244.

Leckie, C.P., McAinsh, M.R., Allen, G.J., Sanders, D. and Hetherington, A.M. (1998) Abscisic acid-induced stomatal closure mediated by cyclic ADP-ribose. *Proc Natl Acad Sci USA*, **95**, 15837–15842.

Lee, S.M., Kim, H.S., Han, H.J., Moon, B.C., Kim, C.Y., Harper, J.F. and Chung, W.S. (2007) Identification of a calmodulin-regulated autoinhibited Ca^{2+}-ATPase (ACA11) that is localized to vacuole membranes in *Arabidopsis*. *FEBS Lett*, **581**, 3943–3949.

Levy, J., Bres, C., Geurts, R., Chalhoub, B., Kulikova, O., Duc, G., Journet ,E.P., Ané, J.M., Lauber, E., Bisseling, T., Dénarié, J., Rosenberg, C. and Debellé, F. (2004) A putative Ca^{2+} and calmodulin-dependent protein kinase required for bacterial and fungal symbioses. *Science*, **303**, 1361–1364.

Li, H., Lin, Y.K., Heath, R.M., Zhu, M.X. and Yang, Z. (1999) Control of pollen tube tip growth by a Rop GTPase-dependent pathway that leads to tip-localized calcium influx. *Plant Cell*, **11**, 1731–1742.

Liao, B., Gawienowski, M.C. and Zielinski, R.E. (1996) Differential stimulation of NAD kinase and binding of peptide substrates by wild-type and mutant plant calmodulin isoforms. *Arch Biochem Biophys*, **327**, 53–60.

Limpens, E., Franken, C., Smit, P., Willemse, J., Bisseling, T. and Geurts, R. (2003) LysM domain receptor kinases regulating rhizobial Nod factor-induced infection. *Science*, **302**, 630–633.

Lin, C.T. and Shalitin, D. (2003) Cryptochrome structure and signal transduction. *Annu Rev Plant Biol*, **54**, 469–496.

Long, S.R. (1996) Rhizobium symbiosis: Nod factors in perspective. *Plant Cell*, **8**, 1885–1898.

Love, J., Dodd, A.N. and Webb, A.A. (2004) Circadian and diurnal calcium oscillations encode photoperiodic information in *Arabidopsis*. *Plant Cell*, **16**, 956–966.

MacAlister, C.A., Ohashi-Ito, K. and Bergmann, D.C. (2007) Transcription factor control of asymmetric cell divisions that establish the stomatal lineage. *Nature*, **445**, 537–540.

MacRobbie, E. (1992) Calcium and ABA-induced stomatal closure. *Phil Trans R Soc Lond B*, **338**, 5–18.

Madsen, E.B., Madsen, L.H., Radutoiu, S., Olbryt, M., Rakwalska, M., Szczyglowski, K., Sato, S., Kaneko, T., Tabata, S., Sandal, N. and Stougaard, J. (2003) A receptor kinase gene of the LysM type is involved in legume perception of rhizobial signals. *Nature*, **425**, 637–640.

Malho, R., Liu, Q., Monteiro, D., Rato, C., Camacho, L. and Dinis, A. (2006) Signalling pathways in pollen germination and tube growth. *Protoplasma*, **228**, 21–30.

Malho, R. and Trewavas, A.J. (1996) Localized apical increases of cytosolic free calcium control pollen tube orientation. *Plant Cell*, **8**, 1935–1949.

McAinsh, M.R., Brownlee, C. and Hetherington, A.M. (1990) Abscisic acid-induced elevation of guard cell cytosolic Ca^{2+} precedes stomatal closure. *Nature*, **343**, 186–188.

McAinsh, M.R., Brownlee, C. and Hetherington, A.M. (1992) Visualizing changes in cytosolic-free Ca^{2+} during the response of stomatal guard cells to abscisic acid. *Plant Cell* **4**, 1113–1122.

McAinsh, M.R., Webb, A.A.R., Taylor, J.E. and Hetherington, A.M. (1995) Stimulus-induced oscillations in guard cell cytosolic free calcium. *Plant Cell*, **7**, 1207–1219.

McCormack, E., Tsai, Y.C. and Braam, J. (2005) Handling calcium signaling: *Arabidopsis* CaMs and CMLs. *Trends Plant Sci*, **10**, 383–389.

Meijer, H.J.G. and Munnik, T. (2003) Phospholipid-based signaling in plants. *Annu Rev Plant Biol*, **54**, 265–306.

Messerli, M.A., Danuser, G. and Robinson, K.P. (1999) Pulsatile influxes of H^+, K^+ and Ca^{2+} tag growth pulses of *Lilium longiflorum* pollen tubes. *J Cell Sci*, **112**, 1497–1509.

Mitra, R.M., Gleason, C.A., Edwards, A., Hadfield, J., Downie, J.A., Oldroyd, G.E.D. and Long, S.R. (2004b) A Ca^{2+}/calmodulin-dependent protein kinase required for symbiotic nodule development: gene identification by transcript-based cloning. *Proc Natl Acad Sci USA*, **101,** 4701–4705.

Mitra, R.M., Shaw, S.L. and Long, S.R. (2004a) Six nonnodulating plant mutants defective for Nod factor-induced transcriptional changes associated with the legume-rhizobia symbiosis. *Proc Natl Acad Sci USA*, **101**, 10217–10222.

Miwa, H., Sun, J., Oldroyd, G.E.D. and Downie, J.A. (2006a) Analysis of nod-factor-induced calcium signaling in root hairs of symbiotically defective mutants of *Lotus japonicus*. *Mole Plant Microbe Interact*, **19**, 914–923.

Miwa, H., Sun, J., Oldroyd, G.E.D. and Downie, J.A. (2006b) Analysis of calcium spiking using a cameleon calcium sensor reveals that nodulation gene expression is regulated by calcium spike number and the developmental status of the cell. *Plant J*, **48**, 883–894.

Neuhaus, G., Bowler, C., Kern, R. and Chua, N.-H. (1993) Calcium/calmodulin-dependent and -independent phytochrome signal transduction pathways. *Cell*, **73**, 937–952.

Ng, C.K., Carr, K., McAinsh, M.R., Powell, B. and Hetherington, A.M. (2001) Drought-induced guard cell signal transduction involves sphingosine-1-phosphate. *Nature*, **410**, 596–599.

Oldroyd, G.E.D. and Downie, J.A. (2004) Calcium, kinases and nodulation signalling in legumes. *Nat Rev Molr Cell Biol*, **5**, 566–576.

Oldroyd, G.E.D. and Downie, J.A. (2006) Nuclear calcium changes at the core of symbiosis signalling. *Curr Opin Plant Biol*, **9**, 351–357.

Patil, S., Takezawa, D. and Poovaiah, B.W. (1995) Chimeric plant calcium/calmodulin-dependent protein-kinase gene with a neural visinin-like calcium-binding domain. *Proc Natl Acad Sci USA*, **92**, 4897–4901.

Pei, Z.-M., Murata, Y., Benning, G., Thomine, S., Klüsener, B., Allen, G.J., Grill, E. and Schroeder, J.I. (2000) Calcium channels activated by hydrogen peroxide mediate abscisic acid signalling in guard cells. *Nature*, **406**, 731–734.

Peiter, E., Maathuis, F.J., Mills, L.N., Knight, H., Pelloux, J., Hetherington, A.M. and Sanders, D. (2005) The vacuolar Ca^{2+}-activated channel TPC1 regulates germination and stomatal movement. *Nature*, **434**, 404–408.

Pillitteri, L.J., Sloan, D.B., Bogenschutz, N.L. and Torii, K.U. (2007) Termination of asymmetric cell division and differentiation of stomata. *Nature*, **445**, 501–505.

Plieth, C. (2005) Calcium: just another regulator in the machinery of life? *Ann Bot (Lond)*, **96**, 1–8.

Plieth, C., Hansen, U.P., Knight, H. and Knight, M.R. (1999) Temperature sensing by plants: the primary characteristics of signal perception and calcium response. *Plant J*, **18**, 491–497.

Pottosin, I.I. and Schönknecht, G. (2007) Vacuolar calcium channels. *J Exp Bot*, **58**, 1559–1569.

Prank, K., Waring, M., Ahlvers, U. Bader, A., Penner, E., Möller, M., Brabant, G. and Schöfl, C. (2005) Precision of intracellular calcium spike timing in primary rat hepatocytes. *Syst Biol (Stevenage)*, **2**, 31–34.

Qi, Z., Stephens, N.R. and Spalding, E.P. (2006) Calcium entry mediated by GLR3.3, an *Arabidopsis* glutamate receptor with a broad agonist profile. *Plant Physiol*, **142**, 963–971.

Radutoiu, S., Madsen, L.H., Madsen, E.B., Felle, H.H., Umehara, Y., Grønlund, M., Sato, S., Nakamura, Y., Tabata, S., Sandal, N. and Stougaard, J. (2003) Plant recognition of symbiotic bacteria requires two LysM receptor-like kinases. *Nature*, **425**, 585–592.

Reddy, V.S. and Reddy, A.S. (2004) Proteomics of calcium-signaling components in plants. *Phytochemistry*, **65**, 1745–1776.

Reddy, V.S., Safadi, F., Zielinski, R.E. and Reddy, A.S. (1999) Interaction of a kinesin-like protein with calmodulin isoforms from *Arabidopsis*. *J Biol Chem*, **274**, 31727–31733.

Riely, B.K., Ane, J.M., Penmetsa, R.V. and Cook, D.R. (2004) Genetic and genomic analysis in model legumes bring Nod-factor signaling to center stage. *Curr Opin Plant Biol*, **7**, 408–413.

Riely, B.K., Lougnon, G., Ane, J.M. and Cook, D.R. (2007) The symbiotic ion channel homolog DMI1 is localized in the nuclear membrane of *Medicago truncatula* roots. *Plant J*, **49**, 208–216.

Roelfsema, M.R.G. and Hedrich, R. (2002) Studying guard cells in the intact plant: modulation of stomatal movement by apoplastic factors. *New Phytologist*, **153**, 425–431.

Saito, K., Yoshikawa, M., Yano, K., Miwa, H., Uchida, H., Asamizu, E., Sato, S., Tabata, S., Imaizumi-Anraku, H., Umehara, Y., Kouchi, H., Murooka, Y., Szczyglowski, K., Downie, J.A., Parniske, M., Hayashi, M. and Kawaguchi, M. (2007) Nucleoporin 85 is required for calcium spiking, fungal and bacterial symbioses, and seed production in *Lotus japonicus*. *Plant Cell*, **19**, 610–624.

Sanders, D., Pelloux, J., Brownlee, C. and Harper, J.F. (2002) Calcium at the crossroads of signaling. *Plant Cell*, **14**, S401– S417.

Sattelmacher, B. (2001) The apoplast and its significance for plant mineral nutrition. *New Phytologist*, **149**, 167–192.

Schroeder, J.I., Kwak, J.M. and Allen, G.J. (2001) Guard cell abscisic acid signalling and engineering drought hardiness in plants. *Nature*, **410**, 327–330.

Scrase-Field, S.A. and Knight, M.R. (2003) Calcium: just a chemical switch? *Curr Opin Plant Biol*, **6**, 500–506.

Shacklock, P.S., Read, N.D. and Trewavas, A.J. (1992) Cytosolic free calcium mediates red light-induced photomorphogenesis. *Nature*, **358**, 753–755.

Shaw, S.L. and Long, S.R. (2003) Nod factor elicits two separable calcium responses in *Medicago truncatula* root hair cells. *Plant Physiol*, **131**, 976–984.

Shigaki, T. and Hirschi, K.D. (2006) Diverse functions and molecular properties emerging for CAX cation/H^+ exchangers in plants. *Plant Biol, (Stuttg)* **8**, 419–429.

Shigaki, T., Sreevidya, C. and Hirschi, K.D. (2002) Analysis of the Ca^{2+} domain in the *Arabidopsis* H^+/Ca^{2+} antiporters CAX1 and CAX3. *Plant Mol Biol*, **50**, 475–483.

Smit, P., Raedts, J., Portyanko, V., Debellé, F., Gough, C., Bisseling, T. and Geurts, R. (2005) NSP1 of the GRAS protein family is essential for rhizobial Nod factor-induced transcription. *Science*, **308**, 1789–1791.

Speksnijder, J.E., Miller, A.L., Weisenseel, M.H., Chen, T.H. and Jaffe, L.F. (1989) Calcium buffer injections block fucoid egg development by facilitating calcium diffusion. *Proc Natl Acad Sci USA*, **86**, 6607–6661.

Stacey, G., Libault, M., Brechenmacher, L., Wan, J.R. and May, G.D. (2006) Genetics and functional genomics of legume nodulation. *Curr Opin Plant Biol*, **9**, 110–121.

Stracke, S., Kistner, C., Yoshida, S., Mulder, L, Sato, S., Kaneko, T., Tabata, S., Sandal, N., Stougaard, J., Szczyglowski, K. and Parniske, M. (2002) A plant receptor-like kinase required for both bacterial and fungal symbiosis. *Nature*, **417**, 959–962.

Sze, H., Liang, F., Hwang, I., Curran, A.C. and Harper, J.F. (2000) Diversity and regulation of plant Ca^{2+} pumps: insights from expression in yeast. *Annu Rev Plant Physiol Plant Mol Biol*, **51**, 433–462.

Szmidt-Jaworska, A., Jaworski, K. and Kopcewicz, J. (2006) The involvement of cyclic ADPR in photoperiodic flower induction of *Pharbitis nil*. *J Plant Growth Regul*, **25**, 233–244.

Szponarski, W., Sommerer, N., Boyer, J.C., Rossignol, M. and Gibrat, R. (2005) Large-scale characterization of integral proteins from *Arabidopsis* vacuolar membrane by two-dimensional liquid chromatography. *Proteomics*, **4**, 397–406.

Tahtiharju, S., Sangwan, V., Monroy, A.F., Dhindsa, R.S. and Borg, M. (1997) The induction of kin genes in cold-acclimating *Arabidopsis thaliana*. Evidence of a role for calcium. *Planta*, **203**, 442–447.

Takeno, K. (1993) Evidence for the involvement of calcium-Ions in the photoperiodic induction of flowering in *Pharbitis nil*. *Plant Cell Physiol*, **34**, 221–225.

Tang, R.-H., Han, S., Zheng, H., Cook, C.W., Choi, C.S., Woerner, T.E., Jackson, R.B. and Pei, Z.-M. (2007) Coupling diurnal cytosolic Ca^{2+} oscillations to the CAS-IP_3 pathway in *Arabidopsis*. *Science*, **315**, 1423–1426.

Taylor, A.R., Manison, N.F.H., Fernandez, C. and Wood, J. (1996) Spatial organization of calcium signaling involved in cell volume control in the Fucus rhizoid. *Plant Cell*, **8**, 2015–2031.

Tirichine, L., Imaizumi-Anraku, H., Yoshida, S., Murakami, Y., Madsen, L.H., Miwa, H., Nakagawa, T., Sandal, N., Albrektsen, A.S., Kawaguchi, M., Downie, A., Sato, S., Tabata, S., Kouchi, H., Parniske, M., Kawasaki, S. and Stougaard, J. (2006) Deregulation of a Ca^{2+}/calmodulin-dependent kinase leads to spontaneous nodule development. *Nature*, **441**, 1153–1156.

Torok, K., Wilding, M., Groigno, L., Patel, R. and Whitaker, M. (1998) Imaging the spatial dynamics of calmodulin activation during mitosis. *Curr Biol*, **8**, 692–699.

Wais, R.J., Galera, C., Oldroyd, G., Catoira, R., Penmetsa, R.V., Cook, D., Gough, C., Denarié, J. and Long, S.R. (2000) Genetic analysis of calcium spiking responses in nodulation mutants of *Medicago truncatula*. *Proc Natl Acad Sci USA*, **97**, 13407–13412.

Walker, S.A., Viprey, V. and Downie, J.A. (2000) Dissection of nodulation signaling using pea mutants defective for calcium spiking induced by Nod factors and chitin oligomers. *Proc Natl Acad Sci USA*, **97**, 13413–13418.

Webb, A.A.R. (2003) The physiology of circadian rhythms in plants. *New Phytologist*, **160**, 281–303.

West, A.E., Chen, W.G., Dalva, M.B., Dolmetsch, R.E., Kornhauser, J.M., Shaywitz, A.J., Takasu, M.A., Tao, X. and Greenberg, M.E. (2001) Calcium regulation of neuronal gene expression. *Proc Natl Acad Sci USA*, **98**, 11024–11031.

White, P.J. (2000) Calcium channels in higher plants. *Biochim Biophys Acta*, **1465**, 171–189.

White, P.J., Bowen, H.C., Demidchik, V., Nichols, C. and Davies, J.M. (2002) Genes for calcium-permeable channels in the plasma membrane of plant root cells. *Biochim Biophys Acta*, **1564**, 299–309.

White, P.J. and Broadley, M.R. (2003) Calcium in plants. *Ann Bot*, **92**, 487–511.

Wood, N.T., Haley, A., Viry-Moussaid, M., Johnson, C.H., van der Luit, A.H. and Trewavas, A.J. (2001) The calcium rhythms of different cell types oscillate with different circadian phases. *Plant Physiol*, **125**, 787–796.

Wu, Z., Liang, F., Hong, B., Young, J.C., Sussman, M.R., Harper, J.F. and Sze, H. (2002) An endoplasmic reticulum-bound Ca^{2+}/Mn^{2+} pump, ECA1, supports plant growth and confers tolerance to Mn^{2+} stress. *Plant Physiol*, **130**, 128–137.

Wymer, C.L., Bibikova, T.N. and Gilroy, S. (1997) Cytoplasmic free calcium distributions during the development of root hairs of *Arabidopsis thaliana*. *Plant J*, **12**, 427–439.

Xiong, T.C., Bourque, S., Lecourieux, D., Amelot, N., Grat, S., Brière, C., Mazars, C., Pugin, A. and Ranjeva, R. (2006) Calcium signaling in plant cell organelles delimited by a double membrane. *Biochim Biophys Acta*, **1763**, 1209–1215.

Yang, T., Segal, G., Abbo, S., Feldman, M. and Fromm, H. (1996) Characterization of the calmodulin gene family in wheat: structure, chromosomal location, and evolutionary aspects. *Mol Gen Genet*, **252**, 684–694.

Zielinski, R.E. (1998) Calmodulin and calmodulin-binding proteins in plants. *Ann Rev Plant Physiol Plant Mol Biol*, **49**, 697–725.

Annual Plant Reviews (2008) **33**, 163–188
doi: 10.1111/b.9781405160025.2008.00006.x www.interscience.wiley.com

Chapter 6

PARADIGMS AND NETWORKS FOR INTRACELLULAR CALCIUM SIGNALING IN PLANT CELLS

Sheng Luan

Department of Plant and Microbial Biology, University of California, Berkeley, CA 94720, USA

Abstract: Calcium (Ca^{2+}) is a ubiquitous second messenger in all eukaryotes. A challenging question is how the cation serves as a messenger for numerous signals and confers specificity to cellular responses. Recent studies have established a concept called "calcium signature" that specifies calcium changes triggered by each signal. Along with this concept comes the "decoding" of calcium signatures—how cells recognize and translate the signatures (codes) into the correct cellular responses. The initial step in the decoding process involves sensor proteins that bind calcium and change their conformation, leading to activation or inactivation of the downstream target molecules thereby regulating the specific biochemical and physiological processes. In this chapter, studies on the major groups of calcium sensors including calcium-dependent protein kinases (CDPKs), calmodulin (CaM), and calcineurin B-like proteins (CBLs) will be summarized and general paradigms and networks of calcium signaling processes will be discussed. I will particularly focus on the more recent studies that have advanced our understanding on calcium signal transduction in plants.

Keywords: calcium; calcium sensors; calcineurin B-like proteins; protein kinase; signal transduction

6.1 Introduction

Despite the lack of a specific nervous system, plants are capable of perceiving external stimuli, processing the signals, calculating specific responses, and sometimes "remembering" the stimulus-response process. This process, often referred to as "signal transduction" or "acclimation," is reminiscent of the "learning" process in animals. In between the signal (input) and response (output), there exists a complicated molecular network for processing the information regardless of the specific organism in question. Within the molecular network for plant signaling, calcium serves as a critical component and plays a role in the signaling of many extracellular stimuli including light, biotic and abiotic stress factors, and developmental cues (Trewavas and Knight, 1994; Bush, 1995; Braam *et al.*, 1997; Felle and Hepler, 1997; Holdaway-Clarke *et al.*, 1997; McAinsh *et al.*, 1997; Wymer *et al.*, 1997; Sanders *et al.*, 1999; Rudd and Franklin-Tong, 2001). One important question in calcium signaling concerns the specificity of signal–response coupling as different signals elicit distinct and specific cellular responses. Recent studies in both animal and plant cells suggest that a Ca^{2+} signal is presented not only by the concentration of Ca^{2+} but also by its spatial and temporal information (Franklin-Tong *et al.*, 1996; Holdaway-Clarke *et al.*, 1997; Dolmetsch *et al.*, 1998; Li *et al.*, 1998; Trewavas, 1999; Allen *et al.*, 2001). A combination of changes in all Ca^{2+} parameters produced by a particular signal is always different from those produced by any other signals and therefore referred to as a "Ca^{2+} signature." If the specificity of the calcium signals is encoded by these signatures, a particular plant cell must be equipped with the mechanisms for decoding various signatures leading to specific responses. Although the decoding process is not well understood, studies indicate that this process starts with the calcium sensors, often calcium-binding proteins, which bind calcium with high affinity and alter their own structural properties. Such structural changes result in functional changes in the sensor proteins (with effector domains) or trigger interaction with the target proteins of the sensors (without effector domains). The sensors or their targets are often regulatory proteins that modulate the function of others and elicit changes in cellular processes (Fig. 6.1, Color plate 7).

Several families of Ca^{2+} sensors have been identified in higher plants. Perhaps the best known is calmodulin (CaM) and CaM-related proteins, which typically contain four EF-hand domains for Ca^{2+} binding (Zielinski, 1998; Snedden and Fromm, 2001; Luan *et al.*, 2002). A more recent addition to Ca^{2+} sensors from *Arabidopsis* is proteins similar to both the regulatory B-subunit of calcineurin and the neuronal Ca^{2+} sensor in animals (Luan *et al.*, 2002). These plant Ca^{2+} sensors are referred to as calcineurin B-like proteins (CBLs) (Kudla *et al.*, 1999). The third major class is exemplified by the Ca^{2+}-dependent protein kinases (CDPKs), which contain CaM-like Ca^{2+}-binding domains and a kinase domain in a single protein (Roberts and Harmon, 1992; Harmon *et al.*, 2000). The CDPK proteins function both as Ca^{2+} sensor and as effector of its Ca^{2+}-sensing activity.

CaM and CBL are small proteins that contain multiple Ca^{2+}-binding domains but lack other effector domains like the kinase domain in CDPKs. To transmit the Ca^{2+} signal, CaMs and CBLs interact with target proteins and regulate their activity. CaM target proteins have been identified in higher plants and include protein kinases, metabolic enzymes, cytoskeleton-associated proteins, and others (Reddy *et al.*, 1996, 2002; Snedden *et al.*, 1996; Zielinski, 1998; Snedden and Fromm, 2001; Luan *et al.*, 2002). A family of SNF1-like protein kinases called CIPKs has been identified as targets for CBL proteins (Shi *et al.*, 1999). The target proteins of these small Ca^{2+} sensors then regulate activities that constitute cellular responses triggered by an external signal. The CDPKs bind calcium and regulate the kinase activity present in the same molecule, more tightly linking calcium sensing and effector activity. Both small (CaM- and CBL-type) and large (CDPK-type) Ca^{2+} sensors are therefore part of a complex signaling network of interconnected pathways. A prime goal of many plant biologists is to understand how this network is established and how it functions to link discrete external signals to specific cellular and physiological responses. In this chapter, I focus on the two families of small Ca^{2+} sensors (CaM and CBL) and CDPKs to explore the general paradigms on how specific signals may be transmitted through the combined action of these proteins.

6.2 CDPKs, plant calcium "sensor responders"

Calcium's role as a second messenger has been identified in all eukaryotes. Before the finding of CaM-domain protein kinases (CDPKs), the general paradigm for action is that calcium binds to a sensor protein (e.g., CaM), altering the conformation of the protein. Conformational change in the sensor protein triggers its interaction with downstream effectors (often enzymes) leading to the modification of that target enzyme (activity). Figure 6.2 (Color plate 8) shows the model for the CaM activation of CaM-dependent protein kinase (CaMK). Identification of CDPK in plants represents a new paradigm for calcium sensing because the sensor protein itself contains a kinase domain that serves as an effector (Harper *et al.*, 1991). Therefore, the CDPKs are also referred to as "sensor responders" as they contain both the sensing domain and the response domain (the kinase) for the calcium signaling process (Harper and Harmon, 2005).

6.2.1 Structural diversity and regulation of CDPK superfamily

The first CDPK to be cloned represents a typical CDPK that contains CaM-like calcium-binding domains (Harper *et al.*, 1991). Subsequent studies identified several subtypes of protein kinases that are highly related to CDPK and may reflect evolutionary diversification of the same type of protein kinases. These include the CDPK-related kinases (CRKs) and calcium and CaM-dependent kinases (CCaMKs). The CRKs have high sequence homology to CDPKs and retain a general similarity in their structural domains (Fig. 6.3, Color plate 9). For example, the kinase domains in the CRKs are followed by a long C-terminal domain corresponding to the CaM-like domains in the CDPKs, although the calcium-binding EF-hand motifs are not conserved anymore in CRKs. The structural features of CCaMK are rather unique in that they contain both calcium-binding and CaM-binding domains in the same protein (Fig. 6.3, Color plate 9). Instead of four EF hands in the CDPKs, CCaMKs usually have three EF hands like those in the other types of calcium-binding proteins called visinin in animals (Patil *et al.*, 1995).

Biochemical studies on CDPKs have identified several regulatory features that represent important models for the regulation of calcium-regulated protein kinases in plants (reviewed by Harper and Harmon, 2005). The structural domains, as they are referred to, carry straightforward features related to their function and regulation. For example, the kinase domain contains the catalytic site of the enzyme; the EF-hand motifs are calcium-binding domains; the autoinhibitory domain is located in between the kinase and the EF hands and functions to repress the kinase activity when calcium signal is absent. Presumably, the autoinhibitory domain serves as a pseudosubstrate that binds to the kinase active site and blocks the access of substrates. Upon calcium binding, the conformational change results in the release of the inhibitory domain from the active site thereby making the kinase site available for substrate access (Fig. 6.4, Color plate 10). In addition to calcium-dependent regulation, some CDPKs have been shown to be modified by myristoylation and palmitoylation (Martin and Busconi, 2000). By attaching a lipid module to the N-terminus of the protein, these modifications can effectively target the protein to the cell membranes. For the regulation of CCaMKs, calcium binding to the visinin-like domain enhances the autophosphorylation that in turn increases CaM-binding affinity, leading to maximal activation of the kinase toward its substrate (Takezawa *et al.*, 1996). The identification of plant CDPKs and CCaMK significantly expands the repertoire of calcium-regulated protein kinases in eukaryotes.

6.2.2 Functional diversity of CDPKs and CCaMKs

Where and when a gene is expressed and subcellular localization of the gene product often determine the function of the gene (product). A number of studies address the temporal and spatial expression patterns of CDPK genes. Recent transcriptional profiling studies further enriched the information at the genome scale on gene expression patterns. Although a comprehensive study of all CDPKs is lacking, some members of CDPK genes have been shown to be ubiquitously expressed, whereas others are expressed with tissue specificity, regulated by various signals such as stress conditions, light, hormones, and pathogens (reviewed by Cheng *et al.*, 2002; Hrabak *et al.*, 2003; Harper *et al.*, 2004). Concerning the subcellular localization, studies have shown that CDPKs can be either soluble or associated with cell membranes. Some members are found to be located throughout the cytoplasm and the nucleus. The subcellular compartments that contain CDPKs range from the plasma membrane (PM), peroxisomes, endoplasmic reticulum, seed oil bodies, and mitochondria (Harper and Harmon 2005). Interestingly, most of the CDPKs contain both myristoylation and palmitoylation sites at their N-termini, which could be responsible for the recruitment to the cell membranes. It is yet to be determined how the subcellular locations of these CDPKs are related to their functions.

Toward the understanding of CDPK function in plant physiology, several approaches have been taken and a number of results are revealing. Using biochemical approaches, a growing list of substrates for CDPKs have been identified, and they are involved in a number of cellular processes. These substrates include but are not limited to enzymes involved in carbon, nitrogen, and sulfur metabolism (Tang *et al.*, 2003; Hardin *et al.*, 2004; Liu *et al.*, 2006), enzymes for secondary metabolism (Cheng *et al.*, 2001), and proteins for ion and water transport (Hwang *et al.*, 2000; Guenther *et al.*, 2003). The phosphorylation of substrates by CDPKs can alter enzyme/transport activity against its substrate (in the case of aquaporin and phenylalanine ammonia lyase) or change the regulatory properties of the substrates (Liu *et al.*, 2006), or protein stability (Tang *et al.*, 2003). These biochemical studies have yet to be connected to the physiological functions of the relevant CDPKs in plants.

Although significant effort has been dedicated to the functional analysis of CDPKs, it has been challenging to assign function to specific CDPKs using either forward or reverse genetics approaches. Available data so far suggest that significant redundancy among CDPK members may account for difficulty in genetic analysis. For example, a recent report (Mori *et al.*, 2006) showed that two CDPKs (CPK3 and CPK6) are involved in the regulation of stomatal response to ABA. Plant hormone ABA is a well-known chemical messenger that is produced upon stress exposure especially under drought conditions. A critical response in plants to drought is closing their stomata to preserve water. Between the drought signal and stomatal closure, a number of signaling components including ABA and calcium have been identified. Furthermore, calcium has been shown to serve as a downstream second messenger for ABA in stomatal closing response. However, little is known regarding the mechanism of calcium action in guard cells except that ion channels responsible for turgor regulation are potential targets for ABA-induced calcium fluctuation (Allen *et al.*, 2000; MacRobbie, 2000; Schroeder *et al.*, 2001; Luan, 2002). As CDPKs are important sensor responders in plants, it is speculated that they may play a role in calcium-regulated stomatal closure. The work by Mori *et al.* (2006) showed that disruption of CPK3 and CPK6 resulted in rather subtle phenotypic changes at the whole plant level despite the changes in the ion channel activities in the guard cells. This study therefore indicates that, in addition to functional redundancy, genetic analyses to identify whole plant phenotype may not be successful due to specific cellular processes that a particular CDPK may be responsible for, and that changes in such processes may not necessarily cause whole plant phenotypic changes.

Calcium signaling is crucial for many aspects of reproductive biology. The earliest evidence of such conclusion was obtained by the finding of a calcium "wave" during the fertilization process in sea urchins. In plants, pollen tube growth has been used as a single-cell model for the study of calcium signaling for decades. The pollen tube elongation is a directional growth that absolutely requires calcium oscillation (Franklin-Tong *et al.*, 1996). It is not known how calcium waves are decoded by the sensors and effectors in the male gametophyte. Some studies indicate involvement of CDPKs. These include finding of a large number of CDPKs expressing in the pollen grains, antisense RNA interference of pollen tube growth (Estruch *et al.*, 1994; Yoon *et al.*, 2006). It is possible that a high degree of functional redundancy may also be found in pollen CDPKs.

Although CCaMK is not found in model plant *Arabidopsis*, studies have demonstrated a critical role of such calcium-regulated protein kinases in plant–microbe symbiosis. One example is the legume–*Rhizobium* symbiosis in nitrogen fixation process. An early signaling event in plant recognition of bacterial partner is calcium oscillation in root hairs (Ehrhardt *et al.*, 1996). Similar calcium signaling has also been identified in the process of plant–fungus symbiosis. A genetic screen identified a mutant defective in nodule formation in the legume *M. truncatula* and the gene affected in the mutant encodes a CCaMK containing typical visinin-like EF hands in the calcium-sensing domain. Together with the identification of several other genes that encode receptor-like kinases and a cation transporter in the nodulation process, it is predicted that the legume–*Rhizobium* interaction signaling may involve generation and decoding of calcium signals (reviewed in Oldroyd and Downie, 2004). The CCaMK-type kinases are clearly candidates for decoding calcium changes during legume–microbe interaction. More recent studies using CCaMK mutants lacking the autoinhibitory domain (making the kinase constitutively active) demonstrate that a CCaMK (DMI3) is required and sufficient for the nodulation-related plant cell morphogenesis (Gleason *et al.*, 2006; Tirichine *et al.*, 2006) highlighting the possibility of transferring nitrogen fixation to nonlegume plants by manipulation of CCaMK and other molecular components in the plant–*Rhizobium* interaction pathway. Because *Arabidopsis* does not seem to have any type of symbiotic relationship with microbes, it is speculated that CCaMKs may be specifically involved in such symbiotic processes.

6.3 CaM: small calcium sensors with a variety of target proteins

6.3.1 Plant genomes encode a large number of CaMs and CaM-related proteins

Perhaps the best-known calcium-binding protein is CaM, a highly conserved protein in all eukaryotic systems. Compared to animal and fungi that contain only a few CaM genes, plants contain an extended superfamily of CaMs and CaM-related proteins with a diverse number of Ca^{2+}-binding EF hands and additional domains (Snedden and Fromm, 1998; Zielinski, 1998; Snedden and Fromm, 2001; Luan *et al.*, 2002). Table 6.1 lists the known and putative CaM genes in *Arabidopsis*. In addition, a large number of CaM-like and

Table 6.1 Conserved (boldface) and divergent CaM genes in *Arabidopsis*

Name	Accession number	CaM function	Expressed sequence tags	Amino acids	Extension	% Identity (% similarity) to CaM2	Closest CaM homolog (% similarity)
CaM1	At5g37780	Yes	7	149	No	96.6 (100)	CaM4 (100)
CaM2	At2g41110	Yes	12	149	No	100 (100)	CaM3, CaM5 (100)
CaM3	At3g56800	Yes	4	149	No	100 (100)	CaM2, CaM5 (100)
CaM4	At1g66410	Yes	18	149	No	96.6 (100)	CaM1 (100)
CaM5	At2g27030	Yes	6	149	No	100 (100)	CaM2, CaM3 (100)
CaM6	Q03509	Yes	4	149	No	98.6 (99.3)	CaM7 (99.3)
CaM7	At3g43810	Yes	9	149	No	99.3 (100)	CaM2, CaM3, CaM5 (100)
CaM8	At4g14640	Yes	3	151	No	72.5 (79.8)	CaM11 (89.3)
CaM9	At3g51920	Yes [a]	5	151	No	49.6 (60.4)	CaM1, CaM4 (60.4)
CaM10 (CaBP-22)	At2g41090	Yes [b]	4	191	C terminus (no homology found with any other database sequences)	66.4 (73.3)	CaM1, CaM4 (74.6)
CaM11	At3g22930	— [c]	2	173	N terminus (22 Gln residues)	74.5 (83.2)	CaM8 (90.0)
CaM12	At2g41100	—	6	324	N terminus (2 EF hands); C terminus (similar to that of CaM10)	63.7 (69.8)	CaM1, CaM4 (70.5)
CaM13	At1g12310	—	11	148	No	50.3 (62.6)	CaM14 (95.9)
CaM14	At1g62820	—	1	148	No	49.6 (61.2)	CaM13 (95.9)

[a] Does not display Ca^{2+}-induced electrophoretic mobility shifting but partially complements a CaM-defective yeast mutant (Zielinski, 2002).
[b] Displays Ca^{2+}-induced electrophoretic mobility shifting (Ling and Zielinski, 1993).
[c] —, unknown.

CaM-related proteins are identified in plant species. In *Arabidopsis*, typical CaM members include CaM1–7 that are highly similar to animal CaM and to each other (>95% identical on amino acid sequence). Other proteins listed in Table 6.1 (CaM8–14) share 50–75% amino acid identity to the typical CaM2 and some of them have been shown to have CaM activity. They are referred to as CaM-like (CaM8, 9, 13, and 14) or, when they have additional non-CaM domains, CaM-related proteins (CaM10–12). For example, *Arabidopsis* CaM8 is a CaM-like protein because of its more divergent sequence. This protein can function as a CaM in Ca^{2+}-binding and yeast complementation experiments, but it appears to interact with a more limited set of target proteins as compared to typical CaM isoforms (Zielinski, 2002). A good example of CaM-related protein is petunia CaM53, which has been demonstrated to have CaM activity but it contains a polybasic C-terminal domain that is not found in a typical CaM (Fig. 6.1, Color plate 7). As discussed later, this extra domain in CaM53 regulates its cellular localization (Rodriguez-Concepcion *et al.*, 1999). It is also interesting that the genes encoding CaM10, CaM12, and CaM2 are organized in a tandem array in this order in chromosome 2. This could result from gene duplication and incorporation of additional domains in a sequence of events from CaM2 to CaM10 to CaM12 (see Fig. 6.1 (Color plate 7) and Table 6.1).

The EF hands in CaM protein are organized into two distinct globular domains, each of which contains one pair of EF hands. Each pair of EF hands is considered to be the basic functional unit. Pairing of EF hands is thought to stabilize the protein and increase its affinity toward Ca^{2+} (Seamon and Kreetsinger, 1983). Although each globular domain binds Ca^{2+} and undergoes conformational changes independently, the two domains act in concert to bind target proteins (Nelson and Chazin, 1998). Upon increase of Ca^{2+} to submicromolar or low micromolar levels, all CaM molecules will are activated. Cooperative binding is required for this "on–off" mechanism to function efficiently. Cooperativity of Ca^{2+} binding ensures that full activation of the CaM occurs in a narrow region of calcium concentrations during a signaling event.

Selectivity of CaM towards Ca^{2+} is also an important factor in effective transduction of the Ca^{2+} signal. CaMs bind Ca^{2+} selectively in the presence of high concentrations of Mg^{2+} and monovalent cations in the cell. The cation selectivity is achieved by optimizations in the structure folds of the binding loop (Fig. 6.5, Color plate 11). For example, the discrimination between Ca^{2+} and Mg^{2+} is accomplished through the reduction in the size of the binding loop. Binding of Mg^{2+} ions would collapse the EF-hand loop, thereby reducing the distance between negatively charged side chains and destabilizing the CaM–Mg^{2+} complex (Falke *et al.*, 1994). Even small changes in the chemical properties of the Ca^{2+}-binding loop (e.g., Glu12→Gln) can drastically reduce the binding affinity to Ca^{2+} (Beckingham, 1991; Haiech *et al.*, 1991). The Glu12→Gln mutation changes the carboxylate side chain into carboxylamide, which removes the oxygen ligand for Ca^{2+} (Nelson and Chazin, 1998). Together, structural analyses in combination with site-directed mutagenesis established that CaMs (and other EF-hand-containing proteins) have evolved as highly specific Ca^{2+} sensors.

Structural analysis of the Ca^{2+}-free and Ca^{2+}-bound states of CaM proteins reveals the conformational changes induced by Ca^{2+} binding (Fig. 6.5, Color plate 11). In the Ca^{2+}-free state, CaM adopts a closed conformation. Ca^{2+} binding triggers a conformational change and the protein now adopts an open conformation with near perpendicular interhelical angles between the globular domains. This open conformation exposes a hydrophobic surface within each globular domain and permits binding of protein targets (Babu *et al.*, 1988; Kuboniwa *et al.*, 1995; Zhang *et al.*, 1995).

6.3.2 CaM targets a large array of proteins with various functions

The diversity of gene expression and protein localization patterns is important for generating functional diversity and specificity. The temporal and spatial expression patterns of CaMs, such as those for CDPKs or any gene family in plants, are diverse. Some are ubiquitously expressed and others are regulated by various factors including light, mechanical stress, heat/cold shock, wounding, osmotic stress, pathogens, and plant hormones. Certain CaM genes are also developmentally regulated and show tissue- and cell-specific expression patterns. Despite extensive analysis of expression patterns, relevant physiological functions are not known. Some touch-induced genes (*TCH*) encode CaM-related proteins, which are rapidly induced by mechanical manipulation, cold- and heat-shock, phytohormones, and Ca^{2+} itself (Braam *et al.*, 1997). The magnitude and kinetics of mRNA induction differ between the different *TCH* genes (Braam *et al.*, 1997). Extensive work with *TCH3* established that the gene is expressed in the shoot apical meristem, vascular tissue, and root pericycle cells during vegetative growth in *Arabidopsis*. Following wind stimuli, *TCH3* becomes abundant in branch points of leaf primordia and stipule, pith parenchyma, and vascular tissues, although the functional consequences of this induction are not understood.

As plants can establish specific cellular Ca^{2+} signatures by restricting Ca^{2+} to a specific compartment of the cell (reviewed in Rudd and Franklin-Tong, 2001), the subcellular location of CaM and other calcium sensors play a role in decoding "local" calcium signals. Certain CaMs are found in different subcellular locations and more importantly such locations can change upon perception of extracellular signals. A good example for this type of regulation is petunia CaM53 (Rodriguez-Concepcion *et al.*, 1999). Similar to rice OsCaM61, CaM53 contains a polybasic 34-residues C-terminal extension ending with a CaaX-box motif for prenylation. CaM53 is efficiently prenylated (Caldelari *et al.*, 2001) and processed (Rodriguez-Concepcion *et al.*, 2000) to be targeted to the PM. When prenylation is blocked, the polybasic domain targets the protein to the nucleus. A similar prenylation-dependent membrane versus nuclear localization has been reported for OsCaM61 (Dong *et al.* 2002). Prenylation and PM targeting of CaM53, however, do not depend on calcium binding. The prenylation status of CaM53 is likely an important aspect of its function, since the set of proteins with which CaM53 could potentially interact upon calcium binding is expected to be very different depending on the subcellular localization of the protein.

An important clue for the function of intracellular calcium sensors is the identity of their target proteins. The Ca^{2+}-bound CaM binds and regulates the activity of a wide range of proteins that are not necessarily related in structure. How can Ca^{2+}-CaMs bind to so many different proteins? More specifically, the plasticity of the Ca^{2+}-CaM structure must accommodate the variation in both the molecular size and composition of the target proteins. This issue has been addressed by structural analyses of Ca^{2+}-CaM and target-bound Ca^{2+}-CaM. Figure 6.2 (Color plate 8) shows that the two globular domains of Ca^{2+}-CaM are interconnected by a flexible tether that can accommodate peptides of varying sizes (Nelson and Chazin, 1998). Upon binding a peptide, the two globular domains fold toward each other to form a hydrophobic channel rich in methionine residues that have flexible hydrophobic side chains. In this channel, Ca^{2+}-CaM interacts with peptides mostly through nonspecific van der Waals interactions that form between the exposed hydrophobic domains of Ca^{2+}-CaM and the target peptides, which explains why Ca^{2+}-CaM can bind many target proteins (O'Neil and DeGrado, 1990; Osawa, *et al.*, 1998; Zhang and Yuan, 1998). Together, the structures of CaM illustrate how this class of proteins can function as extremely efficient Ca^{2+} sensors and on/off switches, allowing them to transduce Ca^{2+} signals with high efficiency and accuracy. Different affinities for Ca^{2+}–CaM interactions with specific target proteins may be sufficient for the differential transduction of the Ca^{2+} signal.

The interaction between CaM and CaMK in animal cells provides a good model that illustrates how Ca^{2+}-CaM regulates the activity of the target. For example, CaMKII contains an autoinhibitory domain, which occludes the active site in the resting state. Ca^{2+}-CaM binds to a site near or overlapping with the autoinhibitory domain, thereby releasing it from the active site and activating the enzyme (reviewed by Hook and Means 2001, discussed earlier in Fig. 6.2, Color plate 8). This model appears to be applicable to interactions between CaMs and their target proteins in plant cells based on the available results. CaM targets in plants have been extensively reviewed (Snedden and Fromm, 1998; Zielinski, 1998; Snedden and Fromm, 2001; Reddy *et al.*, 2002), and therefore we will only introduce the conceptual framework using several examples to explain how CaMs regulate protein target activity in plants.

CaM target proteins can be identified using labeled CaMs to screen expression cDNA libraries. A large number of CaM-binding proteins have been identified from plants. Glutamate decarboxylase (GAD) is one of the best studied (Baum *et al.*, 1993, 1996; Snedden *et al.*, 1996; Zik *et al.*, 1998). The enzyme catalyzes conversion of L-glutamate into gamma-aminobutyric acid (GABA) and is rapidly activated during several stress responses (Snedden and

Fromm, 1998, 2001). GAD is activated by binding either to CaM or to a monoclonal antibody that recognizes the CaM-binding domain of GAD. In analogy to Ca^{2+}-CaM–CaMK interaction, binding of Ca^{2+}-CaM to GAD probably relieves the autoinhibitory effect of the CaM-binding domain, as mutant GAD lacking the CaM-binding domain (GAD-C) is constitutively active. Overexpression of GAD-C in transgenic tobacco induced developmental abnormalities associated with increased GABA levels, concomitant with reduced levels of glutamate (Baum *et al.*, 1996). The activation of GAD by environmental stimuli via the Ca^{2+}-CaM signaling system is very rapid, exemplifying the highly cooperative on/off switch of the CaM response (Snedden and Fromm, 1998).

Ca^{2+}-ATPases are localized in the endomembranes or PM and play a key role in removing Ca^{2+} from the cytoplasm to terminate a signaling event, which is critical for Ca^{2+} homeostasis in all eukaryotic cells (reviewed by Sze *et al.*, 2000). Among the Ca^{2+}- ATPases in higher plants, type IIB Ca^{2+}-ATPases are major targets of Ca^{2+}-CaM regulation. Unlike homologues in animal cells, plant type IIB ATPases are located in both endomembranes (ER and tonoplast) and the PM (Sze *et al.*, 2000). Ca^{2+}-CaM interacts with type IIB ATPases to activate the pump by releasing an autoinhibitory domain from the active site, similar to the Ca^{2+}-CaM–CaMKII interaction in animals. It is noteworthy that plant Ca^{2+}-ATPases are subject to regulation by CDPKs, as briefly described earlier. Interestingly, while Ca^{2+}-CaM activates the pump, CDPK phosphorylation inhibits the pump, demonstrating the complexity in the regulation of Ca^{2+} signal termination by feedback from two different types of Ca^{2+} sensors (Hwang *et al.*, 2000). Several plant nucleotide-gated ion channels may also be regulated by Ca^{2+}-CaM (Schuurink *et al.*, 1998; Arazi *et al.*, 1999, 2000; Kohler *et al.*, 1999; Leng *et al.*, 1999). These channel proteins contain six transmembrane domains and a high-affinity CaM-binding site overlapping with a cyclic nucleotide-binding domain (Arazi *et al.*, 2000).

Ca^{2+} signaling and the role of CaM in the nucleus is drawing increased interest (Snedden and Fromm, 2001; Rudd and Franklin-Tong, 2001). CaMs participate in transcriptional regulation either directly by binding to transcription factors (Szymanski *et al.*, 1996) or indirectly by activating kinases or phosphatases that control transcription factor activity (Marechal *et al.*, 1999). Studies in animal cells demonstrated that CaM localization to the nucleus could be facilitated by differential Ca^{2+} oscillations (Craske *et al.*, 1999; Teruel *et al.*, 2000; Teruel and Meyer, 2000), suggesting additional and complex levels of transcriptional regulation. As discussed earlier, changing the metabolic status of plant cells induced translocation of CaM53 to the nucleus where it appears to activate specific signaling (Rodriguez-Concepcion *et al.*, 1999). Selective Ca^{2+} signals were measured in the cytoplasm and the nucleus of transgenic plants expressing either cytoplasmic or nuclear forms of the Ca^{2+} reporter protein aequorin (van Der Luit *et al.*, 1999; Pauly *et al.*, 2000). Such Ca^{2+} signals may be required for the expression of specific genes. For example, expression of tobacco *NpCaM1* (but not *NpCaM2*, which encodes an identical

CaM protein) in response to wind was stimulated by nuclear Ca^{2+} transients, whereas cold-responsive expression was primarily induced by a cytoplasmic Ca^{2+} transient (van Der Luit *et al.*, 1999). Thus, spatially separated Ca^{2+} signals can also control the function of closely related CaM proteins through the regulation of their genes.

Although many target proteins have been identified for CaMs, relatively little is known about the specific physiological function of each CaM member. Like the situation with CDPKs, the functional redundancy may have hindered the genetic analysis of CaM members in model plants such as *Arabidopsis*.

6.4 The CBL–CIPK network

6.4.1 Plant CBLs are related to calcineurin B but significantly diverged into a group of proteins with new functions

Earlier studies on calcium signaling implicate a calcineurin-like protein in the signaling processes of ion channel regulation and salt tolerance (Luan *et al.*, 1993; Allen and Sanders, 1995; Pardo *et al.*, 1998). Calcineurin is a calcium CaM-dependent protein phosphatase highly conserved in eukaryotes from yeast to mammals (Klee *et al.*, 1998). Like CaM-dependent protein kinase, calcineurin contains a CaM-binding domain in the catalytic subunit (calcineurin A). In addition, another regulatory subunit (calcineurin B) binds to the catalytic subunit and is required for the activation of the phosphatase. Calcineurin B, like CaM, also contains four EF-hand calcium-binding domains, although the overall sequence is not related to CaM. As calcineurin serves as a critical molecular switch to many cellular processes in eukaryotes from yeast to mammals, it was speculated that similar molecules might also exist in plants. Extensive effort focused on the isolation of calcineurin-like proteins and genes from plants and a family of genes encoding calcineurin B-like proteins (CBLs) were eventually identified from *Arabidopsis* (Kudla *et al.*, 1999). Independently, a genetic analysis of salt mutants identified a gene related to calcineurin B (called SOS3, Liu and Zhu, 1998) and is a member of the CBL family (also referred to as CBL4). CBLs are encoded by a multigene family of at least 10 members in *Arabidopsis* (Table 6.2), which have similar structural domains with small variations in the length of the coding regions (Kudla *et al.*, 1999; Kim *et al.*, 2000; Albrecht *et al.*, 2001; Guo *et al.*, 2001b). Their amino acid sequence identity, which ranges from 20 to 90%, would be sufficient for functional redundancy among the closely related members, while allowing for functional specificity among more diverged members. Unlike CaM genes, CBLs have been identified until now only in higher plants, suggesting that CBLs may function in plant-specific signaling processes. Comparing CaM with CBL proteins, the two families do not show significant similarity in their primary amino acid sequences except for the conserved positions in the

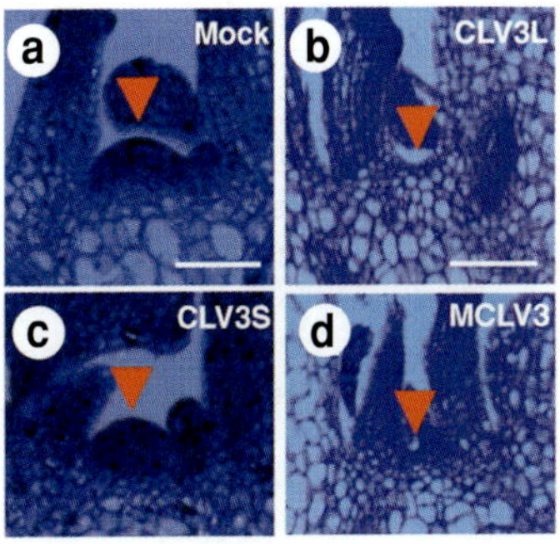

e

```
            CLE motif
CLV3   ..LRTVPSGPDPLHHH..
MCLV3     RTVPSGPDPLHH
CLV3L     RTVPSGPDPLHHHVNPPRQRNNFQP
CLV3S      TVPSGPDPLHHHVNPPRQRNNFQP
```

Plate 1 Exogenous applications of synthetic CLV3 peptides trigger differentiation of the SAM. Shown are the *Arabidopsis* SAM (arrowheads) with mock treatment (a), treated with synthetic CLV3L peptide (b), CLV3S peptide (c), and mature CLV3 peptide (MCLV3) (d). The treatment of CLV3L, which contains the entire CLE motif as well as MCLV3, results in consumption of the SAM, a phenotype resembling that of *wuschel* mutant (b, d). (e) Amino acid sequence of each peptide. (Photographs courtesy of Dr Shinichiro Sawa. Modified from Kondo *et al.* (2006). Copyright, AAAS.)

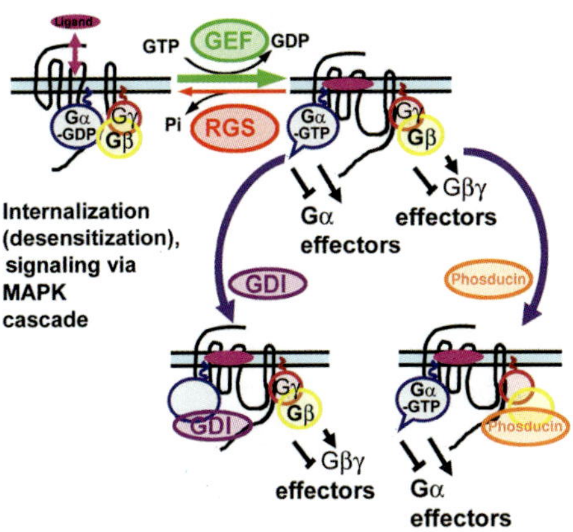

Plate 2 *G-protein-coupled signaling in animals.* G-protein-coupled receptors (GPCRs) interact with a heterotrimeric G-protein complex, composed of Gα, Gβ, and Gγ subunits. In the heterotrimer, Gα is bound to GDP. Activation of GPCRs through ligand occupancy favors guanine-nucleotide exchange factor (GEF) activity of the GPCR and consequently Gα becomes loaded with GTP and dissociates into a free Gα subunit and a free Gβγ dimer. Each of these, in turn, interacts with enzymes (effectors) to alter the amount of secondary messengers produced and/or the movement of ions across the plasma membrane. These interactions can be either positive or negative as depicted by the two types of arrows. The Gα subunit has an intrinsic GTPase that can be accelerated by regulator of G-protein signaling (RGS) proteins. RGS proteins, which have GTPase accelerating protein (GAP) activity, can also block interactions between Gα and its effectors. The Gα subunit and Gβγ dimer can also be modified. Gα in its GDP form while free from the heterotrimer can be bound by guanine dissociation inhibitors (GDI) to prevent reassociation into the heterotrimer. This effectively blocks recycling and prolongs signaling output through the Gβγ dimer. Phosducin can bind the Gβγ dimer and thus block its signaling, preventing reassociation into the heterotrimer, and potentially prolonging signaling output from the activated Gβγ. At the level of the receptor, modification occurs to desensitize signaling and/or facilitate signaling. For some GPCRs, ligand occupancy induces internalization, thus removing receptors from further activation by extracellular signals. Internalization can also be a mechanism for signal propagation because some internalized receptors will recruit other signaling elements such as mitogen activated protein kinases (MAPK) of the MAPK cascade.

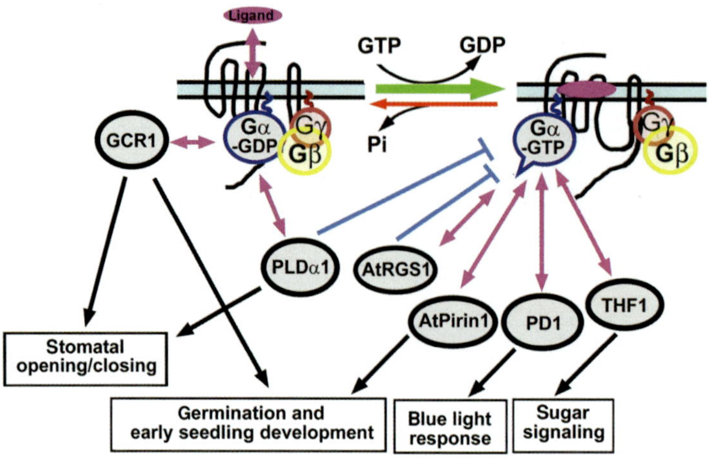

Plate 3 *Heterotrimeric G-protein signaling in plants.* Shown are heterotrimeric G-protein components identified in *Arabidopsis* and proteins demonstrated to physically interact with them. Gα is encoded by the single gene *GPA1*. Gβ is encoded by the single gene *AGB1*. Gγ is encoded by two genes, *AGG1* and *AGG2*. Both GCR1 and AtRGS1 are seven-transmembrane (7TM) proteins and are directly coupled by Gα, but no ligand has been identified for either GCR1 or AtRGS1. AtRGS1 contains an RGS box at its C-terminus, which can accelerate the intrinsic GTPase activity of Gα. PLDα1 is one of the isoforms of phospholipase D. AtPirin1 is a member of the cupin protein superfamily. PD1 is a cytosolic prephenate dehydratase. THF1 is a plastid protein localized to both the outer plastid membrane and the stroma, and does not share significant sequence with any known protein. PLDα1, AtPirin1, PD1, and THF1 have been shown to physically interact with Gα. Double arrows indicate known physical interactions. Blunted arrows indicate an inhibitory effect on Gα presumably through GAP (GTPase accelerating protein) activity.

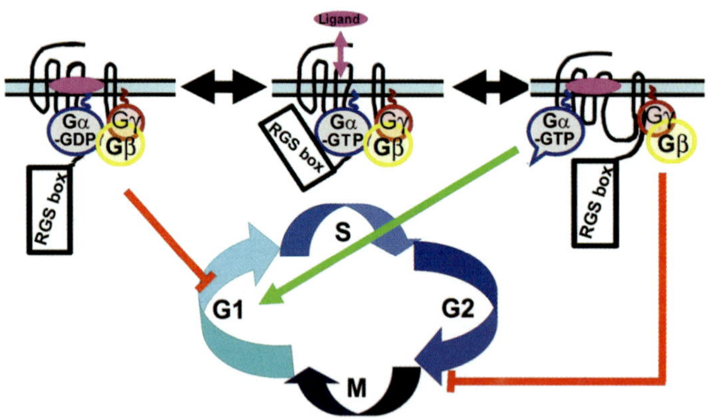

Plate 4 *A model of how AtRGS1 and the G-protein complex modulate cell division in the root.* AtRGS1 is represented by the 7TM domain protein containing an RGS box in its C-terminal domain. AtRGS1 may be a receptor GAP or GEF and the ligand has been suggested to be D-glucose. Shown here is one possibility consistent with D-glucose's positive effect on root growth. In this scenario, D-glucose inhibits the GAP activity of AtRGS1, thus shifting Gα to its GTP-bound form. However, it is also possible that a ligand may induce GEF activity or promote GAP activity. Regardless, the heterotrimer attenuates cell proliferation in the root apical meristem while the Gα-GTP promotes it. In the pericycle, Gβγ attenuates cell division, possibly by blocking reentry into the cell cycle.

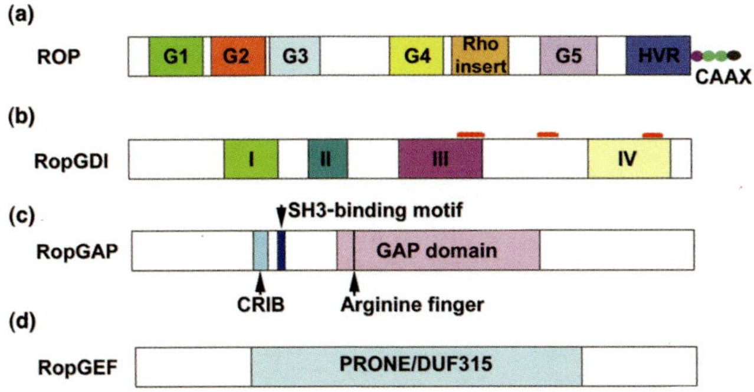

Plate 5 Conserved structures of ROP GTPases, RopGDI, RopGAP, and RopGEF.(a) ROP GTPases contain five conserved domains G1–G5. G2 is the effector-binding domain, and the other four are guanine nucleotide-binding domains. C-terminal hypervariable region (HVR) is highly diverse among ROP members. The Rho insert in ROP GTPases is shorter than that in other members of the Rho family. The CAAX motif at the end of C-terminus is involved in membrane targeting through lipid modification. (b) RopGDIs contain four domains that are highly conserved within all GDIs. Gray bars indicate regions that likely bind to ROP GTPases. (c) RopGAPs have a conserved GAP domain and the arginine finger within the GAP domain. Proximal to the upstream of the GAP domain, there is a CDC42/RAC-interactive binding (CRIB) domain, which is specific to RopGAPs. The CRIB motif and the GAP domain are jointed by an src homology domain 3 (SH3)-binding motif. The N- and C-terminal regions are highly variable, which could play a role in the regulation of RopGAPs and their functional specificity. (d) RopGEFs belong to a plant-specific RhoGEF family. Members of RopGEFs contain a central PRONE domain that is responsible for GEF activity. N- and C-terminal regions are highly variable and are proposed to regulate the GEF activity.

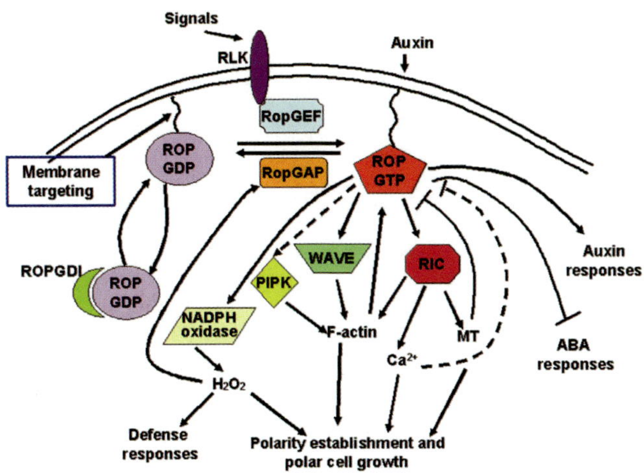

Plate 6 Model for ROP GTPase regulation and function. ROP is targeted to the PM through lipid modification and the HVR domain. Auxin regulates the positioning of ROP in the membrane through an unknown mechanism. GDI association makes ROP sequestered in the cytosol. Unknown extracellular signals may activate RopGEF via RLK to activate ROP. RopGAPs deactivate ROPs by accelerating the intrinsic GTPase activity of ROPs. ROP relays signals to different downstream targets that influence the organization of F-actin and MT and accumulation of second messengers such as H_2O_2 and Ca^{2+}. These downstream signaling events regulate the establishment of polarity, polar cell growth, and defense and hormone responses. F-actin provides a positive feedback mechanism to promote ROP activity, whereas Ca^{2+} and MTs may negatively feedback-regulate ROP activity. H_2O_2 inhibits ROP activation by activating the expression of RopGAPs.

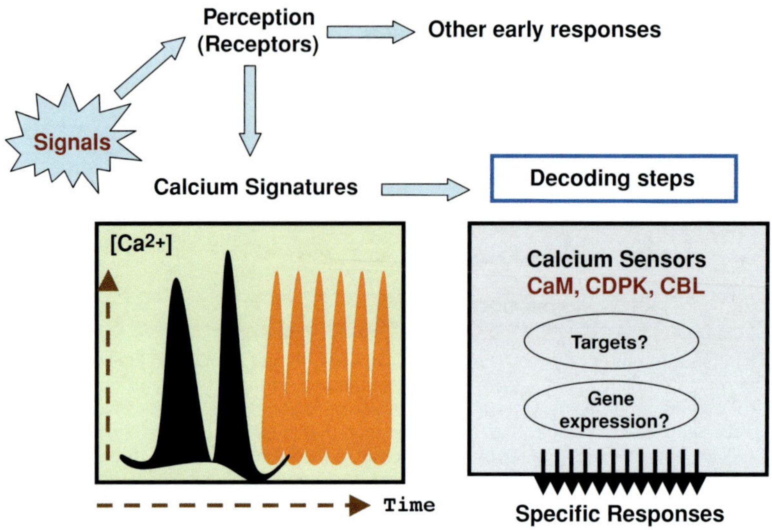

Plate 7 A schematic presentation of calcium decoding in the signal transduction process. Signals are perceived by specific receptors that trigger early responses including calcium fluctuations (signature/code) and other events. Calcium signatures are recognized and decoded by the calcium sensors and their targets followed by biochemical/cellular responses that constitute specific physiological responses in plants.

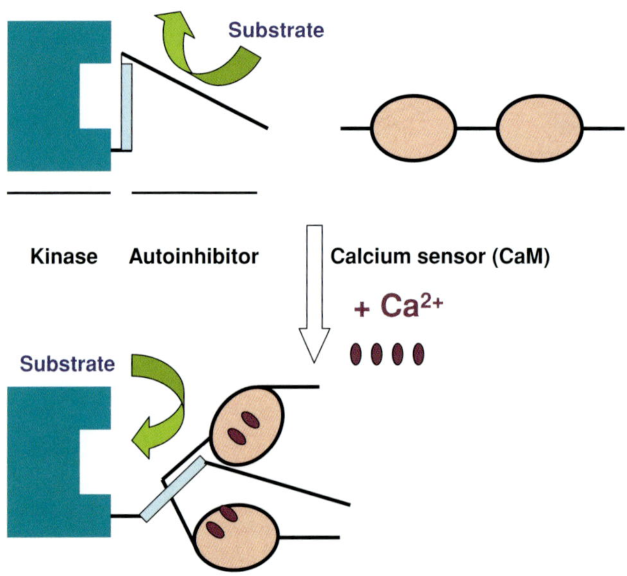

Plate 8 Activation of CaMK by calcium and calmodulin. The basal state of CaMK is not accessible by the substrate due to block by autoinhibitory domain. Calcium binding to CaM changed the conformation of CaM and triggered interaction between CaM and the autoinhibitory domain of CaMK, releasing the kinase active site for substrate access.

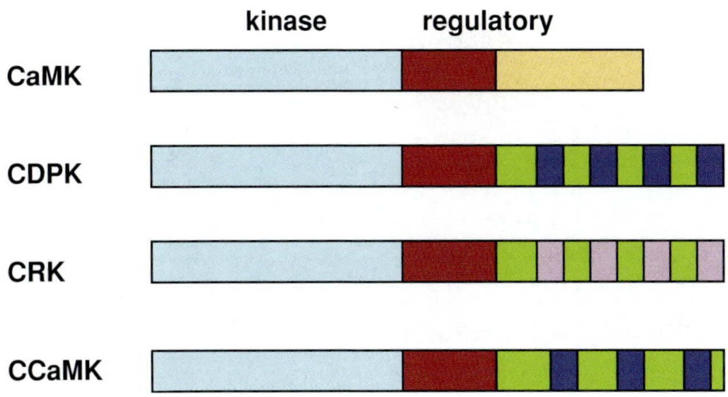

Plate 9 Domain structures of animal CaMK, plant CDPK, and related kinases. The regulatory regions of different kinases contain various domains including the autoinhibitory (crimson), calcium-binding EF hands (blue), and diverged EF hand-like structures (purple).

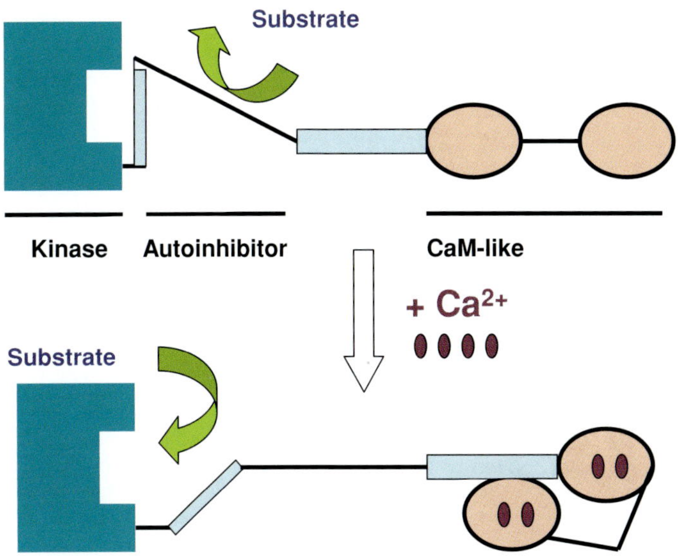

Plate 10 Activation of CDPK by calcium signal. The autoinhibitory domain block the kinase active site under basal conditions. Upon calcium elevation, the CaM-like domain binds calcium and alters its conformation leading to interaction with the junction domain between the CaM-like domain and the autoinhibitory domain. Such interaction releases the block of the kinase active site by autoinhibitory domain allowing substrate access.

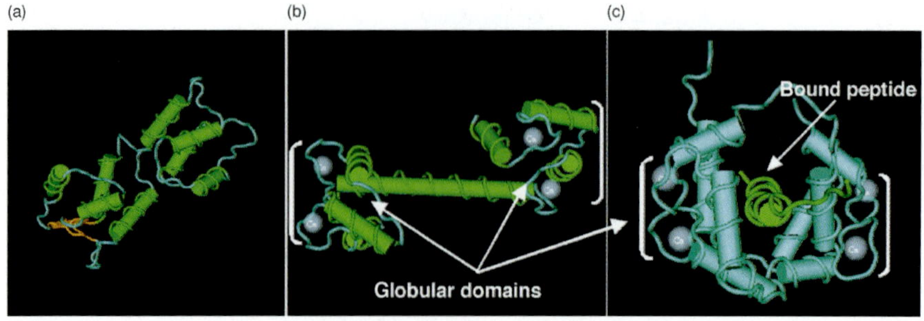

Plate 11 Structural analysis of Apo-CaM, Ca^{2+}-CaM, and the Ca^{2+}-CaM—target complex. Strand-rod presentation of Apo-CaM (a) and Ca^{2+}-CaM (b) showing substantial changes upon Ca^{2+} binding. Part (c) shows a solution structure solved by NMR of peptide-bound Ca^{2+}-CaM. Peptide binding causes disruption of the flexible tether, bringing the globular domains closer to form a channel around the peptide. The majority of contacts between Ca^{2+}-CaM and target peptide are nonspecific van der Waals bonds made by residues in the hydrophobic surfaces. Brackets indicate globular domains.

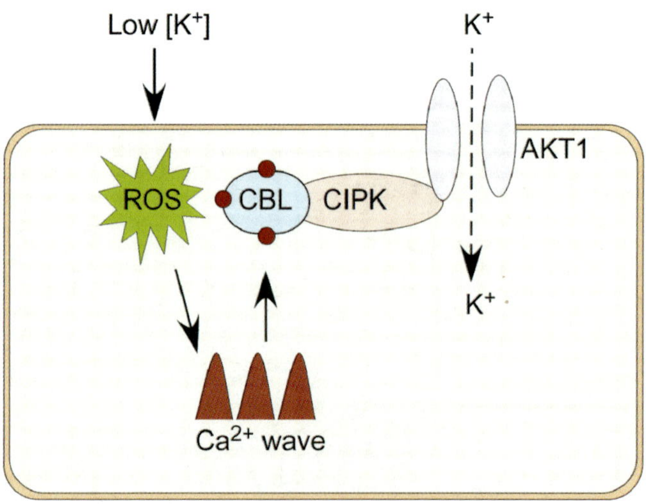

Plate 12 A schematic model of the Ca^{2+}-dependent pathway for low-K response in *Arabidopsis*. Low-K condition triggers the increased level of ROS that induces calcium fluctuations. Calcium binding to CBL1 or CBL9 activates the calcium sensor and leads to CBL–CIPK23 complex formation and activation of CIPK23. CIPK23 physically interacts with AKT1 C-terminus, phosphorylates it and activates the channel resulting in K uptake into the cell.

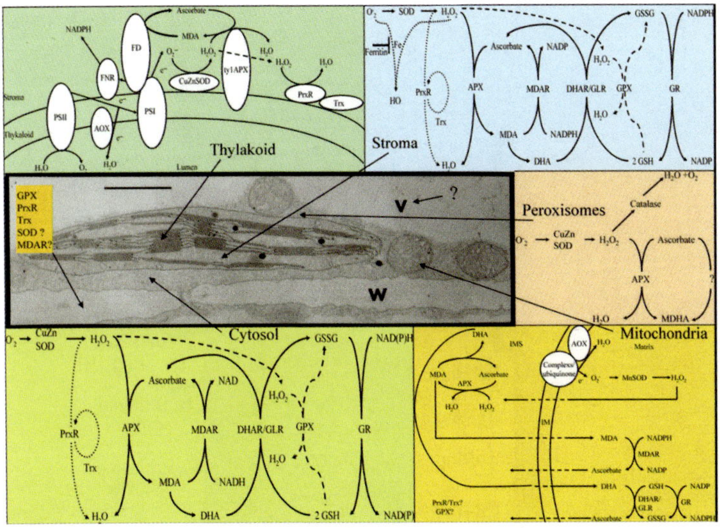

Plate 13 *Localization of ROS-scavenging pathways in plant cells.* A transmission electron micrograph of a portion of a plant cell is used to demonstrate the relative volume of the different cellular compartments and their physical separation (middle, left panel). The water–water cycle and alternative oxidase (AOX) reduces the production of ROS in thylakoids (top, left panel; in some plants FeSOD may replace CuZnSOD). ROS that escape the thylakoid and/or are produced in the stroma undergo detoxification by superoxide dismutase (SOD) and the stromal ascorbate glutathione cycle. Peroxiredoxin (PrxR) and glutathione peroxidase (GPX) are also involved in ROS removal in the stroma (top right panel). ROS produced in peroxisomes during photorespiration, fatty acid oxidation, or other reactions are decomposed by SOD, catalase (CAT), and ascorbate peroxidase (APX) (middle, right panel). SOD and different components of the ascorbate glutathione cycle are also present in mitochondria. In addition, AOX is shown to prevent oxidative damage in mitochondria (bottom, right panel). In principle, the cytosol contains the same set of enzymes found in the stroma (bottom, left panel). However, these are encoded by a different set of genes and the major iron chelating activity in the cytosol responsible for preventing the formation of HO• radicals is unknown. The enzymatic components responsible for ROS detoxification in the apoplast and cell wall (W) are only partially known and the ROS-scavenging pathways at the vacuole (V) are unknown. Additional abbreviations: DHA, dehydroascrobate; DHAR, DHA reductase; FD, ferredoxin; FNR, ferredoxin NADPH reductase; GLR, glutaredoxin; GR, glutathione reductase; GSH and GSSG, reduced and oxidized glutathione, respectively; IM, inner membrane; IMS, IM space; MDA, monodehydroascorbate; MDAR, MDA reductase; PSI and PSII, photosystem I and II, respectively; tyl, thylakoid; Trx, thioredoxin.

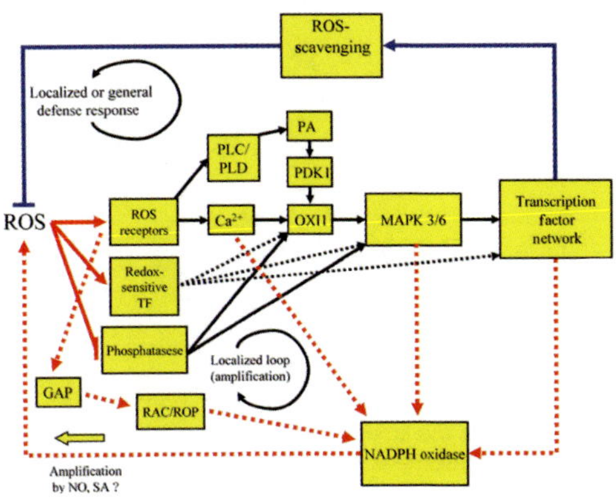

Plate 14 *A hypothetical model for the ROS (O_2^- and H_2O_2) signal transduction pathway of plants.* ROS are shown to be sensed by receptors, redox-sensitive transcriptional regulators, or by inhibition of phosphatases. ROS signaling is shown to be mediated by Ca^{2+}, protein phosphorylation, and G-proteins, and to involve negative (i.e., ROS scavenging), or positive (i.e., Rboh-NADPH oxidase) amplification loops. See text for more details.

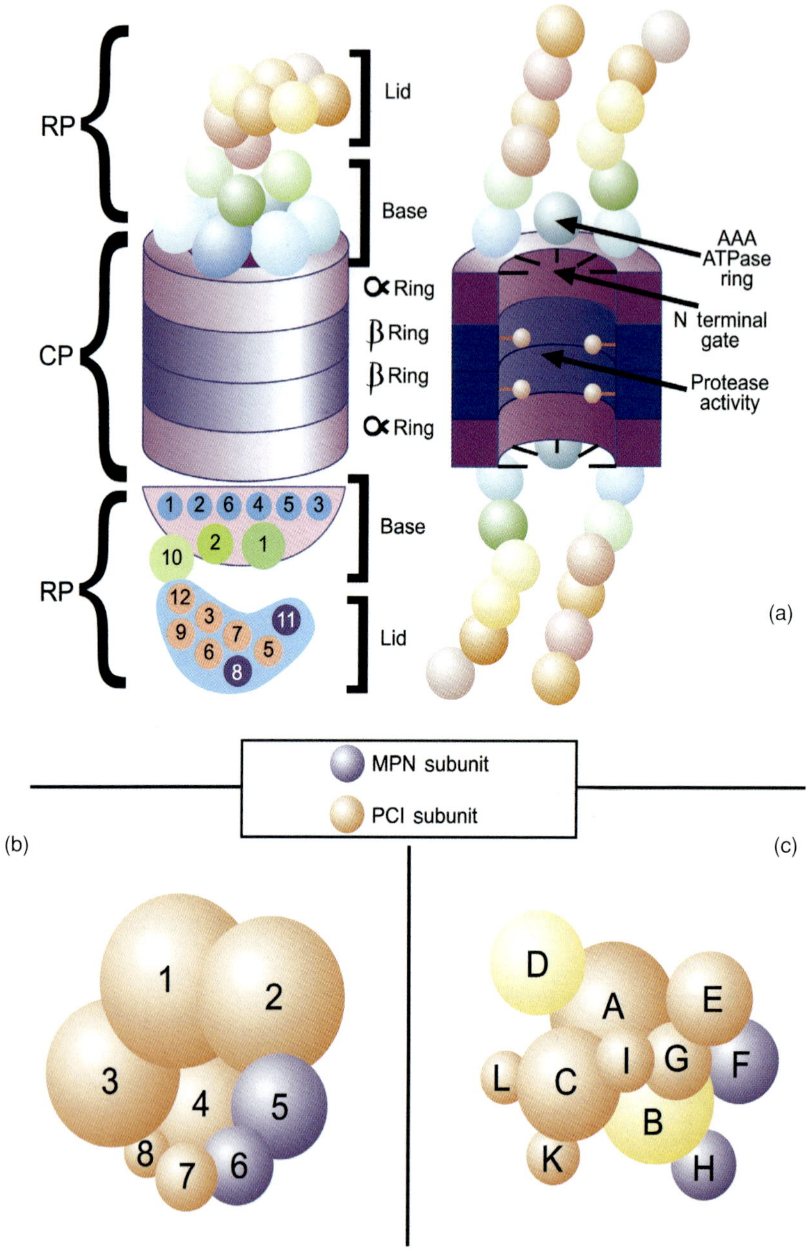

Plate 15 *The PCI complexes.* (a) The proteasome core particle (CP) is capped on both sides by the regulatory particle (RP), which can be further divided in to the lid and the base subparticles (left panel). The right panel shows a section through the proteasome where the α-subunits' N-terminal extensions serve as a gate that controls traffic in and out the CP. The β subunits possess the protease activity. (b) The COP9 signalosome (CSN) has eight subunits. The arrangement and stoichiometry are not accurate. (c) *Arabidopsis* eIF3 has 11 subunits. The arrangement and stochiometry are not accurate.

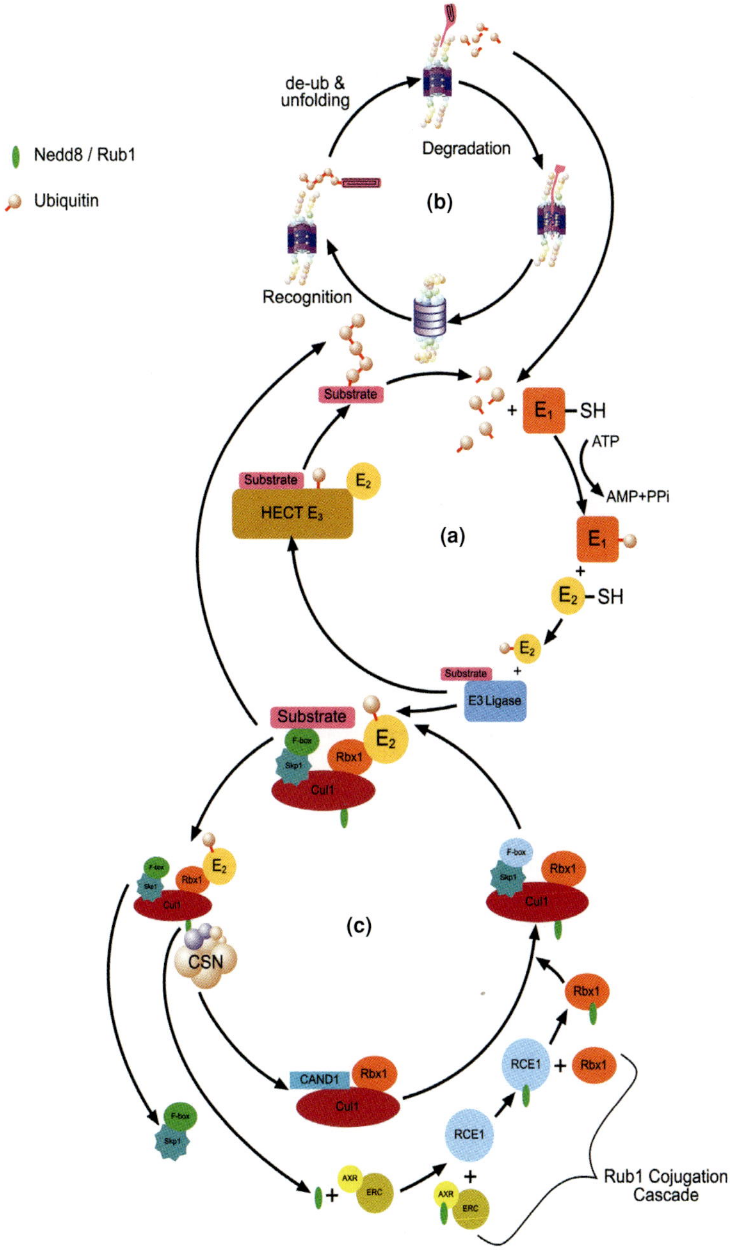

Plate 16 *Overview on the UPS.* (a) *The ubiquitylation cascade:* Ubiquitin is first covalently attached to the E1 activating enzyme in an ATP-dependent manner, then transferred and covalently attached to an E2 conjugating enzyme, and from there directly to the substrate (in the case of RING E3, down arrow) or to the E3 (in the case of HECT, upper arrow). In both cases the E3 defines substrate specify and ubiquitin is finally attached to the substrate. (b) *The degradation process:* Following ubiquitylation, the substrate is recognized by the 26S proteasome lid. The substrate is deubiquitylated, un-folded, the α-subunits' N-terminal extensions are opened and the substrate enters the CP where it is degraded. (c) *Cullin-based E3 ligase regulation*: The active E3 contains a Cullin modified by Rub1. After E3 activity, the CSN removes the Rub1. In the case of SCF E3, this leads to the dissociation of Skp1 and the F-box proteins, and the association of CAND1. The Rub1 cascade restores Rub1 on Cullin, which leads to the dissociation of CAND1 and reassociation of Skp1 with a new F-box protein, and E3 activity.

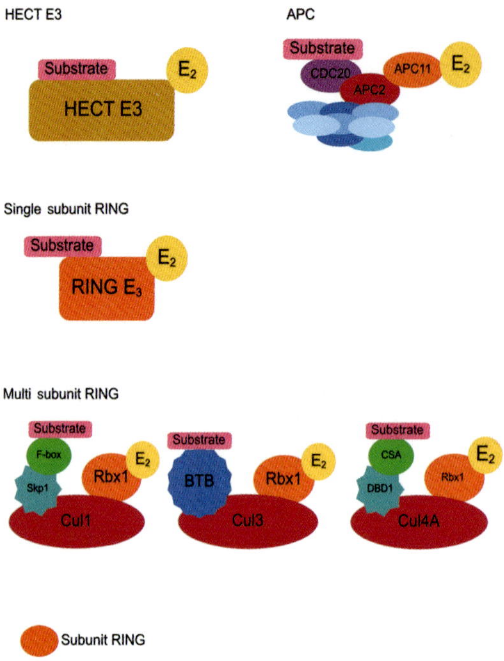

Plate 17 *Overall structures of the E3 ubiquitin ligase family.* The four main E3 families are shown: HECT (homology to E6-AP C terminus), APC (Anaphase-Promoting Complex), and single- and multi-subunit RING (Really Interesting New Gene) ligases. See text for details.

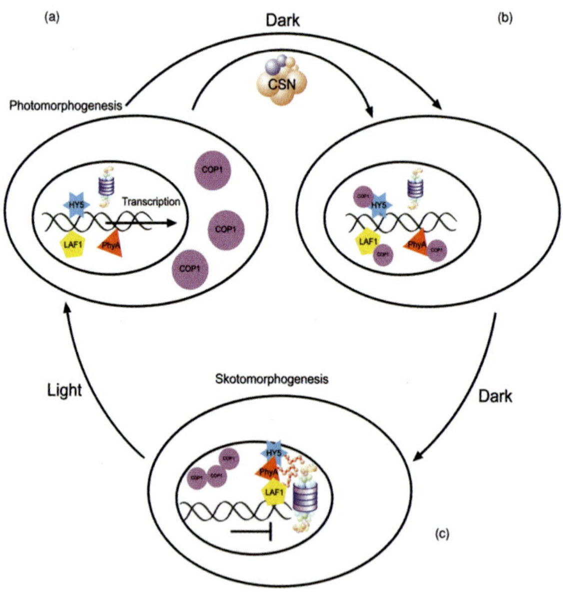

Plate 18 *Model for COP1 function.* (a) Under light conditions, COP1 is located in the cytoplasm and transcription factors such as HY5 and LAF1 are located in nucleus, active in transcription of light induced genes. (b and c) Under dark conditions, COP1 is translocated to the nucleus in a CSN-dependent manner (b) where it is active as E3 ubiquitin ligase toward proteins such as HY5, LAF1, and PhyA (c). Following ubiquitin chain assembly these proteins are degraded by the proteasome and light-induced gene transcription is ceased (c).

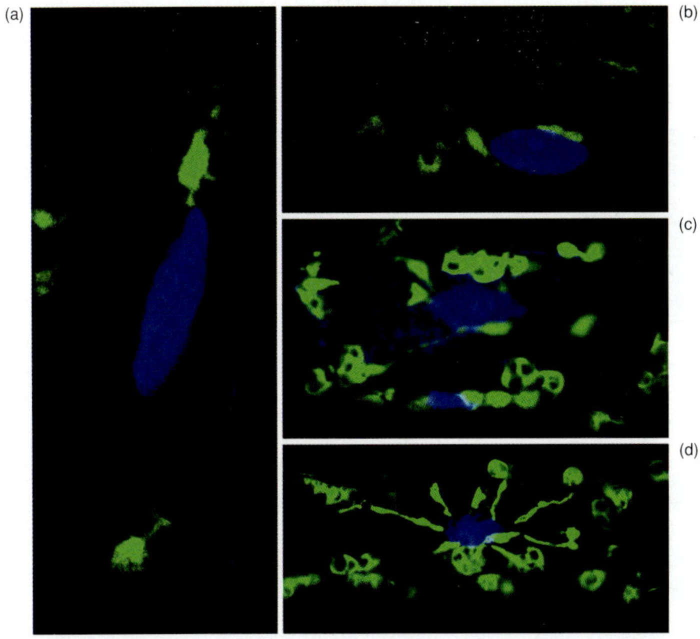

Plate 19 Confocal micrographs of plastids and stromules in norflurazon grown SCO1::GFP *Arabidopsis* seedlings. Confocal image is shown of plastid-tagged GFP (green) and DAPI staining of nuclei and plasma membrane (blue) in root cells (a and b), and hypocotyl cells (c and d).

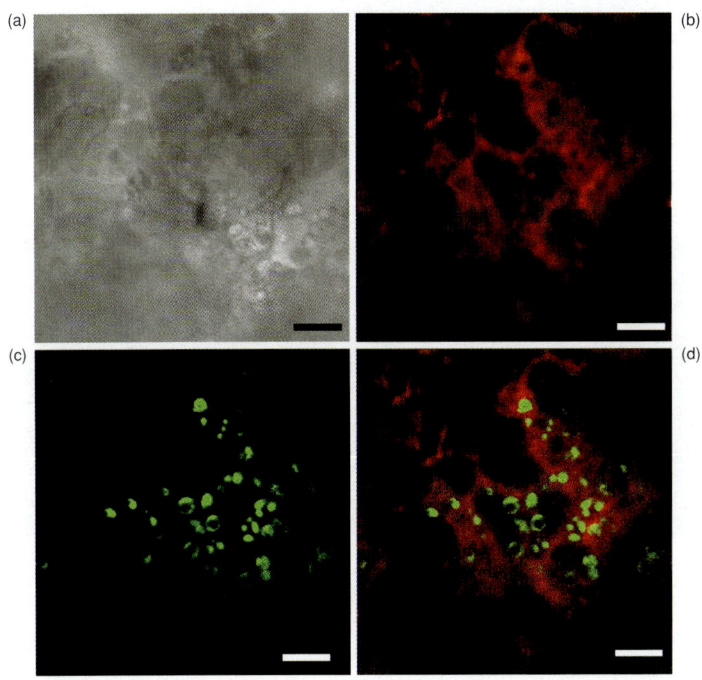

Plate 20 Accumulation of tetrapyrroles visualized using laser-scanning confocal microscopy of norflurazon treated, ALA fed SCO1::GFP *Arabidopsis* seedlings. Emission is shown for the cotyledons. Representative images were retrieved at 507–537 nm and 585–615 nm for specific emission of GFP (c and d) and Mg-ProtoIX (b and d) respectively. Bars = 50 μm.

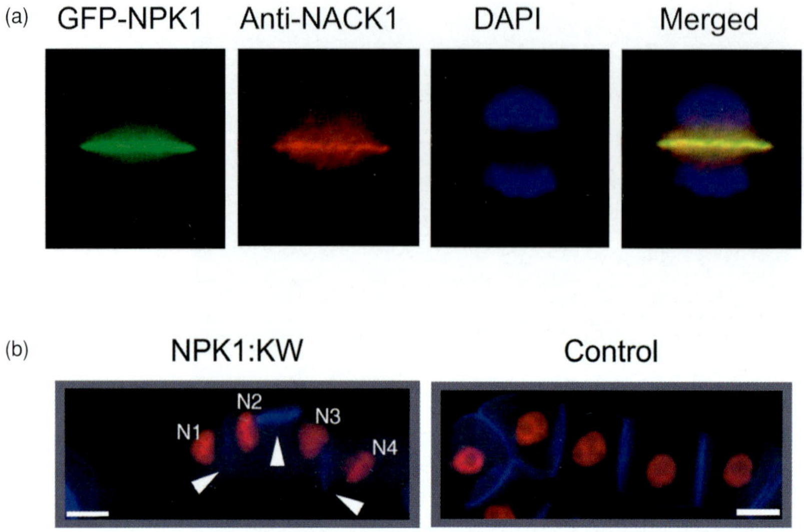

Plate 21 Involvement of the NACK1/NPK1 complex in plant cytokinesis. (a) Co-localization of NPK1 and NACK1 at the equator of the phragmoplast. BY-2 cells expressing GFP-NPK1 (green) were double-stained with rabbit antibodies against NACK1 (red) and 4′,6-diamidino-2-phenylindole (DAPI) for nuclei (blue). (b) Generation of multinucleate cells with incomplete cell plates upon expression of a kinase-defective NPK1 (NPK1:KW; left). Control BY-2 cells are shown in the right panel. Cells were stained with calcofluor (blue; cell wall) and propidium iodide (red; nucleus). Arrowheads, incomplete cell plates; N, nucleus. Bars = 20 μm.

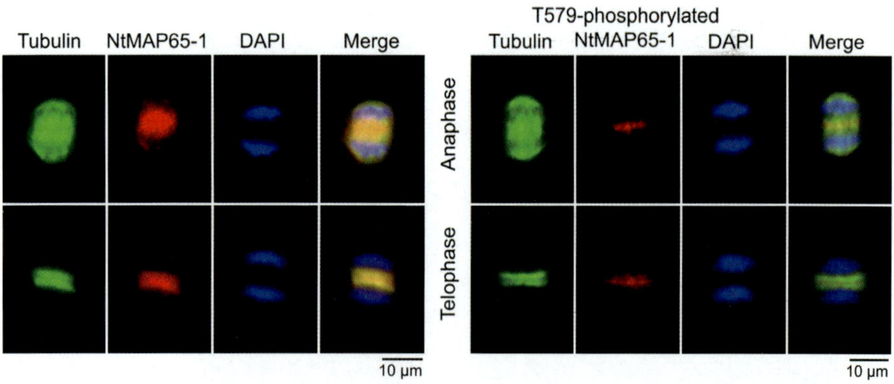

Plate 22 Subcellular localization of NtMAP65-1 (left) and NtMAP65-1 phosphorylated on Thr-579 (right) at anaphase and telophase in BY-2 cells. BY-2 cells were triple-stained with mouse antibodies against a-tubulin (*green*), rabbit antibodies against NtMAP65-1 or NtMAP65-1 phosphorylated on Thr-579 (*red*), and DAPI (*blue*).

Table 6.2 CBL genes and proteins in *Arabidopsis*

Name	Protein accession number	Nucleotide accession number (verified cDNAs)	MIPS/TAIR accession number	Synonyms	Amino acids	Introns
AtCBL1	AAC26008	AF076251	At4g17615		214	Yes
AtCBL2	AAC26009	AF076252	At5g55990		227	Yes
AtCBL3	AAC26010	AF076253	At4g26570		226	Yes
AtCBL4	AAG28402	AF192886	At5g24270	SOS3	223	Yes
AtCBL5	AAG28401	AF192885	At4g01420		214	Yes
AtCBL6	AAG28400	AF192884	At4g16350		227	Yes
AtCBL7	AAG10059	AF290434	At4g26560		214	Yes
AtCBL8	AAL10300	AF411957	At1g64480		214	Yes
AtCBL9	AAL10301	AF411958	At5g47100		213	Yes
AtCBL10	In progress	AF490607	At4g33000		256	Yes

EF-hand motifs. In addition to a general sequence difference, CaMs and CBLs also differ in the number of typical EF-hand motifs in their basic structure. Typically, CaMs contain four EF hands and CBLs contain three canonical EF hands. Recent studies have resolved the 3D structure of two members in the CBL family and in both cases, the fourth "EF hand" appears to diverge into a Mn-binding domain (Nagae *et al.*, 2003; Sanchez-Barrena *et al.*, 2005).

6.4.2 The CBL-type calcium sensors target a family of protein kinases—a shift-of-paradigm from calcineurin in yeast and animals

As discussed earlier, small calcium sensors function by targeting downstream effectors. Unlike CaMs that interact with a large variety of target proteins, CBLs appear to interact with a single family of protein kinases (Shi *et al.*, 1999). These kinases, referred to as CBL-interacting protein kinases (CIPKs), are most similar to sucrose nonfermenting (SNF) protein kinase from yeast and animals in the kinase domain but retain unique C-terminal regulatory domains. The CBL–CIPK interaction represents a major shift-of-paradigm in calcium signaling as compared to yeast and animals where calcineurin B protein interacts and regulates a protein phosphatase. The CBLs interact with CIPKs through the C-terminal nonkinase domain that contains a conserved region among different CIPK members (Shi *et al.*, 1999; Kim *et al.*, 2000; Abrecht 2001; Guo *et al.*, 2001). Interestingly, interaction between CBL1 and CIPK1 requires micromolar levels of Ca^{2+}. This Ca^{2+}-dependent interaction is consistent with the general paradigm established for Ca^{2+}-sensor interactions with target proteins in animals (e.g., Ca^{2+}-CaM–CaMKII interaction). Another study (Halfter *et al.*, 2000) using SOS3 (also referred to as CBL4) as a "bait" also identified several partner proteins that belong to the CIPK family. In particular, SOS3 interaction with SOS2 (also called CIPK24) stimulates kinase activity against a peptide substrate, suggesting that SOS3 serves as a regulatory subunit of SOS2. SOS2 and SOS3 were initially identified by a genetic screen for *Arabidopsis* mutants that are salt-overly sensitive (reviewed by Zhu, 2003).

Regarding the biochemical properties of CIPKs, studies showed that CIPKs have strong substrate specificity with very low activity against generic substrates (Shi *et al.*, 1999). In addition, the CIPK kinase activity prefers Mn^{2+} as a cofactor over Mg^{2+} (Shi *et al.*, 1999). Interaction with CBLs activates the kinase activity of CIPKs. One study suggests that the CBL-interacting domain may serve as an autoinhibitory domain that blocks the kinase active site (like the situation in CDPK or CaMK) (Guo *et al.*, 2001). The CBLs interact with the autoinhibitory domain in CIPKs and by doing so may release the kinase domain for substrate access (see Fig. 6.2, Color plate 8).

The *Arabidopsis* genome contains a large number of genes for putative CIPK proteins. Table 6.3 lists 25 CIPK genes that have been confirmed by cDNA

Table 6.3 CIPK genes and proteins in *Arabidopsis*

Name	Protein accession number	Nucleotide accession number (verified cDNAs)	MIPS/TAIR accession number	Synonyms	Amino acids	Introns
AtCIPK1	AAG28776	AF302112	At3g17510		444	Yes
AtCIPK2	AAF86506	AF286050	At5g07070		456	No
AtCIPK3	AAF86507	AF286051	At2g26980		375	Yes
AtCIPK4	AAG01367	AY007221	At4g14580		426	No
AtCIPK5	AAF86504	AF285105	At5g10930		431	No
AtCIPK6	AAF86505	AF285106	At4g30960		441	No
AtCIPK7	AAK16682	AF290192	At3g23000	AtSR1	429	No
AtCIPK8	AAK16683	AF290193	At4g24400		445	Yes
AtCIPK9	AAK16684	AF295664	At1g01140		449	Yes
AtCIPK10	AAK16685	AF295665	At5g58380		479	No
AtCIPK11	AAK16686	AF295666	At2g30360		435	No
AtCIPK12	AAK16687	AF295667	At4g18700		489	No
AtCIPK13	AAK16688	AF295668	At2g34180		502	No
AtCIPK14	AAK16689	AF295669	At5g01820	AtSR2	442	No
AtCIPK15	AAK16692	AF302111	At5g01810	AtPK10	421	No
AtCIPK16	AAK50348	AY030304	At2g25090		469	Yes
AtCIPK17	AAK64513	AY036958	At1g48260		421	Yes
AtCIPK18	AAK59695	AY034099	At1g29230		520	No
AtCIPK19	AAK50347	AY030303	At5g45810		483	No
AtCIPK20	AAK61493	AY035225	At5g45820		439	No
AtCIPK21	AAK59696	AY034100	At5g57630		417	Yes
AtCIPK22	AAL47845	AF450478	At2g38490		445	Yes
AtCIPK23	AAK61494	AY035226	At1g30270		482	Yes
AtCIPK24	AAK72257	AF395081	At5g35410	SOS2	446	Yes
AtCIPK25	AAL41008	AF448226	At5g25110		487	No

cloning and sequencing. Further experiments have extended the analysis of CBL–CIPK interactions to the entire family of CBLs and a large fraction of the CIPK family in an effort to determine the functional pairs of CBLs and CIPKs. These studies revealed that each CBL interacts with a subset of CIPKs and each CIPK interacts with one or more CBLs. Some CBLs share common CIPK targets and some CIPKs share common CBL-regulatory subunits. Such interaction specificity and overlap among various members in the CBL and CIPK family may well reflect the functional specificity and redundancy (Kim *et al.*, 2000; Albrecht *et al.*, 2001; Guo *et al.*, 2001). It must be noted, however, that these interaction studies were performed using mostly the yeast two-hybrid system and therefore may not necessarily represent the physiological situations in plants. In addition to matching the CBLs with their target kinases, the interaction studies further defined the functional domains of CBLs and CIPKs. For example, the CBL-interacting domain in the C-terminal region of CIPKs was localized to a small region of approximately 20 amino acids (Kim *et al.*, 2000; Albrecht *et al.*, 2001; Guo *et al.*, 2001). This domain may be important in kinase regulation by releasing the autoinhibitory domain (Guo *et al.*, 2001).

Besides regulating the activity of the CIPK kinases, certain structural features of CBLs also suggest these Ca^{2+} sensors can change cellular localization of the CBL–CIPK complexes. Several CBLs have a conserved myristoylation site in their N-terminal region (Liu and Zhu, 1998; Kudla *et al.*, 1999; Kim *et al.*, 2000; Albrecht *et al.*, 2001). It would be expected that these CBLs are localized to cell membranes, which could serve as a regulatory mechanism for establishing a local signal cascade similar to the model discussed for CaM53 above. For example, a significant amount of SOS3/CBL4 is always found associated with the membrane fraction and the myristoylation site is required for the function of the protein (Ishitani *et al.*, 2000). CBL1 and CBL9 are also associated with the membrane (D'Angelo *et al.*, 2006; Xu *et al.*, 2006) and target their common target CIPK1 and CIPK23 to the PM thereby enabling CIPK phosphorylation of membrane associated protein substrate(s) (see details in later sections). Together, the view emerges that in plants certain calcium sensors (including CDPK, CBL, and CaM) have acquired protein domains that restrict their localization, serving as a mechanism to establish local signal transduction pathways that initiate specific cellular responses.

6.4.3 Physiological pathways involving CBL–CIPK signaling modules

The diversity in protein sequence/structure and expression pattern of CBLs and CIPKs suggest that these proteins perform many diverse functions. To date, a physiological function has been established for several CBL–CIPK modules. As discussed above, SOS2/CIPK24 and SOS3/CBL4 have been identified by a genetic screen and play a role in salt tolerance in *Arabidopsis* (reviewed in Hasegawa *et al.*, 2000; Zhu, 2003). Because high salt triggers an increase in Ca^{2+} levels in the cytoplasm (Pauly *et al.*, 2000), salt tolerance could therefore involve Ca^{2+} signaling and the signal could be transmitted via CBL–CIPK pathways for salt detoxification (Hasegawa *et al.*, 2000). In this context, studies have established that SOS3/CBL4–SOS2/CIPK24 may directly regulate the downstream component SOS1, a putative Na^+/H^+ antiporter (Shi *et al.*, 2000), thereby enhancing the salt detoxification process. More recent studies (Kim *et al.*, 2007; Quan *et al.*, 2007) placed another CBL, CBL10, in the salt tolerance pathway. Interestingly, CBL10, like SOS3/CBL4, also interacts with and appears to function through SOS2/CIPK24. However, unlike SOS3/CBL4 that functions mainly in the roots, CBL10 is expressed and functions almost exclusively in the shoots/leaves (Kim *et al.*, 2007; Quan *et al.*, 2007). Perhaps the most unique feature of the *cbl10* mutant is that the mutant plants, despite being more sensitive to salt, accumulate less salt than the wild type (Kim *et al.*, 2007). This feature is consistent with the hypothesis that CBL10 may be required for the transport of salt into vacuole thereby controlling cellular salt homeostasis. This hypothesis is supported by the finding that CBL10 protein interacts with SOS2/CIPK24 and targets the CBL10–SOS2/CIPK24 complex to the vacuole membrane (tonoplast) (Kim *et al.*, 2007).

Biochemical studies have shown that CBL1 expression is highly responsive to a variety of abiotic stress conditions including wounding, cold, drought, and high salt, implicating CBL1 in signaling these stress signals (Kudla *et al.*, 1999). This hypothesis is supported by further genetic analysis of the CBL1 gene. Disruption of CBL1 gene function renders mutant plants hypersensitive to drought, high salt, and hyperosmotic stress (Cheong *et al.*, 2003; Albrecht *et al.*, 2003). These phenotypes suggest that CBL1 gene not only functions in salt tolerance but in other abiotic stress responses as well, which is distinct from SOS3/CBL4 that functions specifically in ionic homeostasis under high-salt condition. Interestingly, CBL9, although highly similar to CBL1 in its amino acid sequence, still displays functional specificity as compared to CBL1 and plays a role in ABA response and ABA biosynthesis in seed germination (Pandey *et al.*, 2004). The CBL1 and CBL9 both interact with CIPK1 forming alternative complexes that functionally diverge in regulating ABA-dependent and independent pathways for osmotic responses (D'Angelo *et al.*, 2006). Furthermore, CBL9 and CIPK3 form a specific complex that functions in ABA response in seed germination (Pandey *et al.*, 2004; Pandey *et al.*, 2008). The functional specificity defined by different CBL–CIPK complexes is a critical feature for the functional versatility of the CBL–CIPK network.

As highly similar CBLs, CBL1 and CBL9 also possess redundant functions. This has been demonstrated by studies showing that CBL1 and CBL9 both target CIPK23 and function redundantly in the regulation of potassium (K) uptake and stomatal movements (Li *et al.*, 2006; Xu *et al.*, 2006; Cheong *et al.*, 2007). A genetic screen for low-K tolerant mutants identified CIPK23 as a critical K-nutrition determinant in *Arabidopsis* (Xu *et al.*, 2006). A reverse genetic analysis on CIPK members identified CIPK23 as critical for both stomatal response and low-K response (Li *et al.*, 2006, Cheong *et al.*, 2007). The K-nutrition phenotype in the mutants was defined as hypersensitivity on low-K medium. Such defect is caused at least in part by the lower K-uptake capacity in the mutants (Li *et al.*, 2006; Xu *et al.*, 2006; Cheong *et al.*, 2007). As CIPK23 interacts with several CBLs including both CBL1 and CBL9 (Xu *et al.*, 2006; Cheong *et al.*, 2007), it was speculated that CBL1 and CBL9 may function in CIPK23-mediated pathways such as K-nutrition and stomatal responses. Because *cbl1* and *cbl9* single mutants did not show phenotypic changes in K-uptake and stomatal movement (Cheong *et al.*, 2003; Pandey *et al.*, 2004), it was predicted that CBL1 and CBL9 may function redundantly in these processes. Indeed, *cbl1cbl9* double mutant displayed defects in the low-K and stomatal response (Li *et al.*, 2006; Xu *et al.*, 2006; Cheong *et al.*, 2007). As both K-uptake and stomatal response involve K-transport across the PM, a potential downstream target of the CBL1/9–CIPK23 pathway may be a K transporter. This is consistent with the fact that both CBL1 and CBL9 are localized to the PM (possibly via myristoylation) and recruit their target CIPK23 to the same location. Using both biochemical and electrophysiological approaches, studies have identified AKT1, a voltage-gated K-channel, as a specific downstream target of CIPK23 (Li *et al.*, 2006; Xu *et al.*, 2006). The CIPK23 protein physically interacts with and phosphorylates the AKT1 channel protein. In the *Xenopus* oocyte model system for AKT1 activity assay, the CBL1 (or CBL9)–CIPK23 complex is required for the activation of the AKT1 channel (Li *et al.*, 2006; Xu *et al.*, 2006). The activation of the AKT1 depends on the presence of calcium and an active kinase domain, indicating that the kinase activity (phosphorylation of AKT1) and calcium messenger are essential for the AKT1 activation (Li *et al.*, 2006). If the CBL1/9–CIPK23–AKT1 pathway operates in vivo, disruption of CBL1/9 or CIPK23 would decrease AKT1 activity. Patch-clamping studies using root hair cells demonstrated that AKT1 activity in the *cipk23* mutant and *cbl1cbl9* double mutant is significantly reduced, connecting the CBL1/9–CIPK23 with the regulation of AKT1 activity and K-uptake *in planta* (Li *et al.*, 2006; Xu *et al.*, 2006, Cheong *et al.*, 2007; Fig. 6.6).

The study on the CBL1/9–CIPK23–AKT1 pathway is significant not only for the functional analysis of the CBL–CIPK network but also for the understanding of nutrient sensing signal transduction (as discussed in Li *et al.*, 2006). It has long been realized that culturing plants in the low-K medium enhances the K-uptake capacity of the plant roots (reviewed by Ashley *et al.*, 2006). However, little is known about the signaling components that connect the low-K signal with the enhanced K uptake. Recent studies suggest that low K increases production of reactive oxygen species (ROS) that trigger calcium fluctuations in root cells (Foreman *et al.*, 2003; Shin and Schachtman, 2005). Studies on the CBL1/9–CIPK23–AKT1 pathway place several components in the signaling process for low-K response (Fig. 6.7, Color plate 12). Further studies show that several more members of CBL and CIPK families can regulate the AKT1 channel. In addition, a PP2C-type protein phosphatase inactivates AKT1, establishing the first comprehensive kinase–phosphatase network that regulates an ion channel in plant cell (Lee *et al.* 2007).

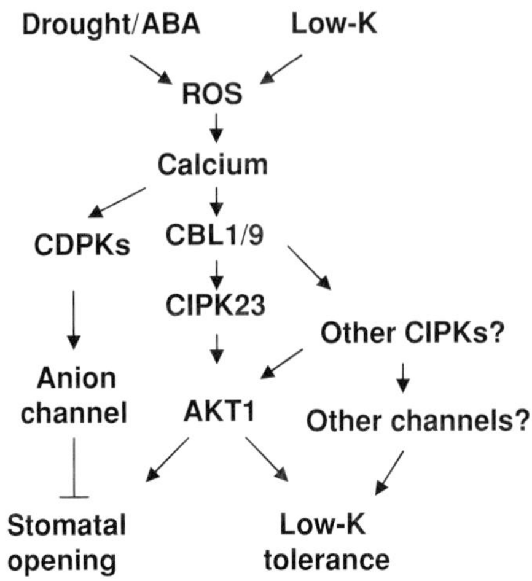

Figure 6.6 CBL–CIPK pathway involved in regulation of both stomatal movement and K uptake. Drought/ABA and low K may utilize ROS as a chemical messenger that triggers calcium changes. CBLs bind calcium and interact and activate CIPK23 that activates AKT1 promoting stomata opening and K uptake/low-K tolerance. Other CIPKs and other channels may also be involved in this "network." The other branch of the calcium signaling activates CDPKs that induces stomatal closure by activating anion channels (see the text).

6.5 Perspectives: complex networks for Ca^{2+} decoding in plant cells

Most if not all of signaling pathways in plants involve Ca^{2+} signal one way or the other (reviewed by Sanders *et al.*, 1999, 2002; Trewavas, 1999; Rudd and Franklin-Tong, 2001). This leads us to the opening question of "how can a cell distinguish different extracellular signals by specific responses if they all use Ca^{2+} as a messenger?" If one examines the nature of Ca^{2+} signals produced by various stimuli in both animals and plants, they vary in their spatial and temporal information, which is referred to as "Ca^{2+} signatures" (reviewed in Rudd and Franklin-Tong, 2001; Sanders *et al.*, 2002). Subtle differences in these signatures may be responsible in part for the "specificity" in cellular responses. There is little doubt that Ca^{2+} sensors and their targets are involved in further defining such signaling "specificity." First, in plants there exist a number of distinct calcium sensors including but not limited to those discussed here (CDPKs, CaMs, and CBLs), their target proteins, and further downstream components in the signaling pathways. What pathway to take for decoding a particular Ca^{2+} signature may depend largely on the presence of these signaling components in a specific cell. This is consistent with the distinct temporal and spatial expression pattern of calcium sensors and their target proteins, which determine the abundance of the signaling proteins in a particular cell under a particular condition. In addition, targeting of certain proteins such as CaM53 and CBL1 to specific subcellular locations will further define their function in the readout of Ca^{2+} signatures. Second, the specificity of calcium sensors in interactions with their targets certainly plays an important role for the diversity of cellular responses. For example, different CBLs interact with different subgroups of CIPKs or different CBLs may interact with the same subgroup of CIPKs with different affinities. The combinatorial potential of these proteins would contribute to the mechanism for a discrete response. Such examples include CBL1 and CBL9, two highly similar calcium sensors that have both redundant and specific functions. Third, substrate specificity and differential cofactor dependence of CIPKs presents an additional level of regulation in the CBL–CIPK system. The availability of a particular substrate for CIPKs in a cell also contributes to specificity. In conclusion, "specificity" in the Ca^{2+} signaling system results from a multifactorial decision process, ranging from a specific Ca^{2+} signature to the availability of a specific set of calcium sensors and their target proteins, which are coupled to downstream components. Each step in this process constrains the Ca^{2+} signal, ultimately leading to "specificity" in cellular responses, yet providing opportunities at every step for potential cross-talk to parallel or competing pathways. To fully understand the Ca^{2+} signaling pathways, we must not only decode the Ca^{2+} signatures but also dissect the "combination code" that consists of calcium sensors and downstream target proteins.

Acknowledgments

The author thanks Dr Sung Chul Lee for assistance in formatting the reference list and colleagues in my laboratory for helpful discussion. Apologies go to colleagues in the field whose work was not discussed because of the focus of this chapter. Related research in the author's laboratory is supported by the National Science Foundation.

References

Albrecht, V., Ritz, O., Linder, S., Harter, K. and Kudla, J. (2001) The NAF domain defines a novel protein-protein interaction module conserved in Ca^{2+}-regulated kinases. *Embo J*, **20**, 1051–1063.

Albrecht, V., Weinl, S., Blazevic, D., D'Angelo, C., Batistic, O., Kolukisaoglu, U., Bock, R., Schulz, B., Harter, K. and Kudla J. (2003) The calcium sensor CBL1 integrates plant responses to abiotic stresses. *Plant J*, **36** (4), 457–470.

Allen, G.J., Chu, S.P., Schumacher, K., Shimazaki, C.T., Vafeados, D., Kemper, A., Hawke, S.D., Tallman, G., Tsien, R.Y., Harper, J.F., Chory, J. and Schroeder, J.I. (2000) Alteration of stimulus-specific guard cell calcium oscillations and stomatal closing in *Arabidopsis* det3 mutant. *Science*, **289**, 2338–2342.

Allen, G.J. and Sanders, D. (1995) Calcineurin, a type 2B protein phosphatase, modulates the Ca^{2+}-permeable slow vacuolar ion channel of stomatal guard cells. *Plant Cell*, **7**, 1473–1483.

Allen, G.J., Chu, S.P., Harrington, C.L., Schumacher, K., Hoffmann, T., Tang, Y.Y., Grill, E. and Schroeder, J.I. (2001) A defined range of guard cell calcium oscillation parameters encodes stomatal movements. *Nature*, **411**, 1053–1057.

Arazi, T., Kaplan, B. and Fromm, H. (2000) A high-affinity calmodulin-binding site in a tobacco plasma-membrane channel protein coincides with a characteristic element of cyclic nucleotide-binding domains. *Plant Mol Biol*, **42**, 591–601.

Arazi, T., Sunkar, R., Kaplan, B. and Fromm, H. (1999) A tobacco plasma membrane calmodulin-binding transporter confers Ni^{2+} tolerance and Pb^{2+} hypersensitivity in transgenic plants. *Plant J*, **20**, 171–182.

Babu, Y.S., Bugg, C.E. and Cook, W.J. (1988) Structure of calmodulin refined at 2.2 Å resolution. *J Mol Biol*, **204**, 191–204.

Baum, G., Chen, Y., Arazi, T., Takatsuji, H. and Fromm, H. (1993) A plant glutamate decarboxylase containing a calmodulin binding domain. Cloning, sequence, and functional analysis. *J Biol Chem*, **268**, 19610–19617.

Baum, G., Lev-Yadun, S., Fridmann, Y., Arazi, T., Katsnelson, H., Zik, M. and Fromm, H. (1996) Calmodulin binding to glutamate decarboxylase is required for regulation of glutamate and GABA metabolism and normal development in plants. *Embo J*, **15**, 2988–2996.

Beckingham, K. (1991) Use of site-directed mutations in the individual Ca2(+)-binding sites of calmodulin to examine Ca2(+)-induced conformational changes. *J Biol Chem*, **266**, 6027–6030.

Braam, J., Sistrunk, M.L., Polisensky, D.H., Xu, W., Purugganan, M.M., Antosiewicz, D.M., Campbell, P. and Johnson, K.A. (1997) Plant responses to environmental stress: regulation and functions of the *Arabidopsis* TCH genes. *Planta*, **203** (Suppl), S35–S41.

Bush, D.S. (1995) Calcium regulation in plant cells and its role in signaling. *Annu Rev Plant Physiol Plant Mol Biol*, **46**, 95–122.

Caldelari, D., Sternberg, H., Rodriguez-Concepcion, M., Gruissem, W. and Yalovsky, S. (2001) Efficient prenylation by a plant geranylgeranyltransferase-I requires a functional CaaL box motif and a proximal polybasic domain. *Plant Physiol*, **126**, 1416–1429.

Cheng, S.H., Sheen, J., Gerrish, C. and Bolwell, G.P. (2001) Molecular identification of phenylalanine ammonia-lyase as a substrate of a specific constitutively active *Arabidopsis* CDPK expressed in maize protoplasts. *FEBS Lett*, **503**, 185–188.

Cheng, S.H., Willmann, M.R., Chen, H.C. and Sheen, J. (2002) Calcium signaling through protein kinases. The *Arabidopsis* calcium-dependent protein kinase gene family. *Plant Physiol*, **129**, 469–485.

Cheong, Y.H., Kim, K.N., Pandey, G.K., Gupta, R., Grant, J.J. and Luan, S. (2003) CBL1, a calcium sensor that differentially regulates salt, drought, and cold responses in *Arabidopsis*. *Plant Cell*, **15**, 1833–1845.

Cheong, Y.H., Pandey, G.K., Grant, J.J., Batistic, O., Li, L., Kim, B.-G., Lee, S.C., Kudla, J. and Luan, S. (2007) Two Calcineurin B-like calcium sensors, interacting with protein kinase CIPK23, regulate leaf transpiration and root potassium uptake in *Arabidopsis*. *Plant J*, **52**, 223–239.

Craske, M., Takeo, T., Gerasimenko, O., Vaillant, C., Torok, K., Petersen, O.H. and Tepikin, A.V. (1999) Hormone-induced secretory and nuclear translocation of calmodulin: oscillations of calmodulin concentration with the nucleus as an integrator. *Proc Natl Acad Sci USA*, **96**, 4426–4431.

D'Angelo, C., Weinl, S., Batistic, O., Pandey, G.K., Cheong, Y.H., Schultke, S., Albrecht, V., Ehlert, B., Schulz, B., Harter, K., Luan, S., Bock, R. and Kudla, J. (2006) Alternative complex formation of the Ca-regulated protein kinase CIPK1 controls abscisic acid-dependent and independent stress responses in *Arabidopsis*. *Plant J*, **48**, 857–872.

Dolmetsch, R.E., Xu, K. and Lewis, R.S. (1998) Calcium oscillations increase the efficiency and specificity of gene expression. *Nature*, **392**, 933–936.

Dong, A., Xin, H., Yu, Y., Sun, C., Cao, K. and Shen, W.H. (2002) The subcellular localization of an unusual rice calmodulin isoform, OsCaM61, depends on its prenylation status. *Plant Mol Biol*, **48**, 203–210.

Ehrhardt, D.W., Wais, R. and Long, S.R. (1996) Calcium spiking in plant root hairs responding to Rhizobium nodulation signals. *Cell*, **85**, 673–681.

Estruch, J.J., Kadwell, S., Merlin, E. and Crossland, L. (1994) Cloning and characterization of a maize pollen-specific calcium-dependent calmodulin-independent protein kinase. *Proc Natl Acad Sci USA*, **91**, 8837–8841.

Falke, J.J., Drake, S.K., Hazard, A.L. and Peersen, O.B. (1994) Molecular tuning of ion binding to calcium signaling proteins. *Q Rev Biophys*, **27**, 219–290.

Felle, H.H. and Hepler, P.K. (1997) The cytosolic Ca^{2+} concentration gradient of *Sinapis alba* root hairs as revealed by Ca^{2+}-selective microelectrode tests and fura-dextran ratio imaging. *Plant Physiol*, **114**, 39–45.

Foreman, J., Demidchik, V., Bothwell, J.H., Mylona, P., Miedema, H., Torres, M.A., Linstead, P., Costa, S., Brownlee, C., Jones, J.D., Davies, J.M. and Dolan, L. (2003) Reactive oxygen species produced by NADPH oxidase regulate plant cell growth. *Nature* 422, 442–426.

Franklin-Tong, V.E., Drobak, B.K., Allan, A.C., Watkins, P. and Trewavas, A.J. (1996) Growth of pollen tubes of papaver rhoeas is regulated by a slow-moving calcium wave propagated by inositol 1,4,5-trisphosphate. *Plant Cell*, **8**, 1305–1321.

Gleason, C., Chaudhuri, S., Yang, T., Munoz, A., Poovaiah, B.W. and Oldroyd, G.E. (2006) Nodulation independent of rhizobia induced by a calcium-activated kinase lacking autoinhibition. *Nature*, **441**, 1149–1152.

Guenther, J.F., Chanmanivone, N., Galetovic, M.P., Wallace, I.S., Cobb, J.A. and Roberts, D.M. (2003) Phosphorylation of soybean nodulin 26 on serine 262 enhances water permeability and is regulated developmentally and by osmotic signals. *Plant Cell*, **15**, 981–991.

Guo, H., Mockler, T., Duong, H. and Lin, C. (2001a) SUB1, an *Arabidopsis* Ca^{2+}-binding protein involved in cryptochrome and phytochrome coaction. *Science*, **291**, 487–490.

Guo, Y., Halfter, U., Ishitani, M. and Zhu, J.K. (2001b) Molecular characterization of functional domains in the protein kinase SOS2 that is required for plant salt tolerance. *Plant Cell*, **13**, 1383–1400.

Haiech, J., Kilhoffer, M.C., Lukas, T.J., Craig, T.A., Roberts, D.M. and Watterson, D.M. (1991) Restoration of the calcium binding activity of mutant calmodulins toward normal by the presence of a calmodulin binding structure. *J Biol Chem*, **266**, 3427–3431.

Halfter, U., Ishitani, M. and Zhu, J.K. (2000) The *Arabidopsis* SOS2 protein kinase physically interacts with and is activated by the calcium-binding protein SOS3. *Proc Natl Acad Sci USA*, **97**, 3735–3740.

Hardin, S.C., Winter, H. and Huber, S.C. (2004) Phosphorylation of the amino terminus of maize sucrose synthase in relation to membrane association and enzyme activity. *Plant Physiol*, **134**, 1427–1438.

Harmon, A.C., Gribskov, M. and Harper, J.F. (2000) CDPKs—a kinase for every Ca^{2+} signal? *Trends Plant Sci*, **5**, 154–159.

Harper, J.F., Breton, G. and Harmon, A. (2004) Decoding Ca(2+) signals through plant protein kinases. *Annu Rev Plant Biol*, **55**, 263–288.

Harper, J.F. and Harmon, A. (2005) Plants, symbiosis and parasites: a calcium signalling connection. *Nat Rev Mol Cell Biol*, **6**, 555–566.

Harper, J.F., Sussman, M.R., Schaller, G.E., Putnam-Evans, C., Charbonneau, H. and Harmon, A.C. (1991) A calcium-dependent protein kinase with a regulatory domain similar to calmodulin. *Science*, **252**, 951–954.

Hasegawa, M., Bressan, R. and Pardo, J.M. (2000) The dawn of plant salt tolerance genetics. *Trends Plant Sci*, **5**, 317–319.

Holdaway-Clarke, T.L., Feijo, J.A., Hackett, G.R., Kunkel, J.G. and Hepler, P.K. (1997) Pollen tube growth and the intracellular cytosolic calcium gradient oscillate in phase while extracellular calcium influx is delayed. *Plant Cell*, **9**, 1999–2010.

Hook, S.S. and Means, A.R. (2001) Ca(2+)/CaM-dependent kinases: from activation to function. *Annu Rev Pharmacol Toxicol*, **41**, 471–505.

Hrabak, E.M., Chan, C.W., Gribskov, M., Harper, J.F., Choi, J.H., Halford, N., Kudla, J., Luan, S., Nimmo, H.G., Sussman, M.R., Thomas, M., Walker-Simmons, K., Zhu, J.K. and Harmon, A.C. (2003) The *Arabidopsis* CDPK-SnRK superfamily of protein kinases. *Plant Physiol*, **132**, 666–680.

Hwang, I., Sze, H. and Harper, J.F. (2000) A calcium-dependent protein kinase can inhibit a calmodulin-stimulated Ca^{2+} pump (ACA2) located in the endoplasmic reticulum of *Arabidopsis*. *Proc Natl Acad Sci USA*, **97**, 6224–6229.

Ishitani, M., Liu, J., Halfter, U., Kim, C.S., Shi, W. and Zhu, J.K. (2000) SOS3 function in plant salt tolerance requires N-myristoylation and calcium binding. *Plant Cell*, **12**, 1667–1678.

Kim, K.N., Cheong, Y.H., Gupta, R. and Luan, S. (2000) Interaction specificity of *Arabidopsis* calcineurin B-like calcium sensors and their target kinases. *Plant Physiol*, **124**, 1844–1853.

Kim, B.G., Waadt, R., Cheong, Y.H., Pandey, G.K., Dominguez-Solis, J.R., Schültke, S., Lee, S.C., Kudla, J. and Luan, S. (2007) The calcium sensor CBL10 mediates salt tolerance by regulating ion homeostasis in *Arabidopsis*. *Plant J*, **52**, 473–484.

Klee, C.B., Ren, H. and Wang, X. (1998) Regulation of the calmodulin-stimulated protein phosphatase, calcineurin. *J Biol Chem*, **273**, 13367–13370.

Kohler, C., Merkle, T. and Neuhaus, G. (1999) Characterisation of a novel gene family of putative cyclic nucleotide- and calmodulin-regulated ion channels in *Arabidopsis thaliana*. *Plant J*, **18**, 97–104.

Kuboniwa, H., Tjandra, N., Grzesiek, S., Ren, H., Klee, C.B. and Bax, A. (1995) Solution structure of calcium-free calmodulin. *Nat Struct Biol*, **2**, 768–776.

Kudla, J., Xu, Q., Harter, K., Gruissem, W. and Luan, S. (1999) Genes for calcineurin B-like proteins in *Arabidopsis* are differentially regulated by stress signals. *Proc Natl Acad Sci USA*, **96**, 4718–4723.

Lee, S.C., Lan, W.Z., Kim, B.G., Li, L., Cheong, Y.H., Pandey, G.K., Lu, G., Buchanan, B.B. and Luan, S. (2007) A protein phosphorylation/dephosphorylation network regulates a plant potassium channel. *Proc Natl Acad Sci USA*, **104**(40), 15959–15964.

Leng, Q., Mercier, R.W., Yao, W. and Berkowitz, G.A. (1999) Cloning and first functional characterization of a plant cyclic nucleotide-gated cation channel. *Plant Physiol*, **121**, 753–761.

Li, L., Kim, B.G., Cheong, Y.H., Pandey, G.K. and Luan, S. (2006) A Ca(2)+ signaling pathway regulates a K(+) channel for low-K response in *Arabidopsis*. *Proc Natl Acad Sci USA*, **103**, 12625–12630.

Li, W., Llopis, J., Whitney, M., Zlokarnik, G. and Tsien, R.Y. (1998) Cell-permeant caged InsP3 ester shows that Ca^{2+} spike frequency can optimize gene expression. *Nature*, **392**, 936–941.

Ling, V. and Zielinski, R.E. (1993) Isolation of an *Arabidopsis* cDNA sequence encoding a 22 kDa calcium-binding protein (CaBP-22) related to calmodulin. *Plant Mol Biol*, **22**, 207–214.

Liu, F., Yoo, B.C., Lee, J.Y., Pan, W. and Harmon, A.C. (2006) Calcium-regulated phosphorylation of soybean serine acetyltransferase in response to oxidative stress. *J Biol Chem*, **281** (37), 27405–27415.

Liu, J., Ishitani, M., Halfter, U., Kim, C.S. and Zhu, J.K. (2000) The *Arabidopsis thaliana* SOS2 gene encodes a protein kinase that is required for salt tolerance. *Proc Natl Acad Sci USA*, **97**, 3730–3734.

Liu, J. and Zhu, J.K. (1998) A calcium sensor homolog required for plant salt tolerance. *Science*, **280**, 1943–1945.

Luan, S. (2002) Signalling drought in guard cells. *Plant Cell Environ*, **25**, 229–237.

Luan, S., Kudla, J., Rodriguez-Concepcion, M., Yalovsky, S. and Gruissem, W. (2002) Calmodulins and calcineurin B-like proteins: calcium sensors for specific signal response coupling in plants. *Plant Cell*, **14** (Suppl), S389–S400.

Luan, S., Li, W., Rusnak, F., Assmann, S.M. and Schreiber, S.L. (1993) Immunosuppressants implicate protein phosphatase regulation of K^+ channels in guard cells. *Proc Natl Acad Sci USA*, **90**, 2202–2206.

MacRobbie, E.A. (2000) ABA activates multiple Ca(2+) fluxes in stomatal guard cells, triggering vacuolar K(+) (Rb(+)) release. *Proc Natl Acad Sci USA*, **97**, 12361–12368.

Marechal, E., Hiratsuka, K., Delgado, J., Nairn, A., Qin, J., Chait, B.T. and Chua, N.H. (1999) Modulation of GT-1 DNA-binding activity by calcium-dependent phosphorylation. *Plant Mol Biol*, **40**, 373–386.

Martin, M.L. and Busconi, L. (2000) Membrane localization of a rice calcium-dependent protein kinase (CDPK) is mediated by myristoylation and palmitoylation. *Plant J*, **24**, 429–435.

McAinsh, M.R., Brownlee, C. and Hetherington, A.M. (1997) Calcium ions as second messengers in guard cell signal transduction. *Physiol Plant*, **100**, 16–29.

Mori, I.C., Murata, Y., Yang, Y., Munemasa, S., Wang, Y.F., Andreoli, S., Tiriac, H., Alonso, J.M., Harper, J.F., Ecker, J.R., Kwak, J.M. and Schroeder, J.I. (2006) CDPKs CPK6 and CPK3 function in ABA regulation of guard cell S-type anion- and Ca(2+)-permeable channels and stomatal closure. *PLoS Biol*, **4**, e327.

Nagae, M., Nozawa, A., Koizumi, N., Sano, H., Hashimoto, H., Sato, M. and Shimizu, T. (2003) The crystal structure of the novel calcium-binding protein AtCBL2 from *Arabidopsis thaliana*. *J Biol Chem*, **278**, 42240–42246.

Nelson, M.R. and Chazin, W. (1998) *Calmodulin as a Calcium Sensor*. Academic Press, San Diego.

O'Neil, K.T. and DeGrado, W.F. (1990) How calmodulin binds its targets: sequence independent recognition of amphiphilic alpha-helices. *Trends Biochem Sci*, **15**, 59–64.

Oldroyd, G.E. and Downie, J.A. (2004) Calcium, kinases and nodulation signalling in legumes. *Nat Rev Mol Cell Biol*, **5**, 566–576.

Osawa, M., Swindells, M.B., Tanikawa, J., Tanaka, T., Mase, T., Furuya, T. and Ikura, M. (1998) Solution structure of calmodulin-W-7 complex: the basis of diversity in molecular recognition. *J Mol Biol*, **276**, 165–176.

Pandey, G.K., Cheong, Y.H., Kim, K.N., Grant, J.J., Li, L., Hung, W., D'Angelo, C., Weinl, S., Kudla, J. and Luan, S. (2004) The calcium sensor calcineurin B-like 9 modulates abscisic acid sensitivity and biosynthesis in *Arabidopsis*. *Plant Cell*, **16**, 1912–1924.

Pandey, G.K., Grant, J.J., Cheong, Y.H., Kim, B.G., Li, L.G. and Luan, S. (2008) Calcineurin-B-like protein CBL9 interacts with target kinase CIPK3 in the regulation of ABA response in seed germination. *Mol Plant*, **1**, 238–248.

Pardo, J.M., Reddy, M.P., Yang, S., Maggio, A., Huh, G.H., Matsumoto, T., Coca, M.A., Paino-D'Urzo, M., Koiwa, H., Yun, D.J., Watad, A.A., Bressan, R.A. and Hasegawa, P.M. (1998) Stress signaling through Ca^{2+}/calmodulin-dependent protein phosphatase calcineurin mediates salt adaptation in plants. *Proc Natl Acad Sci USA*, **95**, 9681–9686.

Patil, S., Takezawa, D. and Poovaiah, B.W. (1995) Chimeric plant calcium/calmodulin-dependent protein kinase gene with a neural visinin-like calcium-binding domain. *Proc Natl Acad Sci USA*, **92**, 4897–4901.

Pauly, N., Knight, M.R., Thuleau, P., van der Luit, A.H., Moreau, M., Trewavas, A.J., Ranjeva, R. and Mazars, C. (2000) Control of free calcium in plant cell nuclei. *Nature*, **405**, 754–755.

Quan, R., Lin, H., Mendoza, I., Zhang, Y., Cao, W., Yang, Y., Shang, M., Chen, S., Pardo, J.M. and Guo, Y. (2007) SCABP8/CBL10, a putative calcium sensor, interacts with the protein kinase SOS2 to protect *Arabidopsis* shoots from salt stress. *Plant Cell*, **19**, 1415–1431.

Reddy, A.S., Narasimhulu, S.B., Safadi, F. and Golovkin, M. (1996) A plant kinesin heavy chain-like protein is a calmodulin-binding protein. *Plant J*, **10**, 9–21.

Reddy, V.S., Ali, G.S. and Reddy, A.S. (2002) Genes encoding calmodulin-binding proteins in the *Arabidopsis* genome. *J Biol Chem*, **277**, 9840–9852.

Roberts, D.M. and Harmon, A.C. (1992) Calcium modulated protein targets of intracellular calcium signals in higher plants. *Annu Rev Plant Physiol Plant Mol Biol*, **43**, 375–414.

Rodriguez-Concepcion, M., Toledo-Ortiz, G., Yalovsky, S., Caldelari, D. and Gruissem, W. (2000) Carboxyl-methylation of prenylated calmodulin CaM53 is required for efficient plasma membrane targeting of the protein. *Plant J*, **24**, 775–784.

Rodriguez-Concepcion, M., Yalovsky, S., Zik, M., Fromm, H. and Gruissem, W. (1999) The prenylation status of a novel plant calmodulin directs plasma membrane or nuclear localization of the protein. *Embo J*, **18**, 1996–2007.

Rudd, J.J. and Franklin-Tong, V.E. (2001) Unravelling response-specificity in Ca^{2+} signaling pathways in plant cells. *New Phytol*, **151**, 7–33.

Sanchez-Barrena, M.J., Martinez-Ripoll, M., Zhu, J.K. and Albert, A. (2005) The structure of the *Arabidopsis thaliana* SOS3: molecular mechanism of sensing calcium for salt stress response. *J Mol Biol*, **345**, 1253–1264.

Sanders, D., Brownlee, C. and Harper, J.F. (1999) Communicating with calcium. *Plant Cell*, **11**, 691–706.

Sanders, D., Pelloux, J., Brownlee, C. and Harper, J.F. (2002) Calcium at the crossroads of signaling. *Plant Cell*, **14** (Suppl), S401–S417.

Schroeder, J.I., Allen, G.J., Hugouvieux, V., Kwak, J.M. and Waner, D. (2001) Guard cell signal transduction. *Annu Rev Plant Physiol Plant Mol Biol*, **52**, 627–658.

Schuurink, R.C., Shartzer, S.F., Fath, A. and Jones, R.L. (1998) Characterization of a calmodulin-binding transporter from the plasma membrane of barley aleurone. *Proc Natl Acad Sci USA*, **95**, 1944–1949.

Seamon, K.B. and Kreetsinger, R.H. (1983) Calcium-modulated proteins. *Met Ions Biol*, **6**, 1–51.

Shi, H., Ishitani, M., Kim, C. and Zhu, J.K. (2000) The *Arabidopsis thaliana* salt tolerance gene SOS1 encodes a putative Na^+/H^+ antiporter. *Proc Natl Acad Sci USA*, **97**, 6896–6901.

Shi, J., Kim, K.N., Ritz, O., Albrecht, V., Gupta, R., Harter, K., Luan, S. and Kudla, J. (1999) Novel protein kinases associated with calcineurin B-like calcium sensors in *Arabidopsis*. *Plant Cell*, **11**, 2393–2405.

Shin, R. and Schachtman, D.P. (2004) Hydrogen peroxide mediates plant root cell response to nutrient deprivation. *Proc Natl Acad Sci USA*, **101**, 8827–8832.

Snedden, W.A. and Fromm, H. (1998) Calmodulin, calmodulin-regulated proteins and plant responses to the environment. *Trends Plant Sci*, **3**, 299–304.

Snedden, W.A. and Fromm, H. (2001) Calmodulin as a versatile calcium signal transducer in plants. *New Phytol*, **151**, 35–66.

Snedden, W.A., Koutsia, N., Baum, G. and Fromm, H. (1996) Activation of a recombinant petunia glutamate decarboxylase by calcium/calmodulin or by a monoclonal antibody which recognizes the calmodulin binding domain. *J Biol Chem*, **271**, 4148–4153.

Sze, H., Liang, F., Hwang, I., Curran, A.C. and Harper, J.F. (2000) Diversity and regulation of plant Ca^{2+} pumps: insights from expression in yeast. *Annu Rev Plant Physiol Plant Mol Biol*, **51**, 433–462.

Szymanski, D.B., Liao, B. and Zielinski, R.E. (1996) Calmodulin isoforms differentially enhance the binding of cauliflower nuclear proteins and recombinant TGA3 to a region derived from the *Arabidopsis* Cam-3 promoter. *Plant Cell*, **8**, 1069–1077.

Takezawa, D., Ramachandiran, S., Paranjape, V. and Poovaiah, B.W. (1996) Dual regulation of a chimeric plant serine/threonine kinase by calcium and calcium/calmodulin. *J Biol Chem*, **271**, 8126–8132.

Tang, G.Q., Hardin, S.C., Dewey, R. and Huber, S.C. (2003) A novel C-terminal proteolytic processing of cytosolic pyruvate kinase, its phosphorylation and degradation by the proteasome in developing soybean seeds. *Plant J*, **34**, 77–93.

Teruel, M.N., Chen, W., Persechini, A. and Meyer, T. (2000) Differential codes for free Ca(2+)-calmodulin signals in nucleus and cytosol. *Curr Biol*, **10**, 86–94.

Teruel, M.N. and Meyer, T. (2000) Translocation and reversible localization of signaling proteins: a dynamic future for signal transduction. *Cell*, **103**, 181–184.

Tirichine, L., Imaizumi-Anraku, H., Yoshida, S., Murakami, Y., Madsen, L.H., Miwa, H., Nakagawa, T., Sandal, N., Albrektsen, A.S., Kawaguchi, M., Downie, A., Sato, S., Tabata, S., Kouchi, H., Parniske, M., Kawasaki, S. and Stougaard, J. (2006) Deregulation of a Ca^{2+}/calmodulin-dependent kinase leads to spontaneous nodule development. *Nature*, **441**, 1153–1156.

Trewavas, A. (1999) How plants learn. *Proc Natl Acad Sci USA*, **96**, 4216–4218.

Trewavas, A. and Knight, M. (1994) Mechanical signalling, calcium and plant form. *Plant Mol Biol*, **26**, 1329–1341.

van Der Luit, A.H., Olivari, C., Haley, A., Knight, M.R. and Trewavas, A.J. (1999) Distinct calcium signaling pathways regulate calmodulin gene expression in tobacco. *Plant Physiol*, **121**, 705–714.

Wymer, C.L., Bibikova, T.N. and Gilroy, S. (1997) Cytoplasmic free calcium distributions during the development of root hairs of *Arabidopsis thaliana*. *Plant J*, **12**, 427–439.

Xu, J., Li, H.D., Chen, L.Q., Wang, Y., Liu, L.L., He, L. and Wu, W.H. (2006) A protein kinase, interacting with two calcineurin B-like proteins, regulates K^+ transporter AKT1 in *Arabidopsis*. *Cell*, **125**, 1347–1360.

Yoon, G.M., Dowd, P.E., Gilroy, S., and McCubbin, A.G. (2006)Calcium-dependent protein kinase isoforms in *Petunia* have distinct functions in pollen tube growth, including regulating polarity. *Plant Cell*, **18**, 867–878.

Zhang, M., Tanaka, T. and Ikura, M. (1995) Calcium-induced conformational transition revealed by the solution structure of apo calmodulin. *Nat Struct Biol*, **2**, 758–767.

Zhang, M. and Yuan, T. (1998) Molecular mechanisms of calmodulin's functional versatility. *Biochem Cell Biol*, **76**, 313–323.

Zhu, J.-K. (2003) Regulation of ion homeostasis under salt stress [review]. *Curr Opin Plant Biol*, **6** (5), 441–445.

Zielinski, R.E. (1998) Calmodulin and calmodulin binding proteins in plants. *Annu Rev Plant Physiol Plant Mol Biol*, **49**, 697–725.

Zielinski, R.E. (2002) Characterization of three new members of the *Arabidopsis thaliana* calmodulin gene family: conserved and highly diverged members of the gene family functionally complement a yeast calmodulin null. *Planta*, **214**, 446–455.

Zik, M., Arazi, T., Snedden, W.A. and Fromm, H. (1998) Two isoforms of glutamate decarboxylase in *Arabidopsis* are regulated by calcium/calmodulin and differ in organ distribution. *Plant Mol Biol*, **37**, 967–975.

Annual Plant Reviews (2008) **33**, 189–201
doi: 10.1111/b.9781405160025.2008.00007.x
www.interscience.wiley.com

Chapter 7

REACTIVE OXYGEN SIGNALING IN PLANTS

Gad Miller,[1] Jesse Coutu,[1] Vladimir Shulaev,[2] and Ron Mittler[1,3]

[1] *Department of Biochemistry and Molecular Biology, University of Nevada, Mail Stop 200, Reno, NV 89557, USA*
[2] *Virginia Bioinformatics Institute, Washington Street, Blacksburg, VA 24061, USA*
[3] *Department of Plant Science, Hebrew University of Jerusalem, Jerusalem 91904, Israel*

Abstract: Reactive oxygen species (ROS) control many different processes in plants, including growth, development, and response to biotic and abiotic stimuli. However, as toxic molecules ROS are also capable of causing cell injury or death. How this conflict is resolved in cells is largely unknown. Nonetheless, it is clear that the steady-state level of ROS in cells needs to be tightly regulated. In the flowering plant *Arabidopsis thaliana*, a network of more than 150 genes is involved in managing the level of ROS in cells. This network is highly dynamic and redundant, and encodes for ROS sensing, scavenging, and producing proteins. Although recent studies unraveled some of the key players of the ROS network, many questions related to its mode of regulation, its protective roles, and its modulation of signaling networks that control growth, development, and stress responses remain unanswered. This chapter gives an overview of the ROS network in plants and discusses its complexity and the challenges it poses for researchers attempting to study basic processes in plant biology.

Keywords: antioxidants; reactive oxygen species (ROS); ROS signal transduction; ROS network; scavenging; oxidative stress

7.1 Introduction to reactive oxygen metabolism

The introduction of molecular oxygen (O_2) into our atmosphere by O_2-evolving photosynthetic organisms, about 2.7 billion years ago, resulted in the production of reactive oxygen species (ROS) that have been, ever since, considered the unwelcome companions of aerobic life (Halliwell and Gutteridge, 1999). In contrast to atmospheric oxygen, these activated or partially reduced derivatives of O_2 (e.g., $^1O_2$, O_2^-, H_2O_2, and $HO^\bullet$) are highly reactive and toxic, and can lead to the oxidative destruction of different cellular components (Asada and Takahashi, 1987; Halliwell, 2006). As a consequence, the successful evolution of aerobic organisms on Earth has been dependent upon the development and/or acquisition of efficient ROS-detoxifying mechanisms and antioxidants. In recent years, a new role for ROS was, however, identified: the control and regulation of biological processes, such as growth, development, programmed cell death, and biotic and abiotic stress responses (Kovtun *et al.*, 2000; Pei *et al.*, 2000; Knight and Knight, 2001; Moller, 2001; Baxter-Burrell *et al.*, 2002; Mullineaux and Karpinski, 2002; Torres *et al.*, 2002, 2005, 2006; Foreman *et al.*, 2003; Kwak *et al.*, 2003; Overmyer *et al.*, 2003; Apel and Hirt, 2004; Shin and Schachtman, 2004; Foyer and Noctor, 2005; Joo *et al.*, 2005; Torres and Dangl, 2005; Asada, 2006; del Rio *et al.*, 2006; Desikan *et al.*, 2006; Gapper and Dolan, 2006; Kwak *et al.*, 2006; Mullineaux *et al.*, 2006; Pavet *et al.*, 2005; Pitzschke and Hirt, 2006; Rhoads *et al.*, 2006; Sagi and Fluhr, 2006; Terada, 2006; Van Breusegem and Dat, 2006; Xing *et al.*, 2006; Zaninotto *et al.*, 2006). These studies extend our understanding of ROS and suggest a dual role for ROS in plant biology as: (i) toxic byproducts of aerobic metabolism and (ii) key regulators of biological processes and pathways.

The use of ROS as signaling molecules by plants suggests that during the course of evolution plant cells were able to achieve a high degree of control over ROS toxicity and are now using ROS as signaling molecules. Controlling ROS toxicity, while enabling ROS such as H_2O_2 or O_2^- to act as signaling molecules, is thought to require a large gene network in plants composed of over 150 genes in the model plants *Arabidopsis thaliana* (Mittler *et al.*, 2004).

7.2 ROS signaling and its modulation by the ROS gene network

While Ca^{2+} signaling is predominantly controlled in plants by storage and release, ROS signaling is thought to be controlled by production and scavenging (Fig. 7.1; Mittler *et al.*, 2004; Bailey-Serres and Mittler, 2006). ROS production is mediated by different cellular pathways, including respiration and photosynthesis, as well as by different proteins and enzymes encoded by the ROS gene network (e.g., NADPH oxidases, amine oxidases, and xanthine oxidase; Table 7.1). In contrast, ROS scavenging is mediated by different ROS-scavenging enzymes and antioxidants that include ascorbate peroxidases, catalases, and superoxide dismutases (Table 7.1; see also a partial list of the ROS gene network of *Arabidopsis* in Table 7.2). These two opposing forces, i.e., ROS scavenging and ROS production, control the level of ROS in cells.

Different environmental or developmental signals feed into the ROS network of plants and alter ROS signaling in a compartment-specific or even a cell-specific manner. ROS signals are perceived by different proteins, enzymes, or receptors, feed into a signal transduction pathway (the ROS signal

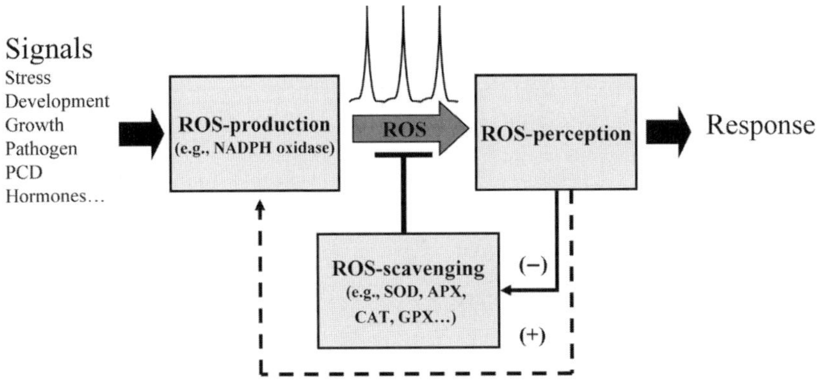

Figure 7.1 *A conceptual model of the ROS gene network.* Different signals alter ROS production in cells. ROS are sensed by ROS perception mechanisms leading to the activation of ROS scavenging (−) and/or the enhanced production of ROS (+). The interplay between ROS production and scavenging determines the steady-state level of ROS in cells, as well as the duration, localization, and amplitude of ROS signals. The decoding of ROS signals, generated by the ROS network, determines in turn the response of cells to the different stimuli. The ROS network is of course integrated into the vast network of signaling pathways of cells and is linked to NO, calcium, and other cellular messengers (not shown).

transduction pathway, described below), and result in enhanced ROS scavenging, thus suppressing the ROS signal (i.e., a feedback inhibition loop), or enhanced ROS production, thereby amplifying the ROS signal (i.e., a feedback amplification loop). The interplay between the ROS-scavenging and the ROS-producing pathways/enzymes of the cell (Tables 7.1 and 7.2), will therefore determine the intensity, duration, and subcellular localization of the ROS signal, and the decoding of this signal will determine the cellular response to the original cue, modulating different developmental, metabolic, and/or defense pathways.

During a plant–pathogen interaction, for example, the identification of a pathogen via a plant receptor will trigger an *R* gene-dependent pathway that will result in the enhanced production of ROS by plasma membrane-localized NADPH oxidases (Torres *et al.*, 2006). This will result in an ROS signal that will activate several different defense pathways. The activity of ROS-scavenging enzymes is important in this case because these enzymes can modulate the ROS signal and determine the intensity, duration, or even the type of response (Mittler *et al.*, 1999). Thus, the interplay between ROS scavenging and production could determine the response of plants to a particular pathogen. An abiotic stress such as osmotic stress, for example, will result in enhanced ROS production due to altered metabolic balance, but also due to the activation of NADPH oxidases and other ROS-producing signals (Mittler, 2002, 2006). The multiple sources of ROS produced in cells during this stress will trigger different ROS-scavenging pathways, and the interplay between the multiple

Table 7.1 Major ROS production and scavenging systems of plants

Mechanism	**Localization**	**Primary ROS**
Production of ROS		
Photosynthesis ET and PSI/II	Chl	O_2^-
Respiration ET	Mit	O_2
Glycolate oxidase	Per	H_2O_2
Excited chlorophyll	Chl	$^1O_2$
NADPH oxidase	PM	O_2
Fatty acid β-oxidation	Per	H_2O_2
Oxalate oxidase	Apo	H_2O_2
Xanthine oxidase	Per	O_2^-
Peroxidases, Mn^{2+} and NADH	CW	H_2O_2, O_2^-
Amine oxidase	Apo	H_2O_2
Scavenging of ROS		
SOD	Chl, Cyt, Mit, Per, Apo	O_2
APX	Chl, Cyt, Mit, Per, Apo	H_2O_2
CAT	Per	H_2O_2
GPX	Cyt, Chl, Mit	H_2O_2, ROOH
Peroxidases	CW, Cyt, Vac	H_2O_2
Thioredoxin Peroxidase	Chl, Cyt, Mit	H_2O_2
Ascorbic acid	Chl, Cyt, Mit, Per, Apo	H_2O_2, O_2
Glutathione	Chl, Cyt, Mit, Per, Apo	H_2O_2
α-tocopherol	Membranes	ROOH, $^1O_2$
Carotenoids	Chl	$^1O_2$
Avoiding ROS production		
Anatomical adaptations	Leaf structure/epidermis	O_2^-, H_2O_2, $^1O_2$
C4/CAM metabolism	Chl, Cyt, Vac	O_2^-, H_2O_2
Chl movement	Cyt	O_2^-, H_2O_2, $^1O_2$
Suppression of Photosynthesis	Chl	O_2^-, H_2O_2
PS/antenna modulations	Chl	O_2^-, $^1O_2$
AOX	Chl, Mit	O_2

Abbreviations: Apo, apoplast; Chl, chloroplast; CW, cell wall; Cyt, cytosol; ET, electron transport; Mit, mitochondria; Per, peroxisome; PM, plasma membrane; PS, photosystem; Vac, vacuole.

ROS-scavenging and ROS-producing pathways during stress will determine the type and intensity of defense response and the capability of plant cells to tolerate the stress (Mittler, 2002, 2006; Mittler *et al.*, 2004).

It is possible that the use of ROS as versatile signaling molecules originated from their proposed use to sense stress. Most forms of stress disrupt the metabolic balance of cells, resulting in altered production of ROS. Simple unicellular organisms, such as bacteria or yeast, sense the altered production of ROS by redox-sensitive transcription factors and other molecular sensors, such as two-component histidine kinases, activate different ROS-defense pathways, and modulate their metabolic pathways to lower the production rate of ROS (Costa and Moradas-Ferreira, 2001; Georgiou, 2002; Mittler *et al.*, 2004; Liu *et al.*, 2005; Kanesaki *et al.*, 2007). Variations on this pathway could

Table 7.2 Major ROS gene network enzymes of *Arabidopsis thaliana*

Enzyme and reaction	Gene name	AGI code	Localization
Superoxide dismutase (SOD)	FeSOD (FSD1)	At4g25100.3	chl
$O_2^- + O_2^- + 2H^+ \rightarrow H_2O_2 + O_2$	FeSOD (FSD2)	At5g51100.1	chl
	FeSOD (FSD3)	At5g23310.1	chl
	Cu/ZnSOD (CSD1)	At1g08830.1	cyt
	Cu/ZnSOD (CSD2)	At2g28190.1	chl
	Cu/ZnSOD (CSD3)	At5g18100.1	per
	MnSOD (MSD1)	At3g10920.1	mit
	MnSOD-like	At3g56350.1	see
Ascorbate peroxidase (APX)	APX1	At1g07890.1	cyt
2 Ase $+H_2O_2 \rightarrow$ 2 MDA + 2 H_2O	APX2	At3g09640.1	cyt
	APX3	At4g35000.1	per/chl
	APX4	At4g09010.1	chl
	APX5	At4g35970.1	per
	APX6	At4g32320.1	cyt/chl/mit
	APX7	At1g33660.1	mit
	stromal-APX	At4g08390.2	chl/mit
	thylakoid-APX	At1g77490.1	chl
Monodehydroascorbate reductase (MDAR)	MDAR1	At1g63940.4	chl/mit
MDA + NAD(P)H + $H^+ \rightarrow$ Ase + NAD(P)	MDAR2	At3g09940.1	cyt
	MDAR3	At3g27820.1	cyt/mit
	MDAR4	At3g52880.1	cyt
	MDAR5	At5g03630.1	cyt
Dehydroascorbate reductase (DHAR)	DHAR1	At5g16710.1	chl/mit
DHA + 2 GSH $\rightarrow$ Asc + GSSG	DHAR2	At5g36270.1	cyt
	DHAR3	At1g75270.1	cyt/chl
	DHAR4	At1g19550.1	cyt/chl
	DHAR5	At1g19570.1	cyt/chl
Glurathlone reductase (GR)	GR1	At3g24170.1	cyt
GSSG + NAD(P)H $\rightarrow$ 2 GSH + NAD(P)	GR2	At3g54660.1	chl/mit
Catalase (Cat)	Cat1	At1g20630.1	per
$2H_2O_2 \rightarrow 2H_2O + O_2$	Cat2	At4g35090.1	per
	Cat3	At1g20620.1	per
Glutathione peroxidase (GPX)	GPX1	At2g25080.1	chl
H_2O_2 + 2 GSH $\rightarrow 2H_2O$ + GSSG	GPX2	At2g31570.1	cyt/chl
	GPX3	At2g43350.1	mit
	GPX4	At2g48150.1	cyt
	GPX5	At3g63080.1	er
	GPX7	At4g31870.1	chl
	GPX8	At1g63460.1	cyt/chl
	Phospholipid GPX6	At4g11600.1	chl/mit
Ferritin	Ferritin 1	At5g01600.1	chl
Fe + P $\rightarrow$ P − Fe	Ferritin 2	At3g56090.1	chl/mit
	Ferritin 3	At2g40300.1	chl/mit
	Ferritin 4	At3g11050.1	chl
NADPH oxidase	NADPH oxidase (RhohA)	At5g07390.1	mem
NADPH + $e^- + O_2 \rightarrow$ $NADP^- + O_2^- + H^+$	NADPH oxidase (RhohB)	At1g09090.2	mem
	NADPH oxidase (RhohC)	At5g51060.1	mem
	NADPH oxidase (RhohD)	At5g47910.1	mem
	NADPH oxidase (RhohE)	At1g19230.1	mem
	NADPH oxidase (RhohF)	At1g64060.1	mem
	NADPH oxidase (RhohG)	At4g25090.1	mem
	NADPH oxidase (RhohH)	At5g60010.1	mem
	NADPH oxidase (RhohI)	At1g11230.1	mem
	NADPH oxidase (RhohJ)	At3g45810.1	mem

(continued)

Table 7.2 *(continued)*

Enzyme and reaction	Gene name	AGI code	Localization
Alternative oxidase (AOX)	AOX putative	At1g32350.1	mit
$2e^- + 2H^+ + O_2 \rightarrow H_2O$	AOX1A	At3g22370.1	mit
	AOX1B	At3g22360.1	mit
	AOX1C	At3g27620.1	mit
	Immutants	At4g22260.1	cld
Peroxiredoxin (PrxR)	1-Cys PrxR	At1g48130.1	nuc
$2P - SH + H_2O_2 \rightarrow P - S - S - P + 2H_2O$	2-cys PrxR A	At3g11630.1	chl
	2-cys PrxR B	At5g06290.1	chl
	2-cys PrxR F	At3g06050.1	mit
	PrxR Q	At3g26060.1	chl
	Type 2 PrxR A	At1g65990.1	mem/chl
	Type 2 PrxR B	At1g65980.1	cyt
	Type 2 PrxR C	At1g65970.1	cyt
	Type 2 PrxR D	At1g60740.1	cyt
	Type 2 PrxR E	At3g52960.1	chl/mit
	Type 2 PrxR-related	At3g03405.1	cyt

A partial list of the ROS gene network is shown. For a full description, see Mittler *et al.*, 2004. Enzyme name and reaction are given on left followed by gene name, locus identifier, and putative cellular localization. The table demonstrates two major principles of the ROS gene network: redundancy and complexity. Thus, multiple ROS-scavenging mechanisms, with a high degree of redundancy per a specific ROS, can be found in each cellular compartment.

have originated during evolution and contributed to the use of ROS as signaling molecules to control more specialized processes such as plant growth and defense, hormonal signaling, and development.

7.3 Subcellular localization and coordination of the ROS network

The different scavenging and producing enzymes encoded by the ROS gene network can be found in many different subcellular compartments (Fig. 7.2, Color plate 13). In addition, usually more than one enzymatic activity per a specific ROS can be found in each of the different compartments (Mittler *et al.*, 2004). Because ROS such as H_2O_2 can diffuse between different cellular compartments (Henzler and Steudle, 2000; Bienert *et al.*, 2007), ROS metabolism in a particular compartment can affect or alter the ROS homeostasis/signaling of a neighboring compartment. Recent studies in *Arabidopsis* suggested that the mode of coordination between the different cellular compartments of plants is complex (Rizhsky *et al.*, 2002; Mittler *et al.*, 2004).

For example, the application of light stress to *Arabidopsis* resulted in the induction of cytosolic and not chloroplastic ROS-defense enzymes (Karpinski *et al.*, 1997, 1999; Pnueli *et al.*, 2003; Davletova *et al.*, 2005), even though most ROS produced during light stress are thought to be generated in chloroplasts and/or peroxisomes. The cytosolic ROS-scavenging pathways were further shown to be required for the protection of chloroplasts during light stress

(Davletova *et al.*, 2005). In a different study, a double mutant deficient in cytosolic asocrbate peroxidase 1 and peroxisomal catalase 1 was found to be more tolerant to light stress compared to wild type or single mutants deficient in ascorbate peroxidase 1 or catalase 1 (Rizhsky *et al.*, 2002). This finding was very surprising because it suggested that different cellular pathways are activated in cells in response to enhanced ROS production in the cytosol or peroxisomes. Activation of both the cytosolic and peroxisomal pathways further results in the generation of a new signal that is different from that activated by the two individual cytosolic or peroxisomal signals. How ROS metabolism and signaling are coordinated between different organelles in cells is largely unknown at present.

7.4 Key components of the ROS gene network identified by reverse genetics

Recent studies of knockout, antisense, and RNAi lines for Cat2, Apx1, chlAOX, mitAOX, chlCuZnSOD (CSD2), peroxiredoxins (PrxR), thio- and gluta-redoxins, and different NADPH oxidases revealed a strong link between ROS and processes such as growth, development, stomatal responses, and biotic and abiotic stress responses (Moller, 2001; Torres *et al.*, 2002; Dietz, 2003; Foreman *et al.*, 2003; Kwak *et al.*, 2003; Pnueli *et al.*, 2003; Rizhsky *et al.*, 2003; Mittler *et al.*, 2004; Davletova *et al.*, 2005; Pavet *et al.*, 2005; Torres and Dangl, 2005; Torres *et al.*, 2005; Cheng *et al.*, 2006; Gadjev *et al.*, 2006; Miao *et al.*, 2006; Perez-Ruiz *et al.*, 2006; Vieira Dos Santos and Rey, 2006). These findings demonstrated the complex nature of the ROS gene network of plants and its modulation of key biological processes. Although all the mutants listed above are viable, demonstrating the redundancy of the ROS gene network, a phenotype was associated with most of the different genes, suggesting that they play a key role in the ROS signaling network of plants. Based on the analysis of the different mutants, proteins such as Cat2, Apx1, CSD2, and 2-cysteine PrxR are involved in the protection of chloroplasts against oxidative damage. By contrast, the absence of the NADPH oxidase genes *RbohD* and *RbohF* alters ROS production in cells and affects the defense response of *Arabidopsis* against pathogen attack (Torres and Dangl, 2005; Torres *et al.*, 2005), and knockout of *RbohC* has an altered root phenotype (Foreman *et al.*, 2003). *RbohD* and *RbohF* are also important for abscisic acid signaling in guard cells (Kwak *et al.*, 2003). Despite these impressive achievements, the function of more than 95% of the genes that compose the ROS gene network of *Arabidopsis* remains unknown at present (Mittler *et al.*, 2004).

7.5 The ROS signal transduction pathway of plants

Recent studies in *Arabidopsis* uncovered some of the key components involved in the ROS signal transduction pathway of plants. Nevertheless, the receptors for ROS are mainly unknown at present. It has been suggested that plant cells sense ROS via at least three different mechanisms (Fig. 7.3, Color plate 14): (i) unidentified receptor proteins, (ii) redox-sensitive transcription factors, and (iii) direct inhibition of phosphatases by ROS (Neill *et al.*, 2002; Mittler *et al.*, 2004; Hancock *et al.*, 2006). Recent studies suggested that the histidine kinase receptor ETR1 and the heat shock transcription factor HSF-A4a play a role in H_2O_2 sensing in *Arabidopsis* (Davletova *et al.*, 2005; Hancock *et al.*, 2006; Miller and Mittler, 2006).

Downstream signaling events associated with ROS sensing involves Ca^{2+} and Ca^{2+}-binding proteins (Bowler and Fluhr, 2000; Coelho *et al.*, 2002; Rentel and Knight, 2004; Evans *et al.*, 2005; Demidchik *et al.*, 2007), G-proteins (Baxter-Burrell *et al.*, 2002; Neill *et al.*, 2002; Joo *et al.*, 2005), and the activation of phospholipid signaling that results in the accumulation of phosphatidic acid (Anthony *et al.*, 2004, 2006). It is possible that the localization of ROS signals in specific cellular sites is similar to that of Ca^{2+} signals in response to stimuli (Coelho *et al.*, 2002). A serine/threonine protein kinase (OXI1) has been shown to play a central role in ROS sensing and the activation of MAPK3/6 by Ca^{2+} (Rentel *et al.*, 2004; Anthony *et al.*, 2006). This kinase is also activated by PDK1 through the phospholipase C or D-phosphatidic acid (PLC/PLD-PA) pathway (Anthony *et al.*, 2004, 2006). A MAPK cascade involving MAPK3/6 acts downstream of OXI1 and controls the activation of different defense mechanisms in response to ROS stress (Kovtun *et al.*, 2000; Apel and Hirt, 2004; Mittler *et al.*, 2004). The expression of different transcription factors is enhanced by ROS and includes members of the WRKY, Zat, RAV, bZIP, GRAS, and Myb families (Davletova *et al.*, 2005; Gadjev *et al.*, 2006; Kaminaka *et al.*, 2006). The possible existence of positive amplification loops, involving NADPH oxidases, in ROS signaling has recently been suggested (Baxter-Burrell *et al.*, 2002; Davletova *et al.*, 2005; Torres *et al.*, 2005). These loops might be activated by low levels of ROS and result in the enhanced production and amplification of ROS signals in specific cellular locations (Fig. 7.3, Color plate 14). Although O_2^- and H_2O_2 have been considered to play a key role as ROS signal transduction molecules, recent studies pointed to the existence of $^1O_2$-specific signaling pathways (Apel and Hirt, 2004; Danon *et al.*, 2006). Taking into account the complex nature of the ROS gene network and its integration into the web of plant signaling networks (Fig. 7.1), we face a major challenge in dissecting the genetic network that controls ROS signaling in plants.

7.6 Summary

Although ROS were initially considered to be toxic byproducts of aerobic metabolism, in recent years it became obvious that plants can cope with ROS toxicity to the degree of using ROS as signal transduction molecules. ROS signaling was shown to be involved in the regulation of basic biological processes such as growth, development, and response to biotic and abiotic stimuli. ROS signaling and ROS toxicity are kept in check by the ROS gene network of plants. This network includes ROS-scavenging and ROS-producing enzymes that modulate the level of ROS in cells (Tables 7.1 and 7.2; Fig. 7.1). Thus, the overall level of ROS is always kept under control and ROS are allowed to accumulate and/or oscillate for the purpose of signaling in a highly controlled manner (Mittler *et al.*, 2004). This process is achieved by a tightly controlled balance between ROS production and ROS scavenging in the different cellular compartments. The interplay between ROS scavenging and production in the different cellular compartments, therefore, determines the intensity, duration, and localization of ROS signals and the decoding of these signals determine the plant's response, or developmental and/or growth adaptations (Fig. 7.1).

References

Anthony, R.G., Henriques, R., Helfer, A., Meszaros, T., Rios, G., Testerink, C., Munnik, T., Deak, M., Koncz, C. and Bogre, L. (2004) A protein kinase target of a PDK1 signalling pathway is involved in root hair growth in *Arabidopsis*. *EMBO J*, **23**, 572–581.

Anthony, R.G., Khan, S., Costa, J., Pais, M.S. and Bogre, L. (2006) The *Arabidopsis* protein kinase PTI1-2 is activated by convergent phosphatidic acid and oxidative stress signaling pathways downstream of PDK1 and OXI1. *J Biol Chem*, **281**, 37536–37546.

Apel, K. and Hirt, H. (2004) Reactive oxygen species: metabolism, oxidative stress, and signal transduction. *Annu Rev Plant Biol*, **55**, 373–399.

Asada, K. (2006) Production and scavenging of reactive oxygen species in chloroplasts and their functions. *Plant Physiol*, **141**, 391–396.

Asada, K. and Takahashi, M. (1987) *Production and Scavenging of Active Oxygen in Photosynthesis*. Elsevier, Amsterdam.

Bailey-Serres, J. and Mittler, R. (2006) The roles of reactive oxygen species in plant cells. *Plant Physiol*, **141**, 311.

Baxter-Burrell, A., Yang, Z., Springer, P.S. and Bailey-Serres, J. (2002) RopGAP4-dependent Rop GTPase rheostat control of *Arabidopsis* oxygen deprivation tolerance. *Science*, **296**, 2026–2028.

Bienert, G.P., Moller, A.L., Kristiansen, K.A., Schulz, A., Moller, I.M., Schjoerring, J.K. and Jahn, T.P. (2007) Specific aquaporins facilitate the diffusion of hydrogen peroxide across membranes. *J Biol Chem*, **282**, 1183–1192.

Bowler, C. and Fluhr, R. (2000) The role of calcium and activated oxygens as signals for controlling cross-tolerance. *Trends Plant Sci*, **5**, 241–246.

Cheng, N.H., Liu, J.Z., Brock, A., Nelson, R.S. and Hirschi, K.D. (2006) AtGRXcp, an *Arabidopsis* chloroplastic glutaredoxin, is critical for protection against protein oxidative damage. *J Biol Chem*, **281**, 26280–26288.

Coelho, S.M., Taylor, A.R., Ryan, K.P., Sousa-Pinto, I., Brown, M.T. and Brownlee, C. (2002) Spatiotemporal patterning of reactive oxygen production and Ca(2+) wave propagation in fucus rhizoid cells. *Plant Cell*, **14**, 2369–2381.

Costa, V. and Moradas-Ferreira, P. (2001) Oxidative stress and signal transduction in *Saccharomyces cerevisiae*: insights into ageing, apoptosis and diseases. *Mol Aspects Med*, **22**, 217–246.

Danon, A., Coll, N.S. and Apel, K. (2006) Cryptochrome-1-dependent execution of programmed cell death induced by singlet oxygen in *Arabidopsis thaliana*. *Proc Natl Acad Sci USA*, **103**, 17036–17041.

Davletova, S., Rizhsky, L., Liang, H., Shengqiang, Z., Oliver, D.J., Coutu, J., Shulaev, V., Schlauch, K. and Mittler, R. (2005) Cytosolic ascorbate peroxidase 1 is a central component of the reactive oxygen gene network of *Arabidopsis*. *Plant Cell*, **17**, 268–281.

del Rio, L.A., Sandalio, L.M., Corpas, F.J., Palma, J.M. and Barroso, J.B. (2006) Reactive oxygen species and reactive nitrogen species in peroxisomes. Production, scavenging, and role in cell signaling. *Plant Physiol*, **141**, 330–335.

Demidchik, V., Shabala, S.N. and Davies, J.M. (2007) Spatial variation in H_2O_2 response of *Arabidopsis thaliana* root epidermal Ca^{2+} flux and plasma membrane Ca^{2+} channels. *Plant J*, **49**, 377–386.

Desikan, R., Last, K., Harrett-Williams, R., Tagliavia, C., Harter, K., Hooley, R., Hancock, J.T. and Neill, S.J. (2006) Ethylene-induced stomatal closure in *Arabidopsis* occurs via AtrbohF-mediated hydrogen peroxide synthesis. *Plant J*, **47**, 907–916.

Dietz, K.J. (2003) Plant peroxiredoxins. *Annu Rev Plant Biol*, **54**, 93–107.

Evans, N.H., McAinsh, M.R., Hetherington, A.M. and Knight, M.R. (2005) ROS perception in *Arabidopsis thaliana*: the ozone-induced calcium response. *Plant J*, **41**, 615–626.

Foreman, J., Demidchik, V., Bothwell, J.H., Mylona, P., Miedema, H., Torres, M.A., Linstead, P., Costa, S., Brownlee, C., Jones, J.D., Davies, J.M. and Dolan, L. (2003) Reactive oxygen species produced by NADPH oxidase regulate plant cell growth. *Nature*, **422**, 442–446.

Foyer, C.H. and Noctor, G. (2005) Redox homeostasis and antioxidant signaling: a metabolic interface between stress perception and physiological responses. *Plant Cell*, **17**, 1866–1875.

Gadjev, I., Vanderauwera, S., Gechev, T.S., Laloi, C., Minkov, I.N., Shulaev, V., Apel, K., Inze, D., Mittler, R. and Van Breusegem, F. (2006) Transcriptomic footprints disclose specificity of reactive oxygen species signaling in *Arabidopsis*. *Plant Physiol*, **141**, 436–445.

Gapper, C. and Dolan, L. (2006) Control of plant development by reactive oxygen species. *Plant Physiol*, **141**, 341–345.

Georgiou, G. (2002) How to flip the (redox) switch. *Cell*, **111**, 607–610.

Halliwell, B. (2006) Reactive species and antioxidants. Redox biology is a fundamental theme of aerobic life. *Plant Physiol*, **141**, 312–322.

Halliwell, B. and Gutteridge, J.M.C. (1999) *Free Radicals in Biology and Medicine*. Clarendon, Oxford.

Hancock, J., Desikan, R., Harrison, J., Bright, J., Hooley, R. and Neill, S. (2006) Doing the unexpected: proteins involved in hydrogen peroxide perception. *J Exp Bot*, **57**, 1711–1718.

Henzler, T. and Steudle, E. (2000) Transport and metabolic degradation of hydrogen peroxide in Chara corallina: model calculations and measurements with the pressure probe suggest transport of H(2)O(2) across water channels. *J Exp Bot*, **51**, 2053–2066.

Joo, J.H., Wang, S., Chen, J.G., Jones, A.M. and Fedoroff, N.V. (2005) Different signaling and cell death roles of heterotrimeric G protein alpha and beta subunits in the *Arabidopsis* oxidative stress response to ozone. *Plant Cell*, **17**, 957–970.

Kaminaka, H., Nake, C., Epple, P., Dittgen, J., Schutze, K., Chaban, C., Holt, B.F., III, Merkle, T., Schafer, E., Harter, K. and Dangl, J.L. (2006) bZIP10-LSD1 antagonism modulates basal defense and cell death in *Arabidopsis* following infection. *EMBO J*, **25**, 4400–4411.

Kanesaki, Y., Yamamoto, H., Paithoonrangsarid, K., Shoumskaya, M., Suzuki, I., Hayashi, H. and Murata, N. (2007) Histidine kinases play important roles in the perception and signal transduction of hydrogen peroxide in the cyanobacterium, Synechocystis sp. PCC 6803. *Plant J*, **49**, 313–324.

Karpinski, S., Escobar, C., Karpinska, B., Creissen, G. and Mullineaux, P.M. (1997) Photosynthetic electron transport regulates the expression of cytosolic ascorbate peroxidase genes in *Arabidopsis* during excess light stress. *Plant Cell*, **9**, 627–640.

Karpinski, S., Reynolds, H., Karpinska, B., Wingsle, G., Creissen, G. and Mullineaux, P. (1999) Systemic signaling and acclimation in response to excess excitation energy in *Arabidopsis*. *Science*, **284**, 654–657.

Knight, H. and Knight, M.R. (2001) Abiotic stress signalling pathways: specificity and cross-talk. *Trends Plant Sci*, **6**, 262–267.

Kovtun, Y., Chiu, W.L., Tena, G. and Sheen, J. (2000) Functional analysis of oxidative stress-activated mitogen-activated protein kinase cascade in plants. *Proc Natl Acad Sci USA*, **97**, 2940–2945.

Kwak, J.M., Mori, I.C., Pei, Z.M., Leonhardt, N., Torres, M.A., Dangl, J.L., Bloom, R.E., Bodde, S., Jones, J.D. and Schroeder, J.I. (2003) NADPH oxidase AtrbohD and AtrbohF genes function in ROS-dependent ABA signaling in *Arabidopsis*. *EMBO J*, **22**, 2623–2633.

Kwak, J.M., Nguyen, V. and Schroeder, J.I. (2006) The role of reactive oxygen species in hormonal responses. *Plant Physiol*, **141**, 323–329.

Liu, H., Colavitti, R., Rovira, II and Finkel, T. (2005) Redox-dependent transcriptional regulation. *Circ Res*, **97**, 967–974.

Miao, Y., Lv, D., Wang, P., Wang, X.C., Chen, J., Miao, C. and Song, C.P. (2006) An *Arabidopsis* glutathione peroxidase functions as both a redox transducer and a scavenger in abscisic acid and drought stress responses. *Plant Cell*, **18**, 2749–2766.

Miller, G. and Mittler, R. (2006) Could heat shock transcription factors function as hydrogen peroxide sensors in plants? *Ann Bot*, **98**, 279–288.

Mittler, R. (2002) Oxidative stress, antioxidants and stress tolerance. *Trends Plant Sci*, **7**, 405–410.

Mittler, R. (2006) Abiotic stress, the field environment and stress combination. *Trends Plant Sci*, **11**, 15–19.

Mittler, R., Herr, E.H., Orvar, B.L., van Camp, W., Willekens, H., Inze, D. and Ellis, B.E. (1999) Transgenic tobacco plants with reduced capability to detoxify reactive oxygen intermediates are hyperresponsive to pathogen infection. *Proc Natl Acad Sci USA*, **96**, 14165–14170.

Mittler, R., Vanderauwera, S., Gollery, M. and Van Breusegem, F. (2004) Reactive oxygen gene network of plants. *Trends Plant Sci*, **9**, 490–498.

Moller, I.M. (2001) Plant mitochondria and oxidative stress: electron transport, NADPH turnover, and metabolism of reactive oxygen species. *Annu Rev Plant Physiol Plant Mol Biol*, **52**, 561–591.

Mullineaux, P. and Karpinski, S. (2002) Signal transduction in response to excess light: getting out of the chloroplast. *Curr Opin Plant Biol*, **5**, 43–48.

Mullineaux, P.M., Karpinski, S. and Baker, N.R. (2006) Spatial dependence for hydrogen peroxide-directed signaling in light-stressed plants. *Plant Physiol*, **141**, 346–350.

Neill, S., Desikan, R. and Hancock, J. (2002) Hydrogen peroxide signalling. *Curr Opin Plant Biol*, **5**, 388–395.

Overmyer, K., Brosche, M. and Kangasjarvi, J. (2003) Reactive oxygen species and hormonal control of cell death. *Trends Plant Sci*, **8**, 335–342.

Pavet, V., Olmos, E., Kiddle, G., Mowla, S., Kumar, S., Antoniw, J., Alvarez, M.E. and Foyer, C.H. (2005) Ascorbic acid deficiency activates cell death and disease resistance responses in *Arabidopsis*. *Plant Physiol*, **139**, 1291–1303.

Pei, Z.M., Murata, Y., Benning, G., Thomine, S., Klusener, B., Allen, G.J., Grill, E. and Schroeder, J.I. (2000) Calcium channels activated by hydrogen peroxide mediate abscisic acid signalling in guard cells. *Nature*, **406**, 731–734.

Perez-Ruiz, J.M., Spinola, M.C., Kirchsteiger, K., Moreno, J., Sahrawy, M. and Cejudo, F.J. (2006) Rice NTRC is a high-efficiency redox system for chloroplast protection against oxidative damage. *Plant Cell*, **18**, 2356–2368.

Pitzschke, A. and Hirt, H. (2006) Mitogen-activated protein kinases and reactive oxygen species signaling in plants. *Plant Physiol*, **141**, 351–356.

Pnueli, L., Liang, H., Rozenberg, M. and Mittler, R. (2003) Growth suppression, altered stomatal responses, and augmented induction of heat shock proteins in cytosolic ascorbate peroxidase (Apx1)-deficient *Arabidopsis* plants. *Plant J*, **34**, 187–203.

Rentel, M.C. and Knight, M.R. (2004) Oxidative stress-induced calcium signaling in *Arabidopsis*. *Plant Physiol*, **135**, 1471–1479.

Rentel, M.C., Lecourieux, D., Ouaked, F., Usher, S.L., Petersen, L., Okamoto, H., Knight, H., Peck, S.C., Grierson, C.S., Hirt, H. and Knight, M.R. (2004) OXI1 kinase is necessary for oxidative burst-mediated signalling in *Arabidopsis*. *Nature*, **427**, 858–861.

Rhoads, D.M., Umbach, A.L., Subbaiah, C.C. and Siedow, J.N. (2006) Mitochondrial reactive oxygen species. Contribution to oxidative stress and interorganellar signaling. *Plant Physiol*, **141**, 357–366.

Rizhsky, L., Hallak-Herr, E., Van Breusegem, F., Rachmilevitch, S., Barr, J.E., Rodermel, S., Inze, D. and Mittler, R. (2002) Double antisense plants lacking ascorbate peroxidase and catalase are less sensitive to oxidative stress than single antisense plants lacking ascorbate peroxidase or catalase. *Plant J*, **32**, 329–342.

Rizhsky, L., Liang, H. and Mittler, R. (2003) The water–water cycle is essential for chloroplast protection in the absence of stress. *J Biol Chem*, **278**, 38921–38925.

Sagi, M. and Fluhr, R. (2006) Production of reactive oxygen species by plant NADPH oxidases. *Plant Physiol*, **141**, 336–340.

Shin, R. and Schachtman, D.P. (2004) Hydrogen peroxide mediates plant root cell response to nutrient deprivation. *Proc Natl Acad Sci USA*, **101**, 8827–8832.

Terada, L.S. (2006) Specificity in reactive oxidant signaling: think globally, act locally. *J Cell Biol*, **174**, 615–623.

Torres, M.A. and Dangl, J.L. (2005) Functions of the respiratory burst oxidase in biotic interactions, abiotic stress and development. *Curr Opin Plant Biol*, **8**, 397–403.

Torres, M.A., Dangl, J.L. and Jones, J.D. (2002) *Arabidopsis* gp91phox homologues AtrbohD and AtrbohF are required for accumulation of reactive oxygen intermediates in the plant defense response. *Proc Natl Acad Sci USA*, **99**, 517–522.

Torres, M.A., Jones, J.D. and Dangl, J.L. (2005) Pathogen-induced, NADPH oxidase-derived reactive oxygen intermediates suppress spread of cell death in *Arabidopsis thaliana*. *Nat Genet*, **37**, 1130–1134.

Torres, M.A., Jones, J.D. and Dangl, J.L. (2006) Reactive oxygen species signaling in response to pathogens. *Plant Physiol*, **141**, 373–378.

Van Breusegem, F. and Dat, J.F. (2006) Reactive oxygen species in plant cell death. *Plant Physiol*, **141**, 384–390.

Vieira Dos Santos, C. and Rey, P. (2006) Plant thioredoxins are key actors in the oxidative stress response. *Trends Plant Sci*, **11**, 329–334.

Xing, S., Lauri, A. and Zachgo, S. (2006) Redox regulation and flower development: a novel function for glutaredoxins. *Plant Biol*, **8**, 547–555.

Zaninotto, F., La Camera, S., Polverari, A. and Delledonne, M. (2006) Cross talk between reactive nitrogen and oxygen species during the hypersensitive disease resistance response. *Plant Physiol*, **141**, 379–383.

Annual Plant Reviews (2008) **33**, 202–243
doi: 10.1111/b.9781405160025.2008.00008.x
www.interscience.wiley.com

Chapter 8

LIPID-MEDIATED SIGNALING

Wendy F. Boss,[1] Daniel V. Lynch,[2] and Xuemin Wang[3]

[1] *Department of Plant Biology, North Carolina State University, Raleigh, NC 27695, USA*
[2] *Department of Biology, Williams College, Williamstown, MA 01267, USA*
[3] *Department of Biology, University of Missouri, St. Louis, MO 63121, USA; Donald Danforth Plant Science Center, St. Louis, MO 63132, USA*

Abstract: Membrane lipids preserve the integrity of cells and organelles in a constantly changing environment. They modulate protein structure and function and can transduce signals from one side of the membrane to the other. For cells to survive the lipid bilayer must maintain a dynamic flux as it responds and adjusts to both chemical and physical cues. Additional burdens are imposed by their sessile lifestyle as plants survive in a constantly changing environment. To understand how plants adjust to their environment it is essential that we understand how they respond and adjust to environmental cues by altering their lipid chemistry and biochemistry. In this chapter, we have focused on the plant-specific features of phospholipid and sphingolipid signaling.

Keywords: lipid; phosphatidylcholine; phospholipid; phosphatidylinositol; sphingolipid; signaling

8.1 Introduction

When Gorter and Grendel first described the bilayer of lipids that defines the outer limits of red blood cells they would not have envisioned lipids as ideal sentries, both initiating and amplifying cell signaling (Gorter and Grendel, 1925). We have come to appreciate that the "lipid world" both senses and communicates changes in the bilayer chemistry and physics. Plasma membrane lipids sense cues such as changes in pH, ionic strength, and tension and pressure that are initiated both on the inside and outside of the cell.

Differences in the chemistry of lipids from different organisms (e.g., plants and animals) were identified early on, and the ability of plants to alter lipid chemistry (e.g., head group composition, fatty acid saturation etc.) as they acclimated to changes in temperature and water availability is well documented (Levitt, 1980; Steponkus, 1984; Lynch and Steponkus, 1987; Thompson, 1989; Wada *et al.*, 1990). Furthermore, biochemists noted some of the distinct characteristics of the lipid metabolizing enzymes in plants. However, it was not until the advent of the genomic era that previously observed differences in kinetic properties of enzymes were accepted as plant-specific features and not the result of differences in lab protocols or secondary metabolites interfering with in vitro assays. Comparisons of DNA sequences provided firm evidence for distinct metabolic pathways.

Comparative genomics also brought more interest in plant lipid-mediated signaling. For example, animal biologists recognized that phospholipase D, which they had once relegated to be a "cabbage enzyme of little interest" (Waite, 1987), was not only present in animals but also initiated important signaling cascades (Exton, 1997). The current challenges for plant biologists are to understand the significance of both the differences and similarities in plant and animal lipid-mediated signaling and to build a comprehensive model which describes the role of plant lipid-mediated signaling in regulating plant growth and development.

In this chapter, we have focused on plant-specific aspects of the fundamental biochemistry of phosphoinositide (PI) signaling, phospholipase D (PLD) signaling, and sphingolipid signaling. The reader is referred to recent reviews that discuss the comparative genetics of the plant and animal lipid signaling pathways (Drøbak *et al.*, 1999; Mueller-Roeber and Pical, 2002; Dunn *et al.*, 2004; Wang, 2004, 2005; Boss *et al.*, 2006; Wang *et al.*, 2006). For other pathways such as N-acyl phospholipids, phospholipase A2, and fatty acid signaling the reader is referred reviews from the laboratories of Chapman (1998; Shrestha *et al.*, 2006), Ryu (2004), and Wang (2001, 2004, 2005).

8.2 Plant-specific features of phosphoinositide signaling

The canonical PI pathway in animal cells is initiated by phospholipase C (PLC) hydrolyzing phosphatidylinositol(4,5)bisphosphate (PtdIns(4,5)P_2) to form inositol(1,4,5)trisphosphate (InsP_3) and diacylglyceride. InsP_3 releases calcium from endoplasmic reticulum stores and diacylglycerol (DAG) recruits and activates protein kinase C. In stimulated animal cells, PLC is recruited to the plasma membrane. Downregulation of PLC-mediated signaling will occur if the PLC is released from the membrane or deactivated and if InsP_3 is either phosphorylated to Ins(1,3,4,5)P_4 by an inositol phosphate (IP) 3-kinase or dephosphorylated to Ins(1,4)P_2 by an IP 5-PTase.

Early studies revealed major differences in plant and animal PI signaling. While the canonical inositol phospholipids were present (Fig. 8.1), InsP_3-mediated calcium release from the ER and DAG-stimulated protein kinases have not been demonstrated (Boss and Massel, 1985; Boss; 1989; Drøbak *et al.*, 1999; Mueller-Roeber and Pical, 2002; Meijer and Munnik, 2003; Boss *et al.*, 2006). Furthermore, although there is preponderance of inositol phosphates and inositol phosphate phosphatases in plants, there is no evidence for an Ins(1,4,5)P_3 3-kinase which in animal cells helps to terminate the

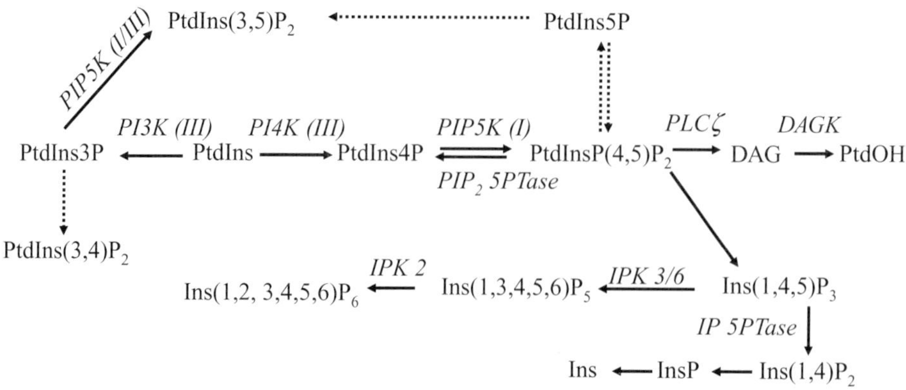

Figure 8.1 The key plant PI pathway intermediates. Solid arrows indicate where the function of the plant enzymes has been verified. Dotted arrows indicate predicted pathways. Abbreviations: InsP, inositol phosphate; IP PTase, inositol phosphate phosphatase; PLC, phospholipase C; DAG, diacylglycerol; DAGK, diacylglycerol kinase; PtdOH, phosphatidic acid; PtdInsP, phosphatidylinositol phosphate; PtdIns4P, phosphatidylinositol-4-phosphate; PtdIns3P, phosphatidylinositol-3-phosphate; PtdIns5P, phosphatidylinositol-5-phosphate; PtdIns(4,5)P_2, phosphatidylinositol-(4,5)-bisphosphate; PIPK, phosphatidylinositol phosphate kinase.

Ins(1,4,5)P_3 signal. In plants, Ins(1,4,5)P_3 is phosphorylated by a 3/6 kinase to Ins(1,3,4,5,6)P_5 (Stevenson-Paulik *et al.*, 2002; Raboy and Bowen, 2006) (Fig. 8.1). In addition, the plant inositol phosphate phosphatases are, for the most part, more promiscuous and less effective in dephosphorylating Ins(1,4,5)P_3 than the mammalian enzymes (Drøbak *et al.*, 1991; Loewus and Murthy, 2000; Perera *et al.*, 2002; Zhong and Ye, 2004; Raboy and Bowen, 2006; Torabinejad and Gillaspy, 2006).

When it comes to $InsP_3$-mediated Ca^{2+}signaling, it is quite possible that in addition to the vacuole, the mitochondria or chloroplast are the $InsP_3$/InsPx-sensitive internal calcium-sensitive stores in plants. Estimates indicate that mitochondria can store up to 60% of the cellular calcium in some plant cells (Subbaiah *et al.*, 1998; Logan and Knight, 2003). Early Ins(1,4,5)P_3 microinjection studies show an increase in cytosolic calcium in regions surrounding chloroplast as well as the vacuole (Gilroy *et al.*, 1990). Alternatively, the major $InsP_3$-mediated ion flux maybe via plasma membrane InsPx-sensitive channels (Lemtiri-Chlieh *et al.*, 2000).

Microinjection studies revealed that $InsP_3$ calcium signaling in plants depended on extracellular calcium (Tucker and Boss, 1996). A requirement for extracellular calcium to propagate the $InsP_3$-mediated calcium signal is supported by recent studies of cells with altered levels of a plasma membrane Ca^{2+} transporter (Tang *et al.*, 2007). Whether the active InsPx is $InsP_3$ or down stream metabolites such as $InsP_6$ as proposed by Lemtiri-Chlieh *et al.* (2000, 2003) and Ali *et al.* (2007) or $InsP_4$ as proposed by Zonia and Munnik (2004, 2006), the data make a compelling argument that PLC-mediated $InsP_3$

biosynthesis either directly or indirectly affects calcium signaling and ion transport differently in plants compared to animals.

8.2.1 Regulation of plant $PtdInsP_2$ biosynthesis

One of the first biochemical differences in plant and animal PI metabolism was the relatively low level of incorporation of *myo* [2-^{3}H]inositol into $PtdInsP_2$ in vivo (Boss and Massel, 1985). This could have been the result of large pools of cold inositol; plants unlike most mammals synthesize inositol de novo (Loewus and Murthy, 2000). However, ^{32}Pi-labeling studies resulted in similar low rates of ^{32}P incorporation into $PtdInsP_2$. One could conclude from the in vivo studies that either there was rapid turnover mediated by PLC or PTases that kept the lipid intermediates low, or that there was a low rate of lipid biosynthesis.

A clue as to what was regulating the biosynthesis of $PtdInsP_2$ in vivo came from in vitro biochemical characterization of phosphatidylinositol lipid kinases. When Sandelius and Sommarin (1990) compared the lipid kinase activity from rat liver and plants, they revealed a 20-fold higher specific activity for the rat plasma membranes relative to soybean hypocotyl or wheat shoot plasma membranes. Furthermore, Augert *et al.* (1989) had shown that rat liver total lipid extracts had a 1:1 ratio of PtdInsP and $PtdInsP_2$ based on mass measurements and that the ratio in isolated plasma membrane was 1:10. These ratios were significantly different from the PtdInsP to $PtdInsP_2$ ratios (from the 10 to 20:1) reported for plants (Boss, 1989).

Clearly, both the in vivo and in vitro data supported the hypothesis that phosphorylation of PtdIns4P by PtdIns(4)P 5-kinase (PIPK) was a flux limiting step in the plant PI pathway (Gross and Boss, 1993; Drøbak *et al.*, 1999; Perera *et al.*, 2002). However, the PIPKs are low abundance proteins that are difficult to purify, and it was not until the genes encoding the enzymes were cloned and recombinant proteins produced that the differences in the enzyme kinetics could be confirmed (Perera *et al.*, 2005; Im *et al.*, 2007b).

Comparative genomics has upheld many of the earlier hypotheses of differences in plant and mammalian PI pathway enzymes. Scientists now are challenged to consider the biochemistry of the plant-specific mechanisms of pathway regulation in order to understand how the PI pathway functions *in planta*.

Kinetic analysis of two isoforms of GST (glutathione S-transferase) tagged recombinant *Arabidopsis* PtdInsP kinases (*At*PIPK1 and *At*PIPK10) indicated that the plant enzymes were significantly less active (the V_{max}/K_m of 20- to 200-fold less) when compared to GST-tagged type I human enzyme, *Hs*PIPKIα (Perera *et al.*, 2005). More recent data indicate that recombinant *At*PIPK3 may be the most active *Arabidopsis* PIPK (Stenzel *et al.*, 2008). While comparative kinetic analyses of *At*PIPK3 with the human kinase have not been done, expressing *Hs*PIPKIα in tobacco cells resulted in a dramatic increase in $PtdInsP_2$ production and clearly demonstrated that none of the plant PIPKs had comparable activity (Im *et al.*, 2007b). The impressive differences in the PtdInsP kinase activities between the wild type and *HsPIPKIα*-expressing cells are reminiscent of the comparative assays of rat liver and plant plasma membranes or microsomes (Sandelius and Sommarin, 1990; Gross and Boss, 1993), respectively.

In vivo labeling studies of the *HsPIPKIα*-expressing cell lines confirmed that PtdInsP kinase was flux limiting in wild type tobacco cells. Mass measurements indicated that the *HsPIPKIα*-expressing cells produced 13-fold more $PtdInsP_2$ per gram fresh weight and thus had ample substrate (PtdIns4P) (Im *et al.*, 2007b). The ratio of PtdInsP to $PtdInsP_2$ decreased to 2:1 in the *HsPIPKIα*-expressing cell lines compared to 12:1 of the wild type cells and was much more comparable to the 1:1 ratio reported for rat liver (Augert *et al.*, 1989). Furthermore, the unstimulated *HsPIPKIα*-expressing cell lines continuously produced 100-fold more $InsP_3$ compared to unstimulated wild type cells based on the $InsP_3$-binding assay (Im *et al.*, 2007b). This synthetic system also confirmed that in plants, more $InsP_6$ would be synthesized in vivo as a result of increased $InsP_3$ production. The fact that the tobacco cells survived the continuous increased rate of PI turnover in the *HsPIPKIα*-expressing cell lines indicated that the cells had somehow compensated for the increased flux through the pathway.

One impact of increasing the PI pathway flux in this system was an increase in respiration. 2ATPs are required to synthesize $PtdInsP_2$ from PI. Potentially, three more ATPs could be required to make $InsP_6$ from $InsP_3$. Thus, increasing PI signaling should in theory affect basal metabolic pathways as well as downstream signaling. This was observed in *HsPIPKIα*-expressing cell lines which depleted the sugar from the culture medium faster (Im *et al.*, 2007b) and were smaller cells (Moran, unpublished results). A recent finding that *At*PIPK9 co-precipitated with *Arabidopsis* cytosolic invertase and repressed invertase activity in vitro also suggests a connection between PtdInsP kinase and energy demand (Lou *et al.*, 2007).

It is well established in animal cells that the PI pathway is very sensitive to inhibition of oxidative phosphorylation (Poggioli *et al.*, 1983; Yeung *et al.*, 2006). In plants, this regulation may involve a change in distribution of the PIPKs. One might imagine that when *At*PIPK9 is membrane bound and functional, invertase is more active and when *At*PIPK9 is not on the membrane, and therefore, presumably less active, *At*PIPK9 might associate with invertase. In vitro biochemical analyses of *At*PIPK9 and its binding proteins are needed to test this hypothesis.

Future studies of cell expressing *HsPIPKIα* should reveal the impact of increased $PtdInsP_2$ on membrane channels and pumps and events downstream of the PI pathway. Undoubtedly, in such a constantly stimulated system compensatory changes will occur in order for cells to survive; however, important insights into the stimulated state will be forthcoming. For example, intermediates which normally might be too low to be detectable might be elevated and pathways which might be transiently elevated in a stimulated cell should be dramatically increased.

Altering the flux in a signaling pathway either by pushing (increasing the signal) or pulling (dampening the signal) will provide quite different outcomes—each yielding important insights (Perera *et al.*, 2002; Im *et al.*, 2007b). It has been hypothesized that the rate of flux through a pathway also is sensed by plants (Paul *et al.*, 2004). If this is true, then altering the flux through a signaling pathway may more closely mimic a stimulated ("pushed") or non-stimulated ("pulled") state than a more classical genetic approach where a disruption in the pathway results in a build up of up stream and complete lack of down stream intermediates. It is important with either approach to appreciate that when it comes to lipid-mediated signaling, discrete microdomains of lipids can regulate different, nonoverlapping functions within a membrane.

8.2.2 Biochemical regulation of plant PIPKs

Both in vivo and in vitro data make a compelling argument that PtdInsP kinase activity is flux limiting in the plants compared to animal cells. It follows that the plant PIPKs must be different from the animal PIPKs and activity must be tightly regulated. There are at least two plant-specific structural features of the major subfamily of PIPKs (subfamily B) that are immediately evident based on sequence comparisons (Fig. 8.2). All subfamily B PIPKs contain membrane occupation recognition nexus (MORN) motifs in the N-terminus and a putative linker region between the N-terminal MORN motifs and the C-terminal catalytic domain (Mueller-Roeber and Pical, 2002). Each of these features contributes important regulatory features to the plant PIPKs. Table 8.1 summarizes some of the features of two representative *At*PIPKs, *At*PIPK1 of subfamily B and *At*PIPK10 of subfamily A, compared to *Hs*PIPKIα.

The MORN motifs are most intriguing. These motifs are found in proteins involved in membrane fission and fusion (Takeshima *et al.*, 2000; Shimada *et al.*, 2004; Gubbels *et al.*, 2006). Of all organisms studied thus far, the only PIPKs that have MORN motifs are those found in plants. Whether this is something was lost from the animal signaling pathway or something gained by plants because the lipids need to be made in close proximity to fusion/fission regions of the membrane remains to be determined. The N-terminal MORN-containing domain of *At*PIPK1 will bind to yeast (Ma *et al.*, 2006) and plant plasma membranes (Im *et al.*, 2007b) suggesting that it may be important for membrane targeting.

The MORN domain, however, seems to do more than just target the enzyme. It binds PtdOH and is essential for PtdOH activation (Im *et al.*, 2007a). One model suggests that when PtdOH binds the MORN domain, a conformational change in the linker region opens up the active site and thereby increases enzyme activity (Im *et al.*, 2007a). Through these studies another unusual feature of the *At*PIPK1 was noted. Unlike most animal PIPKs that are product inhibited, *At*PIPK1 is product activated. Such a feature along with the propensity of the MORN motifs to bind $PtdInsP_2$ would favor the creation of $PtdInsP_2$ microdomains similar to those generated by the

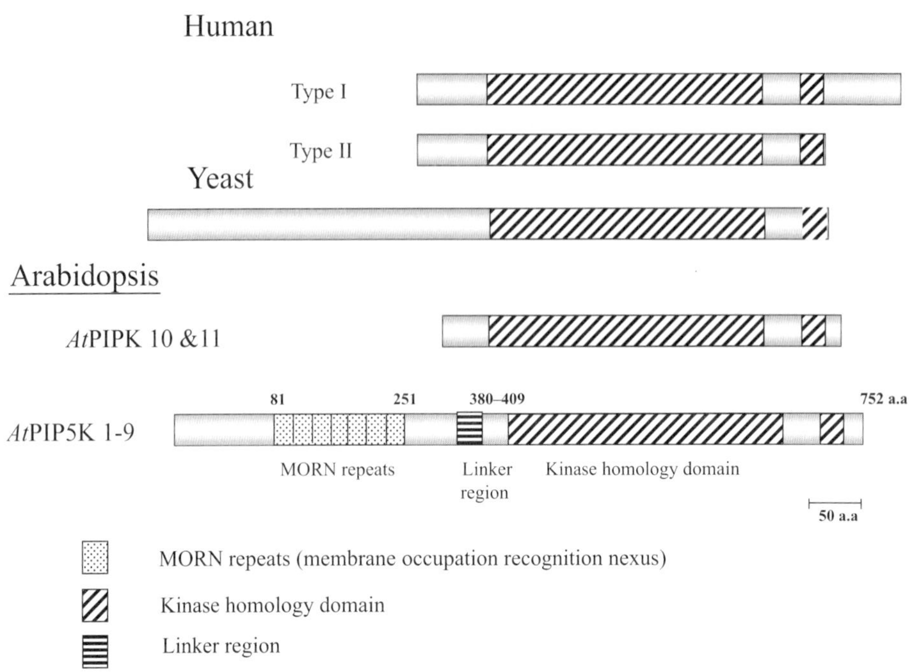

Figure 8.2 There are 11 putative type I AtPtdInsP 5-kinases in *Arabidopsis* arranged in two subfamilies based on size. Subfamily B contains *At*PIPK1-9, all of which contain membrane occupation and recognition nexus (MORN) repeats and a proline-enriched linker region. The MORN domain of *At*PIPK1 is essential for PtdOH activation, which involves a conformational change in the linker region (Im *et al.*, 2007a). The linker region of *At*PIPK1 also is essential for F-actin binding (Davis *et al.*, 2007). *At*PIPK10-11 are in subfamily A with molecular weights less than that of the members of subfamily B and contain no MORN repeats.

myristoylated alanine-rich C kinase substrate (MARKS) proteins (Wang *et al.*, 2002; Zhang *et al.*, 2003).

If the plant-specific PIPKs were stimulated, the creation of regions with large negatively charged head groups such as PtdInsP_2 would enhance membrane curvature. There should also be selective clustering of PtdInsP_2 containing unsaturated fatty acids to fill hydrophobic regions as the hydrophylic region of the bilayer expands. Whether this would favor membrane protrusions, vesicle formation or vesicle fission (Sun *et al.*, 2007) remains to be seen. To our knowledge there are no MARKS analogues in plants. Perhaps the PIPKs are dual-function proteins that generate lipid signaling microdomains as well as lipid-derived second messengers. This would make them the ultimate in "PIP-modulins," PtdInsP_2 modulating proteins (Mclaughlin and Murray, 2005).

The MORN domain is necessary for PtdOH activation and can bind to membrane lipids; however, the linker region of *At*PIPK1 is essential for

Table 8.1 Biochemical differences in plant and mammalian PIPKs

PIPK family	Plant type I (subfamily B)	Plant type I (subfamily A)	Mammalian type I
Representative PIPK	*At*PIPK1 (At1g21980)	*At*PIPK10 (At4g01190)	*Hs*PIPK1 α (NM_003557)
MORN motifs	Yes	No	No
Linker region	Yes	No	No
Actin interaction	Binds directly (requires the linker domain of *At*PIPK1) (Davis *et al.*, 2007)	None detected (neither direct or indirect)	Indirect binding to a scaffold of actin-binding proteins (Yang *et al.*, 2004)
K_m (GST-tagged protein)	69 ± 13 μM	65 ± 6 μM	21 ± 9 μM
V_{max} (GST-tagged protein)	600 ± 15 pmol/min mg protein	67 ± 6 pmol/min mg protein	4100 ± 25 pmol/min mg protein
Phosphatidic acid (PtdOH)-activation	Yes (2-fold) (requires MORN domain; aa #1–251) (Im *et al.*, 2007a)	No	Yes (2-fold) (Jones *et al.*, 2000)
Product activated	Yes (9-fold by PtdIns(4,5)P_2 and 16-fold by PtdIns(4,5)P_2+ PtdOH (Im *et al.*, 2007a)	Not determined	No (product inhibited) except for one report of a soluble PIPK (Imai and Gershengorn, 1986; Berridge and Irvine, 1989)
Phosphorylation decreases activity	Yes (Westergren *et al.*, 2001)	Not determined	Yes (Hinchliffe *et al.*, 1999)
Impact of hyperosmotic stress on plasma membrane-associated PIPK activity	Plasma membrane activity decreases and lower phase membrane activity increases (Im *et al.*, 2007a)	Not determined	Plasma membrane activity increases and cytosolic activity decreases (Yamamoto *et al.*, 2006)

Data are from (Davis *et al.*, 2004; Perera *et al.*, 2005; Im *et al.*, 2007a) unless otherwise indicated.

binding to actin (Davis *et al.*, 2007). This means that PIPKs such as *At*PIPK1 could directly link the actin cytoskeleton to membrane lipids and potentially to PtdInsP_2 microdomains. In contrast, mammalian PIPKs are connected to the cytoskeleton through a scaffold of actin-binding proteins (Yang *et al.*, 2004). Direct binding provides plants with immediate input to the actin cytoskeleton but may reduce "fine control" of cytoskeletal structure that could be conveyed through the multiple actin binding proteins found in mammalian systems.

One of the proteins that *At*PIPK1 recruits to F-actin is *At*PtdIns 4-kinase β1 (*At*PI4Kβ1) (Davis *et al.*, 2007). *At*PI4Kβ1 could supply substrate, PtdIns4P, for the PtdInsP kinase. Presumably, *At*PIPK1 would be less active when bound to actin and become activated when associated with membrane lipids such as PtdOH. It is possible that *At*PI4Kβ1 recruits *At*PIPK1-F-actin to the RabA4b vesicles rather than visa versa (Preuss *et al.*, 2004, 2006). The exact function of this plant-specific connection between the cytoskeleton and $PtdInsP_2$ biosynthesis is a new and exciting avenue for investigation.

Even when some mechanisms of regulation of the PIPKs are similar at a biochemical level, the cell biology in plants and animals can be quite different. For example, in response to hyperosmotic stress, PIPK activity increases in both plants and animals (Heilmann *et al.*, 1999; Pical *et al.*, 1999; Dewald *et al.*, 2001; Yamamoto *et al.*, 2006). However, in HeLa cells, hyperosmotic stress results in activation of the human type I PIPKβ by dephosphorylation and redistributed of the enzyme from the soluble fraction to the plasma membrane (Yamamoto *et al.*, 2006). In plants, the B-subfamily of PIPKs can also be activated by dephosphorylation (Westergren *et al.*, 2001), but $PtdInsP_2$ and PtdInsP kinase activity decrease in the plasma membrane as a result of hyperosmotic stress and PtdInsP kinase activity increases in intracellular membranes or a lower phase fraction isolated by 2-phase partitioning (Im *et al.*, 2007a). It is not known if the differences in targeting and distribution of the PIPKs in plants and animals are a result of differences in cell growth (Heilmann *et al.*, 2001), lipid binding, or protein–protein interactions, but a change in distribution will certainly produce different outcomes in terms of mediating PI signaling as well as membrane biogenesis.

8.2.3 PI metabolism and vesicle trafficking

Much of the data on plant PI metabolism supports a role for PI metabolism in mediating vesicle trafficking. The type III PtdIns 3-kinase (PI3K), which is the only type of PI3K identified in plants, was first characterized as a vacuolar sorting protein (Vps34) in yeast. There is only one gene encoding PI3K in plants. Pharmacological studies suggest that the plant PtdIns 3-kinase is important for vesicle trafficking to the plant vacuole as well (Matsuoka *et al.*, 1995; Jung *et al.*, 2002). Increased activity and expression of PI3K in nitrogen fixing nodules is consistent with an increase in membrane biogenesis (Hong and Verma, 1994), but it could also indicate changes in transcription. PI3K was localized to the putative transcription initiation sites in plant nuclei (Hendrix *et al.*, 1989; Bunney *et al.*, 2000). Aside from one report of the soybean lipid transfer protein (Ssh1) increasing PI3K activity (Monks *et al.*, 2001), little is known about how this lipid kinase is regulated in plant cells and nothing is known about its interacting partners. This is an area where more extensive biochemical studies will surely yield important insights.

Phenotypes of plants with mutations or genetic alterations in other PI pathway enzymes are most prevalent in tip growing or elongating cells where the impact of altering vesicle trafficking would be most evident. While $PtdInsP_2$ biosynthesis is essential for both pollen and root hair growth, too much $PtdInsP_2$ can inhibit growth indicating that turnover is important (Kost *et al.*, 1999; Zhong *et al.*, 2004; Williams *et al.*, 2005; Dowd *et al.*, 2006; Helling *et al.*, 2006).

A common theme in studies of tip growing cells is a need for F-actin-mediated vesicle trafficking. Insights into the mechanisms involved were revealed by Preuss *et al.* who showed that RabA4b containing vesicles bound to PI4Kβ and that both proteins were necessary for actin-mediated trafficking of Golgi-derived vesicles in root hairs (Preuss *et al.*, 2004, 2006). Other evidence supporting PI lipids and Golgi-derived vesicle trafficking in plants comes from studies of the PtdIns5P/PtdIns(4,5)P_2-binding protein Patellin1 (PATL1) (Peterman *et al.*, 2004). PATL1 contains both a Sec14 lipid binding domain and Golgi dynamics domain (GOLD) and is essential for cell plate formation during cytokinesis. Another Golgi protein required for normal growth, SAC1, is functional as a PtdIns(3,5)P_2 phosphatase. A mutation in SAC1, which mistargets the protein but does not affect its function as a lipid PTase, results in decreased cell length and aberrant shapes in fiber cells, pith cells, and trichomes (Zhong *et al.*, 2005). While the exact mechanisms are not known, these examples support a role for the lipids and/or lipid metabolizing enzymes in Golgi-mediated vesicle trafficking. Furthermore, the differences in the phenotype of the SAC family mutants argue for nonoverlapping roles of the lipid PTases and support a model for discrete microdomains of lipids within the membrane (Zhong *et al.*, 2004; Williams *et al.*, 2005).

Pollen tubes are a popular system for studying the PI pathway in vivo (Franklin-Tong *et al.*, 1996; Zonia and Munnik, 2004, 2006; Dowd *et al.*, 2006; Helling *et al.*, 2006). The first report of a role of the PI pathway in pollen tubes indicated that overexpressing RAC increased $PtdInsP_2$ biosynthesis and resulted in tip swelling (Kost *et al.*, 1999). It appears that RhoGAP is essential for the subcellular localization of RAC and normal tip growth (Helling *et al.*, 2006). Increasing $PtdInsP_2$ by inhibiting PLC-mediated turnover of $PtdInsP_2$ also inhibits normal pollen growth and leads to swelling (Dowd *et al.*, 2006; Helling *et al.*, 2006). These experiments make a compelling argument for the need for $PtdInsP_2$ turnover for normal tip growth; however, they have yet to reveal the cause of the tip swelling. While there are many potential reasons for the swollen phenotype, a simple explanation is that a thin cell wall such as was found in FRA3 mutants (Zhong *et al.*, 2004), would be weaker and have a tendency to swell under positive turgor pressure.

If the turnover $PtdinsP_2$ is essential for membrane fusion/fission and vesicle trafficking, then identifying the endogenous PIPKs, PLCs, and InsPx metabolizing enzymes involved and monitoring their distribution will be essential to understand how growth is controlled. With regard to the PIPKs, the genomic data bases indicate that the smaller A-type subfamily of PIPKs (PIPK10/11) as well as some of the B-type PIPKs (PIPK4, PIPK5, PIPK6) are expressed in pollen (https://www.genevestigator.ethz.ch/, Becker *et al.*,

2003). PIPK3 appears to be most prevalent in roots (Stenzel *et al.*, 2008 and Kusano *et al.*, 2008); however, we know little about how these enzymes are regulated.

If $PtdInsP_2$ is essential for pollen growth, are the PIPKs selectively activated as tip growth is initiated? Actin-mediated vesicle trafficking is important for normal pollen tube growth. If the A-type family of PIPKs, e.g., *At*PIPK11, are functioning during tip growth, is this because unlike *At*PIPK1, the PIPKs of the A-subfamily do not bind actin directly? If some of the pollen PIPKs bind actin through a protein scaffold, this may favor more fine control of filament formation through the complex of actin binding proteins. Also, it is possible that the plant-specific PIPKs such as *At*PIPK1 normally suppress vesicle trafficking in plants and that non-actin binding isoforms such as *At*PIPK10 and *At*PIPK11 or a member of the FAB family of PIPKs (PtdIns3P 5-kinases) are essential for rapid vesicle trafficking. Clearly, more information regarding the cellular and subcellular distribution of the PIPKs and their biochemical regulation is needed to understand their role in tip growing cells.

Finally, if $PtdInsP_2$ and the PIPKs are essential for tip growth, what is the source of PtdIns4P? It is possible that PI4Kβ1/2, which produces the PtdIns4P in RabA4b-associated Golgi-derived vesicles (Preuss *et al.*, 2006) also produces the plasma membrane pool of PtdIns4P. There is a large pool of PtdIns4P in plants relative to $PtdInsP_2$ and presumably PtdIns4P is essential for plant membrane structure and function. PI4Kα1 will associate with an F-actin fraction, and thus could also be important for tip growth or as argued above, it may be that the actin-binding lipid kinases may not be prevalent where there is rapid vesicle trafficking. PI4Kα1 is product inhibited and PI4Kβ1 is product activated (Stevenson-Paulik *et al.*, 2003). These differences imply that PI4Kβ1 would be more important for generating high levels of PtdIns4P. This would be consistent with the hypothesis that PI4Kβ1 has a major function in vesicle trafficking (Preuss *et al.*, 2006).

Virtually nothing is known about the putative type II PI4Ks(γ1–8) in plants (Mueller-Roeber and Pical, 2002). The type II family of PI4Ks is thought to be essential for PI signaling in some systems (Minogue *et al.*, 2001; Wei *et al.*, 2002; Guo *et al.*, 2003). At least five of the putative *Arabidopsis* PI4Ks contain ubiquitin-like domains and are more likely involved in protein degradation than lipid signaling (Galvão *et al.*, 2008). More extensive studies of the plant PI4Ks, their subcellular localization, and regulation are needed.

Furthermore, because the ratio of PtdIns4P to $PtdIns(4,5)P_2$ is 10- to 20-fold higher in plants than animals, it is essential to consider the impact of PtdIns4P as a lipid regulator. That is, PtdIns4P is relatively abundant and could bind and regulate membrane-associated proteins. When the PIPKs are activated and PtdIns4P is converted to $PtdIns(4,5)P_2$, would this release PtdIns4P binding proteins and derepress growth? In addition, little is known about the metabolism of PtdIns4P by PLC or PTases. A more complete understanding of PtdIns4P binding proteins is essential to understand how plants utilize inositol phospholipids to regulate growth.

8.2.4 Regulation of $PtdInsP_2$ metabolism by PLC

We know that plant $PtdInsP_2$-PLCs are calcium regulated and that structurally they are most similar to the ζ family of animal PLCs and lack a PH domain (Hunt *et al.*, 2004), but little is known about how or when they bind to membranes or what scaffold of interacting proteins they form in vivo. Early biochemical studies indicated that the activity was membrane associated (Melin *et al.*, 1992). Studies of pollen expressing GFP–PLC constructs indicate punctuate distribution of GFP fluorescence throughout the pollen tube including the subapical region of the plasma membrane (Dowd *et al.*, 2006). While there is no doubt that the PLCs would bind the membrane when activated, to understand how they are regulated, it is essential that we understand the subcellular distribution of the endogenous proteins.

In the classical animal signal transduction scheme, a localized change in Ca^{2+} is propagated throughout the cell and to neighboring cells via the soluble second messenger $InsP_3$. Propagation of the $InsP_3$ signal from cell to cell in plants was shown to require extracellular calcium (Tucker and Boss, 1996). These data are consistent with the in vitro biochemical studies indicating calcium activation of PLC (Hunt *et al.*, 2004). Many of the early experiments used mastoparan or mas-7 to purportedly activate G-proteins, but it is important to appreciate that these amphiphilic peptides will increase the uptake of extracellular calcium. Removing extracellular calcium eliminates the mastoparan effect on $InsP_3$-induced calcium signaling and will even prevent mastoparan-induced cell death (Galanopoulou *et al.*, 1995). Many of these and other earlier studies led to the false assumption that there might be G-protein regulation of PLC in plants (Munnik *et al.*, 1996); however, there is no biochemical or genetic evidence to support this hypothesis. Hopefully, future biochemical studies of the PLCs will characterize more of the plant-specific features of this family of enzymes.

8.2.5 Regulation of $PtdInsP_2$ metabolism by inositol polyphosphate PTases

Several of the plant inositol 5-phosphatases characterized to date can hydrolyze both lipid substrates and soluble inositol phosphates and are therefore not specific for $InsP_3$ hydrolysis (Ercetin and Gillaspy, 2004; Zhong *et al.*, 2004; Torabinejad and Gillaspy, 2006). Mutants identified through mutant screens have subtle developmental phenotypes and sensitivity to growth regulators such as ABA (Berdy *et al.*, 2001; Burnette *et al.*, 2003; Carland and Nelson, 2004; Zhong *et al.*, 2004, 2005; Torabinejad and Gillaspy, 2006). Except for a few studies (Peterman *et al.*, 2004; Zhong *et al.*, 2005), little is known about the subcellular localization and regulation of these enzymes (Torabinejad and Gillaspy, 2006). FRA3 gene is highly expressed in *Arabidopsis* vascular tissues and fiber cells and encodes a type-II inositol polyphosphate (IPP) 5-phosphatase. Biochemical characterizations of FRA3 (Zhong *et al.*, 2004) indicate that FRA3 would hydrolyze either $InsP_3$ or $PtdInsP_2$ depending on the reaction conditions. These data and others characterizing the substrate specificity of the IPP PTases have revealed important mechanisms for fine-tuning PI pathway intermediates.

One of the challenges in using IPP PTase mutants to study the pathway is that the InsPs that build up can be phosphorylated to produce additional signaling molecules. The impact of InsPxs on plant metabolism is an area that needs more research (Stevenson-Paulik *et al.*, 2005; Raboy and Bowen, 2006; Zonia and Munnik, 2006). InsPxs are prevalent in plants and could travel between cells (Tucker and Boss, 1996; Raboy and Bowen, 2006; Zonia and Munnik, 2006). As such, the InsPxs could provide a wealth of information for fine-tuning plant growth and development. One example as yet not explored in plants, is the observation that $InsP_6$ regulates mRNA transport from the nucleus in yeast (Odom *et al.*, 2000). Although the genetics of InsPs in plants is different from yeast (Raboy and Bowen, 2006), the fact that seeds store $InsP_6$ and metabolize it during germination gives the plant a plethora of InsPxs to transmit developmental signals.

In addition to $InsP_6$ being an important means of storing phosphate in seeds, it is now well accepted that $InsP_6$ and potentially $InsP_7$ and $InsP_8$ might be important signaling molecules (Shears, 1998, 2001) and can serve as a source of phosphate for ATP biosynthesis (Phillippy *et al.*, 1994; Raboy *et al.*, 2000; Raboy and Bowen, 2006).

Because $InsP_6$ can be synthesized from Ins(1,4,5)P_3 in plants (Im *et al.*, 2007b), one has to question whether earlier microinjection studies were measuring the impact of $InsP_3$ or $InsP_6$ on guard cell closure or tip growth (Gilroy *et al.*, 1990; Franklin-Tong *et al.*, 1996; Zonia and Munnik, 2006). Furthermore, Zonia showed that in pollen the signal encoded by inositol 3,4,5,6-tetra*kis*phosphate [Ins(3,4,5,6)P_4] antagonized pollen tube growth, induced cell volume increases, and disrupted Cl^- efflux (Zonia and Munnik, 2004). These effects were specific for Ins(3,4,5,6)P_4 and were not mimicked by either Ins(1,3,4,5)P_4 or Ins(1,3,4,5,6)P_5. The data further complicate the interpretation of the role of the PI pathway in pollen tube growth. In addition, it is not known how tip-focused calcium channels impact PI turnover or inositol phosphate production and $InsP_4$-sensitive chloride channels in these rapidly growing structures.

It is quite clear that separating lipid signaling from inositol phosphate signaling in vivo is a challenge. To understand the role of the PI pathway in even the most tractable tip growing system, robust isoform-specific antibodies to the pathway enzymes are needed to identify and localize the proteins in situ.

8.2.6 Nuclear PI signaling in plants

Nuclear PI signaling is well documented in animals and yeast (Irvine, 2003; Gonzales and Anderson, 2006). It has been known for some time that calcium oscillations in plant nuclei are asynchronous with the cytosolic calcium signaling suggesting a nuclear signaling pathway (Pauly *et al.*, 2000). Lipid kinases and inositol phospholipids have been reported in plant nuclei (Hendrix *et al.*, 1989; Bunney *et al.*, 2000); however, little is known about the regulation or the impact of the PI pathway on chromatin structure or nuclear function in plants. In yeast and animals, PI metabolism has been linked to mRNA export (Odom *et al.*, 2000) and chromatin remodeling (Cheng and Shearn, 2004). One of the most dramatic examples of the functional requirements of the pathways enzymes is the embryonic lethal PIPK (skittles) mutation that results in hypercondensation of chromatin in *Drosophila* (Cheng and Shearn, 2004).

In plants, there is good evidence for both PtdIns 3-kinase and PtdIns 4-kinase activity is isolated nuclei (Hendrix *et al.*, 1989; Bunney *et al.*, 2000); however, the data for PtdInsP kinase activity are less convincing. This may result from the low-specific activity of the plant PIPKs because [^{3}H]PtdInsP$_2$ was recovered from isolated nuclei (Hendrix *et al.*, 1989), and recent data suggest that *At*PIPK9 will localize to the nucleus (Lou *et al.*, 2007). It is not clear from the *At*PIPK9 data whether GFP or GFP-peptides were being imaged and there were no biochemical data confirming the presence of a functional recombinant enzyme. Clearly, more work needs to be done to identify the PI pathway enzymes in plant nuclei and to delineate their roles in regulating karyokinesis or DNA synthesis.

Karyokinesis and cytokinesis involve extensive membrane fusion and vesicle trafficking. It is highly likely that PtdInsP or PtdInsP$_2$ and PI metabolism is involved in nuclear membrane restructuring during karyokinesis.

Another exciting areas of investigation are the nuclear PtdInsP- and PtdInsP$_2$-binding proteins. Transcription factors such as PHD domain-containing proteins are potential lipid binding proteins. Virtually nothing is known about the role of the negatively charged inositol phospholipids or of the PIs/PI binding proteins in regulating transcription and translation in plants.

One is reminded of the fact that the first report of changes in PI metabolism by the Hokins was a serendipitous discovery when they precipitated RNA from stimulated cells (Hokin and Hokin, 1953). The similar chemistries of sugar phosphate containing lipids and nucleotides may make them ideal bedfellows.

8.2.7 When it comes to signaling, can plants sense flux?

There is evidence that flux alone can be sensed by plants. Somewhat serendipitously, Paul *et al.* (2004), noticed that when they grow seedlings at low pressure and the seedlings transpire rapidly, even though the seedlings have plenty of water and never wilt, the transcript profile indicates a response characteristic of water loss. That is, transcripts that would have been induced by drought are elevated. Their explanation was that under low pressure there was a very rapid flux of water through the seedlings, and that this rapid *flux* was sensed as lack of water and induced drought-responsive transcripts.

We do not know whether plants sense the turnover of the lipids or simply the second messengers produced, but as indicated above, it is likely that both are important and that in a living cell, both are constantly being sensed. $PtdInsP_2$ is rapidly being turned over even in resting cells and both transient and prolonged increases (15 min to several hours) in $InsP_3$ have been documented in response to gravi-stimulation and osmotic stress (Dewald *et al.*, 2001; Perera *et al.*, 2006).

A synthetic approach to test the impact of the $InsP_3$ signaling on downstream responses was undertaken by Perera *et al.* (2006). Expressing the extremely active and highly specific, plasma membrane localized human type-I InsP 5-PTase in plants dampened the $InsP_3$ signal. Constitutive expression of the human InsP 5-PTase driven by the 35S promoter had no effect on growth under optimal conditions indicating that at normal levels of $InsP_3$ were not essential for growth or development. However, as predicted, the transgenic plants expressing the human InsP 5-PTase had a compromised gravitropic response (Perera *et al.*, 2006) and compromised Ca^{2+} signaling (Perera, unpublished results). The simple interpretation of these data is that because a component of the gravity sensing mechanism (the input from the $InsP_3$ signal) was removed, the sensing mechanism was less robust, and therefore, the response delayed. If so, would other responses to stress (such as drought) be delayed as well and would this make the plants more or less susceptible to the stimulus, e.g., drought? These are intriguing questions that will hopefully be answered in the near future.

With such a complex signaling pathway it is difficult to determine which metabolite is necessary and sufficient to mediate a response. When it comes to flux, one must also consider the rate of turnover of the inositol lipids. If we accept that PtdInsP and $PtdInsP_2$ can be ligands as well as generating second messengers, then every time the lipids turn over, the activity of the proteins to which they are bound will be affected. For example, ATPases, PLDs, actin binding proteins, or transcription factors would directly respond to the rate of $PtdInsP_2$ turnover. Thus, in addition to analyzing InsPx signaling, we must consider lipid-mediated signals and in order to build predicted models of integrated signaling networks and better understand the dynamics of plant PI signaling.

8.2.8 Challenges for studying PI metabolism in plants

A hallmark of the pathway is the specificity with which the polyphosphorylated inositol phospholipids and inositol phosphates are recognized by enzymes and receptors within cells. To be biologically relevant, in vivo and in vitro data must meet this criterion. Studies of calcium-regulated K^+ channels by $InsP_6$ are impressive for this reason. While both *scylo*-$InsP_6$ and *myo*-$InsP_6$ were tested only *myo*-$InsP_6$ affected channel activity (Lemtiri-Chlieh *et al.*, 2000).

One should also appreciate that the lipids by virtue of their very negatively charged head groups are relatively labile especially at neutral pH. Because of cell wall lipases and lipid stability, it is difficult to obtain interpretable results when adding lipids to whole cells. Adequate controls are essential to distinguish between effects of metabolites versus intact lipids. Furthermore, the inhibitors commonly used to study this pathway can have multiple effects, and adequate controls are essential to begin to interpret the results (Boss *et al.*, 2006). Neomycin, for example, interacts with the slow vacuolar channel (Scholz-Starke *et al.*, 2006) and will bind to the cell wall inhibiting the uptake of positively charged molecules (Cho *et al.*, 1995), making it difficult to interpret results when it is added exogenously as an inhibitor of the PI pathway.

Not only are the lipid head groups specifically recognized but recent data indicate that selective pools of polyunsaturated fatty acids identify the plasma membrane signaling $PtdInsP_2$ that responds to hyperosmotic stress (Konig *et al.*, 2007). When plants were exposed to hyperosmotic stress, a pool of PtdIns enriched in unsaturated fatty acids selectively refilled the $PtdInsP_2$ pools and the downstream lipids (DAG and PtdOH). The concept of discrete, nonoverlapping functions of the polyphosphorylated inositol lipids, and inositol phosphates is also supported by molecular genetic studies (for review see Boss *et al.*, 2006; Raboy and Bowen, 2006; Torabinejad and Gillaspy, 2006; Zonia and Munnik, 2006). These data make a compelling argument that adding lipids or inositol phosphates to whole cells cannot replicate in vivo signals.

Approaches to studying the lipids in vivo have involved visualizing lipid domains or endogenous enzymes with GFP-fusion peptides and proteins. Plants present their own challenges with regard to imaging because of their morphology and autofluorescence (Doughman *et al.*, 2003; Vermeer *et al.*, 2006; Balla, 2007), but even in animal cells interpreting expression patterns of GFP-fusion proteins can be equivocal (Doughman *et al.*, 2003; Vermeer *et al.*, 2006; Balla, 2007). Combining the cell biological studies with biochemical analyses and electron microscopy detection of endogenous proteins is essential to interpret the data in a biologically relevant manner. Thus, while molecular genetic studies have led to important insights, until nanocameras become available so that one undertakes a virtual tour inside a cell and visualize the inner workings in real time, combinatorial approaches will be required to identify the scaffold of interacting proteins and connect PI signaling to other known metabolic pathways within the cell.

8.3 Phospholipase D signaling

8.3.1 Plant intracellular phospholipases

Phospholipases produce various lipid and lipid-derived messengers, such as phosphatidic acid (PtdOH), lysophospholipids (lysoPLs), free fatty acids, diacylglycerol (DAG), and inositol 1,4,5 trisphoshates [$Ins(1,4,5)P_3$] (Wang, 2004). The activation of phospholipases often occupies a critical and early

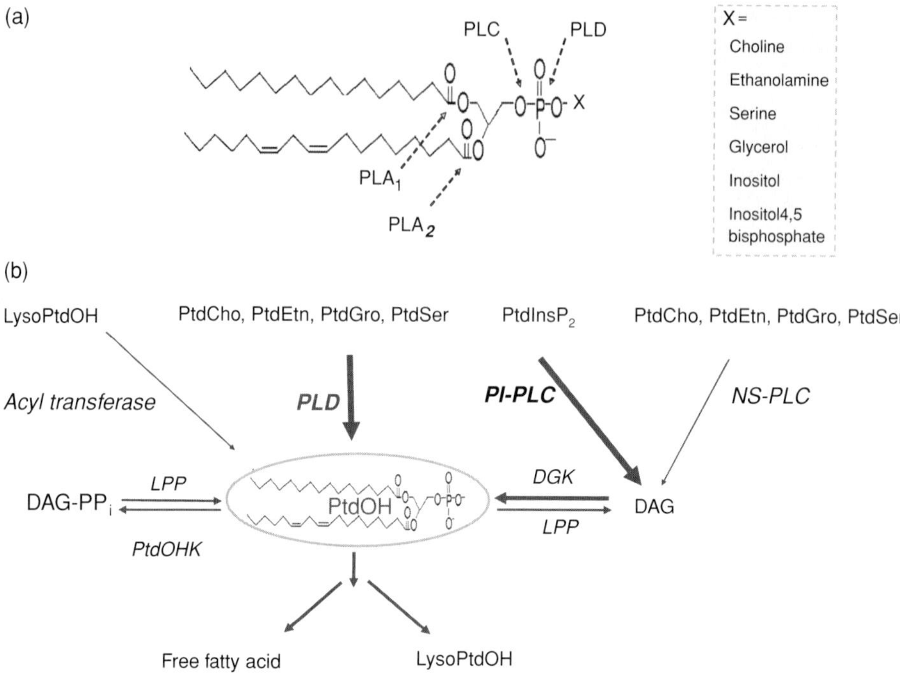

Figure 8.3 Enzymatic reactions lead to PtdOH production and degradation (Wang *et al.*, 2006). (a) Sites of hydrolysis by four types of phospholipases. X denotes the head group that defines different head classes of phospholipids. (b) The enzymatic reactions leading to the PtdOH production (*upper*) and removal (*lower*). Abbreviations: DGK, diacylglycerol kinases; DAG-PP$_i$, diacylglycerol pyrophosphate; LPP, lipid phosphate phosphatase; LysoPtdOH, lysophosphatidic acid; PAK, phosphatidic acid kinase; PtdCho, phosphatidylcholine; PtdEtn, phosphatidylethanolamine; PtdGro, phosphatidylglyerol; PtdInsP$_2$, phosphatidylinositol bisphosphate; PtdSer, phosphatidylserine.

step in a specific signaling cascade. Phospholipases are grouped into four major classes, phospholipase D (PLD), phospholipase C (PLC), phospholipase A_2 (PLA_2), and phospholipase A_1 (PLA_1), according to the site of hydrolysis of glycerophospholipids (Fig. 8.3a). Within each class, the enzymes can be further divided into families or subfamilies accordingly to their sequence similarities, biochemical properties, and biological functions (Wang, 2001; Holk *et al.*, 2002; Ryu, 2004; Nakamura *et al.*, 2005). Distinct differences exist in the occurrence and functions of various phospholipases between plants and animals. Such differences for PtdInsP_2-PLC are highlighted in the earlier section. In addition, plants contain another family of PLCs, nonspecific PLCs that use common membrane lipids, such as phosphatidylcholine (PtdCho) and phosphatidylethnolamine (PtdEtn) as substrate. Some of the nonspecific PLCs are implicated in phospholipid/galactolipid remodeling in plant response to phosphorus deficiency (Nakamura *et al.*, 2005). These enzymes are also potentially important in producing DAG and PtdOH in cell signaling (Fig. 8.3b).

The molecular identity of the intracellular PLA_2 remains elusive in plants. In mammalian cells, intracellular PLA_2 consists of two major types: Ca^{2+}-dependent PLA_2 ($cPLA_2$) and Ca^{2+}-independent PLA_2($iPLA_2$). $cPLA_2$ plays a key role in releasing fatty acids for the production of oxylipins in mammals, but no $cPLA_2$-like enzyme has been found in plants. By comparison, plants have many patatin-like enzymes, some of which share sequence similarities to $iPLA_2$ (Wang, 2001; Holk *et al.*, 2002). However, patatin-like enzymes are better characterized as acyl hydrolases because they utilize phospholipids and galactolipids as substrates and do not show particular preference for sn-2 position (La Camera *et al.*, 2005; Yang *et al.*, 2007). AtPLAI from the *Arabidopsis* genome that is most similar to mammalian $iPLA_2$ is an acyl hydrolase displaying preference to oxygenated galactolipids (Yang *et al.*, 2007). Unique to plants is the presence of high amounts of galactolipids in plastids that harbor the majority of polyunsaturated fatty acids, such as 16:3 and 18:3, used for oxylipin biosynthesis. In addition, the production of oxylipins can begin with 16:3 and 18:3 esterified to phospholipids or galactolipids (Buseman *et al.*, 2006; Yang *et al.*, 2007). These properties underlie a fundamental difference in the fatty acid release and oxylipin biosynthesis between plants and animals. However, the identification of in vivo substrates and products of an acyl-hydrolyzing enzyme is challenging because the substrate selectivity of lipolytic enzymes determined in vitro is strongly influenced by substrate presentation. Thus, intracellular surroundings, membrane environments, and in some cases, heterooligomerization can affect catalytic activity and substrate selectivity.

PLD constitutes a major phospholipase family in plants. PLD was once regarded as a "plant enzyme" because its activity is readily detectable in plant tissues. Because of its high activity, the status of PLD as an enzyme in plants was even once doubted. Now we know that the common PLD activity in *Arabidopsis* comes primarily from only one of a rather complex family of PLDs (Wang, 2004; Wang *et al.*, 2006). Plants contain more genes and types of PLDs than animals do, and different PLDs and their derived PtdOH are involved in a wide range of signaling processes in plant growth, development, and stress response (Wang *et al.*, 2006).

8.3.2 Molecular and biochemical heterogeneity of the PLD family

The *Arabidopsis* genome has 12 identified PLD genes that are grouped into six classes: *PLD*α*(3)*, β*(2)*, γ*(3)*, δ, ε, ζ*(2)* (Wang, 2004; Wang *et al.*, 2006). Ten of the 12 PLDs (αs, βs, γs, δ, and ε) contain the C2 domain, which is a Ca^{2+}-dependent phospholipid binding structural fold found in many proteins involved in signaling, membrane trafficking, and lipid metabolism. The C2-PLDs are unique to plants, and the number of C2-PLDs tends to increase in other plant species, such as 13 C2-PLDs identified in rice. By comparison, the two PLDζs have N-terminal phox homology (PX) and pleckstrin homology

(PH) domains, and two PX/PH-PLDs have also been identified in rice and in mammals. Some of the PLDs also contain a G protein-interacting, DRY motif and a polybasic PI(4,5)P_2-binding motif (Zhao and Wang, 2004). The presence of these different regulatory motifs provides insights to the different modes of activation and functions of PLDs. In addition, individual PLDs can differ in the key amino acid residues in these motifs. In some cases, these differences have been shown to underlie a structural basis for their different biochemical and regulatory properties (Zheng *et al.*, 2000).

PLDα1, β1, γs, δ, and ζ have distinguishable requirements for Ca^{2+}, PtdInsP_2, and free fatty acids (Wang, 2005; and references therein). A recent study on PLDγ1 and γ2 reveals that the two highly homologous PLDs have distinctively different biochemical and molecular properties and thus may have unique regulations and distinguishable functions (Qin *et al.*, 2006). Detailed analyses of PLDβ show that both $Ca^{2+}$and PtdInsP_2 bind to the regulatory C2 domain and the catalytic regions of PLD. Ca^{2+}binding to the two regions differentially affects the interaction of the enzyme with PtdIns(4,5)P_2 and its substrate PC, regulating the membrane association and catalysis of PLD, respectively (Pappan *et al.*, 2004). Besides the above cofactors, PLDs also directly interact with proteins. *Arabidopsis* PLDα1 binds to GDP-bound GPA1, the Gα subunit of the heterotrimeric G-protein in *Arabidopsis* through the DRY motif and the binding inhibits the PLD activity (Zhao and Wang, 2004). PLDα also binds to cardosin A, an RGD-containing aspartic proteinase, through its C2 domain (Simoes *et al.*, 2005). On the other hand, PLDβ is found to bind to actin filaments, whereas PLDδ binds to microtubule cytoskeleton (Kusner *et al.*, 2003). These different molecular interactions indicate that individual PLDs are tightly controlled in the cell and can be activated differentially.

In addition, PLDs display varied substrate preferences. C2-PLDs hydrolyze PC, PE, and PG more efficiently than other lipids, but the preference varies among the enzymes (Wang, 2005; and references therein). However, PX/PH-PLD uses PC as substrate. Thus, activation of individual PLDs could result in the hydrolysis of distinct phospholipids in the cell. PLDs also exhibit unique and overlapping patterns of subcellular association and gene expression. The differences in activation, substrate preferences, gene expression, and subcellular location play an important role in regulating the spatial and temporal hydrolysis of membrane lipids and the production of PtdOH. On the other hand, PLDs also have overlapping biochemical activity and expression patterns. These features contribute to the biochemical and molecular basis for the distinguishable and overlapping functions of the individual PLDs.

8.3.3 Diverse functions of different PLDs

PLDs have been implicated in various processes during plant growth, development, and stress responses. These include effects of *PLD*s on root growth, nodulation, pollination, drought, freezing, salinity, nutrient deficiency, and plant–pathogen interactions (Wang, 2005; Bargmann and Munnik, 2006; Wang *et al.*, 2006; and references therein), and recently on floral abortion (Boyer and Mclaughlin, 2007) and photosensitivity (Kabachevskaya *et al.*, 2007). Analysis of plants altered in specific PLDs has documented unique functions for several PLDs. *PLDα1* and its derived PtdOH decrease transpirational water loss by mediating ABA-promoted stomatal closure (Mishra *et al.*, 2006). Comparison of transcriptomes between *PLDα1*-deficient *Arabidopsis* mutants revealed distinct sets of gene regulated by *PLDα1* under drought stress (Mane *et al.*, 2007). *PLDδ* is involved in reactive oxygen species (ROS) response and cell death to enhance plant stress tolerance, such as freezing (Li *et al.*, 2004). *PLDβ* is implicated in defense responses (Bargmann *et al.*, 2006). *PLDζ1* is a target of the *GLABRA2* transcriptional factor and is proposed to regulate root hair initiation and patterning (Ohashi *et al.*, 2003). *PLDζ2* participates in the regulation of vesicle cycling as well as auxin transport and distribution (Li and Xue, 2007). *PLDζ2* and *ζ1* are involved in primary root growth and lipid remodeling in response to phosphorus deficiency (Cruz-Ramirez *et al.*, 2006; Li *et al.*, 2006a).

The fact that alteration of a PLD leads to a phenotype demonstrates that individual PLDs have unique functions. In some cases, two different PLDs may have opposite roles in the regulation of a physiological process, such as in the case of *PLDα1* and *PLDδ* regulation of freezing tolerance and seed aging (Li *et al.*, 2004; Devaiah *et al.*, 2007). Such effects can be caused, at least in part, by the different roles of PLDs in membrane degradation and cell signaling. However, different PLDs may also have overlapping and additive functions, which have been shown by recent analyses of *PLDζ* single and double knockouts (Li *et al.*, 2006a,b). In addition, *PLDζs* have metabolic and signaling roles, depending upon the severity of phosphorus deficiency. *PLDζs* promote primary root growth under moderate phosphorus limitation (Li *et al.*, 2006a), but they hydrolyze PC to supply phosphorus and diacylglycerol moieties for galactolipid synthesis during severe phosphorus starvation (Li *et al.*, 2006b).

Although detailed mechanisms by which PLDs mediate plant functions are not well understood, a series of recent results have provided interesting mechanistic insights into how a specific PLD regulates plant functions. In stomatal response to ABA, PLDα1 is activated and generates PtdOH. PtdOH recruits the ABI1 PP2C (ABA insensitive protein phosphatase 2C) to the plasma membrane and prevents its translocation to the nucleus (Zhang *et al.*, 2004). This interaction inhibits the negative effect of ABI1 on ABA signaling and promotes stomatal closure (Mishra *et al.*, 2006). The activation of PLDα1 might be mediated by a recently identified ABA receptor GCR2, a G-protein-coupled receptor (Grill and Christmann, 2007; Liu *et al.*, 2007). GDP-bound Gα binds to and inhibits PLDα1 activity, whereas GTP-bound Gα destabilizes the Gα-PLDα1 interaction (Zhao and Wang, 2004). Thus, activation of G protein leads to an increase in PLDα1 activity and, indeed, diminishing the PLDα1 binding to Gα renders plants more sensitive to ABA and decreases water loss

(Mishra *et al.*, 2006). The results indicate that PLD serves as a critical node in assembling regulatory proteins in signaling cascades.

8.3.4 Phosphatidic acid as a class of pivotal messengers

PLDs can mediate cell functions via different modes of action, depending upon the nature of stimuli and the severity of stresses (Wang *et al.*, 2006). One key function of PLD is to produce the lipid mediator PtdOH. PtdOH is a minor membrane lipid, constituting less than 1% of total phospholipids in plants. However, cellular levels of PtdOH in plants are dynamic, increasing rapidly under various conditions, including chilling, freezing, wounding, pathogen elicitation, dehydration, salt, nutrient starvation, nodule induction, and oxidative stress (Bargmann and Munnik, 2006; Wang *et al.*, 2006). The rise is transient, and PtdOH can be removed by multiple reactions, such as dephosphorylation to DAG, phosphorylation to DAG-pyrophosphate, and deacylation (Fig. 8.3b). In addition, PtdOH is composed of different molecular species due to the variation in two fatty acyl chains, and different molecular species differentially affect the interactions of PtdOH with proteins (Zhang *et al.*, 2004; Wang *et al.*, 2006). Moreover, the location and timing of PtdOH production, as dedicated by differential activation, expression, and cellular locales of individual PLDs, are important to its signaling function.

PtdOH has been reported to bind to various proteins, including transcriptional factors, protein kinases, lipid kinases, protein phosphatases, and proteins involved in vesicular trafficking and cytoskeletal rearrangement (Huang *et al.*, 2006; Wang *et al.*, 2006; and references therein). No obvious consensus sequence motifs for PtdOH binding are identified, but a common feature is the requirement of positively charged amino acid residues for PtdOH binding. Lysine and arginine residues enhance the charge of PtdOH via forming hydrogen bonds with the phosphate of PtdOH (Kooijman *et al.*, 2007). The electrostatic/hydrogen bond switch has been proposed to stabilize the protein–lipid interaction and to make the phosphate of PtdOH an effective docking site for positively charged protein domains to interact with membranes (Kooijman *et al.*, 2007). In the case of the ABI1 PP2C, Arg72 is critical to the PtdOH binding, whereas mutation of the two adjacent two basic residues (RK) has no effect on PtdOH binding (Zhang *et al.*, 2004). This indicates that, in addition to the electrostatic interactions, a specific structural fold is required for a PtdOH–effector protein interaction. However, no structure for a PtdOH-bound, functional form has been determined for any protein.

One important mode of action by PtdOH is to tether or recruit proteins to membranes. Signal transduction, vesicular trafficking, and many other critical cellular functions are initiated by the assembly of cytosolic protein complexes to specific sites in cellular membranes. Great advances have been made over the past decade in the understanding of how proteins are recruited to the lipid surface of membranes. Binding to lipid ligands is required for the recruitment and/or regulation of many cytosolic proteins. Lipid mediators, $PtdInsP_2$ and phosphoinositide 3-phosphates, are produced in response to specific stimuli and bind to effector proteins at specific structural folds, such as PH, PX, and FYVE domains. Examples for the membrane-tethering function of PtdOH include animal Raf-1 (Ghosh *et al.*, 2003) and yeast transcriptional repressor Opi1p (Loewen *et al.*, 2004). In plant response to ABA, PtdOH recruits ABI1 to the plasma membrane and decreases the translocation of ABI1 from cytosol to the nucleus (Zhang *et al.*, 2004).

PtdOH can also directly modulate the activity of its effector enzymes. The modulation can be activation or inhibition, depending upon the target proteins (Wang *et al.*, 2006; and references therein). In plants, PtdOH binds to the phosphoinositide-dependent protein kinase 1 (PDK1) and activates the PDK1 and AGC2-1 kinases (Anthony *et al.*, 2004). In contrast, PtdOH decreases ABI1 PP2C activity and is also a potent inhibitor of animal protein phosphatase 1 (Jones and Hannun, 2002). The different effects of PtdOH on protein phosphatases and kinases have led to an intriguing hypothesis: PLD and PtdOH play an important role in the homeostasis of protein phosphorylation by concerted regulation of kinases and phosphatases in a specific signaling response (Wang *et al.*, 2006).

The modulation of enzymatic activity and protein functions should also be considered in the context of tethering and recruiting proteins to the membranes. The basal level of PtdOH in plant cells, 50–150 μM, is considerably above a phospholipid's critical micelle concentration, which is in the subnanomolar range (Wang *et al.*, 2006). Above the critical micelle concentration of a lipid, the concentration of lipid monomer is constant, independent of the total concentration of the lipid. Thus, the accumulation of PtdOH above the critical level during cell activation affects the concentration of membrane-associated, but not monomeric PtdOH. This suggests that PtdOH binding to target proteins occurs at the membrane, but not in solution.

8.3.5 Challenges and perspectives for studying phospholipase-mediated signaling in plants

The current understanding of phospholipase functions in plants is still at an early stage. One complication is that these enzymes have multifaceted functions. For instance, in addition to signaling, activation of PLD may alter membrane lipid composition and result in membrane deterioration. The effect is influenced by the nature of stimuli, severity of stresses, and tissues involved, as shown with PLDα1 in *Arabidopsis* (Mishra *et al.*, 2006; Devaiah *et al.*, 2007). In addition, PtdOH may be involved in cellular processes through different modes of action. Besides protein binding as described above, PtdOH can alter membrane structures and serve as a substrate for the production of other lipid regulators, such as lysoPtdOH, free fatty acids, DAG, and DAG-pyrophosphate (Wang *et al.*, 2006). Furthermore, PtdOH can be produced by other enzymes, such as PLC hydrolysis followed by DAG kinase (Fig. 8.3b) (Bargmann and Munnik, 2006; and references therein). Other lipid mediators, such as lysophospholipids, free fatty acids, and DAG can also result from multiple enzymatic reactions. Discerning the role of lipid mediators and specific lipid signaling reaction has also been confounded by the cross-talk among different lipid signaling processes.

Given the above complications and the multiple members of a phospholipase family, the study of specific genes and enzymes involved in the production of lipid mediators and cellular function is critical to understanding lipid signaling. Current findings suggest that the location and timing of lipid mediator production are tightly regulated by different enzymes (Wang, 2005; Wang *et al.*, 2006; and references therein). A combination of sensitive biochemical and metabolic analyses with genomic manipulation of genes should be effective to unveil the functions of lipid signaling. These enzymes are integral parts of various regulatory cascades in plant growth, development, and stress responses. Appreciation of the differences and similarities between plants and animals will help the investigation and understanding of lipid signaling.

8.4 Sphingolipid signaling

Sphingolipids are present in virtually all eukaryotic cells and have been demonstrated in animals and fungi to serve as structural components of membranes and as participants in signaling pathways involved in cell regulation (reviewed in Dunn *et al.*, 2004; Taha *et al.*, 2006; Cowart and Obeid, 2007). Although sphingolipids were detected in plants in the 1950s, and identified as significant components of the plasma membrane, tonoplast and endomembrane system of plant cells 20 years ago (reviewed in Lynch and Dunn, 2004), our understanding of the function of these lipids in plants, especially in regard to signaling, is meager. While recent studies have implicated sphingolipids in various processes in plants, there is a paucity of mechanistic details regarding their interacting partners or targets. However, increasing convergent interest in plant sphingolipid biology, the availability of *A. thaliana* mutant lines with disrupted sphingolipid metabolism and signal pathways, and methods to analyze sphingolipid profiles and metabolic activities should stimulate progress in the near future.

8.4.1 Plant sphingolipids

Sphingolipids are defined by the presence of a sphingoid long-chain base (LCB). Typically, the LCB sphinganine, synthesized from serine and palmitoyl-CoA, is acylated on the amino group giving rise to ceramide (Cer), which is subsequently modified in plants by addition of glucose or inositol phosphate (which in turn can be further glycosylated) to give rise to the complex sphingolipids glucosylceramide (GlcCer) and glycosylated inositolphosphorylceramide (GIPC) that serve as membrane components. LCB derived from the breakdown of complex sphingolipids and/or from biosynthetic reactions can be phosphorylated, generating long-chain base-1-phosphate (LCBP), the substrate for a lyase that degrades the LCBP and is the sole known route for the destruction of sphingolipids. The structures of these sphingolipids and the enzymatic steps involved in their turnover are shown in Fig. 8.4. A more detailed view of the sphingolipid pathway, including up-to-date information about the *A. thaliana* genes encoding the individual enzymes, can be found at http://www.plantsphingolipids.org/. The sphingolipids most commonly reported to play roles in signaling in eukaryotes include LCB, LCBP, and Cer. It is noteworthy that plants display considerably greater heterogeneity in their LCB profiles than do either yeast or animals (Napier *et al.*, 2002), but sphingosine (4-*trans* sphingenine), the LCB prevalent in many animal sphingolipids and commonly used in studies because of its commercial availability, is virtually absent in plants. Recent advances in extraction methods and mass spectrometry have permitted detailed analyses of sphingolipids from *A. thaliana* (Markham *et al.*, 2006; Markham and Jaworski, 2007). A typical analysis shows that GIPC, GlcCer, Cer, LCB, and LCBP constitute approximately 64%, 34%, 2%, <1%, and <0.1%, respectively, of leaf sphingolipid, proportions consistent with their respective proposed roles.

8.4.2 Sphingolipids and signaling

The discovery that sphingosine is a potent inhibitor of protein kinase C (Hannun *et al.*, 1986) fostered subsequent research on the role of sphingolipids in signal transduction and regulation in mammalian cells and yeast that continues to the present (Taha *et al.*, 2006; Cowart and Obeid, 2007). The sphingolipids most intensively studied in mammalian cells include sphingosine, sphingosine-1-phosphate (S1P), and Cer. As a gross generalization, sphingosine may stimulate growth at lower concentrations but can be toxic or promote cell death at higher concentrations (Merrill, 2002); Cer tends to inhibit cell proliferation and promote apoptosis (Pettus *et al.*, 2002; Zheng *et al.*, 2006) and S1P tends to stimulate cell proliferation (Spiegel and Milstien, 2003; Taha *et al.*, 2006). Thus, Cer and S1P have opposing, even antagonistic, roles and sphingosine is the metabolic intermediate linking the two, since ceramide synthase and ceramidase catalyze the interconversion of sphingosine and Cer, while sphingosine LCB kinase and LCBP phosphatase catalyze the interconversion of sphingosine and S1P (Fig. 8.5). These bioactive sphingolipid molecules are derived from the hydrolysis of complex sphingolipids, especially sphingomyelin, but some sphingolipid signaling molecules may be generated biosynthetically in mammals (Merrill, 2002). Yeast produce only saturated LCBs so, as with plants, are devoid of sphingosine, but produce sphinganine and phytosphingosine (4-hydroxysphinganine). In yeast, LCB and LCBP have been implicated in cell cycle regulation and, along with Cer, in the response to heat stress (Dickson and Lester, 2002; Obeid *et al.*, 2002; Cowart and Obeid, 2007). Phytosphingosine specifically appears to promote protein phosphorylation events involved in growth, cell wall integrity, stress

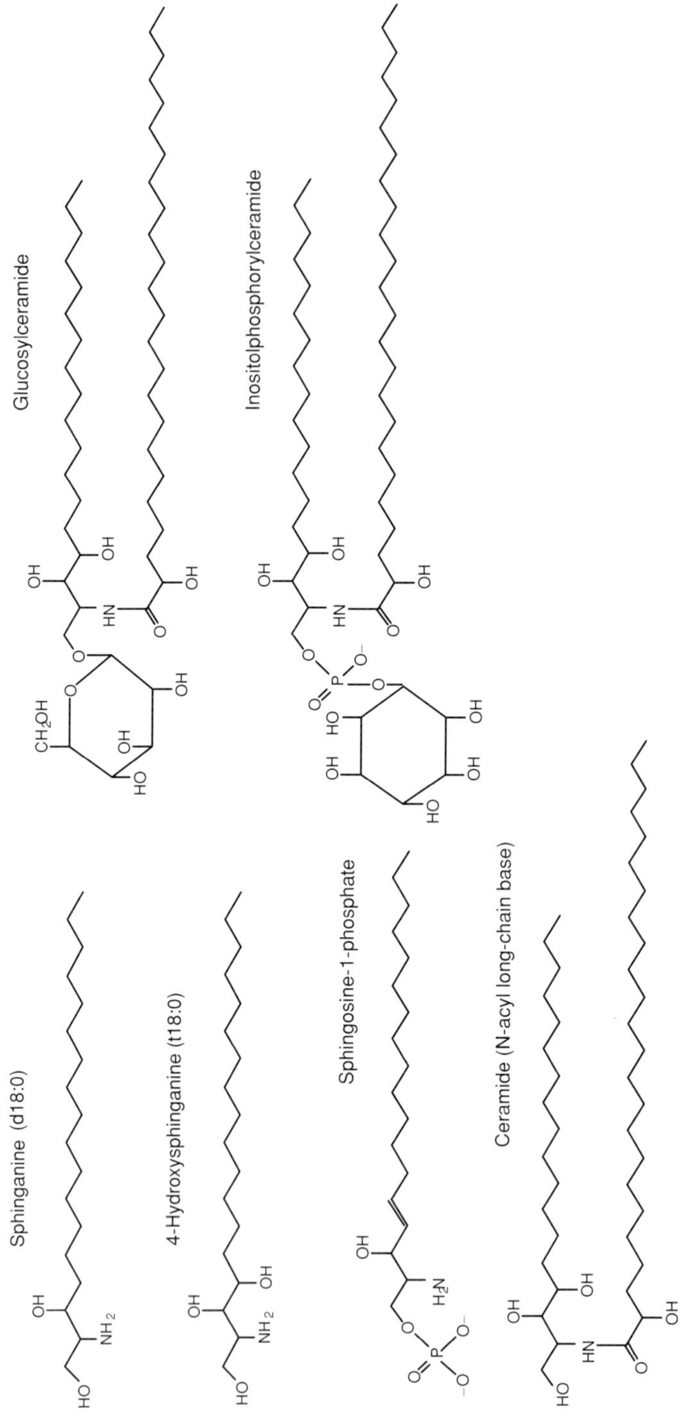

Figure 8.4 Representative sphingolipid structures are shown. Note that while sphingosine 1-phosphate is frequently used in studies of signaling, it has not been unequivocally identified as a significant LCBP in plants, though other LCBPs are present.

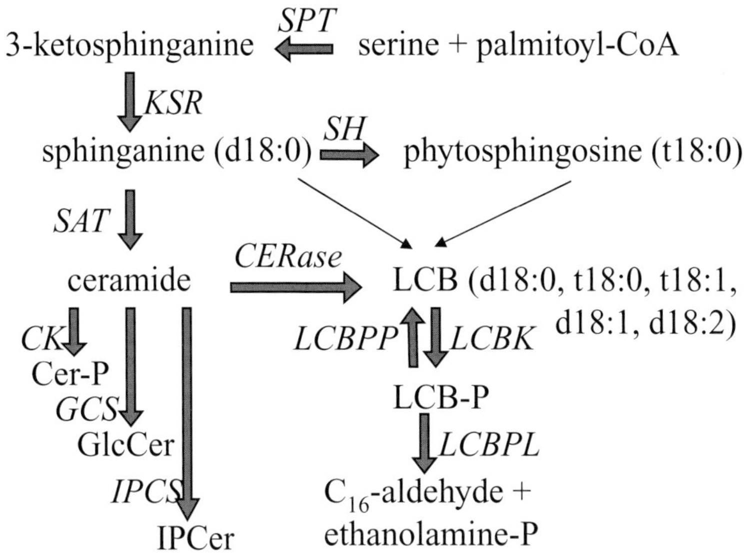

Figure 8.5 A representation of the plant sphingolipid metabolic pathway displaying the enzymatic interconversions. Enzymatic steps modifying long-chain bases and amide-linked acyl chains via hydroxylation and/or desaturation (not shown) are thought to utilize ceramide, glucosylceramide, or inositolphosphorylceramide as substrate, but acyl-chain elongation precedes ceramide formation, and the hydroxylation of free sphinganine has been demonstrated. Inositolphosphorylceramide is thought to be the precursor to complex glycophosphosphingolipids but the glycosyltransferases presumably required have not been identified, nor has the glucosylceramidase or IPC-specific phospholipase C responsible for catabolism of complex sphingolipids, so these reactions are not shown. In addition to free sphinganine (d18:0) and 4-hydroxysphinganine (t18:0, phytosphingosine), other long-chain bases present in plants include *cis* and *trans* isomers of 8-sphingenine (d18:1), 4,8-sphingadienine (d18:2), and 4-hydroxy-8-sphingenine (t18:1). Enzyme abbreviations: CERase, ceramidase; CK, ceramide kinase; GCS, glucosylceramide synthase; IPCS, inositolphophorylceramide synthase; KSR, 3-ketosphinganine reductase; LCBK, long-chain base kinase; LCBPL; long-chain base-phosphate lyase; LCBPP, long-chain base-phosphate phosphatase; SAT, sphinganine acyltransferase; SH, sphinganine hydroxylase; SPT, serine palmitoyltransferase.

resistance, and ageing (Liu *et al.*, 2005). These signaling sphingolipids in yeast apparently are derived primarily from biosynthetic reactions rather than the breakdown of complex lipids (Jenkins and Hannun, 2001; Obeid *et al.*, 2002; Cowart and Hannun, 2007).

8.4.3 Sphingolipid signaling in plants

While relatively few reports have focused on bioactive sphingolipids in plants, several studies have implicated sphingolipids in signaling and regulation. Free LCBs apparently have diverse effects on plants, although in many of the published studies the effects cannot be unambiguously attributed to

LCBs since the levels of related molecules, including LCBP and Cer, were not monitored. Studies of plant–pathogen interactions have implicated disruption of sphingolipid synthesis in programmed cell death (PCD) as part of the hypersensitive response associated with plant defense. The fungal toxins fumonisin B_1 and AAL-toxin are LCB analogs that inhibit sphinganine acyltransferase activity, resulting in an accumulation in plant tissues of sphinganine and 4-hydroxysphinganine (Abbas *et al.*, 1994; Wright *et al.*, 2003) and their phosphorylated intermediates (Lynch *et al.*, unpublished), and promote plant necrosis (Tanaka *et al.*, 1993) and PCD in tissues and protoplasts (Wang *et al.*, 1996; Asai *et al.*, 2000). The response to these toxins appears complex, involving the participation of the ethylene, jasmonate, and salicylate signaling pathways (Asai *et al.*, 2000), but microarray results demonstrated that AAL toxin does not significantly alter the expression of any of the included sphingolipid genes (Gechev *et al.*, 2004). The tomato *Asc1* gene, involved in ceramide synthesis and conferring AAL-toxin resistance, permits sphingolipid synthesis and prevents long-chain base accumulation in tissues exposed to toxin (Brandwagt *et al.*, 2000, 2002; Spassieva *et al.*, 2002). In sensitive tomato plants, AAL-toxin prevents ceramide (and complex sphingolipid) synthesis, resulting in the accumulation of LCB and death. But if long-chain base synthesis (serine palmitoyltransferase) is simultaneously inhibited, LCB accumulation is prevented and the toxic effects of AAL-toxin are at least partially blocked (Spassieva *et al.*, 2002). As well, overexpression of *Asc1* in sensitive plants confers resistance to infection (Brandwagt *et al.*, 2002). These results suggest that LCB may play important regulatory/signaling roles in plant defense, and elevated levels in plant tissues promote cell/tissue death. It must be reiterated, however, that in the above studies LCBP levels were not determined, thus the pathologies cannot be attributed unequivocally to elevated LCBs per se. Indeed, evidence from mutant lines of *A. thaliana* suggests that the fumonisin-induced accumulation of LCBPs (possibly, specific species of LCBP) is associated with death. That is, mutant lines lacking the sole LCBP lyase exhibit hypersensitivity to fumonisin B_1 and a greater accumulation of LCBP in comparison to treated wild type tissues, while mutant lines lacking any one of the three identified LCB kinases are resistant to the toxin (Lynch *et al.*, unpublished).

Additional evidence implicating LCBs in signaling and regulation was provided by a knockout in *A. thaliana* of the *ACD11* gene that results in increased PCD and defense (Brodersen *et al.*, 2002). It was demonstrated that the gene product encodes a putative sphingosine LCB transfer protein. Further studies are needed to better define the specific function of this protein and its sphingolipid LCB specificity in order to understand its role in programmed cell death and defense pathways. Nevertheless, if actual disruption of LCB transfer is responsible for the mutant phenotype, it suggests that the cellular localization of specific sphingolipids may be significant in promoting cell death and other sphingolipid-influenced processes.

A role for Cer in mediating the effects of AAL-toxin exposure was suggested by the partial rescue of AAL-toxin-treated tomato leaves by exogenous ceramide application (Brandwagt *et al.*, 2000). While this does not seem consistent with the proposed role for LCBs (or LCBPs) described above, it was suggested that the relative levels (or ratio) of LCB and Cer may constitute a switch, triggering PCD (Spassieva *et al.*, 2002). Such a "rheostat model" has been proposed to operate in animal cells for LCBP and Cer (Spiegel and Milstien, 2003; Taha *et al.*, 2006). While this merits testing in plants as well, it should include an assessment of the contribution of LCBPs to the rheostat. As well, testing of the rheostat hypothesis should also take into account the levels of specific species of LCB, LCBP, and Cer, in light of recent studies indicating that accumulation of C24 ceramide species induces cell cycle arrest in MCF-7 breast cancer cells while C16 ceramide accumulation is associated with apoptosis (Marchesini *et al.*, 2004).

As with mammalian systems, Cer has been implicated in mediating cell death in plants (Liang *et al.*, 2003; Townley *et al.*, 2005). *A. thaliana acd5* mutants deficient in Cer kinase activity are more sensitive to added Cer, accumulate endogenous Cer (kinase substrate), are more susceptible to pathogen infection, and undergo apoptotic-like cell death late in development. These results point to a role for Cer in promoting PCD in plants and demonstrate a role for the *ACD5* encoded Cer kinase activity in modulating Cer levels in the plant cell. However, it is unclear whether the sole role of Cer kinase is to convert and sequester Cer, or whether Cer-1-phosphate also functions in signaling/cell regulation as demonstrated in mammalian cells where it activates phospholipase A2 (Pettus *et al.*, 2004). As well, open questions pertaining to the fate of Cer-1-phosphate remain: Is there a specific phosphatase capable of dephosphorylating Cer-1-phosphate as demonstrated in brain (Shinghal *et al.*, 1993), or is Cer-1-phosphate used in some as yet uncharacterized pathway? Consistent with the above results, Townley and colleagues demonstrated that exogenous short-chain Cer induces PCD in *A. thaliana* suspension cultures (Townley *et al.*, 2005). Treatment with Cer was accompanied by the generation of a calcium transient and an increase in ROS. Inhibition of the calcium transient was found to prevent cell death, whereas inhibition of ROS had no effect on cell survival. These observations suggest that calcium signaling, but not generation of ROS, is involved in ceramide-induced PCD.

A role for S1P in guard cell signaling and stomatal closure has been investigated (Ng *et al.*, 2001; Ng and Hetherington, 2001; Coursol *et al.*, 2003). S1P was identified in lipid extracts from *C. communis* and increases in S1P content accompanied drought stress (Ng *et al.*, 2001). Stomatal closure occurred following incubation of leaf epidermal strips with exogenous S1P whereas incubation with sphinganine-1-phosphate did not have such an effect, suggesting the significance of the Δ4 double bond in signaling, although phytosphingosine-1-phosphate can also function in signaling stomatal closure (Coursol *et al.*, 2005). It was found that S1P influences calcium mobilization in guard cells and incubation with an inhibitor of sphingosine LCB kinase attenuates the stomatal response to added abscisic acid (ABA) (Ng *et al.*, 2001; Ng and Hetherington, 2001).

This role for S1P in the ABA signaling pathway leading to reduction of guard cell turgor was further investigated in *A. thaliana* (Coursol *et al.*, 2003), where it was found that LCB kinase activity is transiently stimulated by ABA, and inhibition of kinase activity (using inhibitors of the mammalian kinase) diminishes the stomatal response to ABA treatment, as found for *C. communis*. Exogenous S1P is capable of influencing guard cell behavior (both inhibition of stomatal opening and promotion of closure) via inhibition of K^+ influx and stimulation of anion efflux in wild type plant protoplasts but not in protoplasts from knockout plants lacking the heterotrimeric G-protein α-subunit, providing evidence that the G-protein is downstream of S1P in the ABA signaling pathway (Coursol *et al.*, 2003) (see Chapter 2 on heterotrimeric G proteins).

Sphingosine and S1P are virtually absent in plants, so while exogenous S1P may act as a signal molecule, it is doubtful that it is the endogenous signal. However, phytosphingosine-1-phosphate can influence guard cell behavior similar to S1P (Coursol *et al.*, 2005). Since phytosphingosine is relatively abundant in plant tissues (it and sphinganine are the prevalent free LCBs) and it serves as a substrate for the three plant LCB kinases (Coursol *et al.*, 2005; Imai and Nishiura, 2005; Tsegaye *et al.*, unpublished), it is likely that phytosphingosine-1-phosphate, rather than S1P, is the LCBP species involved in a guard cell signaling pathway.

Cumulatively, the above studies support the contention that LCBPs can influence stomatal behavior, but the nature of the interaction between LCBP and the G-protein as well as the identification of the components of the pathway leading from ABA to changes in stomatal aperture need to be investigated further, especially given the recent identification of the ABA receptor as a G-protein-linked receptor (Liu *et al.*, 2007). For example, the location of the putative LCBP receptor merits investigation: that exogenous LCBPs can elicit stomatal closure suggests that LCBP must be either transported into the cell to reach an intracellular target, or act through a receptor/binding protein on the cell surface. If a surface receptor for LCBP exists, then internally generated LCBP (in response to ABA) would need to be exported. In mammals, S1P has been reported to act as both a ligand for certain G-protein-coupled receptors at the plasma membrane and as an intracellular second messenger (Payne *et al.*, 2002; Spiegel and Milstien, 2003). The identification of a plant LCBP receptor, including its location and specificity (and affinity) for LCBP, would improve our understanding of this signaling pathway and its contribution to stomatal behavior in plants.

Although S1P is thought to stimulate proliferation in mammalian cells (Spiegel and Milstien, 2003), and a role for LCBP in "re-entering" the cell cycle following heat stress-induced arrest in yeast has been suggested (Jenkins and Hannun, 2001), such a role for LCBP in plants has not been demonstrated. While the responses to fumonisin described above suggest that LCBP hyperaccumulation is lethal, a basic understanding of the effects of physiological concentrations of LCBP is lacking. The report of G protein involvement in cell proliferation in *A. thaliana* (Ullah *et al.*, 2001) and the evidence from guard cell studies (above) suggesting that the G-protein α-subunit is downstream of LCBP in signaling stomatal closure leads to the suggestion that LCBPs may influence plant cell proliferation (Chalfant and Spiegel, 2005).

8.4.4 Conclusion and a provocation

Looking ahead, the initial studies reviewed here suggest that sphingolipid signaling in plants merits further investigation, and that there are many questions yet to be answered before a coherent understanding of plant sphingolipid function can be realized. Moving forward, there are insights from studies of mammalian and yeast sphingolipid signaling that will be useful in understanding plant signaling, and yet there should not be any expectation that specific aspects of sphingolipid signaling in plants will conform to models developed in other organisms. Many seemingly contradictory results exist in the literature regarding the specific effects attributed to certain sphingolipids in yeast, mammalian cells, and plants. It is not always clear whether these differences are the result of biological plasticity as a consequence of developmental stage or environmental factors, or arise from differences in experimental methods (e.g., methods of supplying exogenous sphingolipids or inhibitors). In either case, they suggest a high level of complexity in the signaling pathways. For example, there is increasing evidence that the actual ratio of LCBP to Cer (the rheostat) is not sufficient to dictate PCD. Assuming it exists at all, could the rheostat "sense" the ratio of only certain molecular species of LCBP or Cer, or could it sense several different types of sphingolipids (e.g., LCB, LCBP, Cer, and Cer-1-phosphate)? In the same vein, could the flux through certain parts of the sphingolipid metabolic pathway, i.e., the rate of interconversion of the sphingolipids and not the static levels in the cell, somehow be sensed by the rheostat? Is the intracellular location of specific sphingolipids crucial to signaling in the plant cell? Evidence indicating that Cer generated in the mitochondria promotes apoptosis in MCF-7 cells but Cer generated in other cellular compartments does not (Birbes *et al.*, 2001) suggests that the location could be important. At another level, how might different cell/tissue types (e.g., guard cell, mesophyll cell, vascular tissue, epidermis) be influenced differentially by sphingolipid signals, and does this impact how we interpret results using entire leaves or roots? Finally, considering future studies of sphingolipid signaling networks in plants, it is evident that the identification of the targets of specific sphingolipid signals will be a critical area of research. In animal cells, several target proteins of Cer, including the protease cathepsin D and protein phosphatase 2A, and of sphingosine, including sphingosine-dependent kinase 1 and PKC, have been implicated in promoting PCD (reviewed in Taha *et al.*, 2006). At present, no protein target of any sphingolipid signal has been identified in plants. In this post-genomic era, the enhanced ability to identify candidates that may function as sphingolipid targets should accelerate progress in this area.

8.5 Summary

Membrane lipids and lipid metabolizing enzymes provide discrete sensors that report a constantly changing environment while maintaining membrane integrity. Innovative approaches being developed both in the lab and in silico will enable future scientists to visualize the dynamic lipid world. It will be exciting to see this knowledge used to improve plant survival and crop productivity in marginal environments.

Acknowledgments

Grant support from the National Science Foundation (MCB-0315869 to WFB; IBN-0454866 to XW; MCB-0455318 to XW; MCB-0312864 to DVL) and the US Department of Agriculture (2005-35308-15253 to XW) are gratefully acknowledged. DVL gratefully acknowledges the assistance and contributions of Janis Bravo and students at Williams College, many of whom were supported by the Howard Hughes Medical Institute. WFB would like to acknowledge Imara Y. Perera, Yang Ju Im of N.C. State University and Ingo Heilmann of Göttingen University & Center for Molecular Biosciences for reading the manuscript, Ingo Heilmann for sharing his unpublished data and Takashi Aoyama of the Institute for Chemical Research, Kyoto University, for sharing his unpublished data.

References

Abbas, H.K., Tanaka, T., Duke, S.O., Porter, J.K., Wray, E.M., Hodges, L., Sessions, A.E., Wang, E., Merrill, A.H., Jr. and Riley, R.T. (1994) Fumonisin and AAL-toxin-induced disruption of sphingolipid metabolism with accumulation of free sphingoid bases. *Plant Physiol*, **106**, 1085–1093.

Ali, R., Ma, W., Lemtiri-Chlieh, F., Tsaltas, D., Leng, Q., Von Bodman, S. and Berkowitz, G.A. (2007) Death don't have no mercy and neither does calcium: *Arabidopsis* cyclic nucleotide gated channel 2 and innate immunity. *Plant Cell*, **19**, 1081–1095.

Anthony, R., Henriques, R., Helfer, A., Meszaros, T., Rios, G., Testerink, C., Munnik, T., Deak, M., Koncz, C. and Bogre, L. (2004) A protein kinase target of a PDK1 signalling pathway is involved in root hair growth in *Arabidopsis*. *EMBO J*, **23**, 572–581.

Asai, T., Stone, J.M., Heard, J.E., Kovtun, Y., Yorgrey, P., Sheen, J. and Ausubel, F.M. (2000) Fumonisin B1-induced cell death in *Arabidopsis* protoplasts requires jasmonate-, ethylene-, and salicylate-dependent signaling pathways. *Plant Cell*, **12**, 1823–1836.

Augert, G., Blackmore, P.F. and Exton, J.H. (1989) Changes in the concentration and fatty acid composition of phosphoinositides induced by hormones in hepatocytes. *J Biol Chem*, **264**, 2574–2580.

Balla, T. (2007) Imaging and manipulating phosphoinositides in living cells. *J Physiol*, **582**, 927–937.

Bargmann, B., Laxalt, A., Riet, B., Schouten, E., Van Leeuwen, W., Dekker, H., De Koster, C., Haring, M. and Munnik, T. (2006) LePLDβ1 activation and relocalization in suspension-cultured tomato cells treated with xylanase. *Plant J*, **45**, 358–368.

Bargmann, B. and Munnik, T. (2006) The role of phospholipase D in plant stress responses. *Curr Opin Plant Biol*, **9**, 515–522.

Becker, J.D., Boavida, L.C., Carneiro, J., Haury, M. and Feijo, J.A. (2003) Transcriptional profiling of *Arabidopsis* tissues reveals the unique characteristics of the pollen transcriptome. *Plant Physiol*, **133**, 713–725.

Berdy, S.E., Kudla, J., Gruissem, W. and Gillaspy, G.E. (2001) Molecular characterization of At5PTase1, an inositol phosphatase capable of terminating inositol trisphosphate signaling. *Plant Physiol*, **126**, 801–810.

Berridge, M.J. and Irvine, R.F. (1989) Inositol phosphates and cell signaling. *Nature*, **341**, 197–205.

Birbes, H., El Bawab, S., Hannun, Y. and Obeid, L. (2001) Selective hydrolysis of a mitochondrial pool of sphingomyelin induces apoptosis. *FASEB J*, **14**, 2669–2679.

Boss, W.F. (1989) *Phosphoinositide Metabolism: Its Relation to Signal Transduction in Plants.* Alan R. Liss, New York.

Boss, W.F., Davis, A.J., Im, Y.J., Galvao, R.M. and Perera, I.Y. (2006) Phosphoinositide metabolism: towards an understanding of subcellular signaling. *Subcell Biochem*, **39**, 181–205.

Boss, W.F. and Massel, M.O. (1985) Polyphosphoinositides are present in plant tissue culture cells. *Biochem Biophys Res Commun*, **132**, 1018–1023.

Boyer, J. and Mclaughlin, J. (2007) Functional reversion to identify controlling genes in multigenic responses: analysis of floral abortion. *J Exp Bot*, **58**, 267–277.

Brandwagt, B., Mesbah, L., Takken, F., Laurent, P., Kneppers, T., Hille, J. and Nijkamp, H. (2000) A longevity assurance gene homolog of tomato mediates resistance to *Alternaria alternata* f. sp.*lycopersici* toxins and fumonisin B1. *Proc Natl Acad Sci USA*, **97**, 4961–4966.

Brandwagt, B.F., Kneeppers, T.J., Nijkamp, H.J. Hillie, J. (2002) Overexpression of the tomato Asc-1 gene mediates high insensitivity to AAL toxins and fumonisin B1 in tomato hairy roots and confers resistance to *Alternaria alternata* f. sp. *lycopersici* in *Nicotiana umbratica* plants. *Mol Plant Microbe Interact*, **15**, 35–42.

Brodersen, P., Petersen, M., Pike, H., Olszak, B., Odum, N., Jorgensen, L., Brown, R. and Mundy, J. (2002) Knockout of *Arabidopsis* accelerated-cell-death11 encoding a sphingosine transfer protein causes activation of programmed cell death and defense. *Genes Dev*, **16**, 490–502.

Bunney, T.D., Watkins, P.A., Beven, A.F., Shaw, P.J., Hernandez, L.E., Lomonossoff, G.P., Shanks, M., Peart, J. and Drøbak, B.K. (2000) Association of phosphatidylinositol 3-kinase with nuclear transcription sites in higher plants. *Plant Cell*, **12**, 1679–1688.

Burnette, R.N., Gunesekera, B.M. and Gillaspy, G.E. (2003) An *Arabidopsis* inositol 5-phosphatase gain-of-function alters abscisic acid signaling. *Plant Physiol*, **132**, 1011–1019.

Buseman, C., Tamura, P., Sparks, A., Baughman, E., Maatta, S., Zhao, J., Roth, M., Esch, S., Shah, J., Williams, T. and Welti, R. (2006) Wounding stimulates the accumulation of glycerolipids containing oxophytodienoic acid and dinor-oxophytodienoic acid in *Arabidopsis* leaves *Plant Physiol*, **142**, 28–39.

Carland, F.M. and Nelson, T. (2004) Cotyledon vascular pattern2-mediated inositol (1,4,5) triphosphate signal transduction is essential for closed venation patterns of *Arabidopsis* foliar organs. *Plant Cell*, **16**, 1263–1275.

Chalfant, C. and Spiegel, S. (2005) Sphingosine 1-phosphate and ceramide 1-phosphate: expanding roles in cell signaling. *J Cell Sci*, **118**, 4605–4612.
Chapman, K.D. (1998) Phospholipase activity during plant growth and development and in response to environmental stress. *Trends Plant Sci*, **3**, 419–426.
Cheng, M.K. and Shearn, A. (2004) The direct interaction between ASH2, a Drosophila trithorax group protein, and SKTL, a nuclear phosphatidylinositol 4-phosphate 5-kinase, implies a role for phosphatidylinositol 4,5-bisphosphate in maintaining transcriptionally active chromatin. *Genetics*, **167**, 1213–1223.
Cho, M.H., Tan, Z., Erneux, C., Shears, S.B. and Boss, W.F. (1995) The effects of mastoparan on the carrot cell plasma membrane polyphosphoinositide phospholipase C. *Plant Physiol*, **107**, 845–856.
Coursol, S., Fan, L., Lestunff, H., Spiegel, S., Gilroy, S. and Assman, S. (2003) Sphingolipid signalling in *Arabidopsis* guard cells involves heterotrimeric G proteins. *Nature*, **423**, 651–654.
Coursol, S., Lestunff, H., Lynch, D.V., Gilroy, S., Assman, S.M. and Spiegel, S. (2005) *Arabidopsis* sphingosine kinase and the effects of phytosphingosine-1-phosphate on stomatal aperture. *Plant Physiol*, **137**, 724–737.
Cowart, L.A. and Hannun, Y.A. (2007) Selective substrate supply in the regulation of yeast de novo sphingolipid synthesis. *J Biol Chem*, **282**, 12330–12340.
Cowart, L.A. and Obeid, L.M. (2007) Yeast sphingolipids: recent developments in understanding biosynthesis, regulation and function. *Biochim Biophys Acta*, **1771**, 421–431.
Cruz-Ramirez, A., Oropeza-Aburto, A., Razo-Hernandez, F., Ramirez-Chavez, E. and Herrera-Estrella, L. (2006) Phospholipase DZ2 plays an important role in extraplastidic galactolipid biosynthesis and phosphate recycling in *Arabidopsis* roots. *Proc Natl Acad Sci USA*, **103**, 6765–6770
Davis, A.J., Im, Y.J., Dubin, J.S., Tomer, K.B. and Boss, W.F. (2007) *Arabidopsis* phosphatidylinositol phosphate kinase 1 binds F-actin and recruits phosphatidylinositol 4-kinase β1 to the actin cytoskeleton. *J Biol Chem*, **282**, 14121–14131.
Davis, A.J., Perera, I.Y. and Boss, W.F. (2004) Cyclodextrins enhance recombinant phosphatidylinositol phosphate kinase activity. *J Lipid Res*, **45**, 1783–1789.
Devaiah, S., Pan, X., Roth, M., Welti, R. and Wang, X. (2007) Enhancing seed quality and viability by suppressing phospholipase D in *Arabidopsis*. *Plant J*, **50**, 950–957.
Dewald, D.B., Torabinejad, J., Jones, C.A., Shope, J.C., Cangelosi, A.R., Thompson, J.E., Prestwich, G.D. and Hama, H. (2001) Rapid accumulation of phosphatidylinositol 4,5-bisphosphate and inositol 1,4,5-trisphosphate correlates with calcium mobilization in salt-stressed *Arabidopsis*. *Plant Physiol*, **126**, 759–769.
Dickson, R.C. and Lester, R.L. (2002) Sphingolipid functions in Saccharomyces cerevisiae. *Biochim Biophys Acta*, **1583**, 13–25.
Doughman, R.L., Firestone, A.J. and Anderson, R.A. (2003) Phosphatidylinositol phosphate kinases put $PI(4,5)P_2$ in its place. *J Membr Biol*, **194**, 77–89.
Dowd, P.E., Coursol, S., Skirpan, A.L., Kao, T.H. and Gilroy, S. (2006) Petunia phospholipase C1 is involved in pollen tube growth. *Plant Cell*, **18**, 1438–1453.
Drøbak, B.K., Dewey, R.E. and Boss, W.F. (1999) Phosphoinositide kinases and the synthesis of polyphosphoinositides in higher plant cells. In: *International Review of Cytology* (ed K.W. Jeon), pp. 95–130. Academic Press, New York.
Drøbak, B.K., Watkins, P.A.C., Chattaway, J.A., Roberts, K. and Dawson, A.P. (1991) Metabolism of inositol (1,4,5)trisphosphate by a soluble enzyme fraction from pea (*Pisum sativum*) roots. *Plant Physiol*, **95**, 412–419.

Dunn, T.M., Lynch, D.V., Michaelson, L.V. and Napier, J.A. (2004) A post-genomic approach to understanding sphingolipid metabolism in *Arabidopsis thaliana*. *Ann Bot*, **93**, 483–497.

Ercetin, M. and Gillaspy, G. (2004) Characterization of 5-phosphatases from *Arabidopsis*. *FASEB J*, **18**, C180.

Exton, J.H. (1997) New developments in phospholipase D. *J Biol Chem*, **272**, 15579–15582.

Franklin-Tong, V.E., Drøbak, B.K., Allan, A.C., Watkins, P.A.C. and Trewavas, A.J. (1996) Growth of pollen tubes of *Papaver rhoeas* is regulated by a slow-moving calcium wave propagated by inositol 1,4,5-trisphosphate. *Plant Cell*, **8**, 1305–1321.

Galanopoulou, D., Moxley, D., Boss, C.B. and Boss, W.F. (1995) Mastoparan induces inositol phospholipid changes and plasmolysis in carrot cells. *Biochem Soc Trans*, **23**, 573S.

Galvão, R.M., Kota, U., Soderblom, E.K., Goshe, M.B. and Boss, W.F. (2008) Characterization of a new family of protein kinases containing phosphoinositide 3/4-kinase and ubiquitin-like domains. *Biochem J*, **409**, 117–127.

Gechev, T.S., Gadjev, I.Z. and Hille, J. (2004) An extensive microarray analysis of AAL-toxin-induced cell death in *Arabidopsis thaliana* brings new insights into the complexity of programmed cell death in plants. *Cell Mol Life Sci*, **61**, 1185–1197.

Ghosh, S., Moore, S., Bell, R. and Dush, M. (2003) Functional analysis of a phosphatidic acid binding domain in human Raf-1 kinase: mutations in the phosphatidate binding domain lead to tail and trunk abnormalities in developing zebrafish embryos. *J Biol Chem*, **278**, 45690–45696.

Gilroy, S., Read, N.D. and Trewavas, A.J. (1990) Elevation of cytoplasmic calcium by caged calcium or caged inositol triphosphate initiates stomatal closure. *Nature*, **346**, 769–771.

Gonzales, M.L. and Anderson, R.A. (2006) Nuclear phosphoinositide kinases and inositol phospholipids. *J Cell Biochem*, **97**, 252–260.

Gorter, E. and Grendel, F. (1925) On bimolecular layers of lipoids on the chromocytes of the blood. *J Exp Med*, **41**, 439–443.

Grill, E. and Christmann, A. (2007) A plant receptor with a big family. *Science*, **315**, 1676–1677.

Gross, W. and Boss, W.F. (1993) Inositol phospholipids and signal transduction. In: *Control of Plant Gene Expression*. CRC Press, Boca Raton.

Gubbels, M.J., Vaishnava, S., Boot, N., Dubremetz, J.F. and Striepen, B. (2006) A morn-repeat protein is a dynamic component of the Toxoplasma gondii cell division apparatus. *J Cell Sci*, **119**, 2236–2245.

Guo, J., Wenk, M.R., Pellegrini, L., Onofri, F., Benfenati, F. and De Camilli, P. (2003) Phosphatidylinositol 4-kinase type II α is responsible for the phosphatidylinositol 4-kinase activity associated with synaptic vesicles. *Proc Natl Acad Sci USA*, **100**, 3995–4000.

Hannun, Y.A., Loomis, C.R., Merrill, A.H. and Bell, R.M. (1986) Sphingosine inhibition of protein kinase C activity and of phorbol dibutyrate binding in vitro and in human platelets. *J Biol Chem*, **261**, 12604–12609.

Heilmann, I., Perera, I.Y., Gross, W. and Boss, W.F. (1999) Changes in phosphoinositide metabolism with days in culture affect signal transduction pathways in Galdieria sulphuraria. *Plant Physiol*, **229**, 1331–1339.

Heilmann, I., Perera, I.Y., Gross, W. and Boss, W.F. (2001) Plasma membrane phosphatidylinositol 4,5-bisphosphate decreases with time in culture. *Plant Physiol*, **126**, 1507–1518.

Helling, D., Possart, A., Cottier, S., Klahre, U. and Kost, B. (2006) Pollen tube tip growth depends on plasma membrane polarization mediated by Tobacco PLC3 activity and endocytic membrane recycling. *Plant Cell*, **18**, 3519–3534.

Hendrix, W., Assefa, H. and Boss, W.F. (1989) The polyphosphoinositides, phosphatidylinositol monophosphate and phosphatidylinositol bisphosphate are present in nuclei isolated from carrot protoplasts. *Protoplasma*, **151**, 62–72.

Hinchliffe, K.A., Ciruela, A., Letcher, A.J., Divecha, N. and Irvine, R.F. (1999) Regulation of type II α phosphatidylinositol phosphate kinase localisation by the protein kinase CK2. *Curr Biol*, **9**, 983–986.

Hokin, M.R. and Hokin, L.E. (1953) Enzyme secretion and the incorporation of ^{32}P into phopholipids of pancreas slices. *J Biol Chem*, **203**, 967–977.

Holk, A., Rietz, S., Zahn, M., Quader, H. and Scherer, G. (2002) Molecular identification of cytosolic, patatin-related phospholipases A from *Arabidopsis* with potential functions in plant signal transduction. *Plant Physiol*, **130**, 90–101.

Hong, Z. and Verma, D.P. (1994) A phosphatidylinositol 3-kinase is induced during soybean nodule organogenesis and is associated with membrane proliferation. *Proc Natl Acad Sci USA*, **91**, 9617–9621.

Huang, S., Gao, L., Blanchoin, L. and Staiger, C. (2006) Heterodimeric capping protein from *Arabidopsis* is regulated by phosphatidic acid. *Mol Biol Cell*, **17**, 1946–1958.

Hunt, L., Otterhag, L., Lee, J.C., Lasheen, T., Hunt, J., Seki, M., Shinozaki, K., Sommarin, M., Gilmour, D.J., Pical, C. and Gray, J.E. (2004) Gene-specific expression and calcium activation of *Arabidopsis thaliana* phospholipase C isoforms. *New Phytol*, **162**, 643–654.

Im, Y.J., Davis, A.J., Perera, I.Y., Johannes, E., Allen, N.S. and Boss, W.F. (2007a) The N-terminal membrane occupation and recognition nexus domain of *Arabidopsis* phosphatidylinositol phosphate kinase 1 regulates enzyme activity. *J Biol Chem*, **282**, 5443–5452.

Im, Y.J., Perera, I.Y., Brglez, I., Davis, A.J., Stevenson-Paulik, J., Phillippy, B.Q., Johannes, E., Allen, N.S. and Boss, W.F. (2007b) Increasing plasma membrane phosphatidylinositol(4,5)bisphosphate biosynthesis increases phosphoinositide metabolism in *Nicotiana tabaccum*. *Plant Cell*, **19**, 1603–1616.

Imai, A. and Gershengorn, M.C. (1986) Phosphatidylinositol 4,5-bisphosphate turnover is transient while phosphatidylinositol turnover is persistent in thyrotropin-releasing hormone-stimulated rat pituitary cells. *Proc Natl Acad Sci USA*, **83**, 8540–8544.

Imai, H. and Nishiura, H. (2005) Phosphorylation of sphingoid long-chain bases in *Arabidopsis*: functional characterization and expression of the first sphingoid long-chain base kinase gene in plants. *Plant Cell Physiol*, **46**, 375–380.

Irvine, R.F. (2003) Nuclear lipid signalling. *Nat Rev Mol Cell Biol*, **4**, 349–360.

Jenkins, G.M. and Hannun, Y.A. (2001) Role for de novo sphingoid base biosynthesis in the heat-induced transient cell cycle arrest of *Saccharomyces cerevisiae*. *J Biol Chem*, **276**, 8574–8581.

Jones, D.H., Morris, J.B., Morgan, C.P., Kondo, H., Irvine, R.F. and Cockcroft, S. (2000) Type I phosphatidylinositol 4-phosphate 5-kinase directly interacts with ADP-ribosylation factor 1 and is responsible for phosphatidylinositol 4,5-bisphosphate synthesis in the Golgi compartment. *J Biol Chem*, **275**, 13962–13966.

Jones, J. and Hannun, Y. (2002) Tight binding inhibition of protein phosphatase-1 by phosphatidic acid: specificity of inhibition by the phospholipid. *J Biol Chem*, **277**, 15530–15538.

Jung, J.Y., Kim, Y.W., Kwak, J.M., Hwang, J.U., Young, J., Schroeder, J.I., Hwang, I. and Lee, Y. (2002) Phosphatidylinositol 3- and 4-phosphate are required for normal stomatal movements. *Plant Cell*, **14**, 2399–2412.

Kabachevskaya, A., Liakhnovich, G., Kisel, M. and Volotovski, I. (2007) Red/far-red light modulates phospholipase D activity in oat seedlings: relation of enzyme photosensitivity to photosynthesis. *J Plant Physiol*, **164**, 108–110.

Konig, S., Mosblech, A. and Heilmann, I. (2007) Stress-inducible and constitutive phosphoinositide pools have distinctive fatty acid patterns in *Arabidopsis thaliana*. *FASEB J*, **21**, 1958–1967.

Kooijman, E., Tieleman, D., Testerink, C., Munnik, T., Rijkers, D., Burger, K. and De Kruijff, B. (2007) An electrostatic/hydrogen bond switch as the basis for the specific interaction of phosphatidic acid with proteins. *J Biol Chem*, **282**, 11356–11364.

Kost, B., Lemichez, E., Spielhofer, P., Hong, Y., Tolias, K., Carpenter, C. and Chua, N.H. (1999) Rac homologues and compartmentalized phosphatidylinositol 4, 5-bisphosphate act in a common pathway to regulate polar pollen tube growth. *J Cell Biol*, **145**, 317–330.

Kusano, H., Testerink, C., Vermeer, J.E.M., Tsuge, T., Shimada, H., Oka, A., Munnik, T. and Aoyama, T. (2008) The arabidopsis phosphatidylinositol phosphate 5-kinase PIP5K3 is a key regulator of root hair tip growth. *Plant Cell*. **20**, 124–141.

Kusner, D.J., Barton, J.A., Qin, C., Wang, X. and Iyer, S.S. (2003) Evolutionary conservation of physical and functional interactions between phospholipase D and actin. *Arch Biochem Biophys*, **412**, 231–241.

La Camera, S., Geoffroy, P., Samaha, H., Ndiaye, A., Rahim, G., Legrand, M. and Heitz, T. (2005) A pathogen-inducible patatin-like lipid acyl hydrolase facilitates fungal and bacterial host colonization in *Arabidopsis*. *Plant J*, **44**, 810–825.

Lemtiri-Chlieh, F., Macrobbie, E.A. and Brearley, C.A. (2000) Inositol hexakisphosphate is a physiological signal regulating the K^+-inward rectifying conductance in guard cells. *Proc Natl Acad Sci USA*, **97**, 8687–8692.

Lemtiri-Chlieh, F., Macrobbie, E.A., Webb, A.A., Manison, N.F., Brownlee, C., Skepper, J.N., Chen, J., Prestwich, G.D. and Brearley, C.A. (2003) Inositol hexakisphosphate mobilizes an endomembrane store of calcium in guard cells. *Proc Natl Acad Sci USA*, **100**, 10091–10095.

Levitt, J. (ed.) (1980) *Chilling, Freezing, and High Temperature Stresses*. Academic Press, New York.

Li, G. and Xue, H.-W. (2007) *Arabidopsis* PLDζ2 regulates vesicle trafficking and is required for auxin response. *Plant Cell*, **19**, 281–295.

Li, M., Qin, C., Welti, R. and Wang, X. (2006a) Double knockouts of phospholipase Dζ1 and ζ2 in *Arabidopsis* affect root elongation during phosphate-limited growth, but do not affect root hair patterning. *Plant Physiol*, **140**, 761–770.

Li, M., Welti, R. and Wang, X. (2006b) Quantitative profiling of *Arabidopsis* polar glycerolipids in response to phosphorus starvation. Roles of phospholipases Dζ1 and Dζ2 in phosphatidylcholine hydrolysis and digalactosyldiacylglycerol accumulation in phosphorus-starved plants. *Plant Physiol*, **142**, 750–761.

Li, W., Li, M., Zhang, W., Welti, R. and Wang, X. (2004) The plasma membrane-bound phospholipase Ddelta enhances freezing tolerance in *Arabidopsis*. *Nature Biotech*, **22**, 427–433.

Liang, H., Yao, N., Song, J.T., Luo, S., Lu, H. and Greenberg, J.T. (2003) Ceramide phosphorylation modulates programmed cell death in plants. *Genes Dev*, **17**, 2636–2641.

Liu, K., Zhang, X., Lester, R.L. and Dickson, R.C. (2005) The sphingoid long-chain base phytosphingosine activates AGC-type protein kinases in *Saccharomyces cerevisiae* including Ypk1, Ypk2, and Sch9. *J Biol Chem*, **280**, 22679–22687.

Liu, X., Yue, Y., Li, B., Nie, Y., Li, W., Wu, W. and Ma, L. (2007) A G protein-coupled receptor is a plasma membrane receptor for the plant hormone abscisic acid. *Science*, **315**, 1712–1716.

Loewen, C., Gaspar, M., Jesch, S., Delon, C., Ktistakis, N., Henry, S. and Levine, T. (2004) Phospholipid metabolism regulated by a transcription factor sensing phosphatidic acid. *Science*, **304**, 1644–1647.

Loewus, F.A. and Murthy, P.P.N. (2000) Myo-inositol metabolism in plants. *Plant Sci*, **150**, 1–19.

Logan, D.C. and Knight, M.R. (2003) Mitochondrial and cytosolic calcium dynamics are differentially regulated in plants. *Plant Physiol*, **133**, 21–24.

Lou, Y., Gou, J.-Y. and Xue, H.-W. (2007) PIP5K9, an *Arabidopsis* phosphatidylinositol monophosphate kinase, interacts with a cytosolic invertase to negatively regulate sugar-mediated root growth. *Plant Cell*, 19, 163–181.

Lynch, D.V. and Dunn, T.M. (2004) An introduction to plant sphingolipids and a review of recent advances in understanding their metabolism and function. *New Phytologist*, **161**, 677–702.

Lynch, D.V. and Steponkus, P.L. (1987) Plasma membrane lipid alterations associated with cold acclimation of winter rye seedlings (*Secale cereale* L. cv Puma). *Plant Physiol*, **83**, 761–767.

Ma, H., Lou, Y., Lin, W.H. and Xue, H.W. (2006) Morn motifs in plant PIPKs are involved in the regulation of subcellular localization and phospholipid binding. *Cell Res*, **16**, 466–478.

Mane, S., Vasquez-Robinet, C., Sioson, A., Heath, L. and Grene, R. (2007) Early PLDα-mediated events in response to progressive drought stress in *Arabidopsis*: a transcriptome analysis. *J Exp Bot*, **58**, 241–252.

Marchesini, N., Osta, W., Bielawski, J., Luberto, C., Obeid, L.M. and Hannun, Y.A. (2004) Role for mammalian neutral sphingomyelinase 2 in confluence-induced growth arrest of MCF7 cells. *J Biol Chem*, **279**, 25101–25111.

Markham, J.E. and Jaworski, J.G. (2007) Rapid measurement of sphingolipids from *Arabidopsis thaliana* by reversed-phase high-performance liquid chromatography coupled to electrospray ionization tandem mass spectrometry. *Rapid Commun Mass Spectrom*, **21**, 1304–1314.

Markham, J.E., Li, J., Cahoon, E.B. and Jaworski, J.G. (2006) Separation and identification of major plant sphingolipid classes from leaves. *J Biol Chem*, **281**, 22684–22694.

Matsuoka, K., Bassham, D.C., Raikhel, N.V. and Nakamura, K. (1995) Different sensitivity to wortmannin of two vacuolar sorting signals indicates the presence of distinct sorting machineries in tobacco cells. *J Cell Biol*, **130**, 1307–1318.

Mclaughlin, S. and Murray, D. (2005) Plasma membrane phosphoinositide organization by protein electrostatics. *Nature*, **438**, 605–611.

Meijer, H.J.G. and Munnik, T. (2003) Phospholipid-based signaling in plants. *Ann Rev Plant Biol*, **54**, 265–306.

Melin, P.M., Pical, C., Jergil, B. and Sommarin, M. (1992) Polyphosphoinositide phospholipase C in wheat root plasma membranes. Partial purification and characterization. *Biochim Biophys Acta*, **1123**, 163–169.

Merrill, A.H., Jr. (2002) De novo sphingolipid biosynthesis: a necessary, but dangerous, pathway. *J Biol Chem*, **277**, 25843–25846.

Minogue, S., Anderson, J.S., Waugh, M.G., Dos Santos, M., Corless, S., Cramer, R. and Hsuan, J.J. (2001) Cloning of a human type II phosphatidylinositol 4-kinase reveals a novel lipid kinase family. *J Biol Chem*, **276**, 16635–16640.

Mishra, G., Zhang, W., Deng, F., Zhao, J. and Wang, X. (2006) A bifurcating pathway directs abscisic acid effects on stomatal closure and opening in *Arabidopsis*. *Science*, **312**, 264–266.

Monks, D.E., Aghoram, K., Courtney, P.D., Dewald, D.B. and Dewey, R.E. (2001) Hyperosmotic stress induces the rapid phosphorylation of a soybean phosphatidylinositol transfer protein homolog through activation of the protein kinases SPK1 and SPK2. *Plant Cell*, **13**, 1205–1219.

Mueller-Roeber, B. and Pical, C. (2002) Inositol phospholipid metabolism in *Arabidopsis*. Characterized and putative isoforms of inositol phospholipid kinase and phosphoinositide-specific phospholipase C. *Plant Physiol*, **130**, 22–46.

Munnik, T., De Vrije, T., Irvine, R.F. and Musgrave, A. (1996) Identification of diacylglycerol pyrophosphate as a novel metabolic product of phosphatidic acid during G-protein activation in plants. *J Biol Chem*, **271**, 15708–15715.

Nakamura, Y., Awai, K., Masuda, T., Yoshioka, Y., Takamiya, K. and Ohta, H. (2005) A novel phosphatidylcholine-hydrolyzing phospholipase C induced by phosphate starvation in *Arabidopsis*. *J Biol Chem*, **280**, 7469–7476.

Napier, J.A., Michaelson, L.V. and Dunn, T.M. (2002) A new class of lipid desaturase central to sphingolipid biosynthesis and signalling. *Trends Plant Sci*, **7**, 475–478.

Ng, C.K., Carr, K., Mcainsh, M.R., Powell, B. and Hetherington, A.M. (2001) Drought-induced guard cell signal transduction involves sphingosine-1-phosphate. *Nature*, **410**, 596–599.

Ng, C.K. and Hetherington, A.M. (2001) Sphingolipid-mediated signaling in plants. *Ann Bot*, **88**, 957–965.

Obeid, L.M., Okamoto, Y. and Mao, C. (2002) Yeast sphingolipids: metabolism and biology. *Biochim Biophys Acta*, **1585**, 163–171.

Odom, A.R., Stahlberg, A., Wente, S.R. and York, J.D. (2000) A role for nuclear inositol 1,4,5-trisphosphate kinase in transcriptional control. *Science*, **287**, 2026–2029.

Ohashi, Y., Oka, A., Rodrigues-Pousada, R., Possenti, M., Ruberti, I., Morelli, G. and Aoyama, T. (2003) Modulation of phospholipid signaling by GLABRA2 in root-hair pattern formation. *Science*, **300**, 1427–1430.

Pappan, K., Zheng, L., Krishnamoorthi, R. and Wang, X. (2004) Evidence for and characterization of Ca^{2+} binding to the catalytic region of *Arabidopsis thaliana* phospholipase Dβ. *J Biol Chem*, **279**, 47833–47839.

Paul, A.L., Schuerger, A.C., Popp, M.P., Richards, J.T., Manak, M.S. and Ferl, R.J. (2004) Hypobaric biology: *Arabidopsis* gene expression at low atmospheric pressure. *Plant Physiol*, **134**, 215–223.

Pauly, N., Knight, M.R., Thuleau, P., Van Der Luit, A.H., Moreau, M., Trewavas, A.J., Ranjeva, R. and Mazars, C. (2000) Control of free calcium in plant cell nuclei. *Nature*, **405**, 754–755.

Payne, S.G., Milstien, S. and Spiegel, S. (2002) Sphingosine-1-phosphate: dual messenger functions. *FEBS Lett*, **531**, 54–57.

Perera, I.Y., Davis, A.J., Galanopoulou, D., Im, Y.J. and Boss, W.F. (2005) Characterization and comparative analysis of *Arabidopsis* phosphatidylinositol phosphate 5-kinase 10 reveals differences in *Arabidopsis* and human phosphatidylinositol phosphate kinases. *FEBS Lett*, **579**, 3427–3432.

Perera, I.Y., Hung, C.Y., Brady, S., Muday, G.K. and Boss, W.F. (2006) A universal role for inositol 1,4,5-trisphosphate-mediated signaling in plant gravitropism. *Plant Physiol*, **140**, 746–760.

Perera, I.Y., Love, J., Heilmann, I., Thompson, W.F. and Boss, W.F. (2002) Up-regulation of phosphoinositide metabolism in tobacco cells constitutively expressing the human type I inositol polyphosphate 5-phosphatase. *Plant Physiol*, **129**, 1795–1806.

Peterman, T.K., Ohol, Y.M., Mcreynolds, L.J. and Luna, E.J. (2004) Patellin1, a novel Sec14-like protein, localizes to the cell plate and binds phosphoinositides. *Plant Physiol*, **136**, 3080–3094.

Pettus, B.J., Bielawska, A., Subramanian, P., Wijesinghe, D.S., Maceyka, M., Leslie, C.C., Evans, J.H., Freiberg, J., Roddy, P., Hannun, Y.A. and Chalfant, C.E. (2004) Ceramide 1-phosphate is a direct activator of cytosolic phospholipase A2. *J Biol Chem*, **279**, 11320–11326.

Pettus, B.J., Chalfant, C.E. and Hannun, Y.A. (2002) Ceramide in apoptosis: an overview and current perspectives.*Biochim Biophys Acta*, **1585**, 114–125.

Phillippy, B., Ullah, A. and Ehrlich, K. (1994) Purification and some properties of inositol 1,3,4,5,6-Pentakisphosphate 2-kinase from immature soybean seeds. *J Biol Chem*, **269**, 28393–28399.

Pical, C., Westergren, T., Dove, S.K., Larsson, C. and Sommarin, M. (1999) Salinity and hyperosmotic stress induce rapid increases in phosphatidylinositol 4,5-bisphosphate, diacylglycerol pyrophosphate, and phosphatidylcholine in *Arabidopsis thaliana* cells. *J Biol Chem*, **274**, 38232–38240.

Poggioli, J., Weiss, S.J., Mckinney, J.S. and Putney, J.W., Jr. (1983) Effects of antimycin A on receptor-activated calcium mobilization and phosphoinositide metabolism in rat parotid gland. *Mol Pharmacol*, **23**, 71–77.

Preuss, M.L., Schmitz, A.J., Thole, J.M., Bonner, H.K.S., Otegui, M.S. and Nielsen, E. (2006) A role for the RabA4b effector protein, PI-4Kβ1, in polarized expansion of root hair cells in *Arabidopsis*. *J Cell Biol*, **172**, 991–998.

Preuss, M.L., Serna, J., Falbel, T.G., Bednarek, S.Y. and Nielsen, E. (2004) The *Arabidopsis* Rab GTPase RabA4b localizes to the tips of growing root hair cells. *Plant Cell*, **16**, 1589–1603.

Qin, C., Li, M., Qin, W., Bahn, S., Wang, C. and Wang, X. (2006) Expression and characterization of *Arabidopsis* phospholipase Dγ2. *Biochim Biophys Acta*, **1761**, 1450–1458.

Raboy, V. and Bowen, D. (2006) Genetics of inositol polyphosphates. *Subcell Biochem*, **39**, 71–102.

Raboy, V., Gerbasi, P.F., Young, K.A., Stoneberg, S.D., Pickett, S.G., Bauman, A.T., Murthy, P.P., Sheridan, W.F. and Ertl, D.S. (2000) Origin and seed phenotype of maize low phytic acid 1-1 and low phytic acid 2-1. *Plant Physiol*, **124**, 355–368.

Ryu, S.B. (2004) Phospholipid-derived signaling mediated by phospholipase A in plants. *Trends Plant Sci*, **9**, 229–235.

Sandelius, A.S. and Sommarin, M. (1990) Membrane-localized reactions involved in polyphosphoinositide turnover in plants. In: *Inositol Metabolism in Plants* (eds D.J. Morre, W.F. Boss. and F.A. Loewus), pp. 139–161. Wiley-Liss, New York.

Scholz-Starke, J., Carpaneto, A. and Gambale, F. (2006) On the interaction of neomycin with the slow vacuolar channel of *Arabidopsis thaliana*. *J Gen Physiol*, **127**, 329–340.

Shears, S.B. (1998) The versatility of inositol phosphates as cellular signals. *Biochim Biophys Acta*, **1436**, 49–67.

Shears, S.B. (2001) Assessing the omnipotence of inositol hexakisphosphate. *Cell Signal*, **13**, 151–158.

Shimada, H., Koizumi, M., Kuroki, K., Mochizuki, M., Fujimoto, H., Ohta, H., Masuda, T. and Takamiya, K. (2004) ARC3, a chloroplast division factor, is a chimera of prokaryotic FtsZ and part of eukaryotic phosphatidylinositol-4-phosphate 5-kinase. *Plant Cell Physiol*, **45**, 960–967.

Shinghal, R., Scheller, R.H. and Bajjalieh, S.M. (1993) Ceramide 1-phosphate phosphatase activity in brain. *J Neurochem*, **61**, 2279–2285.

Shrestha, R., Kim, S.-C., Dyer, J.M., Dixon, R.A. and Chapman, K.D. (2006) Plant fatty acid (ethanol) amide hydrolases. *Biochim Biophys Acta (BBA) – Mol Cell Biol Lipids*, **1761**, 324–334.

Simoes, I., Mueller, E.C., Otto, A., Bur, D., Cheung, A.Y., Faro, C. and Pires, E. (2005) Molecular analysis of the interaction between cardosin A and phospholipase Dα. Identification of RGD/KGE sequences as binding motifs for C2 domains. *FEBS J*, **272**, 5786–5798.

Spassieva, S.D., Markham, J.E. and Hille, J. (2002) The plant disease resistance gene Asc-1 prevents disruption of sphingolipid metabolism during AAL-toxin induced programmed cell death. *Plant J*, **32**, 561–572.

Spiegel, S. and Milstien, S. (2003) Sphingosine 1-phosphate: an enigmatic signalling lipid. *Nat Rev Mol Cell Biol*, **4**, 397–407.

Stenzel, I., Ischebeck, T., Konig, S., Holubowska, A., Sporysz, M., Hause, B. and Heilmann, I. (2008) The type B phosphatidylinositol-4-phosphate 5-kinase 3 is essential for root hair formation in *Arabidopsis thaliana*. *Plant Cell*, February 15, 2008; 10.1105/tpc.107.056119.

Steponkus, P.L. (1984) Role of the plasma membrane in freezing injury and cold acclimation. *Ann Rev Plant Physiol*, **35**, 543–584.

Stevenson-Paulik, J., Bastidas, R.J., Chiou, S.T., Frye, R.A. and York, J.D. (2005) Generation of phytate-free seeds in *Arabidopsis* through disruption of inositol polyphosphate kinases. *Proc Natl Acad Sci USA*, **102**, 12612–12617.

Stevenson-Paulik, J., Love, J. and Boss, W.F. (2003) Differential regulation of two *Arabidopsis* type III phosphatidylinositol 4-kinase isoforms. A regulatory role for the pleckstrin homology domain. *Plant Physiol*, **132**, 1053–1064.

Stevenson-Paulik, J., Odom, A.R. and York, J.D. (2002) Molecular and biochemical characterization of two plant inositol polyphosphate 6-/3-/5-kinases. *J Biol Chem*, **277**, 42711–42718.

Subbaiah, C.C., Bush, D.S. and Sachs, M.M. (1998) Mitochondrial contribution to the anoxic Ca^{2+} signal in maize suspension-cultured cells. *Plant Physiol*, **118**, 759–771.

Sun, Y., Carroll, S., Kaksonen, M., Toshima, J.Y. and Drubin, D.G. (2007) PtdIns(4,5)P2 turnover is required for multiple stages during clathrin- and actin-dependent endocytic internalization. *J Cell Biol*, **177**, 355–367.

Taha, T.A., Mullen, T.D. and Obeid, L.M. (2006) A house divided: ceramide, sphingosine and sphingosine-1-phosphate in programmed cell death. *Biochim Biophys Acta*, **1758**, 2027–2036.

Takeshima, H., Komazaki, S., Nishi, M., Lino, M. and Kangawa, K. (2000) Junctophilins: a novel family of junctional membrane complex proteins. *Mol Cell*, **6**, 11–22.

Tanaka, T., Abbas, H.K. and Duke, S. (1993) Structure-dependent phytotoxicity of fumonisins and related compounds in duckweed bioassay. *Phytochemistry*, **33**, 779–785.

Tang, R.-H., Han, S., Zheng, H., Cook, C.W., Choi, C.S., Woerner, T. E., Jackson, R.B. and Pei, Z.-M. (2007) Coupling diurnal cytosolic Ca^{2+} oscillations to the CAS-IP_3 pathway in *Arabidopsis*. *Science*, **315**, 1423–1426.

Thompson, G.A., Jr. (1989) Membrane acclimation by unicellular organisms in response to temperature change. *J Bioenerg Biomembr*, **21**, 43–60.

Torabinejad, J. and Gillaspy, G.E. (2006) Functional genomics of inositol metabolism. *Subcell Biochem*, **39**, 47–70.

Townley, H.E., Mcdonald, K., Jenkins, G.I., Knight, M.R. and Leaver, C.J. (2005) Ceramides induce programmed cell death in *Arabidopsis* cells in a calcium-dependent manner. *Biol Chem*, **386**, 161–166.

Tucker, E.B. and Boss, W.F. (1996) Mastoparan-induced intracellular Ca^{2+} fluxes may regulate cell-to-cell communication in plants. *Plant Physiol*, **111**, 459–467.

Ullah, H., Chen, J.G., Young, J.C., Im, K.H., Sussman, M.R. and Jones, A.M. (2001) Modulation of cell proliferation by heterotrimeric G protein in *Arabidopsis*. *Science*, **292**, 2066–2069.

Vermeer, J.E., Van Leeuwen, W., Tobena-Santamaria, R., Laxalt, A.M., Jones, D.R., Divecha, N., Gadella, T.W., Jr. and Munnik, T. (2006) Visualization of PtdIns3P dynamics in living plant cells. *Plant J*, **47**, 687–700.

Wada, H., Gombos, Z. and Murata, N. (1990) Enhancement of chilling tolerance of a cyanobacterium by genetic manipulation of fatty acid desaturation. *Nature*, **347**, 200–203.

Waite, M. (ed) (1987) *The Phospholipases.* Plenum Press, New York.

Wang, H., Li, J., Bostock, R.M. and Gilchrist, D.G. (1996) Apoptosis: a functional paradigm for programmed plant cell death induced by a host selective phytotoxin and invoked during development. *Plant Cell*, **8**, 375–391.

Wang, J., Gambhir, A., Hangyas-Mihalyne, G., Murray, D., Golebiewska, U. and Mclaughlin, S. (2002) Lateral sequestration of phosphatidylinositol 4,5-bisphosphate by the basic effector domain of myristoylated alanine-rich C kinase substrate is due to nonspecific electrostatic interactions. *J Biol Chem*, **277**, 34401–34412.

Wang, X. (2001) Plant phospholipases. *Annu Rev Plant Physiol Plant Mol Biol*, **52**, 211–231.

Wang, X. (2004) Lipid signaling. *Curr Opin Plant Biol*, **7**, 329–336.

Wang, X. (2005) Regulatory functions of phospholipase D and phosphatidic acid in plant growth, development, and stress responses. *Plant Physiol*, **139**, 566–573.

Wang, X., Devaiah, S.P., Zhang, W. and Welti, R. (2006) Signaling functions of phosphatidic acid. *Prog Lipid Res*, **45**, 250–278.

Wei, Y.J., Sun, H.Q., Yamamoto, M., Wlodarski, P., Kunii, K., Martinez, M., Barylko, B., Albanesi, J.P. and Yin, H.L. (2002) Type II phosphatidylinositol 4-kinase β is a cytosolic and peripheral membrane protein that is recruited to the plasma membrane and activated by Rac-GTP. *J Biol Chem*, **277**, 46586–46593.

Westergren, T., Dove, S.K., Sommarin, M. and Pical, C. (2001) AtPIP5K1, an *Arabidopsis thaliana* phosphatidylinositol phosphate kinase, synthesizes PtdIns(3,4)P_2 and PtdIns(4,5)P_2 in vitro and is inhibited by phosphorylation. *Biochem J*, **359**, 583–589.

Williams, M.E., Torabinejad, J., Cohick, E., Parker, K., Drake, E.J., Thompson, J.E., Hortter, M. and Dewald, D.B. (2005) Mutations in the *Arabidopsis* phosphoinositide phosphatase gene SAC9 lead to overaccumulation of PtdIns(4,5)P2 and constitutive expression of the stress-response pathway. *Plant Physiol*, **138**, 686–700.

Wright, B.S., Snow, J.W., O'rien, T.C. and Lynch, D.V. (2003) Synthesis of 4-hydroxysphinganine and characterization of sphinganine hydroxylase activity in corn. *Arch Biochem Biophys*, **415**, 184–192.

Yamamoto, M., Chen, M.Z., Wang, Y.J., Sun, H.Q., Wei, Y., Martinez, M. and Yin, H.L. (2006) Hypertonic stress increases phosphatidylinositol 4,5-bisphosphate levels by activating PIP5KIβ. *J Biol Chem*, **281**, 32630–32638.

Yang, S.A., Carpenter, C.L. and Abrams, C.S. (2004) Rho and rho-kinase mediate thrombin-induced phosphatidylinositol 4-phosphate 5-kinase trafficking in platelets. *J Biol Chem*, **279**, 42331–42336.

Yang, W., Devaiah, S., Pan, X., Isaac, G., Welti, R. and Wang, X. (2007) AtPLAI is an acyl hydrolase involved in basal jasmonic acid production and *Arabidopsis* resistance to *Botrytis cinerea*. *J Biol Chem*, **282**, 18116–18128.

Yeung, T., Terebiznik, M., Yu, L., Silvius, J., Abidi, W.M., Philips, M., Levine, T., Kapus, A. and Grinstein, S. (2006) Receptor activation alters inner surface potential during phagocytosis. *Science*, **313**, 347–351.

Zhang, W., Crocker, E., Mclaughlin, S. and Smith, S.O. (2003) Binding of peptides with basic and aromatic residues to bilayer membranes: phenylalanine in the myristoylated alanine-rich C kinase substrate effector domain penetrates into the hydrophobic core of the bilayer. *J Biol Chem*, **278**, 21459–21466.

Zhang, W., Qin, C., Zhao, J. and Wang, X. (2004) Phospholipase Dα1-derived phosphatidic acid interacts with ABI1 phosphatase 2C and regulates abscisic acid signaling. *Proc Natl Acad Sci USA*, **101**, 9508–9513.

Zhao, J. and Wang, X. (2004) *Arabidopsis* phospholipase Dα1 interacts with the heterotrimeric G-protein α-subunit through a motif analogous to the DRY motif in G-protein-coupled receptors. *J Biol Chem*, **279**, 1794–1800.

Zheng, L., Krishnamoorthi, R., Zolkiewski, M. and Wang, X. (2000) Distinct Ca^{2+} binding properties of novel C2 domains of plant phospholipase Dα and β. *J Biol Chem*, **275**, 19700–19706.

Zheng, W., Kollmeyer, J., Symolon, H., Momin, A., Munter, E., Wang, E., Kelly, S., Allegood, J.C., Liu, Y., Peng, Q., Ramaraju, H., Sullards, M.C., Cabot, M. and Merrill, A.H., Jr. (2006) Ceramides and other bioactive sphingolipid backbones in health and disease: lipidomic analysis, metabolism and roles in membrane structure, dynamics, signaling and autophagy. *Biochim Biophys Acta*, **1758**, 1864–1884.

Zhong, R., Burk, D.H., Morrison, W.H., III and Ye, Z.H. (2004) FRAGILE FIBER3, an *Arabidopsis* gene encoding a type II inositol polyphosphate 5-phosphatase, is required for secondary wall synthesis and actin organization in fiber cells. *Plant Cell*, **16**, 3242–3259.

Zhong, R., Burk, D.H., Nairn, C.J., Wood-Jones, A., Morrison, W.H., III and Ye, Z.H. (2005) Mutation of SAC1, an *Arabidopsis* SAC domain phosphoinositide phosphatase, causes alterations in cell morphogenesis, cell wall synthesis, and actin organization. *Plant Cell*, **17**, 1449–1466.

Zhong, R. and Ye, Z.H. (2004) Molecular and biochemical characterization of three WD-repeat-domain-containing inositol polyphosphate 5-phosphatases in *Arabidopsis thaliana*. *Plant Cell Physiol*, **45**, 1720–1728.

Zonia, L. and Munnik, T. (2004) Osmotically induced cell swelling versus cell shrinking elicits specific changes in phospholipid signals in tobacco pollen tubes. *Plant Physiol*, **134**, 813–823.

Zonia, L. and Munnik, T. (2006) Cracking the green paradigm: functional coding of phosphoinositide signals in plant stress responses. *Subcell Biochem*, **39**, 207–237.

Annual Plant Reviews (2008) **33**, 244–272
doi: 10.1111/b.9781405160025.2008.00009.x www.interscience.wiley.com

Chapter 9

THE CYTOSKELETON AND SIGNAL TRANSDUCTION: ROLE AND REGULATION OF PLANT ACTIN- AND MICROTUBULE-BINDING PROTEINS

Patrick J. Hussey[1] and Takashi Hashimoto[2]

[1] *The Integrative Cell Biology Laboratory, School of Biological and Biomedical Sciences, Durham University, Durham DH1 3LE, UK*
[2] *Graduate School of Biological Sciences, Nara Institute of Science and Technology, Nara 630-0192, Japan*

Abstract: The plant cytoskeleton governs plant cell morphogenesis and it is composed of microtubules and actin filaments, and a plethora of associated proteins that serve to anchor, cross-bridge, or otherwise regulate this fibrous network. These associated proteins are involved in competitive and/or cooperative interactions within cells to adjust the dynamics and organization of the cytoskeleton. These associated proteins are often stimulus responsive and are effectors of signaling cascades. This system has evolved so that normally sedentary plant cells can respond to developmental and environmental cues in order to proliferate and grow, to maximize energy production, to take up nutrients from the soil, to reproduce, and to protect from pathogen invasion. In all these cases the cytoskeleton has to respond to signals and reorganize so that cells can divide and expand, generate organelle movement, polarize cell growth, and thicken the cell wall. This chapter will describe the main players in the control of cytoskeletal organization in plant cells and explain their involvement in signal transduction cascades.

Keywords: actin; actin-associated proteins; microtubule; phosphorylation; ROP

9.1 Actin cytoskeleton

The actin network is key to many essential cellular functions. It transports vesicles and organelles such as Golgi stacks and chloroplasts. It is essential for cell expansion and for aspects of cell division such as the positioning of the division spindle and the guidance of the growing cell plate. Consequently many aspects of plant biology are affected by the plant actin cytoskeleton: fertilization through pollen tube extension down the style, hormone transport due to the recycling of membrane-associated transporters, photosynthesis in respect to the movement of chloroplasts, and also management of the transpiration stream in the opening and closing of stomata. In playing these roles the actin network has to be able to respond to internal and external signals. These signals are transmitted through cascades that affect a catalogue of actin-binding proteins that either nucleate, cross-link, tether, or otherwise modulate monomer/polymer dynamics. Most of the work involving studies on actin dynamics and signaling to the actin cytoskeleton has been done using the model cell types: the intercalary growing trichomes and the tip growing cell root hairs and pollen tubes. As trichomes and root hairs are not essential cell types these are excellent models for studying plant cell growth genetically. Pollen tubes on the other hand offer relatively facile micromanipulation.

In this section progress on the inclusion of known actin-binding proteins in signaling cascades and the different modes of regulation will be discussed. For further background in this area also see Hussey *et al.* (2006) and Staiger and Blanchoin (2006), and for a description of the roles of Rho of Plants on actin (ROPs) see Chapter 3 in this book.

9.2 Actin nucleation

9.2.1 The Arp2/3 complex

The Arp2/3 complex nucleates actin by promoting barbed-end actin assembly while capping the pointed end. The complex attaches itself to the flanks of existing filaments and initiates a new F-actin branch at an angle of 70° relative to the parent filament. The Arp2/3 complex consists of seven subunits: Arp2, Arp3, ArpC1/p41, ArpC2/p31, ArpC3/p21, ArpC4/p20, and ArpC5/p16 (Higgs and Pollard, 2001). Homologues of all Arp2/3 complex subunits are present in plants (Mathur *et al.*, 2003b).

The Arp2/3 complex can be activated by several proteins a few of which are conserved in plants, including the components of the SCAR/WAVE complex (Deeks *et al.*, 2004). The SCAR/WAVE complex is an effector of Rac cytoskeletal reorganization in animals and protists. It is composed of five proteins: PIR121, NAP125, ABI, SCAR/WAVE, and HSPC300. SCAR/WAVE control of the Arp2/3 complex is used principally for cell motility in animal cells, but plant cells do not move so it is intriguing that such a pathway exists in plants especially as it is also absent from yeast genomes.

In the mammalian lamellipodium pathway for cell motility, DOCK180 GEF activity is induced when complexed with ELMO. This activates Rac1 which binds the PIR121 subunit of the SCAR/WAVE complex. In vitro, the induced conformational change in this complex can cause the release of an HSP300-WAVE subcomplex which is capable of activating the ARP2/3 complex. In vivo, it has been proposed that Rac1 recruits the complete complex to the lamellipodium leading edge to regulate Arp2/3 complex activity (for review see Deeks and Hussey, 2005).

Yeast 2-hybrid assays and in vitro pull down experiments have demonstrated binary interactions between the plant protein homologues that are equivalent to those characterized in the mammalian complex (Basu *et al.*, 2004, 2005; El-Assal Sel *et al.*, 2004; Frank *et al.*, 2004; Zhang *et al.*, 2005; Uhrig *et al.*, 2007). One member of the putative plant complex, PIR121, binds the active form of ROP2 (Basu *et al.*, 2004) while plant SCAR can activate the Arp2/3 complex in vitro and bind G-actin (Deeks *et al.*, 2004; Frank *et al.*, 2004; Basu *et al.*, 2005; Uhrig *et al.*, 2007). Moreover, *Arabidopsis* SPIKE1 which shares identity with DOCK180, has been shown to interact in yeast 2-hybrid with members of the SCAR and ABI families and that SCAR2 interacts directly with the activated form of ROP7 at the plasma membrane suggesting that SPIKE1 may be a component of the SCAR/WAVE complex and that plant SCAR proteins may act as direct effectors of ROP GTPases (Uhrig *et al.*, 2007).

In *Arabidopsis*, forward and reverse genetic approaches have been used to identify mutants in components of the Arp2/3 and SCAR/WAVE complexes. Four members of the classical *distorted* group of mutants have been shown to encode subunits of the Arp2/3 complex and for the most part further mutant alleles have been identified in insertion line databases; *wurm* is a mutant of the Arp2 subunit, *distorted1* is a mutant of the Arp3 subunit, *distorted2* is a mutant in the ArpC2/p31, and *crooked* represents the ArpC5/p16 subunit (Le *et al.*, 2003; Li *et al.*, 2003; Mathur *et al.*, 2003a; El-Assal Sel *et al.*, 2004). Mutants in ARPC4 have also been identified (Le *et al.*, 2003). Two members of the distorted group have been found to encode components of the SCAR/WAVE complex: *gnarled* which is a mutant of NAP1 (Deeks *et al.*, 2004; El-Assal Sel *et al.*, 2004) and *klunker* a mutant in PIR121 (Basu *et al.*, 2004; Saedler *et al.*, 2004). Alleles of HSPC300 isolated using a TILLING approach also exhibit a strong *distorted* phenotype (Djakovic *et al.*, 2006; Le *et al.*, 2006). The phenotypes of all these mutants are similar with the most dramatic phenotype being observed in trichomes, which develop a distorted phenotype highly similar to that caused by actin depolymerizing drugs, indicating that these mutant phenotypes are caused by defects in the actin cytoskeleton (Mathur *et al.*, 1999; Szymanski *et al.*, 1999). All members of the Arp2/3 and SCAR/WAVE complex are encoded by single copy genes with the exception of ArpC1, ArpC2, ABI, and SCAR. SCAR is encoded by four genes and insertion mutants identified show a degree of redundancy. Weak trichome phenotypes are observed in SCAR2 (Basu *et al.*, 2004; Zhang *et al.*, 2005) but a SCAR2/4 double mutant shows a phenotype very similar to the Arp2/3 mutants (Uhrig *et al.*, 2007). Knocking out components of either the *Arabidopsis* Arp2/3 complex or the SCAR/WAVE complex does not lead to a lethal phenotype as it does in yeast (Winter *et al.*, 1999), *C. elegans* (Sawa *et al.*, 2003), and *Drosophila* (Hudson and Cooley, 2002; Zallen *et al.*, 2002) indicating a less pronounced role for the Arp2/3 complex in actin nucleation in plants.

Several phenotypic studies in animals and Dictyostelium have reported that components of the SCAR/WAVE complex can behave as either activators or repressors of the ARP2/3 complex, causing (at least superficially) a paradox in the understanding of SCAR/WAVE complex function. Quantification of filamentous actin in *Arabidopsis* trichomes mutant for *NAP125* revealed that there was no apparent increase in F-actin in these lines (Deeks *et al.*, 2004). This contrasts tissues carrying loss-of-function mutations of the *Drosophila* NAP125 gene *KETTE* (Hummel *et al.*, 2000). Excessive F-actin accumulation in *kette* mutants has been cited to support the concept of Arp2/3 complex repression by the SCAR/WAVE complex. Phenotypic similarity between loss-of-function Arp2/3 complex and SCAR/WAVE complex mutants in *Arabidopsis* and the absence of F-actin accumulation in *Arabidopsis nap125* mutants suggests a role in Arp2/3 complex activation for the SCAR/WAVE complex in plants. Further studies on Dictyostelium have complicated this debate on SCAR complex function (Blagg and Insall, 2004a). Dictyostelium NAP125 mutants show a reduction in SCAR activity, while Dictyostelium PIR121 mutants show excessive SCAR activity. This is the first significant example of individual SCAR complex core-components behaving differently in genetic models, illustrating that the simple activator and repressor models, which describe the function of the SCAR-complex, are too basic to describe the behavior of the system fully (Blagg and Insall, 2004a).

The situation is further complicated by the fact that deactivation of individual complex members leads to rapid proteolysis of all other members (Blagg *et al.*, 2003; Kunda *et al.*, 2003; Rogers *et al.*, 2003; Innocenti *et al.*, 2004; Schenck *et al.*, 2004; Steffen *et al.*, 2004). For NAP125, PIR121, and ABI mutants, the observed phenotypic variation between organisms could result from the balance between free-SCAR that is inappropriately activating the Arp2/3 complex and increased proteolysis of SCAR in the absence of a complete complex (Blagg and Insall, 2004b). Yet another complication to phenotypic interpretation is a possible relationship between the *D. melanogaster* SCAR complex and the regulation of WASP (Bogdan and Klambt, 2003; Bogdan *et al.*, 2004), another Arp2/3 activating protein that is absent from plants. Despite the continuing debate all the current models that have been developed in animals and protists postulate that the SCAR complex links the activation of the Arp2/3 complex to the polymerization of actin at the cell cortex and the plasma membrane. In plants, however, the subcellular localization of the Arp2/3-SCAR complex interaction remains unidentified. Further work is required in all model organisms to overcome the complexity caused by proteolysis and signaling cross-talk to fully understand the regulating activity of the SCAR complex.

9.2.2 Formins

The second major group of actin nucleators are the formins. The *Arabidopsis* genome harbors 21 formin genes that can be divided into groups I and II based on their sequence similarity and domain organization (Deeks *et al.*, 2002; Cvrckova *et al.*, 2004). All 21 formins have a C-terminal domain that bears significant homology to the contiguous FH1 and FH2 domains present in animal and fungal formins thought to interact with actin, but differences in some of the residues predicted to make contact with the actin monomer between group I and II formins may indicate that members of each group have a distinct biochemistry. The plant formins lack an FH3 domain which has been shown to be required for formin localization (Petersen *et al.*, 1998), yet group I formins have transmembrane domains for tethering to lipid membranes. Two further domains are missing from the plant formins; the GTPase-binding domain (GDB) and Diaphonous-autoregulatory domain (DAD), and these are involved in the autoinhibition of diaphanous-related formins (Watanabe *et al.*, 1997, 1999; Alberts, 2001), which do not have direct homologues in plants. These two domains form an intramolecular interaction, which maintains the formin in an inactive state. The GBD domain can bind to activated Rho small GTPases releasing the DAD domain so that the conformation of the formin is converted into an active state. The absence of these domains indicates that plant Rho-like GTPases (ROPs) may not be direct activators of plant formins.

Formins stimulate de novo actin nucleation and extension from the barbed end (Pruyne *et al.*, 2002; Kovar *et al.*, 2003; Romero *et al.*, 2004). Four plant formins (AtFH1, AtFH4, AtFH5, and AtFH8) have been shown to nucleate purified actin and allow extension from the barbed end of filaments (Deeks *et al.*, 2005; Ingouff *et al.*, 2005; Michelot *et al.*, 2005; Yi *et al.*, 2005). Like other formins, the plant formins appear to bind to the barbed end of F-actin, inhibit actin depolymerization from the barbed end, and partially protect the barbed end from other proteins that otherwise would terminate barbed-end growth. AtFH1 has even been shown to nucleate actin specifically derived from plant tissue (Michelot *et al.*, 2005). One model that explains the mechanism of polymerization of actin by formin is the processive movement of formin attached at the barbed end of an elongating filament (Zigmond *et al.*, 2003). Intriguingly, AtFH1 has recently been shown to be unusual in that it is nonprocessive and moves to the side of the filament after nucleation to facilitate filament bundling (Michelot *et al.*, 2006). In vitro studies have yet to include longer fragments containing transmembrane domains and putative control regions that might influence the interactions between plant formins and actin.

Full length and/or truncated versions of group I formins have been overexpressed in plants or plant cells (Cheung and Wu, 2004; Deeks *et al.*, 2005;

Yi *et al.*, 2005). When full length AtFH1 was overexpressed in tobacco pollen tubes the phenotype of pollen tubes varied from an initial increase in growth rate followed by a growth inhibition, to depolarized pollen tube growth followed by growth inhibition. Overexpressing just the actin nucleating domains increased the number of actin cables indicating that this fragment could promote actin polymerization (Cheung and Wu, 2004). Similarly, overexpression of a comparable fragment of AtFH8 caused an accumulation of fine filamentous actin and the disruption of tip growth in root hairs (Yi *et al.*, 2005). Disruption of root hair growth was also apparent when expression of the N-terminus of AtFH4 was induced, without the actin nucleating C-terminus (Deeks *et al.*, 2005). Taken together these data show that the group I formins at least have the potential to affect growth through F-actin formation. Most likely due to the redundancy in function between members of the formin family, only one isoform AtFH5, has been reported to have any null phenotype (Ingouff *et al.*, 2005). Interestingly, this is a reduction in the rate of cell wall formation indicating that this is involved in cell growth.

9.3 Actin-binding proteins that modulate monomer/polymer dynamics

9.3.1 Capping protein

Heterodimeric capping protein binds tightly to the barbed end of actin filaments. The barbed-end-binding affinity of plant capping protein allows it to act as a nucleator that facilitates pointed-end elongation (Huang *et al.*, 2003). The elongation rate of filaments in vitro is significantly slowed by a combination of capping protein and profilin, as capping protein blocks barbed-end growth and profilin-actin is unable to associate with the pointed end (Huang *et al.*, 2003). Capping protein binds both PIP2 and PA but not to several other phospholipids. Interaction with PA inhibits the actin-binding activity of capping protein. PA was also able to uncap filaments blocked by capping protein on their barbed ends allowing rapid filament assembly from an actin monomer pool buffered with profilin. As exogenously applied PA to *Arabidopsis* tissue culture cells and poppy pollen results in an increase in filamentous actin, it has been proposed that the PA inhibits capping protein activity in these cells resulting in the stimulation of actin polymerization from a pool of actin monomer and profilin (Huang *et al.*, 2006).

9.3.2 Gelsolin, villin, fragmin

Gelsolin and villin form part of a family of actin-binding proteins that share similar domain structure having six gelsolin homology domains and that are regulated by Ca^{2+} (see review by Staiger and Hussey, 2004). Gelsolin is a barbed-end capping protein that is able to nucleate actin filaments so that extension is from the pointed end. Gelsolin also has a Ca^{2+} stimulated filamentous actin severing activity. An 80-kDa actin-binding protein from poppy pollen has been identified that can nucleate actin filaments, has potent Ca^{2+} stimulated severing activity and can regulate assembly dynamics by binding to the barbed end. Sequence analysis revealed that this protein is related to gelsolin (Huang *et al.*, 2004). Villin has an extra subdomain at the C-terminus called the villin headpiece (VHP) which allows it to cross-link adjacent actin filaments to form bundles (Pope *et al.*, 1994). Most villins have an actin-severing and barbed-end capping ability (Glenney *et al.*, 1980; Janmey and Matsudaira, 1988). Two villins of 135 kDa and 115 kDa have been identified biochemically from lily pollen (Yokota and Shimmen, 2000). Both have Ca^{2+}/calmodulin-dependent filamentous actin binding and bundling activity, and the 135-kDa protein has been shown to have Ca^{2+}-dependent capping and depolymerization/fragmentation activity (Yokota *et al.*, 2005). A shorter 41-kDa protein also isolated from lily pollen showing sequence similarity to the villin/gelsolin/fragmin superfamily and has been shown to be a Ca^{2+}-dependent actin-filament-severing protein (Fan *et al.*, 2004). Moreover, a 29-kDa protein arising from an alternatively spliced 135-ABP (*Lillium* villin) transcript has been identified that plays an important role in the organization of the actin cytoskeleton during pollen germination and tube growth (Xiang *et al.*, 2007). There are five villin genes in the *Arabidopsis* genome and the recombinant protein of AtVLN1 has been characterized. This can bind and bundle F-actin in a Ca^{2+} insensitive manner but has no severing or capping activity (Huang *et al.*, 2005).

9.3.3 Profilin

Profilins are small low molecular weight G-actin-binding proteins of approximately 12–15 kDa. When G-actin is bound to profilin, the G-actin cannot incorporate at the pointed end of actin filaments whereas incorporation at the barbed ends continues at the normal rate (Pollard *et al.*, 2000). Animal and fungal profilin can accelerate the rate of G-actin exchange of ADP for ATP, thereby promoting the rate of actin polymerization (Lu and Pollard, 2001). The principal mammalian G-actin sequestering protein is thymosin β4 which is absent from plant genomes (Hussey *et al.*, 2002). The high concentration of profilins in plant pollen would indicate that a major role of these profilins is in G-actin sequestration (Gibbon *et al.*, 1999; Snowman *et al.*, 2002). Surprisingly, plant profilins cannot catalyze ADP to ATP nucleotide exchange on G-actin (Perelroizen *et al.*, 1996; Kovar *et al.*, 2000).

Profilin is believed to function as an intermediate in signaling cascades that affect actin organization. Profilins bind to three major ligands and these are G-actin, polyproline, and phosphoinositides. It has been shown that individual plant profilin isoforms differ in their abilities to sequester G-actin, and differ in their affinities for poly-L-proline and PtdIns(4,5)P_2(Kovar *et al.*, 2000).

This implies differential roles for these isoforms in signaling cascades. Moreover, the actin sequestering activity of pollen profilin is regulated by Ca^{2+} with higher sequestration at higher Ca^{2+} levels, a property which may impact on actin organization in pollen tube tip growth (Kovar *et al.*, 2000; Snowman *et al.*, 2002). Profilin's ability to bind stretches of proline is utilized by several proteins known to transduce signals to or directly affect the actin network. These proteins are the poly-L-proline profilin-binding proteins WASP, SCAR/WAVE, VASP, and FORMIN families (Holt and Koffer, 2001). Within animals and fungi these four classes of protein provide multiple bridges between signaling networks and the actin cytoskeleton. The interactions include numerous small GTPases, kinases, SH3 adaptor proteins, and WW domain proteins. Only members of the SCAR/WAVE and FORMIN families of PLP profilin regulators have been isolated from plants, and searches through the *Arabidopsis* genome sequencing data have not identified any conserved WASP- or VASP-related genes. Another potential control of plant profilin is phosphorylation. It has been shown that tobacco profilin NtProf2 can be phosphorylated by tobacco MAP kinases $p45^{Ntf4}$ and SIPK, when activated by the tobacco MAP kinase kinase NtMEK2. The function of such a phosphorylation remains to be assessed (Limmongkon *et al.*, 2004).

9.3.4 ADF/cofilin and AIP1

ADF/cofilin binds both to G- and F-actin and enhances actin dynamics by severing actin filaments and increasing the depolymerization from the pointed end (Carlier *et al.*, 1997; Gungabissoon *et al.*, 1998). The activity of plant ADFs varies between isoforms. A well-characterized maize vegetative ADF, ZmADF3, is regulated by several factors. Phosphorylation of Ser-6 by calmodulin-domain protein kinase (CDPK) decreases the activity of ZmADF3 (Smertenko *et al.*, 1998; Allwood *et al.*, 2001). The ADF/cofilin phosphorylating LIM/TESK kinases (Arber *et al.*, 1998; Toshima *et al.*, 2001) present in animal cells have not been identified in the plant genomes. The activity of ZmADF3 is also inhibited by phosphatidylinositol 4,5-bisphosphate (PIP2) or phosphatidylinositol 4-monophosphate (PIP) binding (Gungabissoon *et al.*, 1998). Finally, the activity of ZmADF3 is pH dependent. At high pH (8.0), ADF severs actin filaments, whereas it binds F-actin at a lower pH (6.0) (Gungabissoon *et al.*, 1998).

Pollen-specific ADFs from maize and lily show a greater degree of identity than that between ADFs from within the same species, possibly indicating a conserved functional role for pollen ADFs. Lily pollen ADF, LlADF1 has a weak effect on actin dynamics, is not phosphorylated or controlled by phosphorylation on Ser-6 but does interact with PIP2 and PIP and does exhibit a pH sensitive activity (Allwood *et al.*, 2002). Yeast 2-hybrid analysis has shown that LlADF1 interacts with *Arabidopsis* actin interacting protein 1 (AtAIP1) (Allwood *et al.*, 2002). RNAi knockdown of AIP1 in *Arabidopsis* has shown that this protein is essential for normal plant development (Ketelaar *et al.*, 2004,

2007). Immunolocalization studies reveal that pollen ADF and AIP1 colocalize in pollen grains and pollen tubes (Allwood *et al.*, 2002; Lovy-Wheeler *et al.*, 2006). Moreover, in vitro studies have shown that AIP1 enhances the activity of LlADF1 by as much as 60% indicating that this cooperation is a major factor in the control of LlADF1. Also, the pH sensitive activity of LlADF1 correlates with the oscillatory changes in pH in the apical domain of lily pollen tubes and may be a mechanism of control of ADF in this region (Lovy-Wheeler *et al.*, 2006).

9.3.5 Cyclase-associated protein

Cyclase-associated protein (CAP) was first identified in *S. cerevisiae* by co-precipitation with adenylate cyclase (AC) (Field *et al.*, 1990), and as a suppressor (*srv2*) of the activated *RAS2* allele (*RAS2*^{Val19}) (Fedor-Chaiken *et al.*, 1990). Mutations in *CAP/SRV2* affect the regulation of adenylyl cyclase by Ras and affect actin organization. These roles have been attributed to individual domains of the protein: the N-terminal region contains the AC-binding domain responsible for the signaling function, while the C-terminus is responsible for actin binding (reviewed in Hubberstey and Mottillo 2002). CAP has been identified in all eukaryotes analyzed to date including plants (Barrero *et al.*, 2002).

CAP was originally believed to be a monomer sequestering protein because biochemical studies showed that it could suppress spontaneous polymerization of actin (Gottwald *et al.*, 1996; Hubberstey and Mottillo, 2002). More recently, it has been suggested that CAP/Srv2p complexes interact with ADP-actin/cofilin complexes facilitating the conversion to ATP-actin/profilin complexes (Balcer *et al.*, 2003; Mattila *et al.*, 2004). The effect is to recycle cofilin for new rounds of filament depolymerization and to generate profilin bound ATP-actin monomers for polymerization at the barbed ends. In contrast to the signaling activities of the CAP N-terminus, aspects of the *srv2* phenotype dependent upon the actin-binding abilities of CAP/Srv2p can be complemented by CAP isoforms from divergent species. Consequently, the C-terminus of *Arabidopsis* CAP has been shown to complement the cytoskeletal defects of *S. cerevisiae cap* mutants (Barrero *et al.*, 2002). The *Arabidopsis* CAP has been shown to be an actin monomer-binding protein (Barrero *et al.*, 2002; Chaudhry *et al.*, 2007; Deeks *et al.*, 2007) and that it directly accelerates the exchange of ADP for ATP by actin monomers, perhaps filling a functional space in plant actin biochemistry left by the absence of plant profilin nucleotide exchange activity (Staiger and Blanchoin, 2006; Chaudhry *et al.*, 2007). Analysis of null mutant phenotypes in the *Arabidopsis* homologue of CAP (CAP1) shows that this gene is essential for the development of multiple cell types and that its disruption correlates with actin organizational defects. Elongating epidermal cells, tip growing cells, and trichomes show severe morphological abnormalities and unusual aggregation of F-actin (Deeks *et al.*, 2007). A previous study has shown that the overexpression of plant CAP disassembles F-actin arrays in vivo and causes severe growth defects (Barrero *et al.*, 2002), which is surprising when considering that the overexpression of CAP in other organisms does not result in gross phenotypic abnormalities.

The N-terminal region of CAP/Srv2p was previously found to bind AC in yeast (Field *et al.*, 1990), and a 36 amino acid region of the N-terminus is essential for this interaction (Nishida *et al.*, 1998). A motif within this small domain (the "RLE" motif) is highly conserved in yeast, plants, and animals, but no direct interaction with AC has ever been found except in yeast, although indirect interaction has been implicated in Dictyostelium (Noegel *et al.*, 2004). Cell motility in Dictyostelium is dependent on the dual role of CAP in signaling and in actin dynamics indicating that integration of actin dynamics and cell signaling utilizing this protein is essential for normal cell function (Noegel *et al.*, 2004). The N-termini of CAP homologues from different species do not complement one another functionally (Kawamukai *et al.*, 1992), suggesting that the signaling activities of the CAP N-terminus have diversified between species. Plants do not have Ras and only one protein (pollen-specific signaling protein PsiP) with AC activity has been found in plants (Moutinho *et al.*, 2001). The *Arabidopsis cap1* mutants display altered growth behavior of multiple organs resulting in curled inflorescences and meandering roots consistent with CAP1 contributing to the function of plant-specific signaling pathways which remain to be defined (Deeks *et al.*, 2007).

9.4 Microtubule cytoskeleton

During progression of the cell cycle, plant microtubules remodel into distinct organizations that each function at an appropriate stage of the cell cycle (Mineyuki, 2007; Fig. 9.1). Microtubules are organized as cortical arrays in interphase (G_1 and S). Preprophase bands of microtubules (PPB) transiently appear in G_2 and disappear when the nuclear envelope breaks down. During prophase, spindle microtubules are formed. Phragmoplast microtubules appear between the daughter chromosomes in telophase, and function to form new cell walls.

Cortical microtubule arrays are readily remodeled during cell differentiation, and in response to various endogenous and external stimuli. The effects of plant hormones on how arrays are organized are well documented (Shibaoka, 1994). Remodeling has also been observed after attacks by pathogens, following exposure to low temperature, aluminum, or salt stress, and during tropic responses to light and gravity (Nick, 1998; Abdrakhamanova *et al.*, 2003; Sivaguru *et al.*, 2003; Takemoto and Hardham, 2004; Shoji *et al.*, 2006).

These dynamic changes in microtubule organization appear to result from alterations in microtubule dynamics, inter-microtubule association,

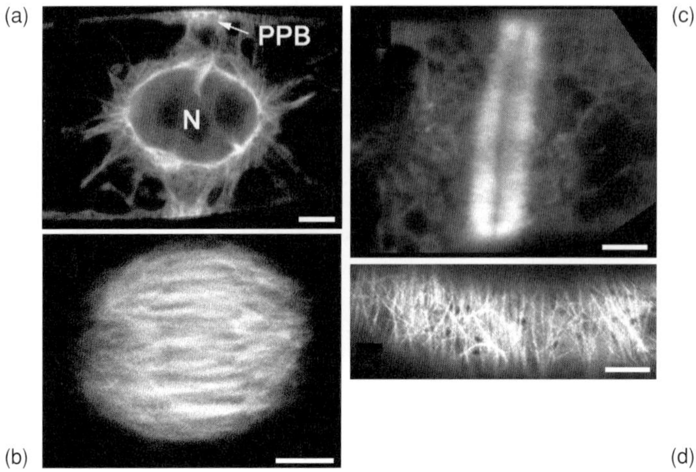

Figure 9.1 Microtubule organization during progression of the cell cycle. Microtubules can be visualized with a GFP:α-tubulin fusion protein stably expressed in tobacco BY-2 cells. (a) The PPB at late G_2, (b) the mitotic spindle, (c) the phragmoplast at telophase, and (d) cortical microtubules at G_1. N; nucleus. Bars: 10 μm. (Reprinted from Yoneda and Hasezawa (2003). Copyright (2003), with permission from Elsevier.)

microtubule nucleation activity, and interactions with the plasma membrane and other cellular components. Activities of various microtubule regulators should be regulated by multiple intracellular signaling pathways, but little is known about the molecules or mechanisms involved. In this review, two signaling pathways that regulate microtubule organization are summarized. As more microtubule regulators are identified and characterized, further progress can be expected in this field.

9.5 ROP

The Rho family small GTPases are highly conserved molecular switches that control morphogenesis, polarity, movement, and cell division in all eukaryotes. Plants have a unique Rho subfamily, ROP (Rho-related GTPase from plants), which mediates the response to a broad range of extracellular stimuli to activate diverse cellular pathways (for recent reviews, see Berken, 2006; Brembu *et al.*, 2006; Nibau *et al.*, 2006; Yang and Fu, 2007) (see Chapter 3). Suppressed expression of two *Arabidopsis* homologs ROP2 and ROP4 in leaf epidermis promoted the formation of thick transverse cortical microtubules. A ROP-interactive CRIB motif-containing RIC1 was shown to mediate the ROP-dependent suppression of microtubule bundles in the early stages of pavement cell morphogenesis (Fu *et al.*, 2005). RIC1 is located to both the plasma membrane and cortical microtubules in young pavement cells, but is redistributed almost exclusively to cortical microtubules in older cells. Overexpression of RIC1-induced microtubule bundling and suppressed lobe formation in pavement cells, whereas *ric1* knockout mutants contained fewer and less bundled cortical microtubules and displayed somewhat wider neck regions. RIC1 binds to the active but not the inactive form of ROPs. Studies on RIC1's localization in cells in which ROP2/4 expression was reduced and in which a constitutively active ROP2 was overexpressed indicated that activated ROP2 removes RIC1 from cortical microtubules, thus suppressing its function. It is not known whether RIC1 directly promotes microtubule organization or acts as a regulator of a bona fide microtubule-associated protein.

9.6 Protein phosphorylation

9.6.1 Inhibitors of protein kinases and phosphatases

Plants contain a huge repertoire of protein kinases that exceed 1000 in number (Wang *et al.*, 2003; Krupa *et al.*, 2006). Serine/threonine protein kinases comprise the largest family, which includes receptor-like kinases, calcium-dependent kinases, cyclin-dependent kinases (CDKs), and kinases in the mitogen-activated protein kinase (MAPK) cascade. Staurosporine inhibits a broad spectrum of kinases through its strong affinity for the ATP-binding site. K-252a is an ATP analog that strongly inhibits calmodulin kinase but at higher concentrations is also an efficient general inhibitor of serine/threonine protein kinases. Olomoucine is a purine derivative that strongly inhibits several CDKs and p44 MAPK, but exhibits reduced sensitivity toward some CDKs.

Plant phosphatases are composed of two large families (Luan, 2003). Conventionally, serine/threonine phosphatases are classified into Type1 (PP1) and Type2 (PP2), and a group of inhibitors have been used to distinguish these types and to further categorize them into subgroups. Both okadaic acid and calyculin A potently inhibit PP1 and PP2A but are not effective against PP2B and PP2C. Cantharidin inhibits PP2A but not the others. Protein tyrosine phosphatases are either tyrosine-specific, or dephosphorylate both tyrosine and serine/threonine residues, as found in mitogen-activated protein kinase phosphatases. Unfortunately, there are no potent and cell-permeable inhibitors readily available for the protein tyrosine phosphatase family.

When synchronized suspension-cultured tobacco cells in the late G_2 phase were treated with 20 μM staurosporine or 2 μM K-252a, the disappearance of PPB, which normally occurs before the breakdown of the nuclear envelope at prophase, was inhibited and cells with PPB accumulated (Katsuta and Shibaoka, 1992). Treatment of synchronized tobacco cells with okadaic acid or calyculin A arrested the cell cycle at specific stages and particular microtubule arrays predominated in these cells (Hasezawa and Nagata, 1992). From these experiments, it is difficult to know whether the reorganization of microtubules is a direct effect of the inhibitors or an indirect consequence of the cell cycle arrest.

Arabidopsis primary roots were used to examine the effects of inhibitors of protein kinases and phosphatases on the morphology and organization of cortical microtubules (Baskin and Wilson, 1997). Microtubule arrays were disrupted and the root cells swelled on treatment with a kinase inhibitor (1 μM staurosporine), or with a phosphatase inhibitor (30 μM cantharidin, 170 nM calyculin A, or 300 nM okadaic acid). Calyculin A and okadaic acid also enhanced the nucleation of microtubules from the nuclear envelope. Cortical microtubules in epicotyl segments of azuki bean shifted from a mainly longitudinal orientation to a transverse orientation when treated with 0.1 mM gibberellin A_3 (GA_3) in the presence of 0.1 mM IAA. This GA_3-induced reorientation was efficiently blocked by 1 mM 6-dimethylaminopurine or other kinase inhibitors (Mizuno, 1994). In lily pollen tubes, microtubules in the cortex and endoplasm disassembled in the presence of 30 nM calyculin A (Foissner *et al.*, 2002).

In vitro, microtubule polymers readily disassemble at low temperatures (e.g., at 0°C), whereas microtubules in living cells often are resistant to disassembly, depending on the cell type and cell-cycle stage. In suspension-cultured tobacco cells, cortical microtubules are labile to cold treatment early on, but acquire stability in the stationary phase. When 1 mM 6-dimethylaminopurine or 10 μM staurosporine was added to the tobacco cells at an early stage of culture, microtubules became resistant to cold treatment (Mizuno, 1992), indicating that kinase-mediated protein phosphorylation results in destabilized microtubules.

Protoplast ghosts, prepared by rupturing protoplasts on coverslips, expose the inner side of the plasma membrane and microtubules associated with it. Cortical microtubules on protoplast ghosts of unsynchronized tobacco cells were depolymerized in the presence of a tobacco cell extract and ATP (plus $MgCl_2$), but not in the presence of either alone (Sonobe, 1990). Phragmoplasts and spindles isolated from cultured tobacco cells were resistant to depolymerization under the same conditions. Interestingly, cortical microtubules and PPB on protoplast ghosts, which were prepared from synchronized tobacco cells in late G_2 or at prophase, depolymerized upon treatment with ATP, and did not require a cytosolic factor (Katsuta and Shibaoka, 1992). Staurosporine at 20 μM suppressed the ATP-induced depolymerization but 2 μM K-252a was ineffective. It has been proposed that a staurosporine-sensitive kinase is associated with microtubules on protoplast ghosts and is directly involved in the depolymerization of microtubules, whereas a K-252a-sensitive kinase in the cytosol is indirectly involved in the depolymerization. These studies also suggest that the stability of microtubules differs with the phase of the cell cycle, and distinct sets of protein regulators that associate with microtubules may contribute to such differences. Primary roots of maize were sectioned without fixation and the stability of cortical microtubules was analyzed in a similar manner (Tian *et al.*, 2004). In this assay, the exogenous ATP readily depolymerized microtubules, which was inhibited by the simultaneous treatment with staurosporine. Conversely, an inhibitor of protein phosphatases, cantharidin, promoted depolymerization in the absence of ATP.

9.6.2 Phosphatase mutants

Visual genetic screens for *Arabidopsis* mutants affecting seedling body organization have isolated the *tonneau2* (*ton2*) mutant (Traas *et al.*, 1995), which is allelic to *fass* (Torres-Ruiz and Jürgens, 1994). *ton2/fass* seedlings of strong alleles are characterized by an extreme compression of all body elements in the apical–basal axis and radial swelling. The mutants can develop into miniature adult plants with all parts. At the cellular level, *ton2/fass* mutants are defective in cell elongation and show random orientation of cell division planes, but are not affected in cell polarity and thus pattern formation. Examination of the microtubule cytoskeleton revealed that the PPB is absent in the mutant cells (Traas *et al.*, 1995; McClinton and Sung, 1997). Cortical microtubules in interphase cells are mostly arranged randomly in the root and have mostly parallel arrays whose orientation is not well fixed, with cells in the same area having a transverse, oblique, or longitudinal array orientation (Camilleri *et al.*, 2002). Interestingly, noncortical microtubules appear to be unaffected; spindles and the phragmoplasts have a normal appearance. The phragmoplasts, however, are often guided to random sites at the cortex, generating the irregularly shaped cells.

The *TON2/FASS* gene encodes a protein similar in its C-terminal part to B″ regulatory subunits of PP2As (Camilleri *et al.*, 2002). Association of the PP2A catalytic type-C subunit with regulatory (type A and type B) subunits produces several species of holoenzymes with distinct properties and functions. B-type subunits associate with the regulatory A subunits and provide targeting and substrate specificity to the enzyme complex (Luan, 2003). In animal cells, a trimeric PP2A complex composed of A and C subunits and of a distinct form of the B subunit associated with microtubules in vivo and in vitro, but a dimeric PP2A complex composed only of A and C subunits did not (Hiraga and Tamura, 2000). The TON2 protein interacts with an *Arabidopsis* type-A subunit of PP2A in the yeast two-hybrid system, indicating that TON2 and its vertebrate homologs define a novel subclass of PP2A B″ subunits. The *Arabidopsis* genome contains three PP2A catalytic C subunit genes, among which *RCN1* exhibits the most obvious morphological defects when mutated alone (Zhou *et al.*, 2004). Double mutants of *rcn1* and either of the two other *pp2a* mutants showed dwarfing and infertility phenotypes that were similar to, but not identical with, those of the moderate alleles of *ton2*. Microtubules were not analyzed in the *pp2a* double mutants. It remains to be seen whether the *pp2a* triple mutants encompass severe cellular phenotypes of the *ton2* null allele.

Genetic screening for *Arabidopsis* mutants that were hypersensitive to a microtubule-destabilizing drug, propyzamide, resulted in the isolation of a unique semidominant mutant allele, *propyzamide-hypersensitive 1-1* (*phs1-1*) (Naoi and Hashimoto, 2004). Seedling roots of *phs1-1* showed left-handed helical growth and became swollen at low doses of microtubule-destabilizing drugs; these phenotypes were synergistically enhanced in the presence of the temperature-sensitive *microtubule organization 1-1* (*mor1-1*) mutation. The cortical microtubules in *phs1-1* appeared to be more destabilized since they formed less ordered and more fragmented arrays, which were more readily depolymerized at low doses of microtubule-destabilizing drugs. PHS1 encodes a MAPK phosphatase-like protein. Recombinant PHS1 showed phosphatase activity toward a bulky polycyclic aryl phosphate, which is a preferred substrate for MAPK phosphatase. Typical mammalian MAPK phosphatases contain a catalytic phosphatase domain at the C-terminus and an N-terminal noncatalytic domain that are divergent among MAPK phosphatases but are important for MAPK-binding specificity. A kinase interaction motif (Arg or Lys)$_{2-3}$-X1-6-ØA-X-ØB, where ØA and ØB are hydrophobic residues such as Leu, Ile, and Val, and X is any amino acid (Liu *et al.*, 2006), is found at the N-terminus of PHS1, and the conserved Arg residue was indeed mutated to Cys in the *phs1-1* allele. Transgenic studies demonstrated that the Arg-to-Cys mutation in *phs1-1* functions dominant negatively. It is not known how this mutation affects docking interactions with cognate MAPKs. A recessive mutant allele, *phs1-3*, whereby *PHS1* expression was reduced by the insertion of T-DNA in the *PHS1* promoter region, is hypersensitive to abscisic acid (ABA), as judged from a strong ABA-induced inhibition of germination, altered expression levels of ABA-responsive genes, and a stronger inhibition of the light-induced opening of stomata by ABA (Quettier *et al.*, 2006). The PHS1 phosphatase may be involved in multiple signaling cascades.

9.6.3 MAP65 phosphorylation

MAPs are obvious candidates for directly phosphorylated substrates that modulate microtubule functions. In plants, phosphorylation and its functional consequences have been demonstrated only for MAP65 proteins. MAP65 is a microtubule-bundling protein conserved in plants, vertebrates, and yeasts, and contains the N-terminal dimerization domain and the highly conserved middle region that binds microtubules (Mollinari *et al.*, 2002; Smertenko *et al.*, 2004). Three isoforms of the tobacco NtMAP65-1 subfamily are the founding members of MAP65 (Smertenko *et al.*, 2000), and are most homologous to the AtMAP65-1 and AtMAP65-2 of *Arabidopsis thaliana* (Hussey *et al.*, 2002). *Arabidopsis* contains nine MAP65 proteins, which likely differ in their subcellular distribution and bundling activities (Hussey *et al.*, 2002; Van Damme *et al.*, 2005; Mao *et al.*, 2005). The fission yeast MAP65 homolog Ase1p was located in regions that contain antiparallel microtubule contacts in vivo, and efficiently promoted the bundling of antiparallel microtubules over the bundling of parallel microtubules in an in vitro assay consisting of bacterially expressed GST-Ase1p and dynamic microtubules grown from purified tubulin (Janson *et al.*, 2007). MAP65-1 proteins from tobacco and *Arabidopsis* were localized to a portion of cortical microtubules, the PPB, the overlap midzone of the two half spindles, and the midzone of the cytokinetic phragmoplast (Smertenko *et al.*, 2000, 2004). The localization at the spindle midzone suggests that MAP65-1 bundles antiparallel microtubules, as found for Ase1p. Electron tomography of cryogenically fixed *Arabidopsis* meristem cells indicated, however, that somatic-type phragmoplast microtubules may and not interdigitate at the cell plate mid-line of the phragmoplast (Austin *et al.*, 2005). Overexpression of GFP-AtMAP65-1 in *Arabidopsis* suspension cells induced bundling of nucleus-associated radial microtubules at the early interphase and the separate halves of the phragmoplast microtubules, indicating MAP65-1 may cross-link parallel microtubules (Mao *et al.*, 2005).

The phosphorylation status of MAP65 changes during the cell cycle peaking during metaphase just when binding of MAP65 to microtubules is significantly reduced. Analysis of phosphorylation sites in the *Arabidopsis* MAP65 homologue AtMAP65-1 demonstrated that several classes of protein kinases including MAPK and CDK regulate MAP65/microtubule interaction (Smertenko *et al.*, 2006). Tobacco MAP65-1s are phosphorylated in vitro by a tobacco MAPK, NRK1/NTF6 on the Thr579 at the C-terminus of NtMAP65-1a, which conforms to the MAPK-phosphorylation motif conserved in MAP65-1 subfamily members (Sasabe *et al.*, 2006). Microtubule-bundling activity of NtMAP65-1 was partially reduced after phosphorylation by NRK1 in vitro. When the subcellular localization of NtMAP65-1 and its phosphorylated form was compared by specific antibodies, the phosphorylated form was found more concentrated at the midzone of the phragmoplast during its lateral expansion. Overexpression of a mutant NtMAP65-1 that cannot be phosphorylated by NRK1 in tobacco BY-2 cells delayed progression of the M phase and phragmoplast expansion. It has been proposed that phosphorylation of MAP65-1 by MAPK reduces its microtubule-bundling activity at the phragmoplast midzone, and facilitates its lateral expansion by facilitating destabilization and turnover of microtubules at the midzone (Sasabe *et al.*, 2006). During cytokinesis in tobacco, kinesin-like proteins (NACK1 and NACK2) bind and the MAPK kinase kinase NPK1, thereby activating it and the subsequent MAPK cascade, consisting of MAPK kinase NQK1/NtMEK1 and the MAPK NRK1/NTF6 (see Chapter 12). Thus, NACK1/2 microtubule motors may concentrate the MAPK activity at the midzone of the phragmoplast, and may spatially regulate activities of MAP65-1 and other microtubule regulators.

In an in vitro phosphorylation assay by M-phase tobacco cell extracts, the CDK-specific inhibitor olomoucine partially inhibited the phosphorylation of AtMAP65-1. When CDK activity in tobacco BY-2 cells were inhibited by olomoucine, GFP-AtMAP65-1 was induced to bind the metaphase spindle microtubules within 10 min of treatment (Smertenko *et al.*, 2006), indicating that phosphorylation of AtMAP65-1 by a CDK-like activity inhibits its microtubule-binding activity. Strong phosphorylation of NtMAP65-1 by CDKs occurred well before the phosphorylation by NRK, and the CDK phosphorylation sites during the M phase were identified as Thr501 and Ser503 of at the C-terminus of NtMAP65-1a. When the NtMAP65-1a form with mutations at these CDK-phosphorylation sites were overexpressed in synchronized tobacco BY-2 cells, progression of the cell cycle from the prometaphase appeared to be unaffected (Sasabe *et al.*, 2006). When the corresponding CDK-non-phosphorylatable mutant of GFP-AtMAP65-1 was overexpressed in *Arabidopsis* suspension cells, microtubules crossing the central region of the metaphase spindle were prematurely decorated by the introduced MAP65 but the subsequent compaction to the midline appeared to be normal (Mao *et al.*, 2005). As expression of a non-phosphorylatable AtMAP65-1 form mutated at nine potential phosphorylation sites of several kinases, including CDKs, delayed the metaphase/anaphase transition in tobacco cells (Smertenko *et al.*, 2006), it is suggested that several kinase pathways are required for the regulation of MAP65-1 activity.

The phosphorylation sites of MAP65-1 by NRK and CDK both fall within the second microtubule interaction region, which is located downstream of the main microtubule region conserved in all MAP65 members. This second interaction region is preferentially coiled-coil and basic. It has been proposed that phosphorylation of critical amino acid residues within this region impairs the charge interaction between the region and the acidic C-terminal region of tubulin, and functions as a potential regulatory mechanism controlling microtubule-binding activities of MAP65 (Smertenko *et al.*, 2006).

9.6.4 Phosphorylation of kinesin-like proteins

Expression of a nondegradable cyclin B1 in tobacco BY-2 cells inhibited transformation of the anaphase midzone microtubule array into a phragmoplast (Weingartner *et al.*, 2004), and CDKA:2-GFP was distributed as a narrow equatorial band, resembling the late PPB, in premitotic cells, on the metaphase spindle, and the midzone of the anaphase spindle and phragmoplast (Weingartner *et al.*, 2001), indicating a CDK-dependent phosphorylation-based control of microtubule dynamics during the M phase. *Arabidopsis* kinesin-like proteins, KCA1 and KCA2, were isolated by their ability to bind CDKA;1 and a geminivirus replication protein in yeast two-hybrid screens (De Veylder *et al.*, 1997; Kong and Hanley-Bowdoin, 2002). These kinesins are unique to plants and possess an N-terminal motor domain and a neck sequence characteristic of minus-end-directed kinesins. A central region downstream of the motor domain interacted with CDKA;1 and a C-terminal tail region of KCA and contained three putative CDK phosphorylation sites (Vanstraelen *et al.*, 2004). When a fragment containing this interaction region was expressed in insect cells, it was phosphorylated, but the phosphorylation was inhibited by treatment with olomoucine (Kong and Hanley-Bowdoin, 2002). When two of these three putative CDK phosphorylatable Ser/Thr residues were mutated simultaneously to a Glu that would mimic the phosphorylated residue, the mutated fragment no longer interacted with either CDKA;1 or the C-terminal region in a two-hybrid assay (Vanstraelen *et al.*, 2004). It has been proposed that the non-phosphorylated KCA has a compact folded conformation. Phosphorylation by CDKA;1 might release the inactive configuration to assume one that is ready for the interaction with a transporting cargo. GFP-KCA1 accumulated at the midline of expanding phragmoplasts in tobacco BY-2 cells, and this localization was not affected by treatment with phosphorylation inhibitors, suggesting that modulation of the phosphorylation status of KCA was not crucial for the interaction with microtubules (Vanstraelen *et al.*, 2004).

A phylogenetic analysis of kinesin motor domains shows that KCA1 and KCA2 form a distinct subclass which includes *Arabidopsis* kinesin-like calmodulin-binding protein (KCBP). KCBP is a minus-end-directed motor (Song *et al.*, 1997), is unique among all known kinesins in having a myosin tail homology-4 region in the N-terminal tail and a calmodulin-binding region following the motor domain, and has a putative CDK consensus phosphorylation site. A plant-specific protein kinase interacted specifically with KCBP mainly at the N-terminal tail region (Day *et al.*, 2000). The catalytic domain of the kinase phosphorylated itself but not KCBP. It is not known whether KCBP or a KCBP-associated protein is regulated by phosphorylation, or whether the interacting kinase is transported by KCBP on the microtubule track.

9.7 Calcium

Free cytoplasmic Ca^{2+} levels fluctuate during the cell cycle and after exposure to environmental stimuli. In stamen hair cells of *Tradescantia*, the free Ca^{2+} level increased transiently after the onset of anaphase, and declined during cytokinesis (Hepler and Callaham, 1987). When carrot protoplasts were exposed to culture media with high levels of Ca^{2+} for 15 min, the free cytoplasmic Ca^{2+} increased from a basal level of approximately 90 nM to a level greater than 600 nM and cortical microtubules largely disassembled (Fisher *et al.*, 1996). Experiments using carrot protoplast ghosts suggest that proteins that were associated with microtubules and were capable of interaction with a Ca^{2+}/calmodulin complex were responsible for the Ca^{2+}-dependent destabilization of cortical microtubules (Cyr, 1991).

Elongation factor-1α (EF-1α) is a highly abundant, ubiquitous G-protein involved in protein translation. EF-1α is also reported to interact with components of signal transduction pathways and the microtubule cytoskeleton. Carrot EF-1α bound, bundled, stabilized, and promoted the assembly of microtubules in vitro in a Ca^{2+}/calmodulin-dependent manner (Durso and Cyr,

1994; Moore *et al.*, 1998). When expressed transiently in fava bean leaf epidermal cells, GFP-EF1 α associated with microtubules only after incubation in weak, lipophilic organic acids; the treatment with a calmodulin inhibitor drug, W7, was ineffective (Moore and Cyr, 2000). Genetic and molecular evidence is required to show whether EF-1 α regulates microtubule functions in plant cells, and whether Ca^{2+}/calmodulin is involved in its regulation.

Kinesin-like KCBP is a member of the Ncd subfamily of minus-end-directed kinesin motor proteins found in all plants and in some animals, and is required for cytokinesis and trichome morphogenesis (Oppenheimer *et al.*, 1997; Vos *et al.*, 2000). In KCBP, a calmodulin-binding region is located at the C-terminal extension of the motor core. Biochemical and structural studies indicated that Ca^{2+}/calmodulin inhibits the binding of KCBP to microtubules by blocking the microtubule-binding sites of KCBP (Narasimhulu and Reddy, 1998; Vinogradova *et al.*, 2004). The calmodulin-binding region in KCBP also associated with KIC, a novel Ca^{2+}-binding protein with one EF-hand motif (Reddy *et al.*, 2004). Interestingly, the microtubule-stimulated ATPase activity of KCBP was inhibited by KIC at lower Ca^{2+} concentrations than by calmodulin, which possessed four EF-hand motifs. Overexpression of KIC in *Arabidopsis* trichomes resulted in reduced branch numbers, as seen in *kcbp* mutants, suggesting that KIC regulates the activity of KCBP in response to changes in cytosolic Ca^{2+}.

9.8 Conclusion

In this chapter, we have described how the proteins of the cytoskeleton are integrated with cytoskeletal organization/dynamics and signal transduction cascades. For the most part the pathways involved are similar to their animal counterparts but significant differences in the way the plant proteins are regulated and utilized are apparent. This is perhaps not too surprising because the modes of development and the environmental cues are different for animals and plants. In particular, plant development is coordinated by hormone signaling pathways. As plant cell morphogenesis is dependent on the cytoskeleton the links between hormone signaling and cytoskeletal regulation, although suggested for auxin in particular, remain unknown. The integration of the cytoskeleton with hormone signaling and plant development is likely to be the next stage in this research.

Acknowledgments

We thank S. Hasezawa and A. Yoneda (University of Tokyo) for providing original artwork for Fig. 9.1. We also thank Andrei Smertenko and Michael Deeks (University of Durham) for comments on this manuscript.

References

Abdrakhamanova, A., Wang, Q.Y., Khokhlova, L. and Nick, P. (2003) Is microtubule disassembly a trigger for cold acclimation? *Plant Cell Physiol*, **44**, 676–686.

Alberts, A.S. (2001) Identification of a carboxyl-terminal diaphanous-related formin homology protein autoregulatory domain. *J Biol Chem*, **276**, 2824–2830.

Allwood, E.G., Anthony, R.G., Smertenko, A.P., Reichelt, S., Drobak, B.K., Doonan, J.H., Weeds, A.G. and Hussey, P.J. (2002) Regulation of the pollen-specific actin-depolymerizing factor LIADF1. *Plant Cell*, **14** (11), 2915–2927.

Allwood, E.G., Smertenko, A.P. and Hussey, P.J. (2001) Phosphorylation of plant actin-depolymerising factor by calmodulin-like domain protein kinase. *FEBS Lett*, **499**, 97–100.

Arber, S., Barbayannis, F.A., Hanser, H., Schneider, C., Stanyon, C.A., Bernard, O. and Caroni, P. (1998) Regulation of actin dynamics through phosphorylation of cofilin by LIM-kinase. *Nature*, **393**, 805–809.

Austin, J.R., II, Seguí-Simarro, J.M. and Staehelin, L.A. (2005) Quantitative analysis of changes in spatial distribution and plus-end geometry of microtubules involved in plant-cell cytokinesis. *J Cell Sci*, **118**, 3895–3903.

Balcer, H.I., Goodman, A.L., Rodal, A.A., Smith, E., Kugler, J., Heuser, J.E. and Goode, B.L. (2003) Coordinated regulation of actin filament turnover by a high-molecular-weight rv2/CAP complex, cofilin, profilin, and Aip1. *Curr Biol*, **13**, 2159–2169.

Barrero, R.A., Umeda, M., Yamamura, S. and Uchimiya, H. (2002) *Arabidopsis* CAP regulates the actin cytoskeleton necessary for plant cell elongation and division. *Plant Cell*, **14**, 149–163.

Baskin, T.I. and Wilson, J.E. (1997) Inhibitors of protein kinases and phosphatases alter root morphology and disorganize cortical microtubules. *Plant Physiol*, **113**, 493–502.

Basu, D., El-Assal Sel, D., Le, J., Mallery, E.L. and Szymanski, D.B. (2004) Interchangeable functions of *Arabidopsis* PIROGI and the human WAVE complex subunit SRA1 during leaf epidermal development. *Development*, **131**, 4345–4355.

Basu, D., Le, J., El-Essal Sel, D., Huang, S., Zhang, C., Mallery, E.L., Koliantz, G., Staiger, C.J. and Szymanski, D.B. (2005) Distorted 3/SCAR2 is a putative *Arabidopsis* WAVE complex subunit that activates the Arp2/3 complex and is required for epidermal morphogenesis. *Plant Cell*, **17**, 502–524.

Berken, A. (2006) ROPs in the spotlight of plant signal transduction. *Cell Motil Life Sci*, **63**, 2446–2459.

Blagg, S.L. and Insall, R.H. (2004a) Control of SCAR activity in Dictyostelium discoideum. *Biochem Soc Trans*, **32**, 1113–1114.

Blagg, S.L. and Insall, R.H. (2004b) Solving the WAVE function. *Nat Cell Biol*, **6**, 279–281.

Blagg, S.L., Stewart, M., Sambles, C. and Insall, R.H. (2003) PIR121 regulates pseudopod dynamics and SCAR activity in Dictyostelium. *Curr Biol*, **13**, 1480–1487.

Bogdan, S., Grewe, O., Strunk, M., Mertens, A. and Klambt, C. (2004) Sra-1 interacts with kette and wasp and is required for neuronal and bristle development in Drosophila. *Development*, **131**, 3981–3989.

Bogdan, S. and Klambt, C. (2003) Kette regulates actin dynamics and genetically interacts with wave and wasp. *Development*, **130**, 4427–4437.

Brembu, T., Winge, P., Bones, A.M. and Yang, Z. (2006) A RHOse by any other name: a comparative analysis of animal and plant Rho GTPases. *Cell Res*, **16**, 435–445.

Camilleri, C., Azimzadeh, J., Pastuglia, M., Bellini, C., Grandjean, O. and Bouchez, D. (2002) The *Arabidopsis TONNEAU2* gene encodes a putative novel protein phosphatase 2A regulatory subunit essential for the control of the cortical cytoskeleton. *Plant Cell*, **14**, 833–845.

Carlier, M.F., Laurent, V., Santolini, J., Melki, R., Didry, D., Xia, G.X., Hong, Y., Chua, N.H. and Pantaloni, D. (1997) Actin depolymerizing factor (ADF/cofilin) enhances the rate of filament turnover: implication in actin-based motility. *J Cell Biol*, **136**, 1307–1322.

Chaudhry, F., Guerin, C., von Witsch, M., Blanchoin, L. and Staiger, C.J. (2007) Identification of *Arabidopsis* cyclase-associated protein1 as the first nucleotide exchange factor for plant actin. *Mol Biol Cell*, **18**, 3002–3014.

Cheung, A.Y. and Wu, H.M. (2004) Overexpression of an *Arabidopsis* formin stimulates supernumerary actin cable formation from pollen tube cell membrane. *Plant Cell*, **16**, 257–269.

Cvrckova, F., Novotny, M., Pickova, D. and Zarsky, V. (2004) Formin homology 2 domains occur in multiple contexts in angiosperms. *BMC Genomics*, **5**, 44.

Cyr, R.J. (1991) Calcium/calmodulin affects microtubule stability in lysed protoplasts. *J Cell Sci*, **100**, 311–317.

Day, I.S., Miller, C., Golovkin, M. and Reddy, A.S.N. (2000) Interaction of a kinesin-like calmodulin-binding protein with a protein kinase. *J Biol Chem*, **275**, 13737–13745.

Deeks, M.J. and Hussey, P.J. (2005) Arp2/3 and SCAR: plants move to the fore. *Nat Rev Mol Cell Biol*, **6** (12), 954–964.

Deeks, M.J., Hussey, P.J. and Davies, B. (2002) Formins: intermediates in signal-transduction cascades that affect cytoskeletal reorganization. *Trends Plant Sci*, **7**, 492–498.

Deeks, M.J., Kaloriti, D., Davies, B., Malho, R. and Hussey, P.J. (2004) *Arabidopsis* NAP1 is essential for Arp2/3-dependent trichome morphogenesis. *Curr Biol*, **14**, 1410–1414.

Deeks, M.J., Rodrigues, C., Dimmock, S., Ketelaar, T., Maciver, S.K., Malho, R. and Hussey, P.J. (2007) *Arabidopsis* CAP1 a key regulator of actin organisation and development. *J Cell Sci*, **120**, 2609–2618.

De Veylder, L., Segers, G., Glab, N., Van Montagu, M. and Inze, D. (1997) Identification of proteins interacting with the *Arabidopsis* Cdc2aAt protein. *J Exp Bot*, **48**, 2113–2114.

Djakovic, S., Dyachok, J., Burke, M., Frank, M.J. and Smith, L.G. (2006) BRICK1/HSPC300 functions with SCAR and the ARP2/3 complex to regulate epidermal cell shape in *Arabidopsis*. *Development*, **133** (6), 1091–1100.

Durso, N.A. and Cyr, R.J. (1994) A calmodulin-sensitive interaction between microtubules and a higher plant homolog of elongation factor-1α. *Plant Cell*, **6**, 893–905.

El-Assal Sel, D., Le, J., Basu, D., Mallery, E.L. and Szymanski, D.B. (2004) *Arabidopsis* GNARLED encodes a NAP125 homolog that positively regulates ARP2/3. *Curr Biol*, **14**, 1405–1409.

Fan, X., Hou, J., Chen, X., Chaudhry, F., Staiger, C.J. and Ren, H. (2004) Identification and characterization of a Ca^{2+}-dependent actin filament-severing protein from lily pollen. *Plant Physiol*, **136**, 3979–3989.

Fedor-Chaiken, M., Deschenes, R.J. and Broach, J.R. (1990) *SRV2*, a gene required for RAS activation of adenylate cyclase in yeast. *Cell*, **61**, 329–340.

Field, J., Vojtek, A., Ballester, R., Bolger, G., Colicelli, J., Ferguson, K., Gerst, J., Kataoka, T., Michaeli, T., Powers, S., Riggs, M., Rodgers, L., Wieland, I., Wheland, B. and

Wigler M. (1990) Cloning and characterization of CAP, the *S. cerevisiae* gene encoding the 70 kD adenylyl cyclase-associated protein. *Cell*, **61**, 319–327.

Fisher, D.D., Gilroy, S. and Cyr, R.J. (1996) Evidence for opposing effects of calmodulin on cortical microtubules. *Plant Physiol*, **112**, 1079–1087.

Foissner, I., Groling, F. and Obermeyer, G. (2002) Reversible protein phosphorylation regulates the dynamic organization of the pollen tubule cytoskeleton: effects of calyculin A and okadaic acid. *Protoplasma*, **220**, 1–15.

Frank, M., Egile, C., Dyachok, J., Djakovic, S., Nolasco, M., Li, R. and Smith, L.G. (2004) Activation of Arp2/3 complex-dependent actin polymerization by plant proteins distantly related to Scar/WAVE. *Proc Natl Acad Sci USA*, **101**, 16379–16384.

Fu, Y., Gu, Y., Zheng, Z., Wasteneys, G. and Yang, Z. (2005) *Arabidopsis* interdigitating cell growth requires two antagonistic pathways with opposing action on cell morphogenesis. *Cell*, **120**, 687–700.

Gibbon, B.C., Kovar, D.R. and Staiger, C.J. (1999) Latrunculin B has different effects on pollen germination and tube growth. *Plant Cell*, **11**, 2349–2363.

Glenney, J.R.J., Bretscher, A. and Weber, K. (1980) Calcium control of the intestinal microvillus cytoskeleton: its implications for the regulation of microfilament organizations. *Proc Natl Acad Sci USA*, **77**, 6458–6462.

Gottwald, U., Brokamp, R., Karakesisoglou, I., Schleicher, M. and Noegel, A.A. (1996) Identification of a cyclase-associated protein (CAP) homologue in Dictyostelium discoideum and characterization of its interaction with actin. *Mol Biol Cell*, **7**, 261–272.

Gungabissoon, R.A., Jiang, C.J., Drobak, B.J., Maciver, S.K. and Hussey, P.J. (1998) Interaction of maize actin-depolymerizing factor with F-actin and phosphoinositides and its inhibition of plant phospholipase C. *Plant J*, **16**, 689–696.

Hasezawa, S. and Nagata, T. (1992) Okadaic acid as a probe to analyze the cell cycle progression in plant cells. *Botanica Acta*, **105**, 63–69.

Hepler, P.K. and Callaham, D.A. (1987) Free calcium increases during anaphase in stamen hair cells of *Tradescantia*. *J Cell Biol*, **105**, 2137–2143.

Higgs, H.N. and Pollard, T.D. (2001) Regulation of actin filament network formation through ARP2/3 complex: activation by a diverse array of proteins. *Ann Rev Biochem*, **70**, 649–676.

Hiraga, A. and Tamura, S. (2000) Protein phosphatase 2A is associated in an inactive state with microtubules through 2A1-specific interaction with tubulin. *Biochem J*, **346**, 433–439.

Holt, M.R. and Koffer, A. (2001) Cell motility: proline-rich proteins promote protrusions. *Trends Cell Biol*, **11** (1), 38–46.

Huang, S., Blanchoin, L., Chaudhry, F., Franklin-Tong, V.E. and Staiger, C.J. (2004) A gelsolin-like protein from Papaver rhoeas pollen (PrABP80) stimulates calcium-regulated severing and depolymerization of actin filaments. *J Biol Chem*, **279**, 23364–23375.

Huang, S., Blanchoin, L., Kovar, D.R. and Staiger, C.J. (2003) *Arabidopsis* capping protein (AtCP) is a heterodimer that regulates assembly at the barbed ends of actin filaments. *J Biol Chem*, **278**, 44832–44842.

Huang, S., Gao, L., Blanchoin, L. and Staiger, C.J. (2006) Heterodimeric capping protein from *Arabidopsis* is regulated by phosphatidic acid. *Mol Biol Cell*, **17** (4), 1946–1958.

Huang, S., Robinson, R.C., Gao, L.Y., Matsumoto, T., Brunet, A., Blanchoin, L. and Staiger, C.J. (2005) *Arabidopsis* VILLIN1 generates actin filament cables that are resistant to depolymerization. *Plant Cell*, **17**, 486–501.

Hubberstey, A.V. and Mottillo, E.P. (2002) Cyclase-associated proteins: capacity for linking signal transduction and actin polymerization. *FASEB J*, **16**, 487–499.
Hudson, A.M. and Cooley, L. (2002) A subset of dynamic actin rearrangements in Drosophila requires the Arp2/3 complex. *J Cell Biol*, **156**, 677–687.
Hummel, T., Leifker, K. and Klambt, C. (2000) The Drosophila HEM-2/NAP1 homolog KETTE controls axonal pathfinding and cytoskeletal organization. *Genes Dev*, **14**, 863–873.
Hussey, P.J., Allwood, E.G. and Smertenko, A.P. (2002a) Actin-binding proteins in the *Arabidopsis* genome database: properties of functionally distinct plant actin-depolymerizing factors/cofilins. *Philos Trans R Soc Lond B*, **357**, 791–798.
Hussey, P.J., Hawkins, T.J., Igarashi, H., Kaloriti, D. and Smertenko, A. (2002b) The plant cytoskeleton: recent advances in the study of the plant microtubule-associated proteins MAP-65, MAP-190 and the *Xenopus* MAP215-like protein, MOR1. *Plant Mol Biol*, **50**, 915–924.
Hussey, P.J., Ketelaar, T. and Deeks M.J. (2006) Control of the actin cytoskeleton in plant cell growth. *Annu Rev Plant Biol*, **57**, 109–125.
Ingouff, M., Fitz Gerald, J.N., Guerin, C., Robert, H., Sorensen, M.B., Van Damme, D., Geelen, D., Blanchoin, L. and Berger, F. (2005) Plant formin AtFH5 is an evolutionarily conserved actin nucleator involved in cytokinesis. *Nat Cell Biol*, **7**, 374–380.
Innocenti, M., Zucconi, A., Disanza, A., Frittoli, E., Areces, L.B., Steffen, A., Stradal, T.E., Di Fiore, P.P., Carlier, M.F. and Scita. G. (2004) Abi1 is essential for the formation and activation of a WAVE2 signalling complex. *Nat Cell Biol*, **6**, 319–327.
Janmey, P.A. and Matsudaira, P.T. (1988) Functional comparison of villin and gelsolin. Effects of Ca^{2+}, KCl, and polyphosphoinositides. *J Biol Chem*, **263**, 16738–16743.
Janson, M.E., Loughlin, R., Loïodice, I., Fu, C., Brunner, D., Nédélec, F.J. and Tran, P.T. (2007) Crosslinkers and motors organize dynamic microtubules to form stable bipolar arrays in fission yeast. *Cell*, **128**, 357–368.
Katsuta, J. and Shibaoka, H. (1992) Inhibition by kinase inhibitors of the development and the disappearance of the PREPROPHASE BAND of microtubules in tobacco BY-2 cells. *J Cell Sci*, **103**, 397–405.
Kawamukai, M., Gerst, J., Field, J., Riggs, M., Rodgers, L., Wigler, M. and Young, D. (1992) Genetic and biochemical analysis of the adenylyl cyclase-associated protein, cap, in Schizosaccharomyces pombe. *Mol Biol Cell*, **3**, 167–180.
Ketelaar, T., Allwood, E.G., Anthony, R., Voigt, B., Menzel, D. and Hussey, P.J. (2004) The actin-interacting protein AIP1 is essential for actin organization and plant development. *Curr Biol*, **14** (2), 145–149.
Ketelaar, T., Allwood, E.G. and Hussey, P.J. (2007) Actin organisation and root hair development are disrupted by ethanol induced overexpression of *Arabidopsis* actin interacting protein 1 (AIP1). *New Phytol*, **174** (1), 57–62.
Kong, L.-J. and Hanley-Bowdoin, L. (2002) A geminivirus replication protein interacts with a protein kinase and a motor protein that display different expression patterns during plant development and infection. *Plant Cell*, **14**, 1817–1832.
Kovar, D.R., Drøbak, B.K. and Staiger, C.J. (April 2000) Maize profilin isoforms are functionally distinct. *Plant Cell*, **12** (4), 583–598.
Kovar, D.R., Kuhn, J.R., Tichy, A.L. and Pollard, T.D. (2003) The fission yeast cytokinesis formin Cdc12p is a barbed end actin filament capping protein gated by profilin. *J Cell Biol*, **161**, 875–887.

Krupa, A., Anamika, and Srinivasan, N. (2006) Genome-wide comparative analyses of domain organization of repertoires of protein kinases of *Arabidopsis thaliana* and *Oryza sativa*. *Gene*, **380**, 1–13.

Kunda, P., Craig, G., Dominguez, V. and Baum, B. (2003) Abi, Sra1 and Kette control the stability and localization of SCAR/WAVE to regulate the formation of actin-based protrusions.*Curr Biol*, **13**, 1867–1875.

Le, J., El-Assal Sel, D., Basu, D., Saad, M.E. and Szymanski, D.B. (2003) Requirements for *Arabidopsis* ATARP2 and ATARP3 during epidermal development. *Curr Biol*, **13**, 1341–1347.

Le, J., Mallery, E.L., Zhang, C., Brankle, S. and Szymanski, D.B. (2006) *Arabidopsis* BRICK1/HSPC300 is an essential WAVE-complex subunit that selectively stabilizes the Arp2/3 activator SCAR2. *Curr Biol*, **16** (9), 895–901.

Li, S., Blanchoin, L., Yang, Z. and Lord, E.M. (2003) The putative *Arabidopsis* arp2/3 complex controls leaf cell morphogenesis. *Plant Physiol*, **132**, 2034–2044.

Limmongkon, A., Giuliani, C., Valenta, R., Mittermann, I., Heberle-Bors, E. and Wilson, C. (2004) MAP kinase phosphorylation of plant profiling. *Biochem Biophys Res Commun*, **324** (1), 382–386.

Liu, S., Sun, J.-P., Zhou, B. and Zhang, Z.-Y. (2006) Structural basis of docking interactions between ERK2 and MAP kinase phosphatase 3. *Proc Natl Acad Sci USA*, **103**, 5326–5331.

Lovy-Wheeler, A., Kunkel, J.G., Allwood, E.G., Hussey, P.J. and Hepler, P.K. (2006) Oscillatory increases in alkalinity anticipate growth and may regulate actin dynamics in pollen tubes of lily. *Plant Cell*, **18** (9), 2182–2193.

Lu, J. and Pollard, T.D. (2001) Profilin binding to poly-L-proline and actin monomers along with ability to catalyze actin nucleotide exchange is required for viability of fission yeast. *Mol Biol Cell*, **12**, 1161–1175.

Luan, S. (2003) Protein phosphatases in plants. *Ann Rev Plant Biol*, **54**, 63–92.

Mao, G., Chan, J., Calder, G., Doonan, J.H. and Lloyd, C.W. (2005) Modulated targeting of GFP-AtMAP65-1 to central spindle microtubules during division. *Plant J*, **43**, 469–478.

Mathur, J., Mathur, N., Kirik, V., Kernebeck, B., Srinivas, B.P. and Hülskamp, M. (2003a) *Arabidopsis* CROOKED encodes for the smallest subunit of the ARP2/3 complex and controls cell shape by region specific fine F-actin formation. *Development*, **130**, 3137–3146.

Mathur, J., Mathur, N., Kernebeck, B. and Hulskamp, M. (2003b) Mutations in actin-related proteins 2 and 3 affect cell shape development in *Arabidopsis*. *Plant Cell*, **15**, 1632–1645.

Mathur, J., Spielhofer, P., Kost, B. and Chua, N. (1999) The actin cytoskeleton is required to elaborate and maintain spatial patterning during trichome cell morphogenesis in *Arabidopsis thaliana*. *Development*, **126**, 5559–5568.

Mattila, P.K., Quintero-Monzon, O., Kugler, J., Moseley, J.B., Almo, S.C., Lappalainen, P. and Goode, B.L. (2004) A high-affinity interaction with ADP-actin monomers underlies the mechanism and in vivo function of Srv2/cyclase-associated protein. *Mol Biol Cell*, **15**, 5158–5171.

Mao, T.L., Jin, L.F., Li, H., Liu, B. and Yuan, M. (2005) Two microtubule-associated proteins of the arabidopsis MAP65 family function differently on microtubules. *Plant Phys* **138**, 654–662.

McClinton, R.S. and Sung, Z.R. (1997) Organization of cortical microtubules at the plasma membrane in *Arabidopsis.Planta*, **201**, 252–260.

Michelot, A., Derivery, E., Paterski-Boujemaa, R., Guerin, C., Huang, S., Parcy, F., Staiger, C.J. and Blanchoin, L. (2006) A novel mechanism for the formation of actin-filament bundles by a nonprocessive formin. *Curr Biol*, **16** (19), 1924–1930.

Michelot, A., Guerin, C., Huang, S., Ingouff, M., Richard, S., Rodiuc, N., Staiger, C.J. and Blanchoin, L. (2005) The formin homology 1 domain modulates the actin nucleation and bundling activity of *Arabidopsis* FORMIN1. *Plant Cell*, **17**, 2296–2313.

Mineyuki, Y. (2007) Plant microtubule studies: past and present. *J Plant Res*, **120**, 45–51.

Mizuno, K. (1992) Induction of cold stability of microtubules in cultured tobacco cells. *Plant Physiol*, **100**, 740–748.

Mizuno, K. (1994) Inhibition of gibberellin-induced elongation, reorientation of cortical microtubules and change of isoform of tubulin in epicotyl segments of azuki bean by protein kinase inhibitors. *Plant Cell Physiol*, **35**, 1149–1157.

Mollinari, C., Kleman, J.-P., Jiang, W., Schoehn, G., Hunter, T. and Margolis, R.L. (2002) PRC1 is a microtubule binding and bundling protein essential to maintain the mitotic spindle midzone. *J Cell Biol*, **157**, 1175–1186.

Moore, R.C. and Cyr, R.J. (2000) Association between elongation factor-1α and microtubules in vivo is domain dependent and conditional. *Cell Motil Cytoskeleton*, **45**, 279–292.

Moore, R.C., Durso, N.A. and Cyr, R.J. (1998) Elongation factor 1α stabilizes microtubules in a calcium/calmodulin-dependent manner. *Cell Motil Cytoskeleton*, **41**, 168–180.

Moutinho, A., Hussey, P.J., Trewavas, A.J. and Malhó, R. (2001) cAMP acts as a second messenger in pollen tube growth and reorientation. *PNAS*, **98** (18), 10481–10486.

Naoi, K. and Hashimoto, T. (2004) A semidominant mutation in an *Arabidopsis* mitogen-activated protein kinase phosphatase-like gene compromises cortical microtubule organization. *Plant Cell*, **16**, 1841–1853.

Narasimhulu, S.B. and Reddy, A.S.N. (1998) Characterization of microtubule binding domains in the *Arabidopsis* kinesin-like calmodulin binding protein. *Plant Cell*, **10**, 957–965.

Nibau, C., Wu, H. and Cheung, A.Y. (2006) RAC/ROP GTPases: "hubs" for signal integration and diversification in plants. *Trends Plant Sci*, **11**, 309–315.

Nick, P. (1998) Signaling to the microtubule cytoskeleton in plants. *Int Rev Cytol*, **184**, 33–80

Nishida, Y., Shima, F., Sen, H., Tanaka, Y., Yanagihara, C., Yamawaki-Kataoka, Y., Kariya, K. and Kataoka, T. (1998) Coiled-coil interaction of N-terminal 36 residues of cyclase-associated protein with adenylyl cyclase is sufficient for its function in *Saccharomyces cerevisiae* ras pathway. *J Biol Chem*, **273** (43), 28019–28024.

Noegel, A.A., Blau-Wasser, R., Sultana, H., Muller, R., Israel, L., Schleicher, M., Patel, H. and Weijer, C.J. (2004) The cyclase-associated protein CAP as regulator of cell polarity and cAMP signaling in Dictyostelium. *Mol Biol Cell*, **15**, 934–945.

Oppenheimer, D.G., Pollock, M.A., Vacik, J., Szymanski, D.B., Ericson, B., Feldmann, K. and Marks, D. (1997) Essential role of a kinesin-like protein in *Arabidopsis* trichome morphogenesis. *Proc Natl Acad Sci USA*, **94**, 6261–6266.

Perelroizen, I., Didry, D., Christensen, H., Chua, N.H. and Carlier, M.F. (1996) Role of nucleotide exchange and hydrolysis in the function of profilin in actin assembly. *J Biol Chem*, **271** (21), 12302–12309.

Petersen, J., Nielson, O., Egel, R. and Hagan, I.M. (1998) FH3, a domain found in formins, targets the fission yeast formin Fus1 to the projection tip during conjugation. *J Cell Biol*, **141**, 1217–1228.

Pollard, T.D., Blanchoin, L. and Mullins, R.D. (2000) Molecular mechanisms controlling actin filament dynamics in nonmuscle cells. *Ann Rev Biophys Biom*, **29**, 545–576.

Pope, B., Way, M., Matsudaira, P.T. and Weeds, A. (1994) Characterisation of the F-Actin binding domains of villin: classification of F-Actin binding proteins into two groups according to their binding sites on actin. *FEBS Lett*, **338**, 58–62.

Pruyne, D., Evangelista, M., Yang, C., Bi, E., Zigmond, S., Bretscher, A. and Boone, C. (2002) Role of formins in actin assembly: nucleation and barbed-end association. *Science*, **297**, 612–615.

Quettier, A.-L., Bertrand, C., Habricot, Y., Miginiac, E., Agnes, C., Jeannette, E. and Maldiney, R. (2006) The *phs1-3* mutation in a putative dual-specificity protein tyrosine phosphatase gene provokes hypersensitive responses to abscisic acid in *Arabidopsis thaliana*.*Plant J*, **47**, 711–719.

Reddy, V.S., Day, I.S., Thomas, T. and Reddy, A.S.N. (2004) KIC, a novel Ca^{2+} binding protein with one EF-hand motif, interacts with a microtubule motor protein and regulates trichome morphogenesis. *Plant Cell*, **16**, 185–200.

Rogers, S.L., Wiedemann, U., Stuurman, N. and Vale, R.D. (2003) Molecular requirements for actin-based lamella formation in Drosophila S2 cells. *J Cell Biol*, **162**, 1079–1088.

Romero, S., Le Clainche, C., Didry, D., Egile, C., Pantaloni, D. and Carlier, M.F. (2004) Formin is a processive motor that requires profilin to accelerate actin assembly and associated ATP hydrolysis. *Cell*, **119**, 419–429.

Saedler, R., Zimmermann, I., Mutondo, M. and Hulskamp, M. (2004) The *Arabidopsis KLUNKER* gene controls cell shape changes and encodes the AtSRA1 homolog. *Plant Mol Biol*, **56**, 775–782.

Sasabe, M. and Machida, Y. (2006) MAP65: a bridge linking a MAP kinase to microtubule turnover. *Curr Opin Plant Biol*, **9**, 563–570.

Sasabe, M., Soyano, T., Takahashi, Y., Sonobe, S., Igarashi, H., Itoh, T.J., Hidaka, M. and Machida, Y. (2006) Phosphorylation of NtMAP65-1 by a MAP kinase downregulates its activity of microtubule bundling and stimulates progression of cytokinesis of tobacco cells. *Genes Dev*, **20**, 1004–1014.

Sawa, M., Suetsugu, S., Sugimoto, A., Miki, H., Yamamoto, M. and Takenawa, T. (2003) Essential role of the *C. elegans* Arp2/3 complex in cell migration during ventral enclosure. *J Cell Sci*, **116**, 1505–1518.

Schenck, A. Qurashi, A., Carrera, P., Bardoni, B., Diebold, C., Schejter, E., Mandel, J.L. and Giangrande, A. (2004) WAVE/SCAR, a multifunctional complex coordinating different aspects of neuronal connectivity. *Dev Biol*, **274**, 260–270.

Shibaoka, H. (1994) Plant hormone-induced changes in the orientation of cortical microtubules. *Ann Rev Plant Physiol Plant Mol Biol*, **45**, 527–544.

Shoji, T., Suzuki, K., Abe, T., Kaneko, Y., Shi, H., Zhu, J.-K., Rus, A., Hasegawa, P.M. and Hashimoto, T. (2006) Salt stress affects cortical microtubule organization and helical growth in *Arabidopsis*. *Plant Cell Physiol*, **47**, 1158–1168.

Sivaguru, M., Pike, S., Gassmann, W. and Baskin, T.I. (2003) Aluminum rapidly depolymerizes cortical microtubules and depolarizes the plasma membrane: evidence that these responses are mediated by a glutamate receptor. *Plant Cell Physiol*, **44**, 667–675.

Smertenko, A., Sleh, N., Igarashi, H., Mori, H., Hauser-Hahn, I., Jiang, C.-J., Sonobe, S., Lloyd, C.W. and Hussey, P.J. (2000) A new class of microtubule-associated proteins in plants. *Nat Cell Biol*, **2**, 750–753.

Smertenko, A.P., Chang, H.-Y., Sonobe, S., Fenyk, S.I., Weingartner, M., Bögre, L. and Hussey, P.J. (2006) Control of the AtMAP65-1 interaction with microtubules through the cell cycle. *J Cell Sci*, **119**, 3227–3237.

Smertenko, A.P., Chang, H.-Y., Wagner, V., Kaloriti, D., Fenyk, S., Sonobe, S., Lloyd, C., Hauser, M.-T. and Hussey, P.J. (2004) The *Arabidopsis* microtubule-associated protein AtMAP65-1: molecular analysis of its microtubule bundling activity. *Plant Cell*, **16**, 2035–2047.

Smertenko, A.P., Jiang, C.J., Simmons, N.J., Weeds, A.G., Davies, D.R. and Hussey, P.J. (1998) Ser6 in the maize actin-depolymerizing factor, ZmADF3, is phosphorylated by a calcium-stimulated protein kinase and is essential for the control of functional activity. *Plant J*, **14**, 187–193.

Snowman, B.N., Kovar, D.R., Shevchenko, G., Franklin-Tong, V.E. and Staiger, C.J. (2002) Signal-mediated depolymerization of actin in pollen during the self-incompatibility response. *Plant Cell*, **14** (10), 2613–2626.

Song, H., Golovkin, M., Reddy, A.S.N. and Endow, S.A. (1997) *In vitro* motility of AtKCBP, a calmodulin-binding kinesin protein of *Arabidopsis*. *Proc Natl Acad Sci USA*, **94**, 322–327.

Sonobe, S. (1990) ATP-dependent depolymerization of cortical microtubules by an extract in tobacco BY-2 cells. *Plant Cell Physiol*, **31**, 1147–1153.

Staiger, C.J. and Blanchoin, L. (2006) Actin dynamics: old friends with new stories. *Curr Opin Plant Biol*, **9** (6), 554–562.

Staiger, C.J. and Hussey P.J. (2004) Actin and actin modulating proteins. In: *The Plant Cytoskeleton* (ed P.J. Hussey), pp. 31–80. Blackwell Publishing, Oxford.

Steffen, A., Rottner, K., Ehinger, J., Innocenti, M., Scita, G., Wehland, J. and Stradal, T.E. (2004) Sra-1 and Nap1 link Rac to actin assembly driving lamellipodia formation. *EMBO J*, **23**, 749–759.

Szymanski, D.B., Marks, M.D. and Wick, S.M. (1999) Organized F-actin is essential for normal trichome morphogenesis in *Arabidopsis*. *Plant Cell*, **11**, 2331–2347.

Takemoto, D. and Hardham, A.R. (2004) The cytoskeleton as a regulator and target of biotic interactions in plants. *Plant Physiol*, **124**, 1493–1506.

Tian, G.-W., Smith, D., Gluck, S. and Baskin, T.I. (2004) Higher plant cortical microtubule array analyzed *in vitro* in the presence of the cell wall. *Cell Motil Cytoskeleton*, **57**, 26–36.

Torres-Ruiz, R.A. and Jürgens, G. (1994) Mutations in the *FASS* gene uncouple pattern formation and morphogenesis in *Arabidopsis* development. *Development*, **120**, 2967–2978.

Toshima, J., Toshima, J.Y., Amano, T., Yang, N., Narumiya, S. and Mizuno, K. (2001) Cofilin phosphorylation by protein kinase testicular protein kinase 1 and its role in integrin-mediated actin reorganization and focal adhesion formation. *Mol Biol Cell*, **12**, 1131–1145.

Traas, J., Bellini, C., Nacry, P., Kronenberger, J., Bouchez, D. and Caboche, M. (1995) Normal differentiation patterns in plants lacking microtubular preprophase bands. *Nature*, **375**, 676–677.

Uhrig, J.F., Mutondo, M., Zimmermann, I., Deeks, M.J., Machesky, L.M., Thomas, P., Uhrig, S., Rambke, C., Hussey, P.J. and Hulskamp, M. (2007) The role of *Arabidopsis*

SCAR genes in ARP2-ARP3-dependent cell morphogenesis. *Development*, **134** (5), 967–977.

Van Damme, D., Van Poucke, K., Boutant, E., Ritzenthaler, C., Inze, D. and Geelen, D. (2005) In vivo dynamic and differential microtubule-binding activities of MAP65 proteins. *Plant Phys* **136**, 3956–3967.

Vanstraelen, M., Acosta, J.A.T., De Veylder, L., Inzé, D. and Geelen, D. (2004) A plant-specific subclass of C-terminal kinesins contains a conserved A-type cyclin-dependent kinase site implicated in folding and dimerization. *Plant Physiol*, **135**, 1417–1429.

Vinogradova, M.V., Reddy, V.S., Reddy, A.S.N., Sablin, E.P. and Fletterick, R.J. (2004) Crystal structure of kinesin regulated by Ca^{2+}-calmodulin. *J Biol Chem*, **279**, 23504–23509.

Vos, J.W., Safadi, F., Reddy, A.S.N. and Hepler, P.K. (2000) The kinesin-like calmodulin binding protein is differentially involved in cell division. *Plant Cell*, **12**, 979–990.

Wang, D., Harper, J.F. and Gribskov, M. (2003) Systematic trans-genomic comparison of protein kinases between *Arabidopsis* and *Saccharomyces cerevisiae*. *Plant Physiol*, **132**, 2152–2165.

Watanabe, N., Madaule, P., Reid, T., Ishizaki, T., Watanabe, G., Kakizuka, A., Saito, Y., Nakao, K., Jockusch, B.M. and Narumiya, S. (1997) p140mDia, a mammalian homolog of Drosophila diaphanous, is a target protein for Rho small GTPase and is a ligand for profiling. *EMBO J*, **16**, 3044–3056.

Watanabe, N., Kato, T., Fujita, A., Ishizaki, T. and Narumiya, S. (1999) Cooperation between mDia1 and ROCK in Rho-induced actin reorganization. *Nat Cell Biol*, **1**, 136–143.

Weingartner, M., Binarova, P., Drykova, D., Schweighofer, A., David, J.-P., Heberle-Bors, E., Doonan, J and Bögre, L. (2001) Dynamic recruitment of Cdc2 to specific microtubule structures during mitosis. *Plant Cell*, **13**, 1929–1943.

Weingartner, M., Criqui, M.-C., Mészáros, T., Binarova, P., Schmit, A.-C., Helfer, A., Derevier, A., Erhardt, M., Bögre, L. and Genschik, P. (2004) Expression of a non-degradable cyclin B1 affects plant development and leads to endomitosis by inhibiting the formation of a phragmoplast. *Plant Cell*, **16**, 643–657.

Winter, D.C., Choe, E.Y. and Li, R. (1999) Genetic dissection of the budding yeast Arp2/3 complex: a comparison of the *in vivo* and structural roles of individual subunits. *Proc Natl Acad Sci USA*, **96**, 7288–7293.

Xiang, Y., Huang, X., Wang, T., Zhang, Y., Liu, Q., Hussey, P.J. and Ren, H. (2007) ACTIN BINDING PROTEIN 29 from Lilium pollen plays an important role in dynamic actin remodeling. *Plant Cell*, **19**, 1930–1946.

Yang, Z. and Fu, Y. (2007) ROP/RAC GTPase signalling. *Curr Opin Plant Biol*, **10**, 490–494.

Yi, K., Guo, C., Chen, D., Zhao, B., Yang, B. and Ren, H. (2005) Cloning and functional characterization of a formin-like protein (AtFH8) from *Arabidopsis*. *Plant Physiol*, **138**, 1071–1082.

Yokota, E. and Shimmen, T. (2000) Characterisation of native actin binding proteins from pollen. In: *Actin, a Dynamic Framework for Multiple Plant Cell Functions* (eds C.J. Staiger, F. Baluska and P.W. Barlow), pp. 103–118. Kluwer Academic Publishers, Dordretch, The Neetherlands.

Yokota, E., Tominaga, M., Mabuchi, I., Tsuji, Y., Staiger, C.J., Oiwa, K. and Shimmen, T. (2005) Plant villin, lily P-135-ABP, possesses G-actin binding activity and accelerates the polymerization and depolymerization of actin in a Ca^{2+}-sensitive manner. *Plant Cell Physiol*, **46** (10), 1690–1703.

Yoneda, A. and Hasezawa, S. (2003) Origin of cortical microtubules organized at M/G_1 interface: recruitment of tubulin from phragmoplast to nascent microtubules. *Eur J Cell Biol*, **82**, 461–471.

Zallen, J.A., Cohen, Y., Hudson, A.M., Cooley, L., Wieschaus, E. and Schejter, E.D. (2002) SCAR is a primary regulator of Arp2/3-dependent morphological events in Drosophila. *J Cell Biol*, **156**, 689–701.

Zhang, X., Dyachok, J., Krishnakumar, S., Smith, L.G. and Oppenheimer, D.G. (2005) Irregular Trichome Branch1 in *Arabidopsis* encodes a plant homolog of the actin-related protein 2/3 complex activator scar/WAVE that regulates actin and microtubule organization. *Plant Cell*, **17**, 2314–2326.

Zhou, H.-W., Nussbaumer, C., Chao, Y. and DeLong, A. (2004) Disparate roles of the regulatory A subunit isoforms in *Arabidopsis* protein phosphatase 2A. *Plant Cell*, **16**, 709–722.

Zigmond, S.H., Evangelista, M., Boone, C., Yang, C., Dar, A.C., Sicheri, F., Forkey, J. and Pring, M. (2003) Formin leaky cap allows elongation in the presence of tight capping proteins. *Curr Biol*, **13**, 1820–1823.

Annual Plant Reviews (2008) **33**, 273–306
doi: 10.1111/b.9781405160025.2008.00010.x
www.interscience.wiley.com

Chapter 10

THE PCI COMPLEXES AND THE UBIQUITIN PROTEASOME SYSTEM (UPS) IN PLANT DEVELOPMENT

Yair Halimi and Daniel A. Chamovitz

Department of Plant Sciences, Tel Aviv University, Tel Aviv 69978, Israel

Abstract: Since the discovery of the ubiquitin proteasome system (UPS) in the 1970s, the UPS field has advanced remarkably, culminating in the 2004 Nobel Prize in Chemistry. The idea that energy should be invested to degrade a substrate, whose own synthesis consumed energy in the first place, was revolutionary (Hershko and Tomkins, 1971). In *Arabidopsis*, ~1400 genes encode components of the UPS, constituting approximately 5% of the *Arabidopsis* proteome (Smalle and Vierstra, 2004), eluding to the great complexity of the UPS system. Describing this complexity is daunting and very much a matter of perspective. This chapter will discuss the UPS in plant development from the point of view of the PCI (*P*roteasome, *C*SN, e*I*F3) or "ZOMES" complexes: the 26S proteasome lid, Cop9 signalosome (CSN), and eIF3 (eukaryote initiation factor 3), and in particular from the point of view of the CSN.

Keywords: Cop9 signalosome; proteasome; ubiquitin; protein degradation

10.1 General overview

Cellular protein content (cellular proteome) is obviously the most important and basic feature that defines every aspect of life. The proteome is a function of numerous regulatory mechanisms functioning at multiple levels. While transcriptional control of the proteome is a major focus of research, over the past decade, we are witnessing a paradigm shift to identifying the central role of posttranscriptional mechanisms in controlling the proteome. The novel idea of precise protein degradation as a mechanism that controls the cellular proteome encompasses both accuracy and sensitivity, and allows coordination of many complicated and parallel processes. Most proteins involved in cellular signaling are known to be subject to proteasome-mediated turnover in responses to specific signals. The PCI (*P*roteasome, *C*SN, e*I*F3) multiprotein complexes, the proteasome regulatory lid, Cop9 signalosome (CSN) and eukaryote initiation factor 3 (eIF3), individually and in concert, affect proteome composition (Kim *et al.*, 2001). These complexes were grouped together on the basis of similarities in their subunit composition and sequence, and predicted structure. While subunits between the complexes share little primary sequence similarity, six subunits in each of all three complexes contain a PCI motif, while two subunits carry an MPN (*M*pr1, *P*ad1, *N*-terminal) motif. This conserved six PCI/two MPN structure strongly suggests a common ancestor for these complexes (Aravind and Ponting, 1998; Hofmann and Bucher, 1998). The PCI motif is found almost exclusively in subunits of these complexes, though the presence of a small number of orphan PCI proteins hints at the possible existence of a fourth PCI complex (Kim *et al.*, 2001).

Both the proteasome lid and CSN have established roles in mediating the ubiquitin-dependent degradation of regulatory proteins such as signaling proteins and transcriptional regulators. Both complexes work in conjunction with another family of multi-subunit protein complexes–E3 ubiquitin ligases. The E3 ligase complex determines specificity of the degradation substrate. While the CSN modulates E3 activity, this activity is necessary for proteasome lid to recognize ubiquitylated proteins destined for degradation in the proteasome. eIF3 obviously controls proteome content through regulating protein synthesis, but eIF3 also impinges on the CSN and proteasome. Interestingly, the CSN resembles on one hand to the regulator of protein synthesis (eIF3) and on the other to the mechanism that degrades them (the lid) and thus is found in the interface between protein synthesis and protein degradation.

10.2 The PCI complexes

10.2.1 Proteasome

The proteasome is a huge 2- to 2.5-MDa molecular machine, found in all eukaryote cells, which carries out nonlysosomal protein degradation (Hershko and Ciechanover, 1998). It is localized mainly in the nucleus but can be found in the cytoplasm as well. The proteasome is composed of two major subcomplexes: a 20S core particle (CP) and a 19S regulatory particle (RP; Fig. 10.1a, Color plate 15).

The CP is a cylindrical structure of four stacked heptameric rings; the two central rings are consisted of seven related β subunits each, and the two distal rings are consisted of seven related α subunits (Fig. 10.1a, Color plate 15). The α-subunits' N-terminal extension probably controls protein traffic in and out the CP (Groll, 2000). Crystallographic structure analysis of the yeast and human CP revealed a cavity within the stacked rings. Protein proteolysis occurs in this cavity through the trypsin-like and chemotrypsin-like protease activity harbored in the β1, β2, and β5 subunits (Groll *et al.*, 1997; Christian *et al.*, 2002). This activity is ATP-independent.

The other major subcomplex of the 26S proteasome is the RP, which can be further divided into two smaller subcomplexes, the base and the lid (Glickman *et al.*, 1998b). The RP is found on one or both sides of the CP, with the RP-base interacting with the α-subunit ring of the CP. The base is consisted of nine subunits, six arranged as a ring of AAA-ATPases (RPT1–RPT6), while the other three (RPN1, 2, and 10), which are not ATPases, are arranged outside the AAA-ATPases ring (Fig. 10.1a, Color plate 15; Zwickl and Baumeister, 1999). Since the RP blocks the CP, proteasome substrate recognition occurs through the RP. Once a substrate binds the RP, the substrate is unfolded by the AAA-ATPase ring, the ubiquitin chain is removed by RPN11 (see below) and recycled, and the "naked" protein is directed through the α-ring gate to the CP for proteolysis (Fig. 10.2b, Color plate 16; Hartmann-Petersen *et al.*, 2003).

The lid is consisted of eight non-ATPase subunits (RPN3, 5–9, and 11–12). While the CP can degrade proteins in the absence of the lid, the lid is necessary for proteolysis of ubiquitinated proteins (Glickman *et al.*, 1998a). The lid probably directly interacts with the polyubiquitin chain attached to the substrate, removes the ubiquitin, and funnels the substrate through to the base and CP. The deubiquitination is carried out not only through associated deubiquitylating enzymes, but directly by the lid itself (Glickman *et al.*, 1999; Lam *et al.*, 2002). RPN11, an MPN subunit, contains a metallo-isopeptidase JAMM motif active site that functions, in conjunction with the entire lid, in deubiquitylation of the substrates (Verma *et al.*, 2002).

10.2.2 CSN (Cop9 signalosome)

The CSN is an eight-subunit complex found in all eukaryotes (Fig. 10.1b, Color plate 15; Chamovitz and Glickman, 2002; Deng and Serino, 2003). The CSN was first identified in plants as a negative regulator of photomorphogenesis (Wei *et al.*, 1994a). *Arabidopsis* seedlings lacking a functional CSN germinate in total darkness as if grown in light, displaying a phenotype referred to as *constitutive photomorphogenesis* (*cop*), *de-etiolated* (*det*), or *fusca* (*fus*). For simplicity sake, this chapter will use the *cop* nomenclature except for mutants widely known by another name. Of the nine *cop* loci identified in *Arabidopsis*, six encode the CSN PCI-containing subunits CSN1, 2, 3, 4, 7, and 8 (Wei *et al.*, 1994a; Staub *et al.*, 1996; Karniol *et al.*, 1999; Serino *et al.*, 1999; Peng *et al.*, 2001b; Serino *et al.*, 2003). Loss-of-function alleles of any of these loci lead to the *cop* phenotype. CSN loss-of-function mutants do not reach reproductive stage and die as a seedling even when growing under optimal light conditions, suggesting that the CSN is necessary for processes other than photomorphogenesis.

The other two CSN subunits, the MPN domain-containing CSN5 and CSN6, are both encoded by duplicated genes in *Arabidopsis*, and thus escaped the saturation *cop* screens. Double mutants for both gene copies for either locus also yield the classic *cop* phenotype, such that lack of any one of the eight CSN subunits leads to an essentially identical *cop* phenotype (Gusmaroli *et al.*, 2004; Dohmann *et al.*, 2005; Gusmaroli *et al.*, 2007).

The known catalytic activity of the CSN is centered in CSN5. This metallo-isopeptidase catalyzes the removal of a small ubiquitin-like molecule termed Rub1 in *Arabidopsis*, and Nedd8 in all other organisms, from the Cullin subunit of certain ubiquitin ligases (Cope *et al.*, 2002) (see Section 10.5.2). Subsequently, it was found that CSN5 possesses a second intrinsic metallo-isopeptidase activity, deubiquitylation. The CSN complex can remove ubiquitin from certain substrates (Gusmaroli *et al.*, 2004). In addition to the intrinsic CSN5 deubiquitylation activity, the deubiquitylating enzyme Ubp12 is active in association with the CSN complex (Zhou *et al.*, 2003; Hetfeld *et al.*, 2005), adding another level of complexity to the involvement of the CSN in regulating substrate ubiquitylation.

Another activity attributed to the CSN is regulation of intracellular partitioning of various key proteins, between the nucleus and cytoplasm. These include CSN5 itself, COP1, $p27^{Kip1}$, and the small subunit of ribonucleotide reductase (the latter three will be discussed below).

While the CSN is often referred to as if it is a uniform complex, CSN subunits are in truth found in multiple configurations. In *Arabidopsis*, CSN4, 5, and 7 are detected by gel filtration analyses in fractions corresponding to smaller weight complexes and/or monomers, as well as the ~500-kDa CSN complex (Karniol *et al.*, 1999; Serino *et al.*, 1999; Dohmann *et al.*, 2005). CSN4 may also be present in an even larger complex. Furthermore, etiolated seedlings appear to have a CSN complex that is larger than that detected in light-grown seedlings (Wei *et al.*, 1994a). Similar results were also found in other organisms. For example, in mammalian cells, the nucleus contains the large CSN complex exclusively, whereas the cytoplasm contains small CSN subcomplexes exclusively. These small subcomplexes included CSN4, 5, 6, 7b, and 8, but not CSN1 (Tomoda *et al.*, 2002). In *Arabidopsis* the CSN also localizes mainly in the nucleus (Chamovitz *et al.*, 1996), though individual subunits, such as CSN7, appear also in the cytoplasm (Yahalom *et al.*, 2001). The role of these mini-complexes is not known. Interestingly, AtCSN5A and AtCSN5B do not integrate to the same complexes, CSN^{CSN5A} complexes are more abundant than CSN^{CSN5B} complexes (Gusmaroli *et al.*, 2004).

CSN integrity is dependent on the expression of almost all of its subunits as loss of an individual subunit usually leads to complex dissociation or prevents complex assembly (Wei and Deng, 1999). One exception to this is CSN5. In *Drosophila*, mutant larvae lacking CSN5 still maintain a CSN complex, though there are changes in the uncomplexed forms of certain subunits (Oron *et al.*, 2002). A similar report was made for *Arabidopsis* (Dohmann *et al.*, 2005), though contradictory results are also claimed (Gusmaroli *et al.*, 2007). If the latter is true then the common *cop* phenotype found in loss-of-function mutants in each subunit is clear. If the former, then it appears that the *cop* phenotype stems from a loss of the CSN-bound form of CSN5.

Very recent work in *Drosophila* has indicated that the CSN complex has roles beyond those mediated by CSN5 and different mutants only partially overlap in their gene expression profiles (Oron *et al.*, 2007). Interestingly, reduction of CSN5 levels prevents the accumulation of CSN1 and CSN8 in *Arabidopsis*, though it is not clear whether this indicates increased degradation of CSN5 or decreased translation/transcription of its gene (Schwechheimer *et al.*, 2001).

Another biochemical activity attributed to the CSN is phosphorylation of key UPS substrates. This activity is not CSN intrinsic, as was initially reported (Seeger *et al.*, 1998), but executed by kinases that work in association with the CSN complex. Purified CSN from human red blood cells copurified with kinase activity directed against c-Jun, IκBα, the NFκB precursor p105, and p53 (Seeger *et al.*, 1998; Bech-Otschir *et al.*, 2001). The phosphorylation by the CSN-associated kinase of p53 is thought to promote its degradation by the 26S proteasome while the phosphorylation of c-Jun is thought to stabilize it (Naumann *et al.*, 1999; Bech-Otschir *et al.*, 2001). These kinases were subsequently identified as CK2, protein kinase D (PKD), and inositol 1,3,4-trisphosphate 5/6-kinase (reviewed in Harari-Steinberg and Chamovitz, 2004). A common feature of these kinase activities is that they are inhibited by curcumin. One CSN-associated kinase has been identified in plants (Malec and Chamovitz, 2006).

Some of the CSN subunits themselves are kinase substrates and can be found in different phosyphorylation states including *Arabidopsis* CSN7 (Karniol *et al.*, 1999) and several human CSN subunits (Henke *et al.*, 1999). The importance of this phosphorylation is not clear.

10.2.3 eIF3

In general, eIF3 is an "outsider" PCI complex in terms of similarity to the other PCI complexes and in connection to the UPS. eIF3 is part of the eukaryote translation machinery (Hinnebusch, 2006). It coordinates the complicated process of translation initiation by stimulating the assembly of the eIF2–GTP–met-tRNAi ternary complex, binding of the ternary complex and other components of the preinitiation complex to the 40S subunit of the ribosome, and scanning the mRNA for the correct AUG start codon.

eIF3 subunit nomenclature is eIF3a to eIF3m in order of descending molecular weight (Browning *et al.*, 2001). eIF3 composition differs among various organisms. *Arabidopsis thaliana* eIF3 contains at least 11 subunits (Table 10.1, Fig. 10.1c (Color plate 15)), compared to the 13 subunits in human and only 6 in budding yeast. Five of these, eIF3a, eIF3b, eIF3c, eIF3g, and eIF3i, are conserved in all organisms and are considered "core" subunits and are essential for translation in vivo (Naranda *et al.*, 1997; Verlhac *et al.*, 1997; Asano *et al.*, 1998; Greenberg *et al.*, 1998; Phan *et al.*, 1998; Valasek *et al.*, 1998; Hanachi

Table 10.1 Nomenclature of PCI complex subunits in *Arabidopsis*

CSN	Proteasome lid	Domain	eIF3	Domain
CSN1	RPN7	PCI	eIF3a	PCI
CSN2	RPN6	PCI	eIF3b	—
CSN3	RPN3	PCI	eIF3c	PCI
CSN4	RPN5	PCI	eIF3d	—
CSN5A			eIF3e	PCI
	RPN11	MPN		
CSN5B			eIF3f	MPN
CSN6A			eIF3g	—
	RPN8	MPN		
CSN6B			eIF3h	MPN
CSN7	RPN9	PCI	eIF3i	PCI
CSN8	RPN12	PCI	eIF3k	PCI
			eIF3l	PCI

The 8 paralogous CSN and proteasome lid subunits, and the 11 eIF3 subunits, are organized in order of descending molecular weights. PCI or MPN domains are noted.

et al., 1999; Vornlocher *et al.*, 1999). Interestingly, only three of these, eIF3a, b, and c, together with eIF3e, f, and h, are sufficient to promote translation in vitro (Masutani *et al.*, 2007).

Most of the other subunits are also conserved among higher eukaryotes, though a few are organism specific. For example, eIF3j is found in human and *Saccharomycescerevisiae* but not in *Arabidopsis* or *Schizosaccharomycespombe* and eIF3m is found in human and *S. pombe* but not in *Arabidopsis* or *S. cerevisiae*.

10.3 PCI/MPN domain

As mentioned above, the subunits of these three complexes contain one of two amino acid motifs—the PCI or MPN motifs (Fig. 10.1, Color plate 15). The PCI motif [also known as the PINT domain (Aravind and Ponting, 1998)] is a predicted α-helical domain of ~200 amino acids, located in the C-terminal region of these proteins. Since it is found almost exclusively in proteins that are part of multiprotein complexes, this motif was suggested to mediate intracomplex protein–protein interactions. In vitro protein interaction assays, protein-binding assays in yeast, and in vivo studies primarily in *Arabidopsis* provided evidence that the PCI motif indeed participates in protein–protein interactions, but in a specific manner. Not all PCI-containing proteins interact with each other. For example, in *Arabidopsis*, CSN2 and CSN8 do not interact in yeast two-hybrid assay although both are PCI proteins (Serino *et al.*, 2003).

The MPN is a ~180 residues motif that has better primary conservation than the PCI motif. It is located in the N-terminus of the protein and is probably more ancient than the PCI domain since it was found also in prokaryotes. All

PCI complexes contain two MPN domain-containing subunits. Two of these, CSN5 and RPN11, contain an MPN variant known as MPN+/JAMM that carries metallo-isopeptidase activity. These proteins cleave Nedd8/Rub1 or ubiquitin, respectively, from conjugated proteins (Cope *et al.*, 2002; Maytal-Kivity *et al.*, 2002). This is the only intrinsic catalytic activity know for the CSN and the proteasome lid complexes. eIF3's MPN subunits have no known catalytic activity.

The crystal structure of the JAMM motif from*Archaeoglobus fulgidus* suggests a zinc binding site similar to many enzymes (Tran *et al.*, 2003). Indeed, the proteasome 19S RP deubiquitylation activity is sensitive to zinc chelators (Cope *et al.*, 2002). The RPN11 JAMM motif removes ubiquitin from substrates designated for degradation in the proteasome. The CSN5 JAMM motif catalyzes the removal of Rub1 (akin to nedd8 in animal and yeast), a ubiquitin-like protein, from Cullin, an SCF E3 ligase subunit (see more in SCF regulation below, Section 10.5.3). Point mutations within the JAMM motif of CSN5 and RPN11 block their activity and led to a lethal phenotype in *Drosophila* (Cope *et al.*, 2002; Verma *et al.*, 2002), in contrast to fission yeast where this mutation had no obvious phenotype other than the block in enzymatic activity (Zhou *et al.*, 2001; Mundt *et al.*, 2002). Interestingly, CSN5 and RPN11 are active only in the context of the complex. This may imply an involvement of the whole complex in interacting with the substrate or other proteins and thus coordinating its metalloprotease activity with the other activities of the complexes.

That the PCI complexes share similar subunit architecture, and a six PCI + two MPN composition, suggests that these complexes evolved from a common ancestor. The CSN and the lid are more closely related to each other than they are to the eIF3, each having only eight subunits. eIF3 is more diverged both at the level of subunit amino acid similarity, and complex structure, as it has additional non-PCI/MPN subunits. The existence of "orphan" PCI proteins may suggest the presence of a fourth PCI complex.

10.4 Inter-PCI-complex relationships

Beyond the structural similarities between the PCI complexes, genetic and biochemical evidence connect them as well. The human CSN was initially discovered as a contamination in proteasome lid preparations (Seeger *et al.*, 1998), implying a close connection of these two complexes. Similarly, three eIF3 subunits, eIF3 c, e, and h, and one lid subunit, Rpn6, copurified with the CSN from cauliflower (Karniol *et al.*, 1998). Yeast two-hybrid screens revealed that AtCSN1 interact with the lid subunit RPN6 through its C-terminus (Kwok *et al.*, 1999). While one explanation for these results could be that similar biochemical properties of similar complexes led to the copurification, subsequent studies have indicated a functional significance to these interactions.

In *Arabidopsis*, the gel filtration fractionation profile of CSN and the proteasome lid partially overlap, and this overlap increases following formaldehyde cross-linking treatment prior to fractionation (Peng *et al.*, 2003). In addition, anti-CSN6 antibodies could pull down some proteasome subunits as well as E3 ubiquitin ligase SCF subunits (Peng *et al.*, 2001a). The interaction between the CSN and the lid is probably direct and does not involve mediating proteins, as purified mammalian CSN and proteasome reciprocally precipitate each other. Interestingly, this interaction was enhanced when ATP was added to the buffer. In this mammalian system flag-tagged CSN2 could also pull down subunits of the SCF E3 ubiquitin ligase (Huang *et al.*, 2005). These data suggest that the CSN, the proteasome, and E3 ubiquitin ligase can form a super complex in order to cooperate in protein degradation. Interestingly, auxin has been shown to induce the formation of a nuclear protein body containing these complexes (Tao *et al.*, 2005), raising the possibility that the super complex formation could be regulation by signaling.

In *S. pombe*, *S. cerevisiae*, and mammalian cells, eIF3e interacts with the 26S proteasome (reviewed in von Arnim and Chamovitz, 2003). In *S. pombe*, this interaction is thought to positively regulate the proteasome and eIF3e mutants accumulate polyubiquitinated proteins just like proteasome mutants. This probably happened due to improper assembly of the proteasome since *eif3e* mutants fail to properly localize the Rpn5 proteasome subunit (Yen and Chang, 2003). In *Arabidopsis* and human cells eIF3e interacts with the CSN, and in *Arabidopsis*, the subcellular localization of eIF3e is tissue dependent (Yahalom *et al.*, 2001; Hoareau Alves *et al.*, 2002). The nature of many of these intercomplex interactions is not clear and further investigation will be needed to understand their biological significance. Nevertheless, an evidence from a diverse array of organisms suggests that these interactions are not coincidental.

10.5 Ubiquitin and ubiquitin-conjugating cascade

Cellular protein degradation occurs in two main pathways: lysosomal and proteasomal. About 80% of the cellular protein degradation occurs through the latter pathway, where protein degradation is dependent on the substrate being first tagged by ubiquitin.

Ubiquitin, a 76 amino acid protein, is, as its name, found in all eukaryotic cells and is very highly conserved (Smalle and Vierstra, 2004). For example, higher plant ubiquitin differs in its amino acid sequence from yeast by only two residues and from animals by three. Ubiquitin has a globular and packed 3D structure, forming a pocket shape called the ub-fold (Vierstra, 1996). A C-terminal extension from the ub-fold terminates with an essential glycine that serves as the covalent interaction site with its substrates and conjugating enzymes. Proteins tagged by a polyubiquitin tail are targeted for degradation by the proteasome. In addition to its famous Nobel-winning role in protein degradation, ubiquitin also participates in many other cellular processes such as cell division, differentiation, signal transduction, protein trafficking, and quality control, which are not considered further in this chapter.

Ubiquitin is only one member of the larger ubiquitin-like family of proteins. In plants these additional ub-like proteins are called Sumo and Rub1 (Nedd8 in nonplants). Interestingly, in plants ubiquitin is transcribed in fusion; it can be fused to other ubiquitins to form a concatomer of ubiquitins, or it can be fused to the ub-like Rub1 or to one of two ribosomal subunits. After translation, a deubiquitylation protease separates the fused proteins to their active form (Callis *et al.*, 1990; Sun and Callis, 1997).

10.5.1 The ubiquitin-conjugating cascade

The specificity and control of the ub-proteasome pathway is dependent on the selective tagging of the targeted protein by polyubiquitin (Rock and Goldberg, 1999). The system that controls this tagging is a cascade of three types of enzymes: the ubiquitin-activating enzyme (E1), ubiquitin-conjugating enzyme (E2), and ubiquitin protein ligase (E3). In general, the E1 activates ubiquitin, E3 determines the specificity of the substrate, and the E2s mediate the interaction between the E1s and the E3s. In the *Arabidopsis* genome there are 2 E1s, 40 E2s, and about 1300 E3s (data from http://plantsubq.genomics.purdue.edu/).

The E1 ubiquitin activating enzyme is a ~1100-residue protein that activates ubiquitin by forming a thioester bond with its C-terminus in an ATP-dependent manner. The bond is created via the E1 cysteine residue that binds ubiquitin in a high-energy bond. The E1 then transfers the activated ubiquitin to an E2 ubiquitin-conjugating enzyme. The E2s are believed to have some part in the determination of the substrate identity. E2s interact with E1s and E3s via the same domain and thus a dynamic binding switch is necessary, with an E2 first binding the E1, and subsequently an E3 (Pickart, 2001). The E3 makes contact both with the E2 and the substrate, and once the E2-E3-substrate complex has assembled, the E2 transfers the activated ubiquitin directly, or via the E3 to the substrate, and this is repeated cyclically. In the first cycle the ubiquitin is conjugated to the substrate, while in subsequent cycles it is conjugated to the previous ubiquitin, forming a linear polyubiquitin chain (Fig. 10.2c, Color plate 16). This chain marks the substrate for degradation by the 26S proteasome (reviewed in Smalle and Vierstra, 2004).

The E3 largely determines the identity of the substrate. The *Arabidopsis* E3 superfamily can be divided into two major families, the HECT (*h*omology to *E*6-AP *C*-*t*erminus) and the Ring/U-box families, with the latter divided into further subclasses (Fig. 10.3, Color plate 17). These two main families differ in the mechanism of ubiquitin transfer. For HECT E3s, the E2 transfers ubiquitin first to the E3, forming a covalent thioester bond between ubiquitin and the E3, and only then is ubiquitin transferred to the substrate. The RING E3s act as a scaffold, mediating direct ubiquitin transfer from the E2 to the substrate (Fig. 10.2, Color plate 16).

10.5.2 E3 ligases

The *Arabidopsis* genome encodes only eight predicted HECT E3s, designated UPL 1–8 (ubiquitin protein ligase). Mammals by contrast have many more HECTs (Schwarz *et al.*, 1998; Downes *et al.*, 2003). HECT E3 ligases are large (up to 400 kDa) single proteins characterized by an HECT domain, a conserved 350 amino acid domain that was initially identified in the human E6AP E3 ligase. Ubiquitin is bound to this domain by the E2. Other motifs in this large protein, such as Armadillo, IQ calmodulin-binding, C-type lectin-binding, transmembrane, ub-interacting motif (UIM), ub-associated (UBA), and ub-like (UBL) domains (Downes *et al.*, 2003), are thought to mediate substrate recognition.

The *RING/U-box* family gets its name from the founding member in humans, *R*eally *I*nteresting *N*ew *G*ene (RING), whose function was initially not clear (Freemont, 1993). The RING family is very diverse; some of its members are single-subunit scaffolds (e.g., RING/U-box) and some are multi-subunit complexes (e.g., Cullin-RING E3 ubiquitin ligases).

The RING/U-box E3s (Fig. 10.3, Color plate 17) are characterized by either a RING finger motif or a structurally related U-box motif. The RING motif, a 70 amino acid zinc-binding motif, acts to dock the ubiquitin-loaded E2. Substrate recognition is mediated by other motifs in the protein such as WD40. *Arabidopsis* contains >400 RING proteins, but it is unclear whether all of them function as E3 ligases. The most conspicuous among the single-subunit E3s are constitutive photomorphogenesis1 (COP1), seven in *Absentia* in *Arabidopsis thaliana* 5 (SINAT5), and arm repeat-containing1 (ARC1).

In *multi-subunit RING E3s*, the RING protein is only one subunit in a larger complex. Substrate recognition is no longer possessed by the RING protein, but by a different subunit. There are two main types of multi-subunit RING E3s, Cullin-RING E3 ubiquitin ligases (CRL E3s) and anaphase promoting complex (APC) E3s (Fig. 10.3, Color plate 17).

CRL E3s contain a Cullin protein backbone that acts as a scaffold to coordinate the RING protein, E2 and substrate-specificity factor subunits. *Arabidopsis* Cullin family contains nine putative members, each Cullin binds to a different family of specificity factors (Fig. 10.3, Color plate 17).

The prototypical CRL is the SCF (Skp1, CDC53, F-Box) complex. The plant SCF contains at least four subunits, the RING protein Rbx1, ASK1 (*A*rabidopsis *Sk*p1), Cullin 1 (Cul1), and one of a plethora of F-box proteins. Cul1 serves as the scaffold of the complex, binding Rbx1 through its carboxyl terminus and the Skp1 adaptor protein through its amino terminus (Zheng *et al.*, 2002b). Skp1 simultaneously interacts with Cul1 and an F-box protein that binds the substrate (Bai *et al.*, 1996).

The F-box family is very diverse in plants and includes about 700 members in *Arabidopsis*. The F-box N-terminus contains a 50–70 amino acid F-box motif that is necessary, but not sufficient, for the interaction with Skp1, while the rest of the protein can contain different protein–protein interaction motifs, including Leu-Rich Repeats (LRR), Kelch, WD-40, and Armadillo (Arm). F-box proteins are classified by these motifs, which are responsible for substrate recognition (Schulman *et al.*, 2000). Besides its role in SCF, F-box proteins may also function in other processes including DNA replication and transcription (Jan *et al.*, 1999; Clifford *et al.*, 2000).

SCF is the best-characterized E3 in plants so far. The huge diversity in *Arabidopsis* F-box proteins, far larger than in yeast or mammals, correlates well with the numerous physiological processes known to be dependent on SCFs. These include photomorphogenesis, phytohormone responses, circadian rhythms, floral development, and senescence, and cellular processes such cell cycle, signal transduction, and gene expression. SCF involvement occurs via regulating degradation of key signaling molecules (Hershko and Ciechanover, 1998; Dharmasiri *et al.*, 2002; Devoto *et al.*, 2003; Itoh *et al.*, 2003; Wang *et al.*, 2003; Pagano, 2004).

Other CRL complexes have different subunit compositions but the core of all of them is the Cullin-Rbx1 dimer. *Arabidopsis* Cul4 interacts with Rbx1 and the CDD complex, and together this complex possesses E3 ligase activity in vitro (Chen *et al.*, 2006). The CDD complex is composed of COP10, damaged DNA binding protein A (DDB1a) and de-etiolated 1 (DET1). The function of these proteins will be discussed later (see Section 10.6). This Cul4-based E3 ligase participates in regulating photomorphogenesis, and mutation in Cul4 led to the classic *cop* phenotype (Chen *et al.*, 2006).

Another type of Cullin-based E3 is the Cul3-BTB complex (Fig. 10.3, Color plate 17). This E3 lacks the Skp1 and F-box subunit, which are replaced by the BTB protein that serves both as the adaptor and as the recognition component of this E3 ligase. *Arabidopsis* genome contains two Cul3s and ~80 BTB proteins. Some of these BTBs were shown to interact with both Cul3s in the yeast two-hybrid assay (Dieterle *et al.*, 2005).

10.5.3 Regulation of Cullin-RING E3 ligases

Since CRL E3 ligases are involved in so many plant physiological processes, their activity must be exquisitely regulated. The CSN takes an important place in this regulation. Copurification of SCF with CSN subunits was the first hint at this regulation (Lyapina *et al.*, 2001). For example, the Cul1 subunit of *Arabidopsis* SCF^{TIR1} immunoprecipitates with the CSN, and vice versa (Schwechheimer *et al.*, 2001). Similarly, the COI1 F-box protein of *Arabidopsis* SCF^{COI1} coprecipitates, in addition to Cul1, with four CSN subunits (Feng *et al.*, 2003). The mammalian F-box protein Skp2 and Cul1 were able to pull down all CSN subunits in 3T3 NIH cells (Yang *et al.*, 2002). CSN binds to Cul1 and Rbx1 via CSN2, CSN6, and CSN1's N-terminal domain. Other CRLs were shown to interact with CSN as well, including Cul4A E3s in human and yeast (Groisman *et al.*, 2003; Liu *et al.*, 2003).

All of the above strongly suggest a conserved interaction between CSN and CRL E3 ligases, but the nature of this interaction was not clear. Cullins can be modified by conjugation by the ubiquitin-like molecule Nedd8 (in *Arabidopsis* Rub1) via a conjugation cascade similar to the ubiquitin E1, E2, and E3 process, called Neddylation (or Rubylation in plants). The *Arabidopsis* Rub8 E1-activating enzyme is the heterodimer Auxin Resistant 1/E1 C-terminal1 (AXR1/ECR1, in human called AppBP1/hUba3); the E2-conjugating enzyme is RUB-conjugating enzyme1 (RCE1, in human called hUbc12); and the only known E3 is the RING protein Rbx1 (Fig. 10.2c, Color plate 16; Liakopoulos *et al.*, 1998; Kamura *et al.*, 1999; Gray *et al.*, 2002).

The conjugation of Rub1 to Cul proteins plays a significant role in plant development (del Pozo *et al.*, 2002a; Dharmasiri *et al.*, 2003; Larsen and Cancel, 2004), and is essential in *S. pombe*, *Caenorhabditis elegans*, and mouse development (Osaka *et al.*, 2000; Tateishi *et al.*, 2001). With the notable exception of p53 (Harper, 2004), Cullins are the only proteins modified by Nedd8. The function of Cullin neddylation/rubbylation has been a matter of intense study and debate. This modification appears to stimulate E3 activity by allowing the recruitment of an E2 enzyme (Wu *et al.*, 2000; Kawakami *et al.*, 2001). CSN, via CSN5, removes Nedd8/Rub1 from Cullins in a process called deneddylation (Cope *et al.*, 2002; Fig. 10.2c, Color plate 16). Evidence from *Arabidopsis*, *S. pombe*, human, and *Drosophila* support the idea that CSN possesses deneddylation activity as *csn* mutants in these organisms accumulate Cullins in their neddylated form (Schwechheimer *et al.*, 2001; Doronkin *et al.*, 2002). An in vitro rescue experiment illustrated that purified pig spleen CSN could restore deneddylase activity to extracts from *csn* mutants (Lyapina *et al.*, 2001). Moreover, a point mutation in the JAMM motif of *S. pombe* CSN5 could not restore the *csn* mutant phenotype of deneddylation, although the CSN complex was intact (Cope *et al.*, 2002). Although the deneddylation activity is centered in the CSN5 subunit, the whole complex is essential; for example, mutants of fission yeast *csn1* do not contain an intact CSN and accumulate monomeric CSN5, but this CSN5 does not function as a deneddylase enzyme by itself (Lyapina *et al.*, 2001).

The function of deneddylation/derubbylation has been a bit of a conundrum. Studies initially carried out in *Arabidopsis*, but later in other organisms, indicate that in *csn* mutants, CRL activity is curtailed, and CRL substrates are stabilized. For example, IAA/AUX proteins, which are substrates of the SCF^{TIR}, are stabilized in *csn5* underexpression lines (Schwechheimer *et al.*, 2001). Another example is the accumulation of Cyclin E in murine *csn2* mutants (Lykke-Andersen and Wei, 2003). This suggests that rubbylation inhibits CRL activity and that CSN-mediated derubbylation promotes activity.

However, in vitro experiments supported an opposing mechanism, that neddylation activates CRLs. Further in vitro work then showed that CSN-mediated deneddylation inhibits SCF activity. For example, Yang *et al.* (2002) showed that cyclin-dependent kinase inhibitor $p27^{kip1}$ degradation was inhibited upon addition of purified CSN, along with a concurrent accumulation of Cul1 in the SCF^{skp2} E3 ligase. This activity extends to other CRLs as

Groisman *et al.* (2003) showed the addition of purified CSN-prevented substrate ubiquitylation by a Cul4 E3 ligase.

The contradiction between the in vivo and in vitro results led to the development of a number of explanations claiming a need for neddylation cycles. Dieter Wolf and colleagues proposed a simple unifying hypothesis whereby neddylation does indeed activate E3 activity (as per the in vitro results), but that proper assembly of the CRL necessitates a deneddylated Cullin (Wolf *et al.*, 2003). If the CRL assembles when the Cullin is neddylated, the substrate specificity factor, such as the F-box, is itself ubiquitylated and degraded before it can bind to its substrate. In this way, the CRL self-destructs before it can act on a substrate. This then can explain the contradiction between in vivo and in vitro results, as substrate stabilization in *csn* mutants would then result from excess CRL activity targeting its own F-box subunit, thus "turning off" the CRL vis-a-vis the F-box substrate (Cope, 2006). Indeed SCF possesses self-ubiquitylation activity (Groisman *et al.*, 2003; Zhou *et al.*, 2003) so it is possible that the cycles of dissociation and reassociation prevent this activity and keep the levels of the SCF subunits.

What is the mechanism by which deneddylation and neddylation activates SCF? Unneddylated Cullin specifically interacts with the protein CAND1 (Cullin-*A*ssociated *N*eddylation *D*issociated/TIP120); this interaction leads to SCF dissociation and only the core SCF subunits Cullin and Rbx1 remain to interact with CAND1 (Fig. 10.2b, Color plate 16). Upon reneddylation of Cullin, CAND1 leaves the complex, allowing reassembly of the SCF with new Skp1 and F-box proteins, as well as the E2 conjugating enzyme (Liu *et al.*, 2002a; Zheng *et al.*, 2002a). This process allows the exchanging of the F-box and subsequently matching of the SCF to a new substrate. What triggers neddylation and deneddylation is not known.

Arabidopsis CAND1 was isolated as an enhancer of the *tir1* mutant. *Tir1* encodes an F-box protein (Chuang *et al.*, 2004). *Arabidopsis* CAND1 positively regulates SCF^{TIR1} because Aux/IAA protein stability is significantly increased when CAN1 is compromised. Interestingly, in vivo studies indicated that liberating CUL1 from CAND1 is not the primary role of the RUB modification pathway in the regulation of SCF activity (Chuang *et al.*, 2004). The *Arabidopsis cand1-1* null mutant displays multiple phenotypes affecting all stages of plant development, including a *cop* phenotype, most likely arising from compromised CRL activity (Feng *et al.*, 2004).

The *anaphase-promoting complex* (APC) is the largest and most complex multi-subunit E3 RING enzyme, composed of 8–13 subunits in various eukaryotes. It was originally discovered in yeast and only recently identified in *Arabidopsis* (Irniger *et al.*, 1995; Eloy *et al.*, 2006). Two APC subunits, APC2 and APC11, are similar to the SCF subunits Cullin and a RING-protein respectively (Tang *et al.*, 2001). In vitro assays showed that recombinant APC11, together with an E1and E2, is sufficient for nonspecific polyubiquitin chain formation (Gmachl *et al.*, 2000). This result raises the question why such a complicated assembly is needed. A possible explanation is that the complex structure is important for substrate recognition and regulation, but the function of most APC subunits is still unclear.

APC is a central cell-cycle regulator, regulating the transition from metaphase to anaphase and mitotic exit by controlling the degradation of many important cell-cycle regulators, including cyclin A, cyclin B, securing (an inhibitor of chromosome separation), and many of the mitotic regulatory kinases (Harper *et al.*, 2002; Peters, 2002). By temporally regulating the degradation of specific cyclins, and the consequent oscillation in cyclin levels, APC can drive or inhibit cell division events in proper order (Murray, 2004). Interestingly, while APC regulates the transition from metaphase to anaphase and exit from mitosis, the SCF E3 complexes control the G_1 to S transition by mediating the degradation of G1 cyclins and CDK inhibitors (Deshaies, 1999). Thus, different classes of multi-subunit E3s control different aspects of the cell cycle regulation, leading to an exquisite control mechanism.

10.6 Other COP/DET proteins

In addition to the *cop/det* mutations in the genes encoding CSN subunits, three more *cop/det* loci have been identified: *Cop1*, *Cop10*, and *Det1*. COP1 was one of the first COP/DET/FUS proteins identified and one of the most studied. Similar to mutants in the CSN, null *cop1* mutants are lethal after the seedling stage and have a strong *cop* phenotype (Deng *et al.*, 1991, 1992). In contrast to the CSN, viable *cop1* mutant alleles also have a *cop* phenotype, and COP1 is not found in all eukaryotes. A clear COP1 ortholog exists in mammals, but has yet to be found in unicellular organisms and invertebrate animals such as *Drosophila* and *C. elegans.*

COP1 is a single-subunit RING domain E3 ligase with seven repeats of WD40 domain, which is probably responsible for substrate recognition and self-dimerization (Holm and Deng, 1999; Holm *et al.*, 2001; Bianchi *et al.*, 2003). COP1 has both nuclear import and nuclear export signals, and in plants this subcellular localization is dynamic and controlled by the light environment, and dependent on CSN (Chamovitz *et al.*, 1996).

COP1 mediates the wide range of plant responses. Microarray analysis demonstrated that 20% of the *Arabidopsis* genome is controlled, directly or indirectly, by COP1, including 20% of the putative *Arabidopsis* transcription factors (Ma *et al.*, 2002). One of the effectors directly downstream of COP1 in plants is HY5, a bZIP transcription factor. Mutations in *HY5* lead to a phenotype that is opposite of *cop*, that is reduced responses to light. COP1 interacts both genetically and physically with HY5 (Ang *et al.*, 1998). The transcript profiling data demonstrated that many of the genes activated by HY5 are repressed by COP1, suggesting opposite function for the two proteins (Ma *et al.*, 2002). In light-grown plants, COP1 localizes to the cytoplasm, while HY5 is in the nucleus, activating light regulated genes (Fig. 10.4a, Color plate 18). In the dark, COP1 relocates from the cytoplasm to the nucleus, where it colocalizes with HY5, and mediates its degradation, repressing HY5-activated expression of light-induced genes (Fig. 10.4, Color plate 18; von Arnim and Deng, 1994). Support for this hypothesis came from the observation that *cop1* mutants accumulate HY5 in high levels in the dark compared with wild type (Osterlund *et al.*, 2000).

COP1 controls many parallel pathways toward light-dependent growth. Another bZIP transcription factor, HYH, which can form a heterodimer with HY5, is also degraded in a COP1-dependent manner (Holm *et al.*, 2002), as are two additional transcription factors, LAF1 and HFR1, which are involved in promoting photomorphogenesis (Seo *et al.*, 2003). While COP1 "looks" like a ring-E3 ubiquitin ligase, only recently has this been demonstrated. COP1 can ubiquitylate LAF1 and HFR1. The in vitro assay demonstrated that COP1 also has an intrinsic E3 activity toward HY5. Interestingly, only one or two ubiquitin molecules were conjugated to HY5 by COP1 in vitro (Saijo *et al.*, 2003), suggesting it is not the only E3 active in HY5 degradation, or that the in vitro conditions were not optimal for this specific assay.

COP1 also plays a role in light signaling via regulation of the photoreceptors Phytochrome A (phyA) and Cryptochrome 2 (cry2). These proteins are degraded rapidly via the UPS under light conditions (Seo *et al.*, 2004; Fig. 10.4, Color plate 18). COP1 physically interacts with both photoreceptors and targets theme for degradation (Yi and Deng, 2005). Interestingly, like with HY5, COP1 localized to the same speckles in the nucleus with PHYA when transiently expresses in epidermis onion cells (Seo *et al.*, 2004) and with LAF1 tagged protein (Seo *et al.*, 2003).

The kinetics of COP1 nuclear export do not jibe with those of light-induction of gene expression. The export of COP1 to the cytoplasm takes about 24 h, while light-induced genes appear within less than 1 h (Stacey *et al.*, 1999). Thus, there must be a rapid mechanism which stabilizes HY5 from COP1 action prior to the latter's nuclear export. The affinity of HY5 toward COP1 is dependant on HY5 phosphorylation state, suggesting another level of complexity, but the identity of this kinase is not known (Hardtke *et al.*, 2000).

In mammalian cells COP1 functions in the degradation of c-Jun and p53, and consequently COP1 influences many cellular processes (Dornan *et al.*, 2004). In the case of c-Jun, COP1 is probably important for c-Jun degradation via interaction with DET1, which is a part of an E3 complex composed of DET1, DDB1 (DNA damage-binding protein 1), Cullin4A, and Roc1 (Wertz *et al.*, 2004). A similar complex containing DET1, DDB1, cullin4, and Roc1 was recently found in *Arabidopsis*, with downregulation of the cullin4 scaffold subunit leading to defects in many developmental aspects including photomorphogenesis .(Bernhardt *et al.*, 2006). These findings suggest a conserved E3 ligase that in different organisms was specialized to different pathways.

det1 was the first of the *cop/det/fus* mutants isolated (Chory *et al.*, 1989), but only recently has the biochemical function of DET1 been elucidated. The essentially identical phenotype to *cop1* mutants hinted that DET1 also functions within the UPS, though its primary sequence gave no hint as to its function. DET1, a 62-KDa protein, is present in a 350-KDa nuclear complex with COP10 and DDB1, probably functioning together as an E3 ligase. On the other hand, some evidence connects DET1 directly to the transcription machinery. Tomato DET1 (also called high pigment 1, HP1) interacts with chromatin via the nonacetylated tail of the histone H2B. Upon light exposure the histone tail is acetylated, leading to the release of DET1 and subsequent derepression of transcription (Benvenuto *et al.*, 2002). The animal DET1 together with DDB1 interacts with histone acetyl transferase (HAT) complexes (Brand, 2001; Martinez *et al.*, 2001). All these data suggest that a role of DET1 complex is to mediate proteolysis of components involved in chromatin structure.

COP10 is nuclear enriched 21-KDa protein that was first identified as an essential mediator of HY5 degradation by COP1 (Wei *et al.*, 1994b; Osterlund *et al.*, 2000). It belongs to a family of ubiquitin E2 variants (UEV) that contains the ubiquitin-conjugating motif (Ubc) but lacks a critical cysteine residue required for ubiquitin conjugation (Sancho *et al.*, 1998; Schwechheimer *et al.*, 2001). UEVs participate in many cellular processes such postreplicative DNA repair and control of the cell cycle (Pickart and Hofmann, 1999), while others UEVs work together with an active E2 enzyme; but it is not clear if this is the case regarding to COP10. Gel filtration analysis of the wild type background showed that COP10 is part of 300 kDa complex and a small fraction can be detected also in the monomer fraction. The most attractive model would have COP10 working with COP1 to directly ubiquitinate substrates, all this regulated by the CSN. Indeed, COP10 directly interacts in yeast with three CSN subunits (CSN3, 4, and 8). In *csn8* (*cop9-1*) and *csn1* (*fus6-1*) mutants, COP10 levels are reduced, and COP10 is present in a complex slightly smaller that the wild type COP10 complex (Suzuki *et al.*, 2002).

10.7 The UPS and plant physiology

Given the vast diversity of plant E3 ligases, it is not surprising that the UPS is involved in a myriad of plant developmental processes. Indeed the list of these processes is growing rapidly. The following attempts to briefly overview the role of the UPS in key plant processes, but should by no means be considered exhaustive.

10.7.1 Response to phytohormones

Auxin participates in almost every aspect of plant life. Despite decades of research, only recently has the elusive auxin receptor been identified. Identification of this receptor, and the prior elucidation of important components of the auxin signaling pathway, was enabled through molecular genetic research mainly employing *Arabidopsis* as a model system.

The Aux/IAA proteins are short-lived regulatory proteins (Abel *et al.*, 1994, 1995), whose rapid turn/over is a major factor in the auxin response. This turnover is dependent on the UPS. Various auxin response (*axr*) mutants in *Arabidopsis* shared a common feature, Aux/IAA stabilization, suggesting the Aux/IAA are the negative regulators of the auxin response. A second group of mutants in the response to auxin included loss of function of the auxin response transcription factors (ARFs). The *Arabidopsis* genome encodes 29 Aux/IAA genes and 23 ARF genes. Active ARFs induce transcription of genes related to the auxin pathway and the Aux/IAA proteins are thought to interact with ARFs to inhibit their activity (Woodward and Bartel, 2005).

The E3 responsible for Aux/IAA degradation is SCF^{TIR1} (Gray *et al.*, 2001) and loss of function mutations in this E3 stabilizes Aux/IAA and leads to an auxin response resistance phenotype (Gray and Estelle, 2000; Worley *et al.*, 2000). As discussed earlier, the CSN regulates E3 ligases via deneddylation of the SCF Cul1 subunit. Not surprisingly then, an in vivo interaction between SCF^{TIR1} and the CSN was found. Likewise, transgenic *Arabidopsis* lines with reduced CSN5 levels showed auxin response phenotypes similar to that of the *tir1* mutants, including stabilization of the Aux/IAA fused to luciferase protein PSIAA6LUC (Schwechheimer *et al.*, 2001). Auxin itself plays an important role in Aux/IAA proteins turnover; on one hand, it induces transcription of the *Aux/IAA* genes in a CSN-dependent manner (Schwechheimer *et al.*, 2001), and on the other hand, auxin pretreatment of TIR1, the SCF F-box component, enhances its affinity to Aux/IAA and subsequently promotes Aux/IAA degradation (Fig. 10.5). Interestingly, pretreatment of Aux/IAA with auxin does not change their affinity to TIR1 (Kepinski and Leyser, 2004). Subsequently, TIR1 was shown to contain a specific auxin binding site and this binding promotes the TIR1 and Aux/IAA interaction (Dharmasiri *et al.*, 2005; Kepinski, 2005). Thus, the TIR1 F-box protein is the elusive auxin receptor, with auxin acting as a cofactor to promote E3-dependent ubiquitylation and subsequent degradation of the Aux/IAA proteins.

Another link to the UPS came from the *axr* mutants that were found on the basis of their auxin resistance response phenotype. It turned out that AXR1 (Auxin Resistance 1) together with ECR1 (E1 c-terminal 1) form a dimer that serves as the E1 for Rub1 (Nedd1) conjugation on Cul1 (see Section 10.5.3 and Fig. 10.2c, Color plate 16).

Jasmonates are potent lipid hormones that mediate defense responses against pathogens, responses to mechanical stimulation, and are essential for stamen and pollen development (Liechti and Farmer, 2002). Responses to this phytohormone are also mediated by an E3 ligase called SCF^{COI1} for coronatine insensitive (Devoto *et al.*, 2002; Xu *et al.*, 2002). Coronatine is a toxin similar to methyl jasmonate in structure; *coi1* mutant are male sterile and JA insensitive.

COI1 is an F-box protein and SCF^{COI1} interacts in vivo with the CSN. *csn1-1*, a weak *csn* mutant, like *coi1*, shows jasmonate insensitivity, suggesting that the CSN plays a roll in jasmonate signaling. The mechanism by which this regulation takes place is not clear but it is possible that it mediates the degradation of some transcriptional repressors.

Gibberellins are potent growth regulators in higher plants, implicated in numerous physiological processes, including seed development and germination, seedling growth, stem and root extension, flower induction and development, pollination, and fruit expansion (Fleet and Sun, 2005). The main active components of the GA signaling pathway are the DELLA proteins, five copies of which are encoded in the *Arabidopsis* genome: GAI (GA insensitive), RGA (Repressor of ga1-3), RGL1, RGL2, and RGL3 (for RGA-like 1, 2, and 3, respectively (Peng *et al.*, 1997; Silverstone *et al.*, 1998; Dill *et al.*, 2001; Chandler *et al.*, 2002). Although the DELLAs are partially redundant in *Arabidopsis*, each has a unique function and mediates a different pathway.

The DELLAs are nuclear localized and inhibit the expression of GA responsive genes, suggesting that they act as transcriptional regulators (Silverstone *et al.*, 1998; Fig. 10.5). GA promotes DELLAs degradation through the E3 ligase $SCF^{SLY1/GID2}$ [for the F-box protein Sleepy 1 (in *Arabidopsis*) or Gibberellin Insensitive Dwarf 2 (in rice)] (Fig. 10.5). Mutations in *SLY1* result in stabilization of RGA and GAI even in the presence of GA. Likewise SLY1 directly binds RGA and GAI and *rga* or *gai* mutants partially suppress the *sly1–10* phenotype. All these data strongly suggest that SLEEPY1 is the F-box protein for DELLAs (Silverstone *et al.*, 2001; McGinnis *et al.*, 2003; Sasaki *et al.*, 2003; Dill *et al.*, 2004; Fu *et al.*, 2004).

DELLAs stability is regulated also via auxin, where auxin promotes GA-dependent degradation of DELLA in the root, whereas ethylene inhibits DELLA protein degradation (Fig. 10.5; Achard *et al.*, 2003; Fu and Harberd, 2003).

The *ethylene* response is mediated by the E3 ligase $SCF^{EBF1/2}$ that is responsible for the degradation of the ethylene insensitive 3 (EIN3) transcription factor. EIN3 normally induces transcription of genes in the ethylene pathway and ethylene is known to inhibit $SCF^{EBF1/2}$ and consequently to stabilize EIN3 (Fig. 10.5; Guo, 2003; Guo and Ecker, 2004).

Recent papers indicated an involvement of the RING E3 keep on going (KEG) in the *ABA* (abscisic acid) response through regulation of ABI5 degradation. ABI5 is stabilized by ABA treatment, and also in *keg* mutants (Stone *et al.*, 2006).

10.7.2 The ubiquitin system and other plant processes

The UPS is implicated in many other plant processes. Proper flower development is dependent on the F-box protein UFO (Levin and Meyerowitz, 1995; Ingram *et al.*, 1997; Samach *et al.*, 1999), and SCF^{UFO} physically interacts with CSN (Wang *et al.*, 2003). Accordingly, *ufo* mutants and weak mutants in the CSN have similar floral phenotypes. This phenotype is also seen in *Arabidopsis* plants that overaccumulate the "e" subunit of eIF3, further supporting a

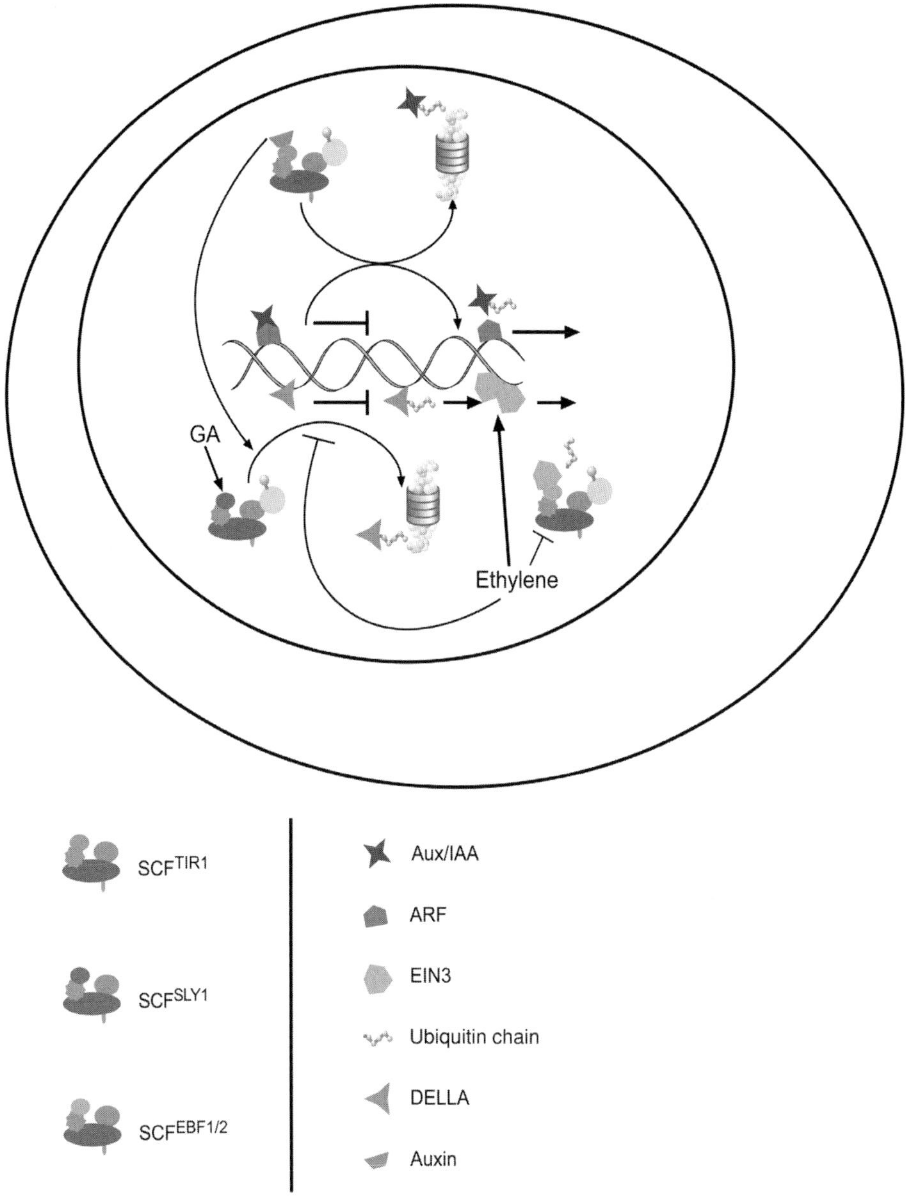

Figure 10.5 *The UPS and phytohormones.* Auxin interacts with the TIR1 F-box protein and induces Aux/IAA polyubiquitylation by SCF^{TIR1} and subsequent proteasomal degradation. Aux/IAA degradation releases ARF inhibition of auxin-induced genes. Gibberellic acid (GA) induces polyubiquitylation of DELLA proteins by SCF^{SLY1}. DELLA inhibits transcription of GA-induced genes and its degradation removes this inhibition. Auxin also induces, while ethylene inhibits, DELLA degradation. Ethylene inhibits SCF$^{EBF1/2}$-dependent EIN3 degradation, leading to EIN3 dimer stabilization and activation of ethylene-induced gene expression.

larger connection between the PCI complexes (A. Yahalom and D. Chamovitz, unpublished). The substrate of UFO has yet to be identified.

Yet another SCF variant, SCF^{ZTL}, regulates circadian rhythms through mediating the degradation of the TOC1 transcription factor (Schultz *et al.*, 2001). The U-box E3 ARC1 (Ushijima *et al.*, 2003) mediates the self-incompatibility response in almond.

There are several indications that the CSN is involved in immune signaling in both plants and animals. The *Nicotiana benthamiana* Rar1 (NbRar1), a zinc-finger protein required for the function of the TMV resistance conferring N gene, interacts with NbSGT1, which in turn interacts with NbSKP1. Thus, NbSGT1 represents a new highly conserved SCF subunit. Suppression of NbSGT1 and NbSKP1 abolishes the N-mediated resistance response to TMV. Both proteins associate with the CSN and thus not surprisingly, repression of the NbCOP9 signalosome also compromises N-mediated resistance to TMV (Liu *et al.*, 2002b).

In *Arabidopsis* the UPS also plays a role in innate immunity. The *snc1 npr1-1* double mutant exhibits high levels of salicylic acid (SA) under nonimmune challenging conditions, and as a result displays enhanced resistance toward some pathogens. UBA1, one of the two AtE1 enzymes, was identified in a suppressor screen for proteins that suppress the SA enhancement in *snc1 npr1-1*plants, suggesting UPS activity is essential for the innate immune response in plants (Goritschnig *et al.*, 2007).

The UPS and CSN are also involved in the *Drosophila* immune system, specifically in the toll pathway, which regulates the *Drosophila* homologue of the NF-κB and IκB, Dorsal and Cactus.*csn5* mutants show constitutive nuclear localization of Dorsal, and overaccumulation of Cactus (Harari-Steinberg *et al.*, 2007). The *Drosophila* Skp1 homologue is a negative regulator of the second immune-signaling pathways, IMD (Khush *et al.*, 2002).

The CSN, and of course the UPS, is also intimately involved in human innate immunity. The CSN copurified with an I-kappaBα kinase activity (Seeger *et al.*, 1998); and CSN3 interacts with IKKγ (Hong *et al.*, 2001).

10.7.3 The UPS and the CSN in cell cycle and DNA damage

Much research has been carried out investigating UPS involvement in cell cycle regulation. One highly studied pathway in plants is the retinoblastoma (RB)/E2F/DP pathway that controls cell division and differentiation in plants and animals (Gutierrez and Boniotti, 2002). E2F dimerizes with DP to form an active heterodimer that induces transcription of genes required for cell cycle progression. E2F can also dimerize with RB and as a result its transcription activity is inhibited. A study carried out by del Pozo and coworkers (2002b) linked this pathway to the UPS in *Arabidopsis*. First, they showed that E2FC degradation is dependent on the SCF^{SKP2A} E3 ligase and that overexpression of an undegradable form of E2FC led to cell division delay. Later they showed that even the degradation of DPB is UPS-dependent and is carried out via the same SCF^{SKP2A} (del Pozo *et al.*, 2006).

A number of key studies have connected the CSN to cell cycle control. This work has been carried out mainly in nonplant systems. However, considering the conservation of CSN function, it is highly likely that CSN impacts cell cycle regulation also in plants. Because of this, the following is a brief overview of what is known mainly from other systems.

CSN is involved in cell cycle control both through regulation of key cell cycle processes, and through a role in the DNA damage response, which is intimately linked with cell cycle control. CSN mediates the degradation of at least two key cell-cycle regulators, $p27^{Kip1}$ and cyclin E. $p27^{Kip1}$ is an inhibitor of cyclin-dependent kinases (Cdks), and through this mechanism controls the G_1 to S transition (Sherr and Roberts, 1999; Slingerland and Pagano, 2000). $p27^{Kip1}$ degradation is carried out following a cascade of events, starting with p27 phosphorylation by the cyclin E-Cdk2 complex (Sheaff *et al.*, 1997; Vlach *et al.*, 1997), transportation from the nucleus to the cytoplasm in a CSN-dependent manner (Tomoda *et al.*, 1999), ubiquitylation mediated by the ubiquitin ligase SCF^{Skp2} complex (Carrano *et al.*, 1999; Sutterluty *et al.*, 1999; Tsvetkov *et al.*, 1999) and subsequent proteolysis by the 26S proteasome.

$p27^{Kip1}$ interacts with CSN5 in the nucleus. This interaction leads to the export of $p27^{Kip1}$ to the cytoplasm. CSN5 itself is exported to the cytoplasm via interaction with CRM1. Point mutations that abolishes the CSN5- $p27^{Kip1}$ interaction results in $p27^{Kip1}$ nuclear accumulation and stabilization while overexpression of CSN5 or CSN3, 6, 7a, 7b, and 8 but not CSN1, 2, and 4 led to $p27^{Kip1}$ destabilization (Tomoda *et al.*, 2002). All of the above strongly suggest that CSN subunits mediate $p27^{Kip1}$ export via the CRM1 mechanism, followed by proteasomal degradation, though it is not clear if the entire CSN takes part in this, or if CSN-related subcomplexes are involved.

Cyclin E is an important regulator of the cell cycle, promoting the G_1 to S transition by activating the cyclin-dependent kinase cdk (Knoblich *et al.*, 1994). Oscillation of cyclin E levels is important for cell cycle progression, and is achieved by its regulated degradation. This degradation is mediated by the SCF E3 ligase (Dealy *et al.*, 1999; Strohmaier *et al.*, 2001). In *Drosophila*, the F-box Archipelago (Ago) participates in cyclin E degradation in the oocyte. Cyclin E accumulates in *ago* mutants and resulting in developmental defects in the oocyte. Similar phenotypes were found in *csn4* and *csn5* mutants suggesting that CSN and the SCF work together in the cyclin E degradation pathway (Doronkin *et al.*, 2003). Similarly, *Murine csn2* null mutants accumulate cyclin E, arrest at the peri-implantation stage and like *Drosophila* mutants, suffered from severe developmental defects (Lykke-Andersen and Wei, 2003).

Ribonucleotide reductase (RNR) is found in all organisms and provides the only mechanism to supply precursors needed for both synthesis and repair of DNA, by reduction of NTPs to dNTPs (Jordan and Reichard, 1998). S-phase necessitates a sufficient pool of dNTPs. In *S. pombe* when there is no need for dNTPs, the small subunit of RNR is anchored to the nucleus by a small protein called Spd1 (S-phase delayed protein), thereby inhibiting RNR activity. Spd1 is degraded upon S-phase or DNA damage signals by the proteasome, and thus the small subunit can be exported to the cytoplasm and form the active holoenzyme with the RNR large subunit. Not surprisingly then, the *S. pombe csn1-d* and *csn2-d* mutants display slow growth and an extended S-phase phenotype (Mundt *et al.*, 2002). DDB1-PCU4 (Cul4) serves as an E3 ligase complex when Cdt2 is its adaptor that recognizes Spd1. The CSN, DDB1, and PCU4 biochemically copurify from *S. pombe*, indicating possible cooperation in SPD1 degradation (Liu *et al.*, 2003, 2005; Bondar *et al.*, 2004). Interestingly, only CSN1 and CSN2, but not the other CSN subunits, are required for SPD1 degradation, suggesting a mechanism distinct from deneddylation by which the CSN controls SPD1 degradation.

The CSN is also involved in UV-induced nucleotide excision repair (NER) in mammalian cells through DDB2 and CSA, which are known as NER proteins. Two distinct mega-complexes are involved in this. Each contains the CSN, DDB1, CUL4A, and ROC1, and they differ by binding either DDB2 or CSA. These two complexes are differently regulated by the CSN. CSN binds to the Ddb2 complex in the absence of UV. Upon UV irradiation, the CSN dissociates from the complex and CUL4A is neddylated. Following DNA repair, the CSN reassociates and CUL4A is deneddylated. Alternatively, the CSN is free from CSA complex under normal conditions, but binds it rapidly after UV irradiation and they both associate with chromatin so that E3 ligase activity is suppressed during UV-induced DNA damage (Groisman *et al.*, 2003).

Acknowledgments

The authors thank all who provided unpublished data for use in this chapter and David Shemesh for the figure artwork.

References

Abel, S., Nguyen, M.D. and Theologis, A. (1995) The PS-IAA4/5-like family of early auxin-inducible mRNAs in *Arabidopsis thaliana*. *J Mol Biol*, **251**, 533–549.

Abel, S., Oeller, P.W. and Theologis, A. (1994) Early auxin-induced genes encode short-lived nuclear proteins. *Proc Natl Acad Sci USA*, **91**, 326–330.

Achard, P., Vriezen, W.H., Van Der Straeten, D. and Harberd, N.P. (2003) Ethylene regulates *Arabidopsis* development via the modulation of DELLA protein growth repressor function. *Plant Cell*, **15**, 2816–2825.

Ang, L.H., Chattopadhyay, S., Wei, N., Oyama, T., Okada, K., Batschauer, A. and Deng, X.W. (1998) Molecular interaction between COP1 and HY5 defines a regulatory switch for light control of *Arabidopsis* development. *Mol Cell*, **1**, 213–222.

Aravind, L. and Ponting, C.P. (1998) Homologues of 26S proteasome subunits are regulators of transcription and translation. *Protein Sci*, **7**, 1250–1254.

Asano, K., Phan, L., Anderson, J. and Hinnebusch, A.G. (1998) Complex formation by all five homologues of mammalian translation initiation factor 3 subunits from yeast *Saccharomyces cerevisiae*. *J Biol Chem*, **273**, 18573–18585.

Bai, C., Sen, P., Hofmann, K., Ma, L., Goebl, M., Harper, J.W. and Elledge, S.J. (1996) SKP1 connects cell cycle regulators to the ubiquitin proteolysis machinery through a novel motif, the F-box. *Cell*, **86**, 263–274.

Bech-Otschir, D., Kraft, R., Huang, X., Henklein, P., Kapelari, B., Pollmann, C. and Dubiel, W. (2001) COP9 signalosome-specific phosphorylation targets p53 to degradation by the ubiquitin system. *EMBO J*, **20**, 1630–1639.

Benvenuto, G., Formiggini, F., Laflamme, P., Malakhov, M. and Bowler, C. (2002) The photomorphogenesis regulator DET1 binds the amino-terminal tail of histone H2B in a nucleosome context. *Curr Biol*, **12**, 1529–1534.

Bernhardt, A., Lechner, E., Hano, P., Schade, V., Dieterle, M., Anders, M., Dubin, M.J., Benvenuto, G., Bowler, C., Genschik, P. and Hellmann, H. (2006) CUL4 associates with DDB1 and DET1 and its downregulation affects diverse aspects of development in *Arabidopsis thaliana*. *Plant J*, **47**, 591–603.

Bianchi, E., Denti, S., Catena, R., Rossetti, G., Polo, S., Gasparian, S., Putignano, S., Rogge, L. and Pardi, R. (2003) Characterization of human constitutive photomorphogenesis protein 1, a RING finger ubiquitin ligase that interacts with Jun transcription factors and modulates their transcriptional activity. *J Biol Chem*, **278**, 19682–19690.

Bondar, T., Ponomarev, A. and Raychaudhuri, P. (2004) Ddb1 is required for the proteolysis of the *Schizosaccharomyces pombe* replication inhibitor Spd1 during S phase and after DNA damage. *J Biol Chem*, **279**, 9937–9943.

Brand, M. (2001) UV-damaged DNA-binding protein in the TFTC complex links DNA damage recognition to nucleosome acetylation. *Curr Opin Neurobiol*, **11**, 34–42.

Browning, K.S., Gallie, D.R., Hershey, J.W., Hinnebusch, A.G., Maitra, U., Merrick, W.C. and Norbury, C. (2001) Unified nomenclature for the subunits of eukaryotic initiation factor 3. *Trends Biochem Sci*, **26**, 284.

Callis, J., Raasch, J.A. and Vierstra, R.D. (1990) Ubiquitin extension proteins of *Arabidopsis thaliana*. Structure, localization, and expression of their promoters in transgenic tobacco. *J Biol Chem*, **265**, 12486–12493.

Carrano, A.C., Eytan, E., Hershko, A. and Pagano, M. (1999) SKP2 is required for ubiquitin-mediated degradation of the CDK inhibitor p27. *Nat Cell Biol*, **1**, 193–199.

Chamovitz, D.A. and Glickman, M.H. (2002) Quick guide: the COP9 signalosome. *Curr Biol*, **12**, R232.

Chamovitz, D.A., Wei, N., Osterlund, M.T., von Arnim, A.G., Staub, J.M., Matsui, M. and Deng, X.W. (1996) The COP9 complex, a novel multisubunit nuclear regulator involved in light control of a plant developmental switch. *Cell*, **86**, 115–121.

Chandler, P.M., Marion-Poll, A., Ellis, M. and Gubler, F. (2002) Mutants at the Slender1 locus of barley cv Himalaya. Molecular and physiological characterization. *Plant Physiol*, **129**, 181–190.

Chen, H., Shen, Y., Tang, X., Yu, L., Wang, J., Guo, L., Zhang, Y., Zhang, H., Feng, S., Strickland, E., Zheng, N. and Deng, X. W. (2006) *Arabidopsis* CULLIN4 forms an E3 Ubiquitin ligase with RBX1 and the CDD complex in mediating light control of development. *Plant Cell*, **18**, 1991–2004.

Chory, J., Peto, C., Feinbaum, R., Pratt, L. and Ausubel, F. (1989) *Arabidopsis thaliana* mutant that develops as a light-grown plant in the absence of light. *Cell*, **58**, 991–999.

Christian, J.F., Sjodin, T., Benson, D.E., Macdonald, I.D., Sligar, S.G., Champion, P.M. and Unno, M. (2002) The structure of the mammalian 20S proteasome at 2.75 A resolution. *J Biol Chem*, **277**, 2547–2553.

Chuang, H.W., Zhang, W. and Gray, W.M. (2004) *Arabidopsis* ETA2, an apparent ortholog of the human cullin-interacting protein CAND1, is required for auxin responses mediated by the SCF(TIR1) ubiquitin ligase. *Plant Cell*, **16**, 1883–1897.

Clifford, R., Lee, M.H., Nayak, S., Ohmachi, M., Giorgini, F. and Schedl, T. (2000) FOG-2, a novel F-box containing protein, associates with the GLD-1 RNA binding protein and directs male sex determination in the *C. elegans* hermaphrodite germline. *Development*, **127**, 5265–5276.

Cope, G.A. and Deshaies, R.J. (2006) Targeted silencing of Jab1/Csn5 in human cells downregulates SCF activity through reduction of F-box protein levels. *BMC Biochem*, **7**, 1.

Cope, G.A., Suh, G.S., Aravind, L., Schwarz, S.E., Zipursky, S.L., Koonin, E.V. and Deshaies, R.J. (2002) Role of predicted metalloprotease motif of Jab1/Csn5 in Cleavage of NEDD8 from CUL1. *Science*, **15**, 15.

Dealy, M.J., Nguyen, K.V., Lo, J., Gstaiger, M., Krek, W., Elson, D., Arbeit, J., Kipreos, E.T. and Johnson, R.S. (1999) Loss of Cul1 results in early embryonic lethality and dysregulation of cyclin E. *Nat Genet*, **23**, 245–248.

del Pozo, J.C., Dharmasiri, S., Hellmann, H., Walker, L., Gray, W.M. and Estelle, M. (2002a) AXR1-ECR1-dependent conjugation of RUB1 to the *Arabidopsis* Cullin AtCUL1 is required for auxin response. *Plant Cell*, **14**, 421–433.

del Pozo, J.C., Diaz-Trivino, S., Cisneros, N. and Gutierrez, C. (2006) The balance between cell division and endoreplication depends on E2FC-DPB, transcription factors regulated by the ubiquitin-SCFSKP2A pathway in *Arabidopsis*. *Plant Cell*, **18**, 2224–2235.

del Pozo, J.C., Ramirez-Parra, E. and Castellano, M.M. (2002b) *Arabidopsis* E2Fc functions in cell division and is degraded by the ubiquitin-SCF(AtSKP2) pathway in response to light. *Curr Opin Plant Biol*, **5**, 480–486.

Deng, X.W., Caspar, T. and Quail, P.H. (1991) cop1: a regulatory locus involved in light-controlled development and gene expression in *Arabidopsis*. *Genes Dev*, **5**, 1172–1182.

Deng, X.W., Matsui, M., Wei, N., Wagner, D., Chu, A.M., Feldmann, K.A. and Quail, P.H. (1992) COP1, an *Arabidopsis* regulatory gene, encodes a protein with both a zinc-binding motif and a G beta homologous domain. *Cell*, **71**, 791–801.

Deng, X.W. and Serino, G. (2003) The COP9 signalosome: regulating plant development through the control of proteolysis. *Annu Rev Cell Dev Biol*, **19**, 261–286.

Deshaies, R.J. (1999) SCF and Cullin/Ring H2-based ubiquitin ligases. *Annu Rev Cell Dev Biol*, **15**, 435–467.

Devoto, A., Muskett, P.R. and Shirasu, K. (2003) Role of ubiquitination in the regulation of plant defence against pathogens. *Curr Opin Plant Biol*, **6**, 307–311.

Devoto, A., Nieto-Rostro, M., Xie, D., Ellis, C., Harmston, R., Patrick, E., Davis, J., Sherratt, L., Coleman, M. and Turner, J.G. (2002) COI1 links jasmonate signalling and fertility to the SCF ubiquitin-ligase complex in *Arabidopsis*. *Plant J*, **32**, 457–466.

Dharmasiri, S., Dharmasiri, N., Hellmann, H. and Estelle, M. (2003) The RUB/Nedd8 conjugation pathway is required for early development in *Arabidopsis*. *EMBO J*, **22**, 1762–1770.

Dharmasiri, S., Estelle, M. and Dharmasiri, N. (2005) The F-box protein TIR1 is an auxin receptor. *Nature*, **435**, 441–445.

Dharmasiri, S., Hellmann, H. and Estelle, M. (2002) The role of regulated protein degradation in auxin response. *Plant Cell*, **14**, 2137–2144.

Dieterle, M., Thomann, A., Renou, J.P., Parmentier, Y., Cognat, V., Lemonnier, G., Muller, R., Shen, W.H., Kretsch, T. and Genschik, P. (2005) Molecular and functional characterization of *Arabidopsis* Cullin 3A. *Plant J*, **41**, 386–399.

Dill, A., Jung, H.S. and Sun, T.P. (2001) The DELLA motif is essential for gibberellin-induced degradation of RGA. *Proc Natl Acad Sci USA*, **98**, 14162–14167.

Dill, A., Thomas, S.G., Hu, J., Steber, C.M., and Sun, T.P. (2004) The *Arabidopsis* F-box protein SLEEPY1 targets gibberellin signaling repressors for gibberellin-induced degradation. *Plant Cell*, **16**, 1392–1405.

Dohmann, E.M., Kuhnle, C. and Schwechheimer, C. (2005) Loss of the CONSTITUTIVE PHOTOMORPHOGENIC9 signalosome subunit 5 is sufficient to cause the cop/det/fus mutant phenotype in *Arabidopsis*. *Plant Cell*, **17**, 1967–1978.

Dornan, D., Wertz, I., Shimizu, H., Arnott, D., Frantz, G.D., Dowd, P., O'Rourke, K., Koeppen, H. and Dixit, V.M. (2004) The ubiquitin ligase COP1 is a critical negative regulator of p53. *Nature*, **429**, 86–92.

Doronkin, S., Djagaeva, I. and Beckendorf, S.K. (2002) CSN5/Jab1 mutations affect axis formation in the Drosophila oocyte by activating a meiotic checkpoint. *Development*, **129**, 5053–5064.

Doronkin, S., Djagaeva, I. and Beckendorf, S.K. (2003) The COP9 signalosome promotes degradation of cyclin E during early *Drosophila* oogenesis. *Dev Cell*, **4**, 699–710.

Downes, B.P., Stupar, R.M., Gingerich, D.J. and Vierstra, R.D. (2003) The HECT ubiquitin-protein ligase (UPL) family in *Arabidopsis*: UPL3 has a specific role in trichome development. *Plant J*, **35**, 729–742.

Eloy, N.B., Coppens, F., Beemster, G.T., Hemerly, A.S. and Ferreira, P.C. (2006) The *Arabidopsis* anaphase promoting complex (APC): regulation through subunit availability in plant tissues. *Cell Cycle*, **5**, 1957–1965.

Feng, S., Ma, L., Wang, X., Xie, D., Dinesh-Kumar, S.P., Wei, N. and Deng, X.W. (2003) The COP9 signalosome interacts physically with SCF COI1 and modulates jasmonate responses. *Plant Cell*, **15**, 1083–1094.

Feng, S., Shen, Y., Sullivan, J.A., Rubio, V., Xiong, Y., Sun, T.P. and Deng, X.W. (2004) *Arabidopsis* CAND1, an unmodified CUL1-interacting protein, is involved in multiple developmental pathways controlled by ubiquitin/proteasome-mediated protein Degradation. *Plant Cell*, **16**, 1870–1882.

Fleet, C.M. and Sun, T.P. (2005) A DELLAcate balance: the role of gibberellin in plant morphogenesis. *Curr Opin Plant Biol*, **8**, 77–85.

Freemont, P.S. (1993) The RING finger. A novel protein sequence motif related to the zinc finger. *Ann N Y Acad Sci*, **684**, 174–192.

Fu, X. and Harberd, N.P. (2003) Auxin promotes *Arabidopsis* root growth by modulating gibberellin response. *Nature*, **421**, 740–743.

Fu, X., Richards, D.E., Fleck, B., Xie, D., Burton, N. and Harberd, N.P. (2004) The *Arabidopsis* mutant sleepy1gar2-1 protein promotes plant growth by increasing the affinity of the SCFSLY1 E3 ubiquitin ligase for DELLA protein substrates. *Plant Cell*, **16**, 1406–1418.

Glickman, M.H., Rubin, D.M., Coux, O., Wefes, I., Pfeifer, G., Cjeka, Z., Baumeister, W., Fried, V.A. and Finley, D. (1998a) A subcomplex of the proteasome regulatory particle required for ubiquitin-conjugate degradation and related to the COP9-signalosome and eIF3. *Cell*, **94**, 615–623.

Glickman, M.H., Rubin, D.M., Fried, V.A. and Finley, D. (1998b) The regulatory particle of the *S. cerevisiae* proteasome. *Mol Cell Biol*, **18**, 3149–3162.
Glickman, M.H., Rubin, D.M., Coux, O., Wefes, I., Pfeifer, G., Cjeka, Z., Baumeister, W., Fried, V.A. and Finley, D. (1999) Functional analysis of the proteasome regulatory particle. *Mol Biol Rep*, **26**, 21–28.
Gmachl, M., Gieffers, C., Podtelejnikov, A.V., Mann, M. and Peters, J.M. (2000) The RING-H2 finger protein APC11 and the E2 enzyme UBC4 are sufficient to ubiquitinate substrates of the anaphase-promoting complex. *Proc Natl Acad Sci USA*, **97**, 8973–8978.
Goritschnig, S., Zhang, Y. and Li, X. (2007) The ubiquitin pathway is required for innate immunity in *Arabidopsis*. *Plant J*, **49**, 540.
Gray, W.M. and Estelle, I. (2000) Function of the ubiquitin-proteasome pathway in auxin response. *Trends Biochem Sci*, **25**, 133–138.
Gray, W.M., Hellmann, H., Dharmasiri, S. and Estelle, M. (2002) Role of the *Arabidopsis* RING-H2 protein RBX1 in RUB modification and SCF function. *Plant Cell*, **14**, 2137–2144.
Gray, W.M., Kepinski, S., Rouse, D., Leyser, O. and Estelle, M. (2001) Auxin regulates SCF(TIR1)-dependent degradation of AUX/IAA proteins. *Nature*, **414**, 271–276.
Greenberg, J.R., Phan, L., Gu, Z., deSilva, A., Apolito, C., Sherman, F., Hinnebusch, A.G. and Goldfarb, D.S. (1998) Nip1p associates with 40 S ribosomes and the Prt1p subunit of eukaryotic initiation factor 3 and is required for efficient translation initiation. *J Biol Chem*, **273**, 23485–23494.
Groisman, R., Polanowska, J., Kuraoka, I., Sawada, J., Saijo, M., Drapkin, R., Kisselev, A.F., Tanaka, K. and Nakatani, Y. (2003) The ubiquitin ligase activity in the DDB2 and CSA complexes is differentially regulated by the COP9 signalosome in response to DNA damage. *Cell*, **113**, 357–367.
Groll, M. (2000) A gated channel into the proteasome core particle [see comment]. *Nat Struct Biol*, **7**, 1062–1067.
Groll, M., Glaeser, P., Osypka, P., Osypka, T., Kunkel, B., Vallbracht, C. and Escherich, A. (1997) Synthesis, kinetic characterization and X-ray analysis of peptide aldehydes as inhibitors of the 20S proteasomes from Thermoplasma acidophilum and *Saccharomyces cerevisiae*. *Comput Biomed Res*, **30**, 403–413.
Guo, H. and Ecker, J.R. (2003) Plant responses to ethylene gas are mediated by SCF(EBF1/EBF2)-dependent proteolysis of EIN3 transcription factor. *Cell*, **115**, 667–677.
Guo, H. and Ecker, J.R. (2004) The ethylene signaling pathway: new insights. *Curr. Opin Plant Biol*, **7**, 40–49.
Gusmaroli, G., Feng, S. and Deng, X.W. (2004) The *Arabidopsis* CSN5A and CSN5B subunits are present in distinct COP9 signalosome complexes, and mutations in their JAMM domains exhibit differential dominant negative effects on development. *Plant Cell*, **16**, 2984–3001.
Gusmaroli, G., Figueroa, P., Serino, G. and Deng, X.W. (2007) Role of the MPN subunits in COP9 signalosome assembly and activity, and their regulatory interaction with *Arabidopsis* Cullin3-Based E3 Ligases. *Plant Cell*, **19**, 564–581.
Gutierrez, C. and Boniotti, M.B. (2002) G(1) to S transition: more than a cell cycle engine switch. *Plant Cell*, **14**, 3057–3071.
Hanachi, P., Hershey, J.W. and Vornlocher, H.P. (1999) Characterization of the p33 subunit of eukaryotic translation initiation factor-3 from *Saccharomyces cerevisiae*. *J Biol Chem*, **274**, 8546–8553.

Harari-Steinberg, O., Cantera, R., Denti, S., Bianchi, E., Oron, E., Segal, D. and Chamovitz, D.A. (2007) COP9 signalosome subunit 5 (CSN5) regulates the development of *Drosophila* immune system: effects on cactus, dorsal and hematopoiesis. *Genes Cells*, **12**, 183–195.

Harari-Steinberg, O. and Chamovitz, D.A. (2004) The COP9 signalosome: mediating between kinase signaling and protein degradation. *Curr Protein Pept Sci*, **5**, 185–189.

Hardtke, C.S., Gohda, K., Osterlund, M.T., Oyama, T., Okada, K. and Deng, X.W. (2000) HY5 stability and activity in *arabidopsis* is regulated by phosphorylation in its COP1 binding domain [in Process Citation]. *EMBO J*, **19**, 4997–5006.

Harper, J.W. (2004) Neddylating the guardian; Mdm2 catalyzed conjugation of Nedd8 to p53. *Cell*, **118**, 2–4.

Harper, J.W., Burton, J.L. and Solomon, M.J. (2002) The anaphase-promoting complex: it's not just for mitosis any more. *Genes Dev*, **16**, 2179–2206.

Hartmann-Petersen, R., Seeger, M. and Gordon, C. (2003) Transferring substrates to the 26S proteasome. *Trends Biochem Sci*, **28**, 26–31.

Henke, W., Ferrell, K., Bech-Otschir, D., Seeger, M., Schade, R., Jungblut, P., Naumann, M. and Dubiel, W. (1999) Comparison of human COP9 signalsome and 26S proteasome lid. *Mol Biol Rep*, **26**, 29–34.

Hershko, A. and Ciechanover, A. (1998) The ubiquitin system. *Annu Rev Biochem*, **67**, 425–479.

Hershko, A. and Tomkins, G.M. (1971) Studies on the degradation of tyrosine aminotransferase in hepatoma cells in culture. Influence of the composition of the medium and adenosine triphosphate dependence. *J Biol Chem*, **246**, 710–714.

Hetfeld, B.K., Helfrich, A., Kapelari, B., Scheel, H., Hofmann, K., Guterman, A., Glickman, M., Schade, R., Kloetzel, P.M. and Dubiel, W. (2005) The zinc finger of the CSN-associated deubiquitinating enzyme USP15 is essential to rescue the E3 ligase Rbx1. *Curr Biol*, **15**, 1217–1221.

Hinnebusch, A.G. (2006) eIF3: a versatile scaffold for translation initiation complexes. *Trends Biochem Sci*, **31**, 553–562.

Hoareau Alves, K., Bochard, V., Rety, S. and Jalinot, P. (2002) Association of the mammalian proto-oncoprotein Int-6 with the three protein complexes eIF3, COP9 signalosome and 26S proteasome. *FEBS Lett*, **527**, 15–21.

Hofmann, K. and Bucher, P. (1998) The PCI domain: a common theme in three multiprotein complexes. *Trends Biochem Sci*, **23**, 204–205.

Holm, M. and Deng, X.W. (1999) Structural organization and interactions of COP1, a light-regulated developmental switch. *Plant Mol Biol*, **41**, 151–158.

Holm, M., Hardtke, C.S., Gaudet, R. and Deng, X.W. (2001) Identification of a structural motif that confers specific interaction with the WD40 repeat domain of *Arabidopsis* COP1. *EMBO J*, **20**, 118–127.

Holm, M., Ma, L.G., Qu, L.J. and Deng, X.W. (2002) Two interacting bZIP proteins are direct targets of COP1-mediated control of light-dependent gene expression in *Arabidopsis*. *Genes Dev*, **16**, 1247–1259.

Hong, X., Xu, L., Li, X., Zhai, Z. and Shu, H. (2001) CSN3 interacts with IKKgamma and inhibits TNF- but not IL-1-induced NF- kappaB activation. *FEBS Lett*, **499**, 133–136.

Huang, X., Hetfeld, B.K., Seifert, U., Kahne, T., Kloetzel, P.M., Naumann, M., Bech-Otschir, D. and Dubiel, W. (2005) Consequences of COP9 signalosome and 26S proteasome interaction. *FEBS J*, **272**, 3909–3917.

Ingram, G.C., Doyle, S., Carpenter, R., Schultz, E.A., Simon, R. and Coen, E.S. (1997) Dual role for fimbriata in regulating floral homeotic genes and cell division in Antirrhinum. *EMBO J*, **16**, 6521–6534.

Irniger, S., Piatti, S., Michaelis, C. and Nasmyth, K. (1995) Genes involved in sister chromatid separation are needed for B-type cyclin proteolysis in budding yeast [erratum appears in Cell 1998 May 1;93(3):487]. *Cell*, **81**, 269–278.

Itoh, H., Matsuoka, M. and Steber, C.M. (2003) A role for the ubiquitin-26S-proteasome pathway in gibberellin signaling. *Trends Plant Sci*, **8**, 492–497.

Jan, E., Motzny, C.K., Graves, L.E. and Goodwin, E.B. (1999) The STAR protein, GLD-1, is a translational regulator of sexual identity in *Caenorhabditis elegans*. *EMBO J*, **18**, 258–269.

Jordan, A. and Reichard, P. (1998) Ribonucleotide reductases. *Annu Rev Biochem*, **67**, 71–98.

Kamura, T., Conrad, M.N., Yan, Q., Conaway, R.C. and Conaway, J.W. (1999) The Rbx1 subunit of SCF and VHL E3 ubiquitin ligase activates Rub1 modification of cullins Cdc53 and Cul2. *Genes Dev*, **13**, 2928–2933.

Karniol, B., Malec, P. and Chamovitz, D.A. (1999) *Arabidopsis FUSCA5* encodes a novel phosphoprotein that is a component of the COP9 complex. *Plant Cell*, **11**, 839–848.

Karniol, B., Yahalom, A., Kwok, S., Tsuge, T., Matsui, M., Deng, X.W. and Chamovitz, D.A. (1998) The *Arabidopsis* homologue of an eIF3 complex subunit associates with the COP9 complex. *FEBS Lett*, **439**, 173–179.

Kawakami, T., Chiba, T., Suzuki, T., Iwai, K., Yamanaka, K., Minato, N., Suzuki, H., Shimbara, N., Hidaka, Y., Osaka, F., Omata, M. and Tanaka, K. (2001) NEDD8 recruits E2-ubiquitin to SCF E3 ligase. *EMBO J*, **20**, 4003–4012.

Kepinski, S. (2005) The *Arabidopsis* F-box protein TIR1 is an auxin receptor [see comment]. *Nature* **435**, 446–451.

Kepinski, S. and Leyser, O. (2004) Auxin-induced SCFTIR1-Aux/IAA interaction involves stable modification of the SCFTIR1 complex. *Proc Natl Acad Sci USA*, **101**, 12381–12386.

Khush, R.S., Cornwell, W.D., Uram, J.N. and Lemaitre, B. (2002) A ubiquitin-proteasome pathway represses the Drosophila immune deficiency signaling cascade. *Curr Biol*, **12**, 1728–1737.

Kim, T., Hofmann, K., von Arnim, A.G. and Chamovitz, D.A. (2001) PCI complexes: pretty complex interactions in diverse signaling pathways. *Trends Plant Sci*, **6**, 379–386.

Knoblich, J.A., Sauer, K., Jones, L., Richardson, H., Saint, R. and Lehner, C.F. (1994) Cyclin E controls S phase progression and its down-regulation during Drosophila embryogenesis is required for the arrest of cell proliferation. *Cell*, **77**, 107–120.

Kwok, S.F., Staub, J.M. and Deng, X.W. (1999) Characterization of two subunits of *Arabidopsis* 19S proteasome regulatory complex and its possible interaction with the COP9 complex. *J Mol Biol*, **285**, 85–95.

Lam, Y.A., Lawson, T.G., Velayutham, M., Zweier, J.L. and Pickart, C.M. (2002) A proteasomal ATPase subunit recognizes the polyubiquitin degradation signal. *Nature*, **416**, 763–767.

Larsen, P.B. and Cancel, J.D. (2004) A recessive mutation in the RUB1-conjugating enzyme, RCE1, reveals a requirement for RUB modification for control of ethylene biosynthesis and proper induction of basic chitinase and PDF1.2 in *Arabidopsis*. *Plant J*, **38**, 626–638.

Levin, J.Z. and Meyerowitz, E.M. (1995) UFO: an *Arabidopsis* gene involved in both floral meristem and floral organ development. *Plant Cell*, **7**, 529–548.

Liakopoulos, D., Doenges, G., Matuschewski, K. and Jentsch, S. (1998) A novel protein modification pathway related to the ubiquitin system. *EMBO J*, **17**, 2208–2214.

Liechti, R. and Farmer, E.E. (2002) The jasmonate pathway. *Science*, **296**, 1649–1650.

Liu, C., Poitelea, M., Watson, A., Yoshida, S.H., Shimoda, C., Holmberg, C., Nielsen, O. and Carr, A.M. (2005) Transactivation of *Schizosaccharomyces pombe* cdt2+ stimulates a Pcu4-Ddb1-CSN ubiquitin ligase. *EMBO J*, **24**, 3940–3951.

Liu, C., Powell, K.A., Mundt, K., Wu, L., Carr, A.M. and Caspari, T. (2003) Cop9/signalosome subunits and Pcu4 regulate ribonucleotide reductase by both checkpoint-dependent and -independent mechanisms. *Genes Dev*, **14**, 14.

Liu, J., Furukawa, M., Matsumoto, T. and Xiong, Y. (2002a) NEDD8 modification of CUL1 dissociates p120(CAND1), an inhibitor of CUL1-SKP1 binding and SCF ligases. *Mol Cell*, **10**, 1511–1518.

Liu, Y., Schiff, M., Serino, G., Deng, X.-W. and Dinesh-Kumar, S.P. (2002b) Role of SCF ubiquitin-ligase and the COP9 signalosome in the N gene-mediated resistance response to tobacco mosaic virus. *Plant Cell*, **14**, 1483–1496.

Lyapina, S., Cope, G., Shevchenko, A., Serino, G., Tsuge, T., Zhou, C., Wolf, D.A., Wei, N. and Deshaies, R.J. (2001) Promotion of NEDD-CUL1 conjugate cleavage by COP9 signalosome. *Science*, **292**, 1382–1385.

Lykke-Andersen, K. and Wei, N. (2003) Disruption of the COP9 signalosome Csn2 subunit in mice causes deficient cell proliferation, accumulation of p53 and cyclin E, and early embryonic death. *Gene*, **321**, 65–72.

Ma, L., Gao, Y., Qu, L., Chen, Z., Li, J., Zhao, H. and Deng, X.W. (2002) Genomic evidence for COP1 as a repressor of light-regulated gene expression and development in *Arabidopsis*. *Plant Cell*, **14**, 2383–2398.

Malec, P. and Chamovitz, D.A. (2006) Characterization and purification of kinase activities against *Arabidopsis* COP9 signalosome subunit 7. *Isr J Chem*, **46**, 239–246.

Martinez, E., Moggs, J.G., Oulad-Abdelghani, M., Lejeune, F., Dilworth, F.J., Stevenin, J., Almouzni, G. and Tora, L. (2001) Human STAGA complex is a chromatin-acetylating transcription coactivator that interacts with pre-mRNA splicing and DNA damage-binding factors in vivo. *EMBO J*, **20**, 3187–3196.

Masutani, M., Sonenberg, N., Yokoyama, S. and Imataka, H. (2007) Reconstitution reveals the functional core of mammalian eIF3. *EMBO J*, **26**, 3373–3383.

Maytal-Kivity, V., Reis, N., Hofmann, K. and Glickman, M.H. (2002) MPN+, a putative catalytic motif found in a subset of MPN domain proteins from eukaryotes and prokaryotes, is critical for Rpn11 function. *BMC Biochem*, **3**, 28.

McGinnis, K.M., Thomas, S.G., Soule, J.D., Strader, L.C., Zale, J.M., Sun, T.P. and Steber, C.M. (2003) The *Arabidopsis* SLEEPY1 gene encodes a putative F-box subunit of an SCF E3 ubiquitin ligase. *Plant Cell*, **15**, 1120–1130.

Mundt, K.E., Liu, C. and Carr, A.M. (2002) Deletion mutants in COP9/signalosome subunits in fission yeast *Schizosaccharomyces pombe* display distinct phenotypes. *Mol Biol Cell*, **13**, 493–502.

Murray, A.W. (2004) Recycling the cell cycle: cyclins revisited. *Cell*, **116**, 221–234.

Naranda, T., Kainuma, M., MacMillan, S.E. and Hershey, J.W. (1997) The 39-kilodalton subunit of eukaryotic translation initiation factor 3 is essential for the complex's integrity and for cell viability in *Saccharomyces cerevisiae*. *Mol Cell Biol*, **17**, 145–153.

Naumann, M., Bech-Otschir, D., Huang, X., Ferrell, K. and Dubiel, W. (1999) COP9 signalosome-directed c-Jun activation/stabilization is independent of JNK. *J Biol Chem*, **274**, 35297–35300.

Oron, E., Mannervik, M., Rencus, S., Harari-Steinberg, O., Neuman-Silberberg, S., Segal, D. and Chamovitz, D.A. (2002) COP9 signalosome subunits 4 and 5 regulate multiple pleiotropic pathways in Drosophila melanogaster. *Development*, **129**, 4399–4409.

Oron, E., Tuller, T., Li, L., Yekutieli, D., Segal, D., Chor, B., Edgar, B.A. and Chamovitz, D.A. (2007) Genomic analysis of COP9 signalosome function in Drosophila melanogaster reveals a role in temporal regulation of gene expression. *Mol Sys Biol*, *Mol Syst Biol*, **3**, 108.

Osaka, F., Saeki, M., Katayama, S., Aida, N., Toh, E.A., Kominami, K., Toda, T., Suzuki, T., Chiba, T., Tanaka, K. and Kato, S. (2000) Covalent modifier NEDD8 is essential for SCF ubiquitin-ligase in fission yeast. *EMBO J*, **19**, 3475–3484.

Osterlund, M.T., Hardtke, C.S., Wei, N. and Deng, X.W. (2000) Targeted destabilization of HY5 during light-regulated development of *Arabidopsis*. *Nature*, **405**, 462–466.

Pagano, M. (2004) Control of DNA synthesis and mitosis by the Skp2-p27-Cdk1/2 axis. *Mol Cell*, **14**, 414–416.

Peng, J., Carol, P., Richards, D.E., King, K.E., Cowling, R.J., Murphy, G.P. and Harberd, N.P. (1997) The *Arabidopsis* GAI gene defines a signaling pathway that negatively regulates gibberellin responses. *Genes Dev*, **11**, 3194–3205.

Peng, Z., Serino, G. and Deng, X.W. (2001a) Molecular characterization of subunit 6 of the COP9 signalosome and its role in multifaceted developmental processes in *Arabidopsis*. *Plant Cell*, **13**, 2393–2407.

Peng, Z., Serino, G. and Deng, X.W. (2001b) A role of *Arabidopsis* COP9 signalosome in multifaceted developmental processes revealed by the characterization of its subunit 3. *Development*, **128**, 4277–4288.

Peng, Z., Shen, Y., Feng, S., Wang, X., Chitteti, B.N., Vierstra, R.D. and Deng, X.W. (2003) Evidence for a physical association of the COP9 signalosome, the proteasome, and specific SCF E3 ligases in vivo. *Curr Biol*, **13**, 504–505.

Peters, J.M. (2002) The anaphase-promoting complex: proteolysis in mitosis and beyond. *Mol Cell*, **9**, 931–943.

Phan, L., Zhang, X., Asano, K., Anderson, J., Vornlocher, H.P., Greenberg, J.R., Qin, J. and Hinnebusch, A.G. (1998) Identification of a translation initiation factor 3 (eIF3) core complex, conserved in yeast and mammals, that interacts with eIF5. *Mol Cell Biol*, **18**, 4935–4946.

Pickart, C.M. (2001) Ubiquitin enters the new millennium. *Annu Rev Biochem*, **70**, 503–533.

Pickart, C.M. and Hofmann, R.M. (1999) Noncanonical MMS2-encoded ubiquitin-conjugating enzyme functions in assembly of novel polyubiquitin chains for DNA repair. *Cell*, **96**, 645–653.

Rock, K.L. and Goldberg, A.L. (1999) Degradation of cell proteins and the generation of MHC class I-presented peptides. *Annu Rev Immunol*, **17**, 739–779.

Saijo, Y., Sullivan, J.A., Wang, H., Yang, J., Shen, Y., Rubio, V., Ma, L., Hoecker, U. and Deng, X.W. (2003) The COP1-SPA1 interaction defines a critical step in phytochrome A-mediated regulation of HY5 activity. *Genes Dev*, **17**, 2642–2647.

Samach, A., Klenz, J.E., Kohalmi, S.E., Risseeuw, E., Haughn, G.W. and Crosby, W.L. (1999) The UNUSUAL FLORAL ORGANS gene of *Arabidopsis thaliana* is an F-box protein required for normal patterning and growth in the floral meristem. *Plant J*, **20**, 433–445.

Sancho, E., Vila, M.R., Sanchez-Pulido, L., Lozano, J.J., Paciucci, R., Nadal, M., Fox, M., Harvey, C., Bercovich, B., Loukili, N., Ciechanover, A., Lin, S.L., Sanz, F., Estivill, X., Valencia, A. and Thomson, T.M. (1998) Role of UEV-1, an inactive variant of the E2 ubiquitin-conjugating enzymes, in in vitro differentiation and cell cycle behavior of HT-29-M6 intestinal mucosecretory cells. *Mol Cell Biol*, **18**, 576–589.

Sasaki, A., Itoh, H., Ueguchi-Tanaka, M., Ashikari, M., Kitano, H. and Matsuoka, M. (2003) Accumulation of phosphorylated repressor for gibberellin signaling in an F-box mutant [see comment]. *Science*, **299**, 1896–1898.

Schulman, B.A., Carrano, A.C., Jeffrey, P.D., Bowen, Z., Kinnucan, E.R., Finnin, M.S., Elledge, S.J., Harper, J.W., Pagano, M. and Pavletich, N.P. (2000) Insights into SCF ubiquitin ligases from the structure of the Skp1-Skp2 complex. *Nature*, **408**, 381–386.

Schultz, T.F., Kiyosue, T., Yanovsky, M., Wada, M. and Kay, S.A. (2001) A role for LKP2 in the circadian clock of *Arabidopsis*. *Plant Cell*, **13**, 2659–2670.

Schwarz, S.E., Rosa, J.L. and Scheffner, M. (1998) Characterization of human hect domain family members and their interaction with UbcH5 and UbcH7. *J Biol Chem*, **273**, 12148–12154.

Schwechheimer, C., Serino, G., Callis, J., Crosby, W.L., Lyapina, S., Deshaies, R.J., Gray, W.M., Estelle, M. and Deng, X.W. (2001) Interactions of the COP9 signalosome with the E3 ubiquitin ligase SCFTIR1 in mediating auxin response. *Science*, **3**, 3.

Seeger, M., Kraft, R., Ferrell, K., Bech-Otschir, D., Dumdey, R., Schade, R., Gordon, C., Naumann, M. and Dubiel, W. (1998) A novel protein complex involved in signal transduction possessing similarities to 26S proteasome subunits. *FASEB J*, **12**, 469–478.

Seo, H.S., Watanabe, E., Tokutomi, S., Nagatani, A. and Chua, N.H. (2004) Photoreceptor ubiquitination by COP1 E3 ligase desensitizes phytochrome A signaling. *Genes Dev*, **18**, 617–622.

Seo, H.S., Yang, J.Y., Ishikawa, M., Bolle, C., Ballesteros, M.L. and Chua, N.H. (2003) LAF1 ubiquitination by COP1 controls photomorphogenesis and is stimulated by SPA1. *Nature*, **423**, 995–999.

Serino, G., Su, H., Peng, Z., Tsuge, T., Wei, N., Gu, H. and Deng, X.W. (2003) Characterization of the last subunit of the *Arabidopsis* COP9 signalosome: implications for the overall structure and origin of the complex. *Plant Cell*, **15**, 719–731.

Serino, G., Tsuge, T., Kwok, S., Matsui, M., Wei, N. and Deng, X.W. (1999) *Arabidopsis* cop8 and fus4 mutations define the same gene that encodes subunit 4 of the COP9 signalosome. *Plant Cell*, **11**, 1967–1980.

Sheaff, R.J., Groudine, M., Gordon, M., Roberts, J.M. and Clurman, B.E. (1997) Cyclin E-CDK2 is a regulator of p27Kip1. *Genes Dev*, **11**, 1464–1478.

Sherr, C.J., and Roberts, J.M. (1999) CDK inhibitors: positive and negative regulators of G1-phase progression. *Genes Dev*, **13**, 1501–1512.

Silverstone, A.L., Ciampaglio, C.N. and Sun, T. (1998) The *Arabidopsis* RGA gene encodes a transcriptional regulator repressing the gibberellin signal transduction pathway. *Plant Cell*, **10**, 155–169.

Silverstone, A.L., Jung, H.S., Dill, A., Kawaide, H., Kamiya, Y. and Sun, T.P. (2001) Repressing a repressor: gibberellin-induced rapid reduction of the RGA protein in *Arabidopsis*. *Plant Cell*, **13**, 1555–1566.

Slingerland, J. and Pagano, M. (2000) Regulation of the cdk inhibitor p27 and its deregulation in cancer. *J Cell Physiol*, **183**, 10–17.

Smalle, J. and Vierstra, R.D. (2004) The ubiquitin 26S proteasome proteolytic pathway. *Annu Rev Plant Biol*, **55**, 555–590.

Stacey, M.G., Hicks, S.N. and von Arnim, A.G. (1999) Discrete domains mediate the light-responsive nuclear and cytoplasmic localization of *Arabidopsis* COP1. *Plant Cell*, **11**, 349–364.

Staub, J.M., Wei, N. and Deng, X.W. (1996) Evidence for FUS6 as a component of the nuclear-localized COP9 complex in *Arabidopsis*. *Plant Cell*, **8**, 2047–2056.

Stone, S.L., Williams, L.A., Farmer, L.M., Vierstra, R.D. and Callisa, J. (2006) KEEP ON GOING, a RING E3 ligase essential for *Arabidopsis* growth and development, is involved in abscisic acid signaling. *Plant Cell*, **18**, 3415–3428.

Strohmaier, H., Spruck, C.H., Kaiser, P., Won, K.A., Sangfelt, O. and Reed, S.I. (2001) Human F-box protein hCdc4 targets cyclin E for proteolysis and is mutated in a breast cancer cell line [see comment]. *Nature*, **413**, 316–322.

Sun, C.W. and Callis, J. (1997) Independent modulation of *Arabidopsis thaliana* polyubiquitin mRNAs in different organs and in response to environmental changes. *Plant J*, **11**, 1017–1027.

Sutterluty, H., Chatelain, E., Marti, A., Wirbelauer, C., Senften, M., Muller, U. and Krek, W. (1999) p45SKP2 promotes p27Kip1 degradation and induces S phase in quiescent cells. *Nat Cell Biol*, **1**, 207–214.

Suzuki, G., Yanagawa, Y., Kwok, S.F., Matsui, M. and Deng, X.W. (2002) *Arabidopsis* COP10 is a ubiquitin-conjugating enzyme variant that acts together with COP1 and the COP9 signalosome in repressing photomorphogenesis. *Genes Dev*, **16**, 554–559.

Tang, Z., Li, B., Bharadwaj, R., Zhu, H., Ozkan, E., Hakala, K., Deisenhofer, J. and Yu, H. (2001) APC2 Cullin protein and APC11 RING protein comprise the minimal ubiquitin ligase module of the anaphase-promoting complex. *Mol Biol Cell*, **12**, 3839–3851.

Tao, L.Z., Cheung, A.Y., Nibau, C. and Wu, H.M. (2005) RAC GTPases in tobacco and *Arabidopsis* mediate auxin-induced formation of proteolytically active nuclear protein bodies that contain AUX/IAA proteins. *Plant Cell*, **17**, 2369–2383.

Tateishi, K., Omata, M., Tanaka, K. and Chiba, T. (2001) The NEDD8 system is essential for cell cycle progression and morphogenetic pathway in mice. *J Cell Biol*, **155**, 571–579.

Tomoda, K., Kubota, Y., Arata, Y., Mori, S., Maeda, M., Tanaka, T., Yoshida, M., Yoneda-Kato, N. and Kato, J.Y. (2002) The cytoplasmic shuttling and subsequent degradation of p27Kip1 mediated by Jab1/CSN5 and the COP9 signalosome complex. *J Biol Chem*, **277**, 2302–2310.

Tomoda, K., Kubota, Y. and Kato, J. (1999) Degradation of the cyclin-dependent-kinase inhibitor p27Kip1 is instigated by Jab1. *Nature*, **398**, 160–165.

Tran, H.J., Allen, M.D., Lowe, J. and Bycroft, M. (2003) Structure of the Jab1/MPN domain and its implications for proteasome function. *Biochemistry*, **42**, 11460–11465.

Tsvetkov, L.M., Yeh, K.H., Lee, S.J., Sun, H. and Zhang, H. (1999) p27(Kip1) ubiquitination and degradation is regulated by the SCF(Skp2) complex through phosphorylated Thr187 in p27. *Curr Biol*, **9**, 661–664.

Ushijima, K., Sassa, H., Dandekar, A.M., Gradziel, T.M., Tao, R. and Hirano, H. (2003) Structural and transcriptional analysis of the self-incompatibility locus of almond: identification of a pollen-expressed F-box gene with haplotype-specific polymorphism. *Plant Cell*, **15**, 771–781.

Valasek, L., Trachsel, H., Hasek, J. and Ruis, H. (1998) Rpg1, the *Saccharomyces cerevisiae* homologue of the largest subunit of mammalian translation initiation factor 3, is required for translational activity. *J Biol Chem*, **273**, 21253–21260.

Verlhac, M.H., Chen, R.H., Hanachi, P., Hershey, J.W. and Derynck, R. (1997) Identification of partners of TIF34, a component of the yeast eIF3 complex, required for cell proliferation and translation initiation. *EMBO J*, **16**, 6812–6822.

Verma, R., Aravind, L., Oania, R., McDonald, W.H., Yates, J.R., III, Koonin, E.V. and Deshaies, R.J. (2002) Role of Rpn11 metalloprotease in deubiquitination and degradation by the 26S proteasome. *Science*, **298**, 611–615.

Vierstra, R.D. (1996) Proteolysis in plants: mechanisms and functions. *Plant Mol Biol*, **32**, 275–302.

Vlach, J., Hennecke, S. and Amati, B. (1997) Phosphorylation-dependent degradation of the cyclin-dependent kinase inhibitor p27. *EMBO J*, **16**, 5334–5344.

von Arnim, A.G. and Chamovitz, D.A. (2003) Protein homeostasis: a degrading role for Int6/eIF3e. *Curr Biol*, **13**, R323–325.

von Arnim, A.G. and Deng, X.W. (1994) Light inactivation of *Arabidopsis* photomorphogenic repressor COP1 involves a cell-specific regulation of its nucleocytoplasmic partitioning. *Cell*, **79**, 1035–1045.

Vornlocher, H.P., Hanachi, P., Ribeiro, S. and Hershey, J.W. (1999) A 110-kilodalton subunit of translation initiation factor eIF3 and an associated 135-kilodalton protein are encoded by the *Saccharomyces cerevisiae* TIF32 and TIF31 genes. *J Biol Chem*, **274**, 16802–16812.

Wang, X., Feng, S., Nakayama, N., Crosby, W.L., Irish, V., Deng, X.W. and Wei, N. (2003) The COP9 signalosome interacts with SCF(UFO) and participates in *Arabidopsis* flower development. *Plant Cell*, **15**, 1071–1082.

Wei, N., Chamovitz, D.A., and Deng, X.W. (1994a) *Arabidopsis* COP9 is a component of a novel signaling complex mediating light control of development. *Cell*, **78**, 117–124.

Wei, N. and Deng, X.W. (1999) Making sense of the COP9 signalosome. A regulatory protein complex conserved from *Arabidopsis* to human. *Trends Genet*, **15**, 98–103.

Wei, N., Kwok, S.F., von Arnim, A.G., Lee, A., McNellis, T.W., Piekos, B. and Deng, X.W. (1994b) *Arabidopsis* COP8, COP10, and COP11 genes are involved in repression of photomorphogenic development in darkness. *Plant Cell*, **6**, 629–643.

Wertz, I.E., O'Rourke, K.M., Zhang, Z., Dornan, D., Arnott, D., Deshaies, R.J. and Dixit, V.M. (2004) Human de-etiolated-1 regulates c-Jun by assembling a CUL4A ubiquitin ligase. *Science*, **303**, 1371–1374.

Wolf, D.A., Zhou, C. and Wee, S. (2003) The COP9 signalosome: an assembly and maintenance platform for cullin ubiquitin ligases? *Nat Cell Biol*, **5**, 1029–1033.

Woodward, A.W. and Bartel, B. (2005) Auxin: regulation, action, and interaction. *Ann Bot*, **95**, 707–735.

Worley, C.K., Zenser, N., Ramos, J., Rouse, D., Leyser, O., Theologis, A. and Callis, J. (2000) Degradation of Aux/IAA proteins is essential for normal auxin signalling. *Plant J*, **21**, 553–562.

Wu, K., Chen, A. and Pan, Z.Q. (2000) Conjugation of Nedd8 to CUL1 enhances the ability of the ROC1-CUL1 complex to promote ubiquitin polymerization. *J Biol Chem*, **275**, 32317–32324.

Xu, L., Liu, F., Lechner, E., Genschik, P., Crosby, W.L., Ma, H., Peng, W., Huang, D. and Xie, D. (2002) The SCF(COI1) ubiquitin-ligase complexes are required for jasmonate response in *Arabidopsis*. *Plant Cell*, **14**, 1919–1935.

Yahalom, A., Kim, T.H., Winter, E., Karniol, B., von Arnim, A.G. and Chamovitz, D.A. (2001) *Arabidopsis* eIF3e (INT-6) associates with both eIF3c and the COP9 signalosome subunit CSN7. *J Biol Chem*, **276**, 334–340.

Yang, X., Menon, S., Lykke-Andersen, K., Tsuge, T., Di, X., Wang, X., Rodriguez-Suarez, R.J., Zhang, H. and Wei, N. (2002) The COP9 signalosome inhibits p27(kip1) degradation and impedes G1-S phase progression via deneddylation of SCF Cul1. *Curr Biol*, **12**, 667–672.

Yen, H.C. and Chang, E.C. (2003) INT6: a link between the proteasome and tumorigenesis. *Cell Cycle*, **2**, 81–83.

Yi, C. and Deng, X.W. (2005) COP1 – from plant photomorphogenesis to mammalian tumorigenesis. *Trends Cell Biol*, **15**, 618–625.

Zheng, J., Yang, X., Harrell, J.M., Ryzhikov, S., Shim, E.H., Lykke-Andersen, K., Wei, N., Sun, H., Kobayashi, R. and Zhang, H. (2002a) CAND1 binds to unneddylated CUL1 and regulates the formation of SCF ubiquitin E3 ligase complex. *Mol Cell*, **10**, 1519–1526.

Zheng, N., Schulman, B.A., Song, L., Miller, J.J., Jeffrey, P.D., Wang, P., Chu, C., Koepp, D.M., Elledge, S.J., Pagano, M., Conaway, R.C., Conaway, J.W., Harper, J.W. and Pavletich, N.P. (2002b). Structure of the Cul1-Rbx1-Skp1-F-boxSkp2 SCF ubiquitin ligase complex. *Nature*, **416**, 703–709.

Zhou, C., Seibert, V., Geyer, R., Rhee, E., Lyapina, S., Cope, G., Deshaies, R.J. and Wolf, D.A. (2001) The fission yeast COP9/signalosome is involved in cullin modification by ubiquitin-related Ned8p. *BMC Biochem*, **2**, 7.

Zhou, C., Wee, S., Rhee, E., Naumann, M., Dubiel, W. and Wolf, D.A. (2003) Fission yeast COP9/signalosome suppresses cullin activity through recruitment of the deubiquitylating Enzyme Ubp12p. *Mol Cell*, **11**, 927–938.

Zwickl, P. and Baumeister, W. (1999) AAA-ATPases at the crossroads of protein life and death. *Nat Cell Biol*, **1**, E97–E98.

Annual Plant Reviews (2008) **33**, 307–335
doi: 10.1111/b.9781405160025.2008.00011.x
www.interscience.wiley.com

Chapter 11

SIGNALING BETWEEN THE ORGANELLES AND THE NUCLEUS

Aurora Piñas Fernández and Åsa Strand

Umeå Plant Science Centre, Department of Plant Physiology, Umeå University, S-901 87 Umeå, Sweden

Abstract: The function of the eukaryotic cell depends on regulated and reciprocal communication between its different compartments. This includes the exchange of metabolic intermediates and energy equivalents but also exchange of information. The presence of genes encoding organellar proteins in both the nucleus and the organelle necessitates tight coordination of expression by the different genomes, and this has led to the evolution of retrograde control. Retrograde (organelle-to-nucleus) signaling coordinates the expression of nuclear genes encoding organellar proteins with the metabolic and developmental state of the organelle. Complex networks of retrograde signals coordinate cellular activities and assist the cell during plant development and stress response.

Keywords: retrograde communication; plastids; mitochondria; ROS; GUN

11.1 Introduction

The eukaryotic cell is subdivided into functionally distinct, membrane-enclosed compartments or organelles and proper function of the cell depends on tightly regulated exchange of metabolic intermediates, energy equivalents, and information between these different compartments. The chloroplasts and mitochondria, which evolved from free-living prokaryotic organisms that entered the eukaryotic cell through endosymbiosis, have retained their own distinct genomes. However, while the genome of the cyanobacterium *Synechocystis*, which shares the same ancestor as the chloroplast, encodes more than 3100 genes (Kotani and Tabata, 1998), the plastid genome codes for less than 100 of the estimated 3000 proteins in the chloroplast (Leister, 2003). The gradual conversion from endosymbiont to organelle during the course of evolution has clearly been accompanied by a dramatic reduction in genome size as the organelles lost most of their genes to the nucleus and became dependent on their eukaryotic host. The genes that remained in the chloroplast and the mitochondrial genome are photosynthesis- or respiration-related, respectively, or encode components of the organelle gene expression machinery (rRNA, tRNA, and some ribosomal proteins; Unseld *et al.*, 1997; Wakasugi *et al.*, 2001). Thus, the majority of the organellar proteins are encoded in the nucleus and the presence of genes encoding organellar proteins in different cellular compartments presents the complex problem to coordinate the activities of the different genomes of the plant cell (Surpin and Chory, 1997; Surpin *et al.*, 2002; Richly *et al.*, 2003). In order to achieve this coordination, mechanisms to orchestrate nuclear and organellar gene expression have evolved and these include both anterograde (nucleus-to-organelles) and retrograde (organelles-to-nucleus) controls (Rodermel and Park, 2003). Anterograde mechanisms coordinate gene expression in the organelle with cellular and environmental cues that are perceived and choreographed by genes in the nucleus. This type of traffic includes nuclear-encoded proteins that regulate the transcription and translation of organellar genes. Retrograde (organelle-to-nucleus) signaling, on the other hand, coordinates the expression of nuclear genes encoding plastid proteins with the metabolic and developmental state of the plastid and mitochondria (Fig. 11.1; Susek *et al.*, 1993).

11.2 Plastid-to-nucleus communication

The photosynthetic apparatus is composed of proteins encoded by genes from both the nucleus and the chloroplast. For example, in the photosynthetic electron transport complexes of the thylakoid membrane, the core subunits are encoded by the chloroplast genome and the peripheral subunits are encoded by the nuclear genome. In the stroma, the large subunit of Rubisco is chloroplastically encoded whereas the small subunit is nuclear encoded. To ensure that all these photosynthetic complexes are assembled stoichiometrically, and to enable their rapid reorganization in response to a changing environment, the plastids emit signals that regulate nuclear gene expression to match the status of the plastids (Mayfield and Taylor, 1984; Burgess and Taylor, 1988; Taylor, 1989; Susek *et al.*, 1993; Beck, 2005). It is now clear that several different plastid processes produce signals influencing nuclear photosynthetic gene expression (Beck, 2005; Nott *et al.*, 2006) and it has been demonstrated that different signals are produced at different developmental stages (Sullivan and Gray, 2002). To date, the best-characterized plastid signals are mediated through (1) changes of the redox state of the chloroplast, (2) reactive oxygen species, (3) the organellar transcriptional and translational activity (PGE), and (4) accumulation of the tetrapyrrole Mg-protoporphyrinIX (Mg-ProtoIX; Figs. 11.1 and 11.2).

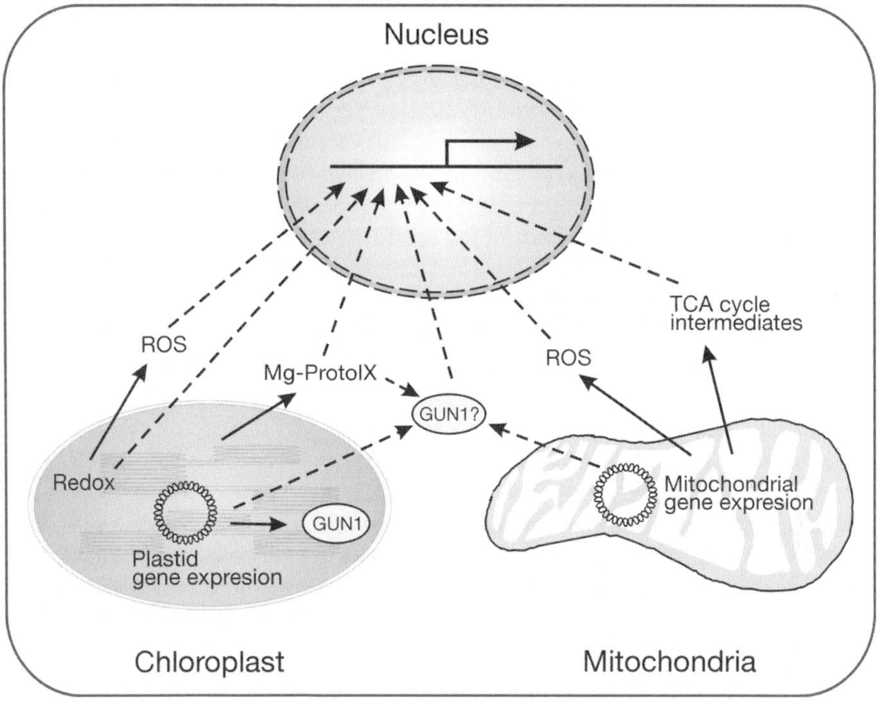

Figure 11.1 Model of retrograde signaling between the organelles and the nucleus. The organelles produce multiple signals at different times of their development, and in response to changes in the environment, that orchestrate major changes in nuclear gene expression. To date, the best-characterized plastid signals are mediated through (1) changes of the redox state of the chloroplast, (2) reactive oxygen species, (3) plastid transcriptional and translational activity (PGE), and (4) accumulation of the tetrapyrrole Mg-ProtoIX. The signals described involved in mitochondria-to-nucleus communication are (1) reactive oxygen species, (2) TCA intermediates, and (3) mitochondrial transcriptional and translational activity.

11.2.1 Chloroplast "redox signals"

In photosynthesis, light energy is absorbed by the light harvesting antennae and converted into chemical energy by the reaction centers. However, when photon intensity exceeds the photon utilization capacity of the chloroplast, photosynthesis becomes photoinhibited and the reaction centers, particularly photosystem II, can become irreversibly damaged and require repair (Aro *et al.*, 1993a,b). Furthermore, elevated excitation pressure has been demonstrated to increase the production of reactive oxygen species (ROS; Karpinski *et al.*, 1997; Huner *et al.*, 1998; Foyer and Allen, 2003) and the damaging effects of ROS include oxidation of lipids, proteins, and enzymes necessary for proper function of the chloroplast and the cell as a whole (Foyer and Allen, 2003). In order to avoid such damage, plants are able to sense when photon intensity exceeds the photon utilization capacity of the chloroplast and communicate this information to stimulate changes in nuclear and chloroplast gene expression. Recent microarray experiments have revealed that the expression of a large number of nuclear-encoded genes is affected by exposure to high light (Rossel *et al.*, 2002; Kimura *et al.*, 2003; Richly *et al.*, 2003; Vanderauwera *et al.*, 2005) and in a study investigating the response to excess light of the different photoreceptors, phyA, phyB, cry1, and cry2, the blue light receptor cry1 was shown to act as a mediator of excess light stress (Kleine *et al.*, 2007). As a consequence the *cry1* mutant displayed a light-sensitive phenotype with a significant photoinactivation of PSII following exposure to excess light (Kleine *et al.*, 2007). This indicates that light stress can also be sensed in a chloroplast-independent manner by a cytosolic/nucleic component. However, the vast majority of the high light-regulated genes are probably regulated by signals independent of cry1 and most likely originating from redox changes in the chloroplast.

11.2.1.1 Plastid signals associated with the rate of electron transport

The redox state of the plastoquinone electron carrier pool (PQ) was shown to be correlated with the expression of photosynthetic genes encoded in both the chloroplast and the nucleus (Fig. 11.2; Escoubas *et al.*, 1995; Huner *et al.*, 1998; Pfannschmidt *et al.*, 1999, 2001; Pfannschmidt, 2003). However, subsequent detailed analysis in cyanobacteria and higher plants using inhibitors of electron transport and wavelengths that preferentially excited either PSII or PSI, demonstrated that the redox state of the PQ pool was not the major source of the high-light chloroplast-to-nucleus communication signal (Hihara *et al.*, 2003; Fey *et al.*, 2005; Piippo *et al.*, 2006). Light shift experiments, combined with DCMU, demonstrated that only 54 *Arabidopsis* genes were "ideal redox regulated genes" or regulated directly by the reduction state of PQ (Fig. 11.2; Fey *et al.*, 2005). Among those 54 genes, none encoded components directly associated with photosynthesis. In a more recent study, the origin of the chloroplast signal was investigated in *Arabidopsis* by modulating the redox state of the PQ pool using wavelengths of light that preferentially excited either PSII or PSI (Piippo *et al.*, 2006). Elements on the reducing side of PSI were shown to be of greater importance in light-regulated modulation of nuclear gene expression than was the redox state of PQ (Fig. 11.2; Piippo *et al.*, 2006). Furthermore, the steady state CO_2-fixation rate was clearly reflected in the orchestration of the expression of nuclear-encoded photosynthesis-related genes (Fig. 11.2), suggesting that the metabolic activity of the chloroplast could also be a source of plastid signals (Piippo *et al.*, 2006). Thus, the more recent work suggests that rather than the reduction state of the PQ itself, the generation of metabolites or signaling molecules during photosynthesis is more likely to be involved in the relay of information from chloroplasts to the nucleus. This new model is attractive because the redox state of the down stream components of the photosynthetic electron transport chain can be directly affected by the energy balance of the cell.

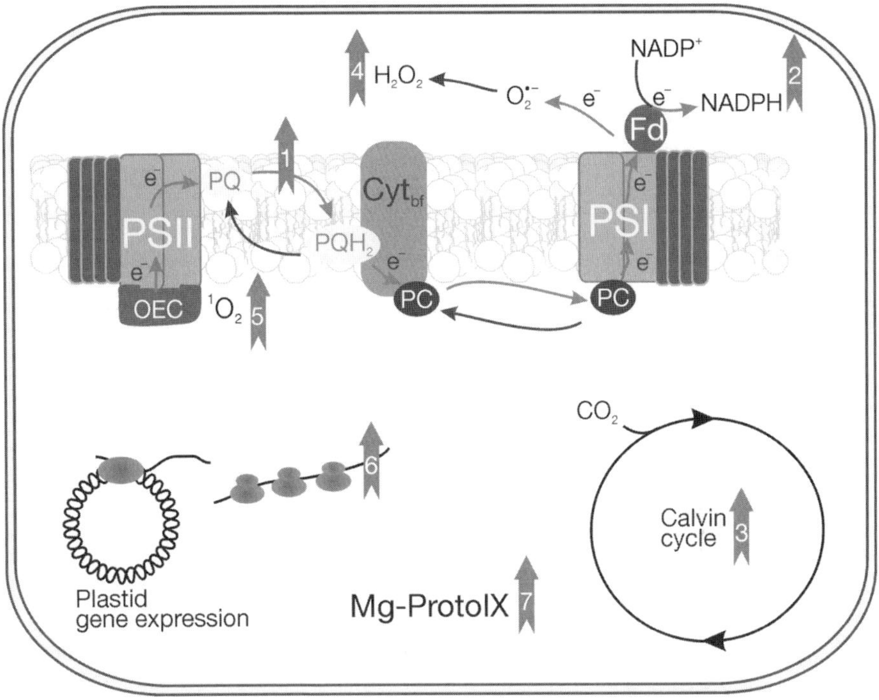

Figure 11.2 Model showing the potential outputs from the plastid that may initiate plastid-to-nucleus communication. (1) The reduction state of the plastoquinone pool; (2) the reducing side of PSI; (3) CO_2 fixation rate; (4) accumulation of hydrogen peroxide (H_2O_2); (5) accumulation of singlet oxygen ($^1O_2$); (6) inhibition of plastid transcription or translation; and (7) accumulation of the tetrapyrrole Mg-ProtoIX.

The actual mechanism(s) of how the plant cell can convey the changes in redox status of the chloroplast to the nucleus is still unknown. Nakamura and Hihara (2006) recently reported that a small LuxR-type regulator in *Synechocystis*, PedR, could work as a sensor for the availability of reducing equivalents supplied from the photosynthetic electron transport chain. The mutant demonstrated mis-regulation of several genes encoding components associated with chlorophyll biosynthesis (Nakamura and Hihara, 2006). The reducing equivalents from electron transport triggered a conformational change critical for the activity of the PedR protein and when the supply of reducing equivalents from photosynthetic electron transport increases upon elevated irradiation PedR was found to be inactivated through a conformational change (Nakamura and Hihara, 2006). This mechanism enables a transient induction or repression of the target genes in response to sudden changes in the light environment.

Furthermore, a unique light- and redox-controlled protein phosphorylation system has evolved in plant thylakoid membranes where intrinsic protein kinases are activated by light or reducing conditions and subsequently phosphorylate the membrane proteins of photosystem II and its light-harvesting antenna LHCII (Vener *et al.*, 1998). The phosphorylation state of these proteins has been suggested to be involved in the regulation of LHC expression in the nucleus (Rintamaki *et al.*, 1997). A small (9 kDa) plant-specific soluble phosphoprotein (TSP9) was found to be released from the thylakoid membrane into the stroma in response to phosphorylation in the light by the redox-dependent protein kinases (Carlberg *et al.*, 2003). The TSP9 protein was suggested to play a role in cell signaling and the regulation of gene expression in response to changing light conditions (Carlberg *et al.*, 2003). However, such a role for TSP9 has not yet been established.

11.2.1.2 Reactive oxygen species

At high irradiances, and other stress conditions such as exposure to low temperature, the equilibrium between production and scavenging of ROS is perturbed, resulting in a transient increase in ROS levels (Karpinski *et al.*, 2003). The ROS generated in chloroplasts are singlet oxygen ($^1O_2$) by PSII and the superoxide anion ($O_2^{\bullet -}$) formed at PSI due to an over-reduction of electron carriers, leading to the reduction of oxygen (the Mehler reaction). The accumulating $O_2^{\bullet -}$ can be metabolized to H_2O_2 (Mullineaux and Karpinski, 2002; Apel and Hirt, 2004). The damaging effects of ROS are oxidation of lipids, proteins, and enzymes necessary for the proper function of the chloroplast and the cell as a whole (Foyer and Allen, 2003). Plants have developed several strategies to protect themselves against excess ROS. Carotenoids, tocopherols, glutathione, and ascorbate are all ROS scavengers (Niyogi, 1999). ROS-converting enzymes such as superoxide dismutase (SOD), ascorbate peroxidase (APX), catalase (CAT), glutathione peroxidases (GPX), and peroxiredoxin (PrxR) that can dismutate O_2 radicals and scavenge H_2O_2 play an important role in ROS defense mechanism (Niyogi, 1999; Mittler *et al.*, 2004; Chapter 7). In correlation with the increased production of ROS, there is an induction of nuclear genes encoding proteins involved in the antioxidant defense system (Karpinski *et al.*, 1997, 1999). Changes in concentrations or rates of ROS production could be initiators of signaling pathways originating in the chloroplast (Mullineaux and Karpinski, 2002). Increases in foliar H_2O_2 concentrations have been shown to be important for the induction of the ascorbate peroxidase gene *APX2*, and for the expression of a number of genes involved in plant development, stress responses, and programmed cell death (Fig. 11.2; Pnueli *et al.*, 2003; Vanderauwera *et al.*, 2005; Gadjev *et al.*, 2006).

A specific function for $^1O_2$ in retrograde communication was discovered by the conditional fluorescent (*flu*) mutant of *Arabidopsis* (Fig. 11.2).FLU is a nuclear-encoded chloroplast protein that plays a key role during negative feedback control of chlorophyll biosynthesis, through interaction with GluTR (Meskauskiene *et al.*, 2001; Meskauskiene and Apel, 2002). Inactivation of this protein in the *flu* mutant leads to the over accumulation of free protochlorophyllide (Pchlide). Following light excitation of Pchlide, $^1O_2$ is generated in the plastid and the *flu* mutant can therefore be used to induce the release of $^1O_2$ in a controlled and noninvasive manner. Transcriptome analysis of the *flu* mutant demonstrated that the release of $^1O_2$-activated genes involved in the cell death-regulating pathway (Danon *et al.*, 2005). Interestingly, the most strongly induced genes in the *flu* mutant are genes encoding ethylene-responsive element-binding proteins that are indicative of ethylene signaling (Gadjev *et al.*, 2006). The identification of a suppressor mutant of the *flu* mutant, *executer1*, contributed to further understanding of the mechanism of cell death triggered by $^1O_2$ (Wagner *et al.*, 2004). The *executer1 flu* double mutant over accumulates $^1O_2$ but abrogates the stress responses of the *flu* mutant. EXECUTER1 has homologs in all higher plants but is unrelated to any known protein. The N-terminal of EXECUTER1 resembles the import signal sequences of plastid proteins. Hence, EXECUTER1 represents a highly conserved plastid protein that seems to enable higher plants to perceive the release of $^1O_2$ as a stress signal and to activate a genetically determined stress response program. Results from Wagner *et al.* (2004) suggested that, depending on the $^1O_2$ levels, two different cell death programs are triggered. First, when the $^1O_2$ levels are low EXECUTER1-dependent cell death program is triggered. Second, when $^1O_2$ levels are high the EXECUTER1-independent cell death takes place. This second cell death reaction seems to be caused primarily by the toxicity of elevated levels of $^1O_2$.

Different stress conditions might provoke the production of specific ROS, which may modulate nuclear gene expression specifically (Gadjev *et al.*, 2006; Laloi *et al.*, 2007). Transcriptome analysis using the *flu* mutant and the herbicide paraquat demonstrated that $^1O_2$activates a distinct set of genes that are different from those induced by superoxide ($O_2^{\bullet-}$) and/or H_2O_2, suggesting that different types of reactive oxygen species activate distinct signaling pathways (Laloi *et al.*, 2007). Furthermore, by using a transgenic line overexpressing the thylakoid bound ascorbate peroxidase (tAPX) gene, the level of H_2O_2 in the plastids could be modulated noninvasively. Overexpression of the H_2O_2-specific scavenger strongly reduced the activation of nuclear genes in plants treated with the herbicide paraquat, which in the light leads to the enhanced generation of $O_2^{\bullet-}$ and H_2O_2. In the *flu* mutant background, overexpression of tAPX resulted in significantly higher $^1O_2$-mediated induction of nuclear-encoded genes, compared to wild-type plants overexpressing tAPX. These results suggest that H_2O_2 antagonizes the $^1O_2$-mediated stress responses observed in the *flu* mutant. This cross-talk between H_2O_2- and $^1O_2$-dependent signaling pathways may contribute to the fine-tuning of the response to environmental stresses.

11.2.2 The response to inhibition of plastid gene expression

Some of the earliest evidence for the existence of a "plastid signal" came from studies of mutants with morphologically aberrant plastids. These mutants are either albinos, have a pale phenotype, or show white–green variegations. Several mutants within these categories demonstrated reduced expression of nuclear-encoded plastid components, including mutants with defective plastid protein synthesis such as the plastid ribosome-deficient *albostrians* barley mutant and the *Brassica napus al* mutant (Bradbeer *et al.*, 1979; Hess *et al.*, 1994; Zubko and Day, 1998). These mutants show reduced expression of nuclear-encoded plastid components suggesting that functional plastids are required for induction of nuclear-encoded photosynthetic genes (Fig. 11.2). The phenotype of these mutants can be mimicked by treating wild-type seedlings with inhibitors of organellar protein translation such as lincomycin, erythromycin, and chloramphenicol (Oelmuller and Mohr, 1986; Sullivan and Gray, 1999, 2000, 2002). Interestingly, inhibitors of plastid transcription and translation have no effect on the induction of LHC and RBCS expression if applied later than 48–72 h after germination. This result suggests that the plastid signal necessary for the induction of nuclear-encoded photosynthesis genes must involve a product of early plastid gene expression (PGE; Sullivan and Gray, 1999, 2000).

The decrease in expression of nuclear-encoded photosynthesis genes in pea seedlings treated with lincomycin and erythromycin (Sullivan and Gray, 1999) was phenocopied in the *prors1-1* and *prors1-2* mutants. The *prors1* mutants were isolated as mutants with a decrease in effective PSII quantum yield (Pesaresi *et al.*, 2006) and the mutations were found in the 5′-untranslated region of the nuclear gene *PROLYL-tRNA SYNTHETASE1* (*PRORS1*). PRORS1 is necessary for translation of the plastidic- and mitochondrial-encoded proteins. The mutation leads to transcriptional downregulation of nuclear genes encoding proteins involved in the photosynthetic light reactions, whereas genes encoding other chloroplast proteins, such as those related to carbohydrate metabolism, were found to be upregulated (Pesaresi *et al.*, 2006). To discriminate the influence of PRORS1 on the regulation of nuclear photosynthetic gene expression, against the influence of PRORS1 on the plastid and the mitochondria, two mutants were used with altered organellar translation: *prpl11-1* with a defect plastid ribosome and *mrpl11* with a defect mitochondrial 50S subunit (Pesaresi *et al.*, 2006). Impaired protein synthesis in either mitochondria or chloroplast failed to convey a transcriptional response such as that observed in the *prors1* mutants. Thus, only when protein synthesis of both organelles was affected, as in the *mrpl11 prpl11-1* double mutant, was a *prors1* like transcriptional response observed. The results of Pesaresi *et al.* (2006) imply that both organelles cooperate in regulating the expression of nuclear photosynthetic genes (Fig. 11.3).

11.2.3 The tetrapyrrole-mediated pathway

Higher plants synthesize four major tetrapyrrole molecules via a common branched pathway: chlorophyll, heme, sitoheme, and phytochromobilin. Many tetrapyrroles are excited by light and if left unquenched they can form

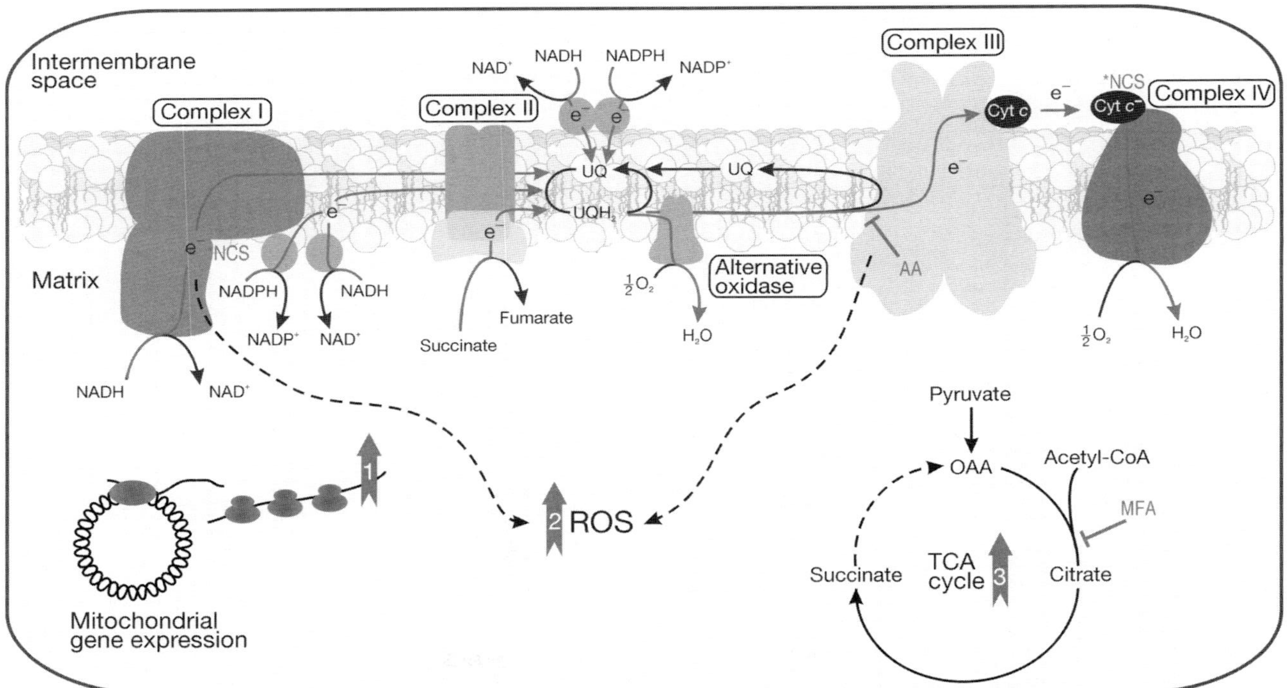

Figure 11.3 Model showing the potential outputs of the mitochondria that may initiate mitochondria-to-nucleus communication. (1) Inhibition of mitochondrial transcription or translation; (2) disruption of respiratory electron transport and accumulation of ROS; and (3) accumulation of TCA cycle intermediates.

highly toxic radicals. Tetrapyrrole synthesis is therefore tightly regulated to prevent the accumulation of intermediates that may endanger the plant cell. Perturbations in the tetrapyrrole pathway have been shown to affect nuclear gene expression in both green algae and higher plants (Beck, 2005).

11.2.3.1 The *gun*2-5 mutants revealed Mg-ProtoIX as a plastid signal

The *g*enome *un*coupled mutants, or *gun1-5* mutants are *Arabidopsis* mutants where the communication between the chloroplast and the nucleus has been disrupted (Susek *et al.*, 1993). The *gun* mutants express nuclear-encoded photosynthetic genes when grown on norflurazon (Susek *et al.*, 1993; Mochizuki *et al.*, 2001; Koussevitzky *et al.*, 2007). Norflurazon inhibits phytoene desaturase in carotenoid biosynthesis. The photooxidation caused by norflurazon treatment is limited to the plastid and results in complete destruction of the thylakoid membrane but does not affect the envelope membrane (Oelmuller and Mohr, 1986; Puente *et al.*, 1996). The resulting inhibition of chloroplast function leads to decreased transcription of nuclear-encoded photosynthetic genes, as demonstrated by nuclear run-on assays (Burgess and Taylor, 1988). Four of the *GUN* genes encode components closely associated with the tetrapyrrole biosynthesis. The *gun2* and *gun3* mutations are alleles of *hy1* and *hy2*, respectively. *HY1/GUN2* encodes heme oxygenase and *HY2/GUN3* encodes phytochromobilin synthase. These enzymes are required for the synthesis of phytochromobilin, the chromophore of phytochrome (Mochizuki *et al.*, 2001). Repression of chlorophyll synthesis in these mutants is thought to be mediated through allosteric inhibition by heme accumulation of glutamyl tRNA reductase, which catalyzes the conversion of glutamate to ALA (Beale, 1999; Terry and Kendrick, 1999). The *GUN5* gene encodes the H-subunit of Mg-chelatase (Mochizuki *et al.*, 2001). Mg-chelatase catalyzes the first reaction in the "chlorophyll branch" of tetrapyrrole biosynthesis, inserting Mg^{2+} into the protoporphyrin ring and is composed of three subunits referred to as ChlH, ChlD, and ChlI (Jensen *et al.*, 1996). *GUN4* encodes a novel chloroplast protein demonstrated to bind both protoporphyrinIX (ProtoIX) and Mg-protoporphyrin (Mg-ProtoIX) and to activate Mg-chelatase *in vitro* (Larkin *et al.*, 2003). GUN4 is predicted to be a soluble protein but has been localized to the chloroplast stroma, thylakoids, and envelopes. GUN4 could also play a role in photoprotection by binding the photooxidizing ProtoIX and Mg-ProtoIX. Consistent with this proposal is the observation that *gun4* seedlings bleach under high light (Larkin *et al.*, 2003). Thorough characterization of the genome uncoupled mutants, *gun2* and *gun5*, with restrictions in defined steps in tetrapyrrole biosynthesis, provided conclusive evidence that Mg-ProtoIX acts as a signaling molecule initiating retrograde communication between the chloroplast and the nucleus (Fig. 11.2; Strand *et al.*, 2003; Nott *et al.*, 2006). Mg-ProtoIX has been shown to accumulate under stress conditions affecting the structure and function of the thylakoid membrane and acts as a negative regulator of nuclear-encoded photosynthesis genes (Strand *et al.*, 2003; Wilson *et al.*, 2003). In the genome uncoupled mutants *gun2* and *gun5*, Mg-ProtoIX does not accumulate to the same critical amount during stress due to reduced flux through the tetrapyrrole pathway. As a result, the plastid signal is lost in these mutants and expression of a large number of nuclear genes encoding chloroplastic proteins directly associated with the photosynthetic reaction is maintained (Strand *et al.*, 2003; Koussevitzky *et al.*, 2007).

In addition to exerting control over nuclear gene expression, stress-induced accumulation of Mg-ProtoIX also affects the expression of the plastid-encoded photosynthesis genes. Transcription of chloroplast-encoded genes in higher plants is performed by two RNA polymerases, plastid-encoded RNA polymerase (PEP), and nuclear-encoded RNA polymerase (NEP). Similar to the nuclear-encoded photosynthesis genes, expression of the PEP-dependent plastid-encoded photosynthesis genes *psbA,psbD,psaA,psaC,* and *rbcL* was mis-regulated following norflurazon treatment in the *gun5* mutant (Ankele *et al.*, 2007). In contrast, expression levels of the NEP-dependent genes *RpoB, AccD, ClpP,* and *Rpl33* (Hess *et al.*, 1993; Hajdukiewicz *et al.*, 1997; Liere *et al.*, 2004) was not repressed by the norflurazon treatment and no difference in expression levels could be found between the *gun5* mutant and wild type (Ankele *et al.*, 2007). The de-repression of the plastid-encoded photosynthesis genes in the *gun5* mutant could be matched with the maintenance of expression of the nuclear-encoded plastidic sigma factors necessary for the function of the multi-subunit enzyme PEP. Thus, in addition to exerting control over nuclear-encoded photosynthesis genes, stress-induced accumulation of Mg-ProtoIX also affects the expression of the plastid-encoded photosynthesis genes by controlling the expression of the sigma factors necessary for the function of the multi-subunit enzyme PEP.

11.2.4 GUN1 and ABI4 act in concert in response to plastid-derived signals

The fifth of the *GUN* genes, *GUN1* was recently cloned and found to encode a chloroplast-localized pentatricopeptide-repeat (PPR) containing protein (Koussevitzky *et al.*, 2007). *GUN1* encodes a member of the P-subfamily of the PPR proteins. A majority of the PPR proteins are targeted to the mitochondria or the plastids where they have been proposed to function in processing, editing, stability, and translation of RNA molecules (Lurin *et al.*, 2004). Among the *gun* mutants, *gun1* is the only mutant that shows a *gun* phenotype when grown in the presence of lincomycin as well as of norflurazon. Furthermore, the *gun1* mutant was demonstrated to be insensitive to the accumulation of Mg-ProtoIX induced by treatment with dipyridyl (DP) in addition to norflurazon. Unlike the GUN2-GUN5, GUN1 does not seem to be involved in chlorophyll biosynthesis, indicating that it is acting downstream of Mg-ProtoIX accumulation. Thus, the *gun1* mutant was proposed to be affected in both the Mg-ProtoIX mediated and the PGE pathway (Koussevitzky *et al.*, 2007). However, treatment with norflurazon and resulting accumulation of Mg-ProtoIX was recently demonstrated to also affect the transcription of the plastid-encoded photosynthesis genes through an effect on expression of the nuclear-encoded sigma factors (Ankele *et al.*, 2007). Thus, norflurazon treatment may also trigger the PGE pathway as a secondary effect.

The promoter regions of the genes mis-regulated in the *gun1* mutant when grown on norflurazon were found to be enriched in an abscisic acid (ABA) response element (ABRE). When mutants affected in ABA synthesis or responses were tested for a *gun* phenotype, only one ABA-related mutant, *abi4* showed *LHCB* mRNA accumulation when chloroplast function was impaired (Koussevitzky *et al.*, 2007). *ABI4* encodes an AP2-type transcription factor and several lines of evidence presented by Koussevitzky *et al.* (2007) indicated that GUN1 and ABI4 act in the same signaling pathway. The *gun1* mutant is epistatic to *abi4*; *LHCB* expression levels in the *gun1abi4* double mutant were similar to the *gun1* single mutant. Yeast one-hybrid assays indicated that ABI4 binds the *LHCB* promoter. From these data Koussevitzky *et al.* (2007) suggest that multiple signals from damaged chloroplasts converge on a common pathway to regulate nuclear gene expression. The existence of a major "Master switch" that coordinates several plastid signals and regulates the expression of the nuclear-encoded plastid proteins as a whole has been suggested (Richly *et al.*, 2003) and ABI4 may be the proposed "Master switch." The *gun1* and *abi4* mutants are affected in both the PGE- and the Mg-ProtoIX-mediated pathways and a model is proposed by Koussevitzky *et al.* (2007) where several plastid signals are integrated within the plastids, and GUN1 is required to either generate or transmit a second, common signal to the nucleus following growth on norflurazon or lincomycin (Fig. 11.1). To further understand the possible role of GUN1 as a coordinator of multiple plastid signals it would be interesting to investigate, without treatment with any inhibitors, the relationship between GUN1, PRORS1, and ABI4 (Fig. 11.1).

11.3 Mitochondria-to-nucleus communication

Plant mitochondria contain a minimum of 325–450 proteins and of those only about 60 are encoded by the mitochondrial genome (Unseld *et al.*, 1997; Notsu *et al.*, 2002). Thus, similar to the chloroplast, most of the mitochondrial proteins are nuclear encoded and mitochondria-to-nucleus communication, referred to as *m*itochondria *r*etrograde *r*egulation (MRR), is necessary for proper mitochondrial function. Much less is known about MRR compared to the plastid-to-nucleus communication in plants, and most of our knowledge of MRR comes from yeast or mammalian systems. Experimental results from budding yeast demonstrate a connection between retrograde communication and nutrient sensing, TOR signaling and aging (Liu and Butow, 2006). Although little is known about MRR in plant cells, several studies indicate that plant mitochondrial dysfunction such as disruption of the electron transport or the TCA cycle, leads to changes in nuclear gene expression (Fig. 11.1; Vanlerberghe and McIntosh, 1992, 1994, 1997).

11.3.1 Tetrapyrroles involved in mitochondria-to-nucleus communication

In yeast (*Sacharomyces cerevisiae*) heme synthesized in the mitochondria has been shown to regulate transcription of nuclear genes encoding mitochondrial proteins (Forsburg and Guarente, 1989). Heme synthesis is directly correlated with oxygen levels in the cellular environment and when cells are grown aerobically heme is synthesized in the mitochondria. Thus, heme functions as a regulatory switch between anaerobic to aerobic growth. The oxygen-dependent regulation of the expression of the genes encoding components of the mitochondrial electron transport chain (mtETC) involves heme-dependent transcription factors such as HAP1p (heme-activating protein) and the HAP2/3/4/5p complex. HAP1p is a transcriptional activator that initiates the expression of several genes encoding mtETC proteins (Kwast *et al.*, 1998) and also activates expression of the ROX1 gene, encoding the aerobic repressor that represses genes encoding proteins required for anaerobic growth (Zhang and Hach, 1999). A current model suggests that heme binds to the "sensory domain" of HAP1p that is part of a DNA-bound, repressed complex (Kwast *et al.*, 1998). The result is the formation of a smaller HAP1p-containing complex that is transcriptionally active (Kwast *et al.*, 1998). The *Arabidopsis* genome has not revealed any homologs of the yeast HAP1 protein, so it remains to be clarified whether plant mitochondria also communicate with the nucleus via heme-HAP1p. The HAP2/3/4/5p complex is also involved in oxygen-induced activation of several nuclear genes encoding yeast mtETC proteins (Schuller, 2003). However, the mode of action of the HAP2/3/4/5p complex and the potential involvement of heme in its regulation is not clear (Kwast *et al.*, 1998). In the *Arabidopsis* genome there are multiple genes encoding homologs of HAP2, 3, and 5 (Gusmaroli *et al.*, 2002) and the *Arabidopsis* genes were able to complement yeast mutants (Edwards *et al.*, 1998). The presence of multiple forms of each HAP homolog in *Arabidopsis*, in contrast to the single genes in yeast and vertebrates, suggests that the HAP2, 3, 5 complex may play diverse roles in gene transcription in higher plants. The HAP2/3/4/5p complex binds to the CCAAT box and two Hap2/3/4/5p-binding motifs have been identified in the promoter region of one of the better-characterized plant MRR reporter gene, *AtAOX1a* (Dojcinovic *et al.*, 2005).

11.3.2 Inhibition of the mitochondrial electron transport chain

The disruption of mtETC, either chemically or genetically, results in the production of mitochondrial reactive oxygen species (mtROS) and the accumulation of mtROS is correlated with the induction of the genes encoding the terminal oxidase of the mitochondria alternative pathway of electron transport,*AOX* (Fig. 11.3; Vanlerberghe and McIntosh, 1992). The *Arabidopsis* genome encodes five AOX genes classified as the *AOX1* type, named from *a* to *d*, and the *AOX2* type (Saisho *et al.*, 1997). Treatment of *Arabidopsis* with inhibitors of mETC, such as Antimycin A (AA; an inhibitor of complex III of the cytochrome pathway), results in increased *AtAOX1a* expression (Fig. 11.3; Clifton *et al.*, 2006). However, when AA was combined with antioxidants in yeast cells (*H. anomala*) and cultured tobacco cells, the accumulation of *AOX* transcript was blocked, suggesting that ROS is the MRR signal controlling *AOX1a* expression (Maxwell *et al.*, 1999). Thus, inhibition of the cytochrome pathway results in increased ROS production, which in turn triggers the signal leading to the induction of *AOX* expression (Wagner, 1995; Maxwell *et al.*, 1999). One important role of AOX is to protect the mitochondria during oxidative stress by ameliorating ROS production (Maxwell *et al.*, 1999, 2002). In addition to the induction of AOX, AA treatment of *Arabidopsis* leaves induces expression of another mtETC component, the CYTC-encoding cytochrome *c* (Yu *et al.*, 2001) and many genes encoding proteins involved in stress responses. The stress-related genes include genes encoding glutathione S-transferase, monodehydroascorbate reductase, and a mitochondrial uncoupling protein (Yu *et al.*, 2001). Analysis of the *Arabidopsis* transcriptome following AA treatment shows a unique pattern of nuclear gene induction, and it is likely that mtROS production triggers a more complex set of distinct signals than that initiated by simple experimental addition of ROS such as hydrogen peroxide (Gadjev *et al.*, 2006).

The MRR has also been studied using maternally inherited mitochondrial DNA deletion mutants of different complexes of mtETC, such as the nonchromosomal stripe (*ncs*) mutants in maize. The *ncs* mutants have diverse defects, ranging from lesions in complexes I and IV of the mitochondrial electron transport chain to defects in mitochondrial translation (Lauer *et al.*, 1990; Marienfeld and Newton, 1994; Newton *et al.*, 1996; Karpova *et al.*, 2002). The various mutants all develop characteristic yellow stripes on the leaves and stems. The striping phenotype results from somatic segregation of cells containing mutant mitochondria from heteroplasmic progenitor cells. Although homoplasmic and near-homoplasmic mutant leaf sectors appear to grow, chloroplasts do not develop fully and do not function photosynthetically (Gu *et al.*, 1993). The maize genome encodes three AOX: *AOX1*, *AOX2*, and *AOX3* (Karpova *et al.*, 2002). Interestingly, deficiencies on each respiratory complex trigger induction of individual AOX. Thus, CI- and CIV-deficient mutants were found to induce *AOX2* and *AOX3*, respectively (Karpova *et al.*, 2002). In contrast, the translation-defective *ncs* mutant demonstrated induction of both *AOX2* and *AOX3* (Karpova *et al.*, 2002).

11.3.3 Accumulation of intermediates of the TCA cycle trigger changes in nuclear gene expression

Inhibition of the tricarboxylic acid cycle by monofluoroacetate (MFA), which inhibits the enzyme aconitase, results in the accumulation of citrate and in the accumulation *AOX* transcript in tobacco cell cultures and *Arabidopsis* (Vanlerberghe and McLntosh, 1996; Zarkovic *et al.*, 2005). Furthermore, accumulation of intermediates of the TCA cycle, such as malate, 2-oxoglutarate and citrate, induces *AOX1* expression even at very low concentrations in tobacco or soybean cells (Fig. 11.3), but does not cause a measurable increase in the cellular ROS concentration (Vanlerberghe and McLntosh, 1996; Djajanegara *et al.*, 2002; Gray *et al.*, 2004). The effects on nuclear gene expression resulting from increased amounts of citrate cannot strictly be assigned MRR because it is not clear whether the effect of MFA is solely in the mitochondria, although effects on mitochondria function and resulting MRR are likely the most prominent. However, increased citrate from MFA treatment could be the cause of *AOX* induction and the action of citrate would represent an MRR pathway that is separate from AA-induced MRR (Rhoads and Subbaiah, 2007).

11.4 Emission of organellar signals

For the signal to be emitted from the organelles to affect nuclear gene expression it must overcome certain obstacles. First, the signal must exit the organelle. Second, the signal needs to be transmitted through the cytoplasm to finally convey the information to the nucleus to initiate a genetic response. The distance that separates the plastid and the nucleus can vary because plastids can move within the cytoplasm (Gray *et al.*, 2001). Studies with the centric diatom *Pleurosira leavis* have demonstrated that chloroplasts move from cell cortex to the nucleus upon illumination as a mechanism of photo avoidance (Makita and Shihira-Ishikawa, 1997; Furukava, 1998). In addition, recent studies with tobacco and *Arabidopsis* cells demonstrated that the ability of the plastids to move toward the nuclei is dependent on cell type and stage of development (Kwok and Hanson, 2004a,b; Natesan *et al.*, 2005). Plastids also have the capacity to form stromules (Kohler *et al.*, 1997). Stromules are highly dynamic structures emanating from the plastid surface and are enclosed by the outer and inner plastid envelope membranes (Gray *et al.*, 2001). The formation of stromules provides a means to increase the plastid surface area with only relatively small changes in plastid volume. In general, stromules are more abundant in tissues containing non-green plastids and in cells containing smaller plastids (Kumar *et al.*, 2005). From studies in tobacco seedlings and adult plants, stromules were shown to be more abundant at the seedling stage compared with similar cell types in mature plants (Kumar *et al.*, 2005). Plastids and stromules lie within grooves and invaginations of the nuclear envelope membrane, which may facilitate the exchange of signals. The formation of stromules appears to be induced by exposure to stress conditions, when the exchange of signals can be crucial to the cellular response and subsequent survival. Seedlings grown on norflurazon, the condition used to initiate accumulation of the signaling metabolite Mg-ProtoIX, have stromules expanding out from the chloroplast toward the nucleus (Fig. 11.4, Color plate 19). This stromule formation on norflurazon was much more striking in hypocotyl cells, where Mg-ProtoIX accumulates, compared to root cells that do not have active chlorophyll biosynthesis (Fig. 11.4, Color plate 19; Ankele *et al.*, 2007). The stress caused by the norflurazon treatment appears to stimulate the formation of stromules, possibly to facilitate the communication between the plastids and the nucleus.

The ROS H_2O_2 is thought to diffuse as easily as water across biological membranes (Karpinski *et al.*, 1999). Chloroplast-derived H_2O_2 therefore could directly influence the functions of cytosolic signaling components. The potential for H_2O_2 to act as an intracellular signaling molecule was demonstrated by its role in the systemic response of plants exposed to excess light (Karpinski *et al.*, 1999), pathogens (Bolwell, 1999), and physical damage (Orozco-Cardenas *et al.*, 2001). It is widely believed that, in biological systems, $^1O_2$ has a very short half-life of around 200 ns (Gorman and Rodgers, 1992) which would preclude its direct involvement in signaling responses. In this case, the hydroperoxides and endoperoxides formed when $^1O_2$ oxidizes biological molecules are more likely to be the actual signaling molecules. Alternatively, chlorophyll degradation products may act as signaling molecules (Krieger-Liszkay, 2005). However, recent observations suggest that the half-life of $^1O_2$ in cells may be much longer than 200 ns, thus allowing diffusion over appreciable distances and across cell membranes (Skovsen *et al.*, 2005; Snyder *et al.*, 2005) in which case $^1O_2$ may itself act as a signal molecule.

The signaling molecule Mg-ProtoIX is exported from the chloroplast (Ankele *et al.*, 2007). Mg-ProtoIX accumulation in norflurazon grown seedlings could be visualized using a laser scanning confocal microscope by taking advantage of the molecules photoreactive properties. The fluorescence images obtained demonstrated that Mg-ProtoIX accumulated in the cytosol (Fig. 11.5, Color plate 20). The relative cytoplasmic accumulation of Mg-ProtoIX was greater in cotyledons, compared to hypocotyls, suggesting that the export mechanisms is more active in leaf tissue and that the export of tetrapyrroles is an active and regulated process. Supporting this conclusion, Beck and colleagues demonstrated in *Chlamydomonas* that the light responsive gene *HSP70*, encoding a heat shock protein, could be induced in the dark by feeding Mg-ProtoIX to the cells (Kropat *et al.*, 1995, 1997, 2000). However, expression of *HSP70* was not induced when the cells were fed the precursor ProtoIX in the dark, with resulting accumulation of Mg-ProroIX in the plastids (Kropat *et al.*, 2000). These data suggest that the plastid export mechanism for Mg-ProtoIX is light regulated. Efflux of related molecules such as heme, heme precursors, phytochromobilin, and chlorophyll-degradation products has been observed from chloroplasts (Thomas and Weinstein, 1990; Matile *et al.*, 1992; Jacobs and Jacobs, 1993; Terry *et al.*, 1993). It is possible that the export mechanism is promiscuous and that the same transport route(s) could be used by Mg-ProtoIX and ProtoIX. Mutant analysis of *Arabidopsis* suggests that a plastid localized ABC-transporter like-protein may be involved in the translocation of ProtoIX across the envelope membrane (Moller *et al.*, 2001). Furthermore, a mitochondrial peripheral-type benzodiazepine receptor from *Arabidopsis* was demonstrated to transport ProtoIX (Lindemann *et al.*, 2004).

The GUN4 protein, which was shown to bind both ProtoIX and Mg-ProtoIX and to be localized to the envelope of the chloroplast (Larkin *et al.*, 2003), may also be involved in the export of tetrapyrroles from the chloroplast.

Identification of signaling metabolites emitted from the organelles used in retrograde communication is novel in plants. However, in yeast (*Sacharomyces cerevisiae*) heme synthesized in the mitochondria regulate transcription of nuclear genes encoding mitochondrial proteins (Forsburg and Guarente, 1989). Heme synthesis is directly correlated to oxygen levels in the cellular environment and when cells are grown aerobically heme is synthesized in the mitochondria, exported from the mitochondria and imported to the nucleus where it activates transcription factors (Zhang and Hach, 1999). Mg-ProtoIX was shown to accumulate evenly through the cytosol and confocal images do not imply that Mg-ProtoIX accumulated specifically in the nucleus (Fig. 11.5, Color plate 20). These data suggest that Mg-ProtoIX likely binds a regulatory protein in the cytosol, and modifies the activity and/or the translocation of this protein, perhaps through a photoreactive action. As a consequence, expression of nuclear-encoded photosynthetic genes is inhibited.

11.5 Targets of retrograde communication

The cytoplasmic and nuclear proteins that participate in the organellar-to-nucleus signaling pathways in plants are poorly understood. Nevertheless, progress has been made on the identification of the *cis* elements in the promoter regions of nuclear genes responding to signals originating in the plastids and the mitochondria. Nuclear genes that encode organellar components are regulated by a diverse group of *cis* regulatory elements that act in combination. Promoter::reporter gene fusions have been used to successfully identify light- and plastid-response elements in the promoters. However, so far it has been impossible to uncouple the plastid- from the light-responsive *cis* elements (Kusnetsov *et al.*, 1996; Puente *et al.*, 1996; McCormac *et al.*, 2001; Strand *et al.*, 2003; Koussevitzky *et al.*, 2007).

11.5.1 Targets of plastid signals

The chlorophyll intermediates Mg-ProtoIX and Mg-ProtoIX methyl ester was found to induce the nuclear-encoded heat shock genes *HSP70A* and *HSP70B* of *Chlamydomonas reinhardtii* mimicking the normal light induction of these genes (Kropat *et al.*, 1995, 1997, 2000). Analysis of the *HSP70A* promoter revealed two regulatory regions that each confer responsiveness to Mg-ProtoIX and light (von Gromoff *et al.*, 2006). These regions were cloned into promoter::reporter gene fusions to verify their biological activity. Mutational analysis of one of those regulatory regions and an alignment with promoters of other Mg-ProtoIX inducible genes uncovered the sequence motif (G/C)CGA(C/T)N(A/G)N_{15}(T/C/A)(A/T/G) that may confer Mg-ProtoIX responsiveness (von Gromoff *et al.*, 2006). A shift of *C. reinhardtii* cultures from dark to light induces a transient accumulation of Mg-ProtoIX in the chloroplast and a fraction of the accumulated intermediate is transported to the cytosol. In the cytosol, Mg-ProtoIX is bound by a regulatory protein, possibly a transcription factor. As a consequence this protein, or a factor activated by the Mg-ProtoIX binding protein, may interact with the *cis* element to stimulate induction of *HSP70A* expression. Thus, light induction of *HSP70A* is mediated via Mg-ProtoIX. With a *cis*-element in hand that is a target for the plastid derived Mg-ProtoIX signal, strategies can be designed to identify the protein(s) that interact with this element upon Mg-ProtoIX accumulation (von Gromoff *et al.*, 2006).

In *Arabidopsis,* promoter::reporter gene fusions revealed that the Mg-ProtoIX signal is mediated by one the best-defined binding sites involved in light-regulated transcription of *LHCB* genes, the G-box (CUF1) element (Strand *et al.*, 2003). A true CUF1 (CACGTA) is present in the promoter region of 18 of the genes mis-regulated in *gun5*, and an additional 24 genes have the closely related CACGTG (Strand *et al.*, 2003). The finding that both light and plastid signals act via CUF1 is consistent with other experiments showing that the *cis*-elements cannot be separated from each other and that light and plastid signaling pathways interact and converge. It has been proposed that this convergence occurs downstream from COP1, a factor that is required for normal photomorphogenesis (Sullivan and Gray, 1999, 2000; Osterlund *et al.*, 2000a,b). Yeast one-hybrid assays indicated that ABI4 binds in close proximity of the CUF1 element of the *LHCB* promoter (Koussevitzky *et al.*, 2007). ABI4 is an AP2-type transcription factor and when the CUF1 element was mutated to TTACGT, ABI4 could not facilitate growth on His-deficient media in the yeast one-hybrid assay. In contrast, when the CUF1 element was mutated to CCACAA, ABI4-enabled growth suggesting that ABI4 binds the CCAC motif (Koussevitzky *et al.*, 2007). In the CUF1 element required for retrograde signaling, two cytosines precede the G-box resulting in a CCAC core. It is conceivable that by binding the CCAC motif, ABI4 inhibits the G-box-mediated light induced expression of photosynthetic genes when chloroplast development is arrested (Koussevitzky *et al.*, 2007).

11.5.2 Targets of mitochondrial signals

The transcriptional induction of *AtAOX1a* by inhibition of the mitochondrial electron transport by AA and of the TCA cycle by MFA has made the *AtAOX1a* a marker gene for MRR in plants. The promoter region of *AtAOX1a* responding to MRR was narrowed down to a 93 bp region (Dojcinovic *et al.*, 2005). Through sequence analysis of this 93 bp region, five different putative transcription factor binding motif were identified: (1) two CCAT-boxes, (2) two G-boxes, (3) a GC-motif, (4) a W-box, and (5) a Dof transcription factor-binding motif (Dojcinovic *et al.*, 2005). The CCAT box is a *cis*-regulatory element widely present in eukaryotes and yeast and in *Saccharomyces cerevisae*, the Hap2/3/4/5p complex bind to this box and regulates the expression of the mitochondrial respiratory complex subunit genes (Schuller, 2003). Proteins that bind G-boxes are basic leucine zipper (bZIP) transcription factors that are implied in developmental processes, light and stress responses, and responses to pathogen attack (Siberil *et al.*, 2001; Jakoby *et al.*, 2002). The WRKY proteins are zing-finger-type transcription factors that bind to W-box sequences (Ishiguro and Nakamura, 1994) that are involved in pathogen resistance and stress responses, development, and senescence processes (Wang *et al.*, 1998; Eulgem *et al.*, 1999; Du and Chen, 2000). Dof transcription factors are involved in plant-specific processes such as responding to light, phytohormones and pathogens, and in seed development and germination (Yanagisawa, 2002). Mutational analysis of each of these motifs demonstrated that all of them are necessary for full induction of *AtAOX1a* in response to mtETC and TCA cycle inhibition. Nevertheless, detailed analysis of mutated forms of the 93 bp promoter::reporter construct showed that for some of the mutations *AtAOX1a* did not respond to AA inhibition of the mtETC but did respond to MFA inhibition of TCA cycle, suggesting that there are at least two separate pathways regulating AOX expression.

11.6 Organelle-to-organelle communication

11.6.1 Plastid-to-plastid communication

Plastids can communicate directly with each other via physical interaction using stromules. The role of stromules in transfer between plastids was observed by visualizing GFP-tagged chloroplast proteins. The movement of either foreign small proteins, such as GFP (Kohler *et al.*, 1997; Tirlapur *et al.*, 1999), or GFP-tagged plastid proteins, such as aspartate aminotransferase and Rubisco (Kwok and Hanson, 2004b), suggests that small metabolites and large macromolecules are able to move through stromules between the plastids. Stromule formation would enhance the exchange of metabolite or signaling molecules between plastids in different parts of the cell and recent studies that have located proteins such as the plastidic hexokinase (Kandel-Kfir *et al.*, 2006) and an ABA-responsive RNA-binding protein (Raab *et al.*, 2006) within stromules that interconnect plastids. Stromules also harbor some normal plastid metabolism, as indicated by the observation of long chain acyl-CoA synthase LACS9 in the envelope membrane of the stromules (Schnurr *et al.*, 2002), and it is possible that stromules contain all normal metabolic activities of the plastids with the exception of the activity associated with the thylakoid membranes, which are not found in stromules. It is still unknown when stromules are formed and become interconnected with other plastids.

However, because they are observed in all plastid types it appears as though stromules are a natural feature of plastid morphology. Stromule morphology and occurrence are variable between cell types (Natesan *et al.*, 2005) and the frequency of stromule formation appears to be related to the developmental stage of plants (Waters *et al.*, 2004; Natesan *et al.*, 2005). Stromules may be considered an ancestral feature that plastids have conserved in order to communicate between themselves since the endosymbiont event (Waters *et al.*, 2004).

11.6.2 Mitochondria-to-chloroplast communication

The mitochondria and chloroplast depend on each other for the exchange of metabolites and energy equivalents. Mitochondrial metabolism is essential for sustaining photosynthetic carbon assimilation and mitochondrial respiration protects photosynthesis against photoinhibition by dissipating excess redox equivalents from chloroplast (Niyogi, 1999, 2000). By using inhibitors of mitochondrial electron transport it was demonstrated that mitochondrial metabolism is essential for photosynthesis (Raghavendra and Padmasree, 2003). Furthermore, genetic defects on mitochondrial function such as in the potato and barley *gdc* mutants, which both have impaired photorespiration, result in an excessive reduction and energization of the chloroplast (Heineke *et al.*, 2001; Igamberdiev *et al.*, 2001). The same effect was also observed in the tobacco *cmsII* mutant, which lacks the major mitochondrial NADH dehydrogenase. In this mutant, the rate of photosynthesis is reduced during the dark-light transition or when carbon fixation and photorespiration are simultaneously active (Sabar *et al.*, 2000; Dutilleul *et al.*, 2003). Photosynthesis provides substrates for mitochondrial respiration and transporters located in the envelope membranes of mitochondria and chloroplasts mediate the exchange of metabolites, and this generates an important channel of communication between the organelles (Raghavendra and Padmasree, 2003). To facilitate this exchange, plastid stromules and mitochondria are closely associated, as was demonstrated by DIC microscopy in sub-epidermal cells of *Iris unguicularis* (Gunning, 2005) and by fluorescence microscopy with GFP-labeled plastid in tobacco (Kwok and Hanson, 2004a,b).

The communication between chloroplast and mitochondria may also be involved in programmed cell death (Yao *et al.*, 2004). ACD2, which encodes the red chlorophyll catabolite reductase, plays a role in protecting cells from Protoporphyrin IX-induced cell death. ACD2 is located in the chloroplast but when *acd2* mutant cells were treated with Protoporphyrin IX to induce light triggered cell death, a change in mitochondrial function was observed almost immediately (Yao *et al.*, 2004). Thus, a stress signal must be released by chloroplast and received by the mitochondria and the intercommunication established in program cell death is an example of close interaction between the organelles (Yao *et al.*, 2004).

11.7 Concluding remarks

The eukaryotic cell depends on the interaction between its different organelles and the nucleus. This includes the exchange of metabolic intermediates and energy equivalents but also information. While it is true that the chloroplast and the mitochondria are dependent on the nucleus to supply much of the genetic information necessary for their function, it is also becoming clear that the organelles produce multiple signals at different times of their development and in response to changes in the environment, and orchestrate major changes in nuclear gene expression. Thus, although the chloroplast and mitochondria are dependent on the nucleus, the organelles exert significant control over the running of the cell. To date, multiple sources of retrograde signals have been identified, such as the accumulation of tetrapyrroles and ROS, for example, and some of the responding elements in the promoter of the nuclear-encoded genes regulated by retrograde signals have been identified. However, the components transducing the signal to the nucleus and the*trans*-acting factors controlling the expression of nuclear-encoded genes remain elusive and identifying these players are challenging tasks for the future.

References

Ankele, E., Pesquet, E., Kindgren, P. and Strand, Å. (2007) *In vivo* visualization of Mg-ProtoporphyrinIX, a coordinator of photosynthetic gene expression in the nucleus and the chloroplast. *Plant Cell*, **19**, 1964–1979.

Apel, K. and Hirt, H. (2004) Reactive oxygen species: metabolism, oxidative stress, and signal transduction. *Annu Rev Plant Biol*, **55**, 373–399.

Aro, E.M., McCaffery, S. and Anderson, J.M. (1993a) Photoinhibition and D1 protein degradation in peas acclimated to different growth irradiances. *Plant Physiol*, **103**, 835–843.

Aro, E.M., Virgin, I. and Andersson, B. (1993b) Photoinhibition of Photosystem II. Inactivation, protein damage and turnover. *Biochim Biophys Acta*, **1143**, 113–134.

Beale, S. (1999) Enzymes of chlorophyll biosynthesis. *Photosynth Res*, **60**, 43–73.

Beck, C.F. (2005) Signaling pathways from the chloroplast to the nucleus. *Planta*, **222**, 743–756.

Bolwell, G.P. (1999) Role of active oxygen species and NO in plant defence responses. *Curr Opin Plant Biol*, **2**, 287–294.

Bradbeer, J., Atkinson, Y., Börner, T. and Hagemann, R. (1979) Cytoplasmic synthesis of plastid polypeptides may be controlled by plastid-synthesised RNA. *Nature*, **279**, 816–817.

Burgess, D. and Taylor, W. (1988) The chloroplast affects the transcription of a nuclear gene family. *Mol Gen Genet*, **214**, 89–96.

Carlberg, I., Hansson, M., Kieselbach, T., Schroder, W.P., Andersson, B. and Vener, A.V. (2003) A novel plant protein undergoing light-induced phosphorylation and release from the photosynthetic thylakoid membranes. *Proc Natl Acad Sci USA*, **100**, 757–762.

Clifton, R., Millar, A.H. and Whelan, J. (2006) Alternative oxidases in *Arabidopsis*: a comparative analysis of differential expression in the gene family provides new insights into function of non-phosphorylating bypasses. *Biochim Biophys Acta*, **1757**, 730–741.

Danon, A., Miersch, O., Felix, G., Camp, R.G. and Apel, K. (2005) Concurrent activation of cell death-regulating signaling pathways by singlet oxygen in *Arabidopsis thaliana*. *Plant J*, **41**, 68–80.

Djajanegara, I., Finnegan, P.M., Mathieu, C., McCabe, T., Whelan, J. and Day, D.A. (2002) Regulation of alternative oxidase gene expression in soybean. *Plant Mol Biol*, **50**, 735–742.

Dojcinovic, D., Krosting, J., Harris, A.J., Wagner, D.J. and Rhoads, D.M. (2005) Identification of a region of the *Arabidopsis* AtAOX1a promoter necessary for mitochondrial retrograde regulation of expression. *Plant Mol Biol*, **58**, 159–175.

Du, L. and Chen, Z. (2000) Identification of genes encoding receptor-like protein kinases as possible targets of pathogen- and salicylic acid-induced WRKY DNA-binding proteins in *Arabidopsis*. *Plant J*, **24**, 837–847.

Dutilleul, C., Driscoll, S., Cornic, G., De Paepe, R., Foyer, C.H. and Noctor, G. (2003) Functional mitochondrial complex I is required by tobacco leaves for optimal photosynthetic performance in photorespiratory conditions and during transients. *Plant Physiol*, **131**, 264–275.

Edwards, D., Murray, J.A. and Smith, A.G. (1998) Multiple genes encoding the conserved CCAAT-box transcription factor complex are expressed in *Arabidopsis*. *Plant Physiol*, **117**, 1015–1022.

Escoubas, J.M., Lomas, M., LaRoche, J. and Falkowski, P.G. (1995) Light intensity regulation of cab gene transcription is signaled by the redox state of the plastoquinone pool. *Proc Natl Acad Sci USA*, **92**, 10237–10241.

Eulgem, T., Rushton, P.J., Schmelzer, E., Hahlbrock, K. and Somssich, I.E. (1999) Early nuclear events in plant defence signalling: rapid gene activation by WRKY transcription factors. *EMBO J*, **18**, 4689–4699.

Fey, V., Wagner, R., Brautigam, K., Wirtz, M., Hell, R., Dietzmann, A., Leister, D., Oelmuller, R. and Pfannschmidt, T. (2005) Retrograde plastid redox signals in the expression of nuclear genes for chloroplast proteins of *Arabidopsis thaliana*. *J Biol Chem*, **280**, 5318–5328.

Forsburg, S.L. and Guarente, L. (1989) Communication between mitochondria and the nucleus in regulation of cytochrome genes in the yeast *Saccharomyces cerevisiae*. *Annu Rev Cell Biol*, **5**, 153–180.

Foyer, C.H. and Allen, J.F. (2003) Lessons from redox signaling in plants. *Antioxid Redox Signal*, **5**, 3–5.

Furukawa, T., Watanabe, M. and Shihira-Ishikawa, I. (1998) Green- and blue-light mediate chloroplast migration in the centric diatom *Pleurosira laevis*. *Protoplasma*, **203**, 214–220.

Gadjev, I., Vanderauwera, S., Gechev, T.S., Laloi, C., Minkov, I.N., Shulaev, V., Apel, K., Inze, D., Mittler, R. and Van Breusegem, F. (2006) Transcriptomic footprints disclose specificity of reactive oxygen species signaling in *Arabidopsis*. *Plant Physiol*, **141**, 436–445.

Gorman, A.A. and Rodgers, M.A. (1992) Current perspectives of singlet oxygen detection in biological environments. *J Photochem Photobiol B*, **14**, 159–176.

Gray, G.R., Villarimo, A.R., Whitehead, C.L. and McIntosh, L. (2004) Transgenic tobacco (*Nicotiana tabacum* L.) plants with increased expression levels of mitochondrial

NADP+-dependent isocitrate dehydrogenase: evidence implicating this enzyme in the redox activation of the alternative oxidase. *Plant Cell Physiol*, **45**, 1413–1425.

Gray, J.C., Sullivan, J.A., Hibberd, J.M. and Hansen, M.R. (2001) Stromules: mobile protrusions and interconnections between plastids. *Plant Biol*, **3**, 223–233.

Gu, J., Miles, D. and Newton, K.J. (1993) Analysis of leaf sectors in the NCS6 mitochondrial mutant of maize. *Plant Cell*, **5**, 963–971.

Gunning, B.E.S. (2005) Plastid stromules: video microscopy of their outgrowth, retraction, tensioning, anchoring, branching, bridging and tip-shedding. *Protoplasma*, **225**, 33–42.

Gusmaroli, G., Tonelli, C. and Mantovani, R. (2002) Regulation of novel members of the *Arabidopsis thaliana* CCAAT-binding nuclear factor Y subunits. *Gene*, **283**, 41–48.

Hajdukiewicz, P.T., Allison, L.A. and Maliga, P. (1997) The two RNA polymerases encoded by the nuclear and the plastid compartments transcribe distinct groups of genes in tobacco plastids. *EMBO J*, **16**, 4041–4048.

Heineke, D., Bykova, N., Gardestrom, P. and Bauwe, H. (2001) Metabolic response of potato plants to an antisense reduction of the P-protein of glycine decarboxylase. *Planta*, **212**, 880–887.

Hess, W.R., Muller, A., Nagy, F. and Borner, T. (1994) Ribosome-deficient plastids affect transcription of light-induced nuclear genes: genetic evidence for a plastid-derived signal. *Mol Gen Genet*, **242**, 305–312.

Hess, W.R., Prombona, A., Fieder, B., Subramanian, A.R. and Borner, T. (1993) Chloroplast rps15 and rpoB/C1/C2 gene cluster are strongly transcribed in ribosome-deficient plastids: evidence for a functioning non-chloroplast-encoded RNA polymerase. *EMBO J*, **12**, 563–571.

Hihara, Y., Sonoike, K., Kanehisa, M. and Ikeuchi, M. (2003) DNA microarray analysis of redox-responsive genes in the genome of the cyanobacterium Synechocystis sp. strain PCC 6803. *J Bacteriol*, **185**, 1719–1725.

Huner, N.P.A., Öquist, G. and Sarhan, F. (1998) Energy balance and acclimation to light and cold. *Trends Plant Sci*, **3**, 224–230.

Igamberdiev, A.U., Bykova, N.V., Lea, P.J. and Gardestrom, P. (2001) The role of photorespiration in redox and energy balance of photosynthetic plant cells: a study with a barley mutant deficient in glycine decarboxylase. *Physiol Plant*, **111**, 427–438.

Ishiguro, S. and Nakamura, K. (1994) Characterization of a cDNA encoding a novel DNA-binding protein, SPF1, that recognizes SP8 sequences in the 5′ upstream regions of genes coding for sporamin and beta-amylase from sweet potato. *Mol Gen Genet*, **244**, 563–571.

Jacobs, J. and Jacobs, N. (1993) Porphyrin accumulation and export by isolated barley (*Hordum vulgare*) plastids. *Plant Physiol*, **101**, 1181–1187.

Jakoby, M., Weisshaar, B., Droge-Laser, W., Vicente-Carbajosa, J., Tiedemann, J., Kroj, T. and Parcy, F. (2002) bZIP transcription factors in *Arabidopsis*. *Trends Plant Sci*, **7**, 106–111.

Jensen, P.E., Willows, R.D., Petersen, B.L., Vothknecht, U.C., Stummann, B.M., Kannangara, C.G., von Wettstein, D. and Henningsen, K.W. (1996) Structural genes for Mg-chelatase subunits in barley: Xantha-f, -g and -h. *Mol Gen Genet*, **250**, 383–394.

Kandel-Kfir, M., Damari-Weissler, H., German, M.A., Gidoni, D., Mett, A., Belausov, E., Petreikov, M., Adir, N. and Granot, D. (2006) Two newly identified membrane-associated and plastidic tomato HXKs: characteristics, predicted structure and intracellular localization. *Planta*, **224**, 1341–1352.

Karpinski, S., Escobar, C., Karpinska, B., Creissen, G. and Mullineaux, P.M. (1997) Photosynthetic electron transport regulates the expression of cytosolic ascorbate peroxidase genes in *Arabidopsis* during excess light stress. *Plant Cell*, **9**, 627–640.

Karpinski, S., Gabrys, H., Mateo, A., Karpinska, B. and Mullineaux, P.M. (2003) Light perception in plant disease defence signaling. *Curr Opin Plant Biol*, **6** (4), 390–396.

Karpinski, S., Reynolds, H., Karpinska, B., Wingsle, G., Creissen, G. and Mullineaux, P. (1999) Systemic signaling and acclimation in response to excess excitation energy in *Arabidopsis*. *Science*, **284**, 654–657.

Karpova, O.V., Kuzmin, E.V., Elthon, T.E. and Newton, K.J. (2002) Differential expression of alternative oxidase genes in maize mitochondrial mutants. *Plant Cell*, **14**, 3271–3284.

Kimura, M., Yamamoto, Y.Y., Seki, M., Sakurai, T., Sato, M., Abe, T., Yoshida, S., Manabe, K., Shinozaki, K. and Matsui, M. (2003) Identification of *Arabidopsis* genes regulated by high light-stress using cDNA microarray. *Photochem Photobiol*, **77**, 226–233.

Kleine, T., Kindgren, P., Benedict, C., Hendricksen, L. and Strand, Å. (2007) Genome-wide gene expression analysis reveals a critical role for CRY1 in the response of *Arabidopsis* to high irradiance. *Plant Physiol*, **144**, 1391–1406

Kohler, R.H., Cao, J., Zipfel, W.R., Webb, W.W. and Hanson, M.R. (1997) Exchange of protein molecules through connections between higher plant plastids. *Science*, **276**, 2039–2042.

Kotani, H. and Tabata, S. (1998) Lessons from sequencing of the genome of a unicellular cyanobacterium, *Synechocystis Sp. Pcc6803*. *Annu Rev Plant Physiol Plant Mol Biol*, **49**, 151–171.

Koussevitzky, S., Nott, A., Mockler, T.C., Hong, F., Sachetto-Martins, G., Surpin, M., Lim, J., Mittler, R. and Chory, J. (2007) Signals from chloroplasts converge to regulate nuclear gene expression. *Science*, **316**, 715–719.

Krieger-Liszkay, A. (2005) Singlet oxygen production in photosynthesis. *J Exp Bot*, **56**, 337–346.

Kropat, J., Oster, U., Rudiger, W. and Beck, C.F. (1997) Chlorophyll precursors are signals of chloroplast origin involved in light induction of nuclear heat-shock genes. *Proc Natl Acad Sci USA*, **94**, 14168–14172.

Kropat, J., Oster, U., Rudiger, W. and Beck, C.F. (2000) Chloroplast signalling in the light induction of nuclear HSP70 genes requires the accumulation of chlorophyll precursors and their accessibility to cytoplasm/nucleus. *Plant J*, **24**, 523–531.

Kropat, J., von Gromoff, E.D., Muller, F.W. and Beck, C.F. (1995) Heat shock and light activation of a Chlamydomonas HSP70 gene are mediated by independent regulatory pathways. *Mol Gen Genet*, **248**, 727–734.

Kumar, S., Naresan, A., Sullivan, J.A. and Gray, J.C. (2005) Stromules: a characteristic cell-specific feature of plastid morphology. *J Exp Bot*, **56**, 787–797.

Kusnetsov, V., Bolle, C., Lubberstedt, T., Sopory, S., Herrmann, R.G. and Oelmuller, R. (1996) Evidence that the plastid signal and light operate via the same cis-acting elements in the promoters of nuclear genes for plastid proteins. *Mol Gen Genet*, **252**, 631–639.

Kwast, K.E., Burke, P.V. and Poyton, R.O. (1998) Oxygen sensing and the transcriptional regulation of oxygen-responsive genes in yeast. *J Exp Biol*, **201**, 1177–1195.

Kwok, E.Y. and Hanson, M.R. (2004a) Stromules and the dynamic nature of plastid morphology. *J Microsc*, **214**, 124–137.

Kwok, E.Y. and Hanson, M.R. (2004b) Plastids and stromules interact with the nucleus and cell membrane in vascular plants. *Plant Cell Rep*, **23**, 188–195.

Laloi, C., Stachowiak, M., Pers-Kamczyc, E., Warzych, E., Murgia, I. and Apel, K. (2007) Cross-talk between singlet oxygen- and hydrogen peroxide-dependent signaling of stress responses in *Arabidopsis thaliana*. *Proc Natl Acad Sci USA*, **104**, 672–677.

Larkin, R.M., Alonso, J.M., Ecker, J.R. and Chory, J. (2003) GUN4, a regulator of chlorophyll synthesis and intracellular signaling. *Science*, **299**, 902–906.

Lauer, M., Knudsen, C., Newton, K.J., Gabay-Laughnan, S. and Laughnan, J.R. (1990) A partially deleted mitochondrial cytochrome oxidase gene in the NCS6 abnormal growth mutant of maize. *New Biol*, **2**, 179–186.

Leister, D. (2003) Chloroplast research in the genomic age. *Trends Genet*, **19**, 47–56.

Liere, K., Kaden, D., Maliga, P. and Borner, T. (2004) Overexpression of phage-type RNA polymerase RpoTp in tobacco demonstrates its role in chloroplast transcription by recognizing a distinct promoter type. *Nucleic Acids Res*, **32**, 1159–1165.

Lindemann, P., Koch, A., Degenhardt, B., Hause, G., Grimm, B. and Papadopoulos, V. (2004) A novel *Arabidopsis thaliana* protein is a functional peripheral-type benzodiazepine receptor. *Plant Cell Physiol*, **45**, 723–733.

Liu, Z. and Butow, R.A. (2006) Mitochondrial retrograde signaling. *Annu Rev Genet*, **40**, 159–185.

Lurin, C., Andres, C., Aubourg, S., Bellaoui, M., Bitton, F., Bruyere, C., Caboche, M., Debast, C., Gualberto, J., Hoffmann, B., Lecharny, A., Le Ret, M., Martin-Magniette, M.L., Mireau, H., Peeters, N., Renou, J.P., Szurek, B., Taconnat, L. and Small, I. (2004) Genome-wide analysis of *Arabidopsis* pentatricopeptide repeat proteins reveals their essential role in organelle biogenesis. *Plant Cell*, **16**, 2089–2103.

Makita, N. and Shihira-Ishikawa, I. (1997) Chloroplast assemblage by mechanical stimulation and its intercelular transmission in diatom cells. *Protoplasma*, **197**, 86–95.

Marienfeld, J.R. and Newton, K.J. (1994) The maize NCS2 abnormal growth mutant has a chimeric nad4-nad7 mitochondrial gene and is associated with reduced complex I function. *Genetics*, **138**, 855–863.

Matile, P., Schellenberg, M. and Peisker, C. (1992) Production and release of a chlorophyll catabolite in isolated senescent chloroplasts. *Planta*, **187**, 230–235.

Maxwell, D.P., Nickels, R. and McIntosh, L. (2002) Evidence of mitochondrial involvement in the transduction of signals required for the induction of genes associated with pathogen attack and senescence. *Plant J*, **29**, 269–279.

Maxwell, D.P., Wang, Y. and McIntosh, L. (1999) The alternative oxidase lowers mitochondrial reactive oxygen production in plant cells. *Proc Natl Acad Sci USA*, **96**, 8271–8276.

Mayfield, S. and Taylor, W. (1984) Carotenoid-deficient maize seedlings fail to accumulate light harvesting chlorophyll a/b binding protein (LHCP) mRNA. *Eur J Biochem*, **144**, 79–84.

McCormac, A.C., Fischer, A., Kumar, A.M., Soll, D. and Terry, M.J. (2001) Regulation of HEMA1 expression by phytochrome and a plastid signal during de-etiolation in *Arabidopsis thaliana*. *Plant J*, **25**, 549–561.

Meskauskiene, R. and Apel, K. (2002) Interaction of FLU, a negative regulator of tetrapyrrole biosynthesis, with the glutamyl-tRNA reductase requires the tetratricopeptide repeat domain of FLU. *FEBS Lett*, **532**, 27–30.

Meskauskiene, R., Nater, M., Goslings, D., Kessler, F., op den Camp, R. and Apel, K. (2001) FLU: a negative regulator of chlorophyll biosynthesis in *Arabidopsis thaliana*. *Proc Natl Acad Sci USA*, **98**, 12826–12831.

Mittler, R., Vanderauwera, S., Gollery, M. and Van Breusegem, F. (2004) Reactive oxygen gene network of plants. *Trends Plant Sci*, **9**, 490–498.

Mochizuki, N., Brusslan, J.A., Larkin, R., Nagatani, A. and Chory, J. (2001) *Arabidopsis* genomes uncoupled 5 (GUN5) mutant reveals the involvement of Mg-chelatase H subunit in plastid-to-nucleus signal transduction. *Proc Natl Acad Sci USA*, **98**, 2053–2058.

Moller, S.G., Kunkel, T. and Chua, N.H. (2001) A plastidic ABC protein involved in intercompartmental communication of light signaling. *Genes Dev*, **15**, 90–103.

Mullineaux, P. and Karpinski, S. (2002) Signal transduction in response to excess light: getting out of the chloroplast. *Curr Opin Plant Biol*, **5**, 43–48.

Nakamura, K. and Hihara, Y. (2006) Photon flux density-dependent gene expression in *Synechocystis sp. PCC 6803* is regulated by a small, redox-responsive, LuxR-type regulator. *J Biol Chem*, **281**, 36758–36766.

Natesan, S.K., Sullivan, J.A. and Gray, J.C. (2005) Stromules: a characteristic cell-specific feature of plastid morphology. *J Exp Bot*, **56**, 787–797.

Newton, K.J., Mariano, J.M., Gibson, C.M., Kuzmin, E. and Gabay-Laughnan, S. (1996) Involvement of S2 episomal sequences in the generation of NCS4 deletion mutation in maize mitochondria. *Dev Genet*, **19**, 277–286.

Niyogi, K.K. (1999) Photoprotection revisited: genetic and molecular approaches. *Annu Rev Plant Physiol Plant Mol Biol*, **50**, 333–359.

Niyogi, K.K. (2000) Safety valves for photosynthesis. *Curr Opin Plant Biol*, **3**, 455–460.

Notsu, Y., Masood, S., Nishikawa, T., Kubo, N., Akiduki, G., Nakazono, M., Hirai, A. and Kadowaki, K. (2002) The complete sequence of the rice (*Oryza sativa* L.) mitochondrial genome: frequent DNA sequence acquisition and loss during the evolution of flowering plants. *Mol Genet Genomics*, **268**, 434–445.

Nott, A., Jung, H.S., Koussevitzky, S. and Chory, J. (2006) Plastid-to-nucleus retrograde signaling. *Annu Rev Plant Biol*, **57**, 739–759.

Oelmuller, R. and Mohr, H. (1986) Photo-oxidative destruction of chloroplasts and its consequences for expression of nuclear genes. *Planta*, **167**, 106–113.

Orozco-Cardenas, M.L., Narvaez-Vasquez, J. and Ryan, C.A. (2001) Hydrogen peroxide acts as a second messenger for the induction of defense genes in tomato plants in response to wounding, systemin, and methyl jasmonate. *Plant Cell*, **13**, 179–191.

Osterlund, M.T., Hardtke, C.S., Wei, N. and Deng, X.W. (2000a) Targeted destabilization of HY5 during light-regulated development of *Arabidopsis*. *Nature*, **405**, 462–466.

Osterlund, M.T., Wei, N. and Deng, X.W. (2000b) The roles of photoreceptor systems and the COP1-targeted destabilization of HY5 in light control of *Arabidopsis* seedling development. *Plant Physiol*, **124**, 1520–1524.

Pesaresi, P., Masiero, S., Eubel, H., Braun, H.P., Bhushan, S., Glaser, E., Salamini, F. and Leister, D. (2006) Nuclear photosynthetic gene expression is synergistically modulated by rates of protein synthesis in chloroplasts and mitochondria. *Plant Cell*, **18**, 970–991.

Pfannschmidt, T. (2003) Chloroplast redox signals: how photosynthesis controls its own genes. *Trends Plant Sci*,**8**, 33–41.

Pfannschmidt, T., Nilsson, A. and Allen, J.F. (1999) Photosynthetic control of chloroplast gene expression. *Nature*, **397**, 625–628.

Pfannschmidt, T., Schutze, K., Brost, M. and Oelmuller, R. (2001) A novel mechanism of nuclear photosynthesis gene regulation by redox signals from the chloroplast during photosystem stoichiometry adjustment. *J Biol Chem*, **276**, 36125–36130.

Piippo, M., Allahverdiyeva, Y., Paakkarinen, V., Suoranta, U.M., Battchikova, N. and Aro, E.M. (2006) Chloroplast-mediated regulation of nuclear genes in *Arabidopsis thaliana* in the absence of light stress. *Physiol Genomics*, **25**, 142–152.

Pnueli, L., Liang, H., Rozenberg, M. and Mittler, R. (2003) Growth suppression, altered stomatal responses, and augmented induction of heat shock proteins in cytosolic ascorbate peroxidase (Apx1)-deficient *Arabidopsis* plants. *Plant J*, **34**, 187–203.

Puente, P., Wei, N. and Deng, X.W. (1996) Combinatorial interplay of promoter elements constitutes the minimal determinants for light and developmental control of gene expression in *Arabidopsis*. *EMBO J*, **15**, 3732–3743.

Raab, S., Toth, Z., de Groot, C., Stamminger, T. and Hoth, S. (2006) ABA-responsive RNA-binding proteins are involved in chloroplast and stromule function in *Arabidopsis* seedlings. *Planta*, **224**, 900–914.

Raghavendra, A.S. and Padmasree, K. (2003) Beneficial interactions of mitochondrial metabolism with photosynthetic carbon assimilation. *Trends Plant Sci*, **8**, 546–553.

Rhoads, D.M. and Subbaiah, C.C. (2007) Mitochondrial retrograde regulation in plants. *Mitochondrion*, **7**, 177–194.

Richly, E., Dietzmann, A., Biehl, A., Kurth, J., Laloi, C., Apel, K., Salamini, F. and Leister, D. (2003) Covariations in the nuclear chloroplast transcriptome reveal a regulatory master-switch. *EMBO Rep*, **4**, 491–498.

Rintamaki, E., Salonen, M., Suoranta, U.M., Carlberg, I., Andersson, B. and Aro, E.M. (1997) Phosphorylation of light-harvesting complex II and photosystem II core proteins shows different irradiance-dependent regulation in vivo. Application of phosphothreonine antibodies to analysis of thylakoid phosphoproteins. *J Biol Chem*, **272**, 30476–30482.

Rodermel, S. and Park, S. (2003) Pathways of intracellular communication: tetrapyrroles and plastid-to-nucleus signaling. *Bioessays*, **25**, 631–636.

Rossel, J.B., Wilson, I.W. and Pogson, B.J. (2002) Global changes in gene expression in response to high light in *Arabidopsis*. *Plant Physiol*, **130**, 1109–1120.

Sabar, M., De Paepe, R. and de Kouchkovsky, Y. (2000) Complex I impairment, respiratory compensations, and photosynthetic decrease in nuclear and mitochondrial male sterile mutants of *Nicotiana sylvestris*. *Plant Physiol*, **124**, 1239–1250.

Saisho, D., Nambara, E., Naito, S., Tsutsumi, N., Hirai, A. and Nakazono, M. (1997) Characterization of the gene family for alternative oxidase from *Arabidopsis thaliana*. *Plant Mol Biol*, **35**, 585–596.

Schnurr, J.A., Shockey, J.M., de Boer, G.J. and Browse, J.A. (2002) Fatty acid export from the chloroplast. Molecular characterization of a major plastidial acyl-coenzyme A synthetase from *Arabidopsis*. *Plant Physiol*, **129**, 1700–1709.

Schuller, H.J. (2003) Transcriptional control of nonfermentative metabolism in the yeast *Saccharomyces cerevisiae*. *Curr Genet*, **43**, 139–160.

Siberil, Y., Doireau, P. and Gantet, P. (2001) Plant bZIP G-box binding factors. Modular structure and activation mechanisms. *Eur J Biochem*, **268**, 5655–5666.

Skovsen, E., Snyder, J.W., Lambert, J.D. and Ogilby, P.R. (2005) Lifetime and diffusion of singlet oxygen in a cell. *J Phys Chem B Condens Matter Mater Surf Interfaces Biophys*, **109**, 8570–8573.

Snyder, J.W., Skovsen, E., Lambert, J.D. and Ogilby, P.R. (2005) Subcellular, time-resolved studies of singlet oxygen in single cells. *J Am Chem Soc*, **127**, 14558–14559.

Strand, Å., Asami, T., Alonso, J., Ecker, J.R. and Chory, J. (2003) Chloroplast to nucleus communication triggered by accumulation of Mg-protoporphyrinIX. *Nature*, **421**, 79–83.

Sullivan, J.A. and Gray, J.C. (1999) Plastid translation is required for the expression of nuclear photosynthesis genes in the dark and in roots of the pea lip1 mutant. *Plant Cell*, **11**, 901–910.

Sullivan, J.A. and Gray, J.C. (2000) The pea light-independent photomorphogenesis1 mutant results from partial duplication of COP1 generating an internal promoter and producing two distinct transcripts. *Plant Cell*, **12**, 1927–1938.

Sullivan, J.A. and Gray, J.C. (2002) Multiple plastid signals regulate the expression of the pea plastocyanin gene in pea and transgenic tobacco plants. *Plant J*, **32**, 763–774.

Surpin, M. and Chory, J. (1997) The co-ordination of nuclear and organellar genome expression in eukaryotic cells. *Essays Biochem*, **32**, 113–125.

Surpin, M., Larkin, R.M. and Chory, J. (2002) Signal transduction between the chloroplast and the nucleus. *Plant Cell*, **14** (Suppl), S327–S338.

Susek, R.E., Ausubel, F.M. and Chory, J. (1993) Signal transduction mutants of *Arabidopsis* uncouple nuclear CAB and RBCS gene expression from chloroplast development. *Cell*, **74**, 787–799.

Taylor, W. (1989) Regulatory interactions between nuclear and plastid genomes. *Annu Rev Plant Physiol Plant Mol Biol*, **40**, 211–233.

Terry, M.J. and Kendrick, R.E. (1999) Feedback inhibition of chlorophyll synthesis in the phytochrome chromophore-deficient aurea and yellow-green-2 mutants of tomato. *Plant Physiol*, **119**, 143–152.

Terry, M.J., Maines, M.D. and Lagarias, J.C. (1993) Inactivation of phytochrome- and phycobiliprotein-chromophore precursors by rat liver biliverdin reductase. *J Biol Chem*, **268**, 26099–26106.

Thomas, J. and Weinstein, J. (1990) Measurement of heme efflux and heme content in isolated developing chloroplasts. *Plant Physiol*, **94**, 1414–1423.

Tirlapur, U.K., Dahse, I., Reiss, B., Meurer, J. and Oelmuller, R. (1999) Characterization of the activity of a plastid-targeted green fluorescent protein in *Arabidopsis. Eur J Cell Biol*, **78**, 233–240.

Unseld, M., Marienfeld, J.R., Brandt, P. and Brennicke, A. (1997) The mitochondrial genome of *Arabidopsis thaliana* contains 57 genes in 366,924 nucleotides. *Nat Genet*, **15**, 57–61.

Vanderauwera, S., Zimmermann, P., Rombauts, S., Vandenabeele, S., Langebartels, C., Gruissem, W., Inze, D. and Van Breusegem, F. (2005) Genome-wide analysis of hydrogen peroxide-regulated gene expression in *Arabidopsis* reveals a high light-induced transcriptional cluster involved in anthocyanin biosynthesis. *Plant Physiol*, **139**, 806–821.

Vanlerberghe, G.C. and McIntosh, L. (1992) Coordinate regulation of cytochrome and alternative pathway respiration in tobacco. *Plant Physiol*, **100**, 1846–1851.

Vanlerberghe, G.C. and McIntosh, L. (1994) Mitochondrial electron transport regulation of nuclear gene expression. Studies with the alternative oxidase gene of tobacco. *Plant Physiol*, **105**, 867–874.

Vanlerberghe, G.C. and McLntosh, L. (1996) Signals regulating the expression of the nuclear gene encoding alternative oxidase of plant mitochondria. *Plant Physiol*, **111**, 589–595.

Vanlerberghe, G.C. and McIntosh, L. (1997) Alternativeoxidase: from gene to function. *Annu Rev Plant Physiol Plant Mol Biol*, **48**, 703–734.

Vener, A.V., Ohad, I. and Andersson, B. (1998) Protein phosphorylation and redox sensing in chloroplast thylakoids. *Curr Opin Plant Biol*, **1**, 217–223.

von Gromoff, E.D., Schroda, M., Oster, U. and Beck, C.F. (2006) Identification of a plastid response element that acts as an enhancer within the *Chlamydomonas* HSP70A promoter. *Nucleic Acids Res*, **34**, 4767–4779.

Wagner, A.M. (1995) A role for active oxygen species as second messengers in the induction of alternative oxidase gene expression in Petunia hybrida cells. *FEBS Lett*, **368**, 339–342.

Wagner, D., Przybyla, D., Op den Camp, R., Kim, C., Landgraf, F., Lee, K.P., Wursch, M., Laloi, C., Nater, M., Hideg, E. and Apel, K. (2004) The genetic basis of singlet oxygen-induced stress responses of *Arabidopsis thaliana*. *Science*, **306**, 1183–1185.

Wakasugi, T., Tsudzuki, T. and Sugiura, M. (2001) The genomics of land plant chloroplasts: gene content and alterations of genomic information by RNA editing. *Photosynth Res*, **70**, 107–118.

Wang, Z., Yang, P., Fan, B. and Chen, Z. (1998) An oligo selection procedure for identification of sequence-specific DNA-binding activities associated with the plant defence response. *Plant J*, **16**, 515–522.

Waters, M.T., Fray, R.G. and Pyke, K.A. (2004) Stromule formation is dependent upon plastid size, plastid differentiation status and the density of plastids within the cell. *Plant J*, **39**, 655–667.

Wilson, K.E., Sieger, S.M. and Huner, N.P.A. (2003) The temperature-dependent accumulation of Mg-protoporphyrin IX and reactive oxygen species in *Chlorella vulgaris*. *Physiol Plant*, **119**, 126–136.

Yanagisawa, S. (2002) The Dof family of plant transcription factors. *Trends Plant Sci*, **7**, 555–560.

Yao, N., Eisfelder, B.J., Marvin, J. and Greenberg, J.T. (2004) The mitochondrion – an organelle commonly involved in programmed cell death in *Arabidopsis thaliana*. *Plant J*, **40**, 596–610.

Yu, J., Nickels, R. and McIntosh, L. (2001) A genome approach to mitochondrial–nuclear communication in *Arabidopsis*. *Plant Physiol Biochem*, **39**, 345–353.

Zarkovic, J., Anderson, S.L. and Rhoads, D.M. (2005) A reporter gene system used to study developmental expression of alternative oxidase and isolate mitochondrial retrograde regulation mutants in *Arabidopsis*. *Plant Mol Biol*, **57**, 871–888.

Zhang, L. and Hach, A. (1999) Molecular mechanism of heme signaling in yeast: the transcriptional activator Hap1 serves as the key mediator. *Cell Mol Life Sci*, **56**, 415–426.

Zubko, M.K. and Day, A. (1998) Stable albinism induced without mutagenesis: a model for ribosome-free plastid inheritance. *Plant J*, **15**, 265–271.

Annual Plant Reviews (2008) **33**, 336–361
doi: 10.1111/b.9781405160025.2008.00012.x www.interscience.wiley.com

Chapter 12

SIGNALING BY PROTEIN PHOSPHORYLATION IN CELL DIVISION

Michiko Sasabe and Yasunori Machida

Division of Biological Science, Graduate School of Science, Nagoya University Chikusa-ku, Nagoya 464-8602, Japan

Abstract: Regulation of cell division in eukaryotes including higher plants is crucial for growth, differentiation, development, and cell death. Protein phosphorylation plays a major role in signaling to control the cell division, and is driven by protein kinases called mitotic kinases, including a cyclin-dependent kinase (CDK), an Aurora kinase, and a mitogen-activated protein kinase (MAPK). In contrast to animals and yeast, our knowledge on the regulatory mechanism of the cell division in plants is very limited. Although plant cells seem to share some basic processes during the cell cycle progression with animal and yeast cells, they have acquired unique molecules orchestrating the cell division, reflecting the plant-specific fashion in cell division such as cytokinesis. This chapter focuses on recent advances in the understanding of signaling molecules, so-called mitotic kinases (CDKs, Auroras, MAPKs) during mitosis and cytokinesis in plant cells.

Keywords: mitosis; cyclin-dependent kinase (CDK); Aurora kinase; mitogen-activated protein kinase (MAPK); cytokinesis; phragmoplast; cell plate

12.1 Introduction

Cell division in eukaryotes is operated by protein kinases that control the progression of cell cycle at various phases. Cyclin-dependent kinases (CDKs) regulate DNA synthesis and mitosis onset. Members of the Aurora kinase and Polo-like kinase families participate in the segregation of chromosomes, modulation of spindle function, and cytokinesis. Mitogen-activated protein kinase (MAPK) cascades are involved in various aspects of cell cycle, including entry into the cell cycle, transition from G2 to M phase, and spindle assembly checkpoint in yeast and animals (Pagès *et al.*, 1993; Minshull *et al.*, 1994; Takenaka *et al.*, 1998; Wright *et al.*, 1999), and they are involved in the progression of cytokinesis in plants (Nishihama *et al.*, 2001, 2002; Krysan *et al.*, 2002; Strompen *et al.*, 2002; Soyano *et al.*, 2003; Yang *et al.*, 2003). Although plants share the basic mechanisms of cell division with other eukaryotes, they have unique molecules modulating the cell cycle. For example, plants have a unique class of CDKs, whereas CDC25 phosphatase appears to lack this function (Boudolf *et al.*, 2006). The ortholog of Polo-like kinase, a mitotic kinase in animals, has not been reported in plants until now. Such differences possibly reflect the distinctive intercellular structures of plants and the noteworthy aspects of plants in the cell cycle system. Many plant cells can replicate DNA in the absence of mitosis, which is known as endoreduplication. The mode of cytokinesis is strikingly different between plants and animals. Plants have rigid cell walls, which disturb the "outside-in" cytokinesis by constriction, unlike in animal cells. Plants have the "inside-out" system for cytokinesis to generate cell plates, including cell wall and cell membrane. In this chapter, we describe the recent advances in the understanding of signaling regulated by the so-called mitotic kinases (CDKs, Auroras, MAPKs) during mitosis in plant cells.

12.2 Progression of mitosis by cyclin-dependent kinases in plants

12.2.1 Plant CDKs and cyclins

Cell cycle is unidirectional and consists of a G1 phase (Gap1), S phase (DNA synthesis), G2 phase (Gap2), and M phase (Mitosis). CDKs are the engines that drive the events of the cell cycle and have also been implicated in the control of gene transcription and other processes. The CDK activity depends on a noncatalytic partner, a cyclin. In yeast, a single CDK (CDC2 in *Schizosaccharomyces pombe*, CDC28 in *Saccharomyces cerevisiae*) is associated with different cyclins that drive the various aspects of the cell cycle. In animals, several CDKs and cyclins regulate the progression of the cell cycle at distinct phases (Morgan, 1997). Similarly, plants have a large number of different *CDK* and *cyclin* genes (Inzé and De Veylder, 2006; Francis, 2007).

Plant CDKs are classified into seven types, designated CDKA to CDKG, and each family has some members, although the detailed functions of CDKC, CDKE, and CDKG in the cell cycle remain unclear (Francis, 2007). *CDKA* is a single gene in *Arabidopsis* and an ortholog of the mammalian CDK1 and yeast CDC2 that conserve PSTAIRE motif, an important motif for binding with cyclins (Ferreira *et al.*, 1991; Hirayama *et al.*, 1991; Imajuku *et al.*, 1992). CDKAs play a central role at both the G1/S and G2/M transitions. CDKBs are plant-specific, and they are divided into two subfamilies according to the differences in the sequence of the cyclin-binding motifs, CDKB1 (PPTALRE) and CDKB2 (P[P/S]TTLRE) (Joubès *et al.*, 2000; Dewitte and Murray, 2003).

The transcripts of CDKB1 and CDKB2 show slightly different accumulation patterns, those of CDKB1 accumulate from S phase until mid-M phase, and those of CDKB2 accumulate during G2 phase and M phase (Fobert *et al.*, 1996; Segers *et al.*, 1996). The protein accumulation and kinase activity of CDKB1 and CDKB2 follow the transcription of these genes, suggesting that the CDKBs function at the G2/M transition and M phase (Magyar *et al.*, 1997; Umeda *et al.*, 1999; Mészáros *et al.*, 2000; Porceddu *et al.*, 2001; Sorrell *et al.*, 2001; Breyne *et al.*, 2002). CDK activity is regulated by phosphorylation, which is responsible for CDK-activating kinases (CAKs). CAKs phosphorylate CDKs and induce the conformational change of CDKs allowing the interaction between CDKs and substrates. CDKDs and CDKF are known to function as CAKs (Umeda *et al.*, 1998, 2005; Vandepoele *et al.*, 2002). CDKDs are divided into three subfamilies and are functional orthologs of vertebrate CAKs (CDK7), whereas CDKF is a plant-specific CAK that has unique enzymatic properties (Umeda *et al.*, 2005). CDKF seems to be able to phosphorylate and activate CDKDs (Shimotohno *et al.*, 2004).

There are seven groups of cyclins in plants, A to D, H, P, and T. In contrast, there are 13 groups in animals (A to L and T). In plants, each group has many members, e.g., *Arabidopsis* has at least 32 putative cyclins. Among them, A-, B-, and D-type cyclins are assumed to be involved in cell cycle control. In summary, A- and B-type cyclins are expressed from S phase to M phase, and they control DNA replication, G2/M transition, and mitosis (Doerner *et al.*, 1996; Magnard *et al.*, 2001; Wang *et al.*, 2004; Weingartner *et al.*, 2004). D-type cyclins appear to regulate G1/S through interaction with CDKA (Nakagami *et al.*, 1999; Boniotti and Gutierrez, 2001; Healy *et al.*, 2001; Menges *et al.*, 2006) and G2/M transition and mitosis, probably with CDKB and CDKA (Schnittger *et al.*, 2002; Kono *et al.*, 2003, 2007). D-type cyclins of *Arabidopsis* (CYCDs) have ten members classified into seven groups (CYCD1 to CYCD7). CYCD3 has three members, CYCD4 has two, and the other groups consist of one member each (Vandepoele *et al.*, 2002). In animals, D-type cyclins are known to form complexes with CDK4 and CDK6, and these complexes phosphorylate and inactivate the retinoblastoma proteins (Rb). This is followed by the release of active E2F transcription factors from Rb to induce the transcription of S-phase-related genes (Harbour and Dean, 2000). In plants, a similar pathway has been reported (Nakagami *et al.*, 1999, 2002; Boniotti and Gutierrez, 2001), suggesting that the regulatory mechanisms for G1/S transition are conserved between animals and plants. However, the large number of D-type cyclins in plants might possess distinct expression patterns and different CDK partners, and they play a variety of roles in the progression of cell cycle. In vertebrates, cyclin H has been identified as the regulatory subunit of CAK (CDK7) and stimulates CDK7 activity (Fisher and Morgan, 1994; Labbé *et al.*, 1994; Mäkelä *et al.*, 1994). Plant H-type cyclins also interact with CDKDs that are classified into CAKs and elevate the kinase activity of CDKDs in vitro (Shimotohno *et al.*, 2004), suggesting that plant cyclin H is a regulatory subunit of CAKs similar to that in animals and yeast.

Recently, the roles and regulatory mechanisms of CDK/cyclin complexes in the plant cell cycle and plant development have been reviewed in detail (Umeda *et al.*, 2005; Inzé and De Veylder, 2006; Francis, 2007). In this chapter, we describe the functions of CDKs involved in the progression of mitosis.

12.2.2 Initiation and progression of mitosis by CDKs

The peaks of kinase activity of CDKA are detected at G1/S and G2/M in *Arabidopsis*, from S phase to early M phase in tobacco, and that of kinase activity of CDKB1;1 are detected at G2/M in *Arabidopsis* and in mid-to-late G2 in tobacco (Joubès *et al.*, 2000; Porceddu *et al.*, 2001; Sorrell *et al.*, 2001).

Overexpression of a dominant-negative CDKA of *Arabidopsis* in tobacco plants produced smaller plants, which is a result of reduced rate of cell division (Hemerly *et al.*, 1995). Recently, it has been reported that the mutant of *CDKA* in *Arabidopsis* exhibited a defect in pollen mitosis during male gametogenesis (Iwakawa *et al.*, 2006; Nowack *et al.*, 2006). These results show that plant CDKA is required for entry into mitosis. Overexpression of a dominant-negative of *Arabidopsis* CDKB1;1 caused delayed G2/M transition in tobacco cells, suggesting that CDKB1 is also involved in mitotic entry (Porceddu *et al.*, 2001).

The regulatory subunits of CDKA and CDKB at G2/M transition are unclear. It has been reported that alfalfa A-type cyclin interacts with CDKA in the yeast two-hybrid system (Roudier *et al.*, 2000), and *Arabidopsis* CYCD4;1 and CYCB1;1 interact with and activate CDKA;1 in vitro (Kono *et al.*, 2003; Weingartner *et al.*, 2004). CYCB1;1 is assumed to interact with and activate G2/M-specific CDKB2;1 in vitro (Weingartner *et al.*, 2004). CDKB2;1 also interacts with CYCA2;2, CYCB2;1, CYCD1;1, and CYCD4;1, and it is activated by the association with CYCD4;1 in vitro (Kono *et al.*, 2003). Although CYCD1;1 binds to both CDKA;1 and CDKB1;1 in addition to CDKB2;1, the functions of the binding are unknown (Kono *et al.*, 2003). In tobacco, the overexpression of nondegradable tobacco cyclin B1 resulted in defective phragmoplast formation, suggesting that CDK/cyclin B complexes regulate the initiation of cytokinesis (Weingartner *et al.*, 2004). In summary, CDKs make complexes with A-, B-, and D4-type cyclins at the proper timing and might regulate the entry into and progression of mitosis. Downstream factors of CDKs in plants remain to be identified.

12.3 Aurora kinases in plants

12.3.1 Aurora kinases in animals

Accurate segregation of chromosomes is one of the major events during mitosis. This event depends on the capture of chromosomes by spindle microtubules during prometaphase and the dynamics of these microtubules during anaphase. In plants, however, the signaling molecules of the progression of chromosome separation are not clear. In animals, it has been reported that there are many proteins essential for chromosome motility and spindle checkpoint, and these proteins are composed of the mitotic spindle, centrosomes (spindle poles), centromeres (kinetochores), and pairing chromosomes (Karsenti and Vernos, 2001; Scholey *et al.*, 2003). The Aurora kinase family belongs to the serine/threonine protein kinase family and is a key regulator to control precise and timely localization and functions of the above-mentioned proteins (Cheeseman *et al.*, 2002; Andrews *et al.*, 2003; Carmena and Earnshaw, 2003). This family has three members—Aurora A, Aurora B, and Aurora C—in vertebrates and two members in *Drosophila* and *Caenorhabditis elegans* (*C. elegans*). All Aurora kinases share similar structures, with catalytic domains flanked by short C-terminal tails and N-terminal domains of variable lengths (Carmena and Earnshaw, 2003). Budding yeast as well as fission yeast have a single Aurora kinase, Ipl1p (Chan and Botstein, 1993), and Ark1p (Petersen *et al.*, 2001). They are structurally and functionally classified as Aurora B.

12.3.1.1 Aurora A

In animals, Aurora A is localized at the spindle poles from prophase to telophase and seems to mainly regulate the assembly of the mitotic spindles through the maturation of centrosomes. The centrosome is known as the microtubule-nucleating center in most animal cells. At the initiation of mitosis, many proteins, including the tubulin ring complex, transforming acidic coiled-coil (TACC) protein family, and other microtubule-associated proteins (MAPs), are recruited to centrosomes. This process is called the centrosome maturation. During the progression of centrosome maturation, microtubules grow out from the centrosomes and form the bipolar spindle. Recent studies show that Aurora A kinase phosphorylates the conserved centrosomal protein TACC, which promotes microtubule assembly (Barros *et al.*, 2005; Kinoshita *et al.*, 2005; Peset *et al.*, 2005). Phosphorylation of TACC by Aurora A seems to ensure the localization of TACC to centrosomes and the assembly of microtubules at the centrosome during mitosis, because mutation of Aurora A phosphorylation sites in TACC dramatically reduced the centrosome localization of TACC and the number of microtubules emanating from the centrosome (Barros *et al.*, 2005; Kinoshita *et al.*, 2005; Peset *et al.*, 2005). It is reported that *Drosophila* TACC (D-TACC) physically interacts with mini spindles (Msps), a *Drosophila* homolog of *Xenopus* XMAP215 (Cullen and Ohkura, 2001; Lee *et al.*, 2001), required for the recruitment of Msps to the centrosomes (Lee *et al.*, 2001). Msps and XMAP215 are microtubule-associated proteins and are known to play critical roles as regulators of microtubule assembly in mitosis (Kinoshita *et al.*, 2002). Since *Xenopus* TACC (TACC3/Maskin) can alter the activity of XMAP215 to polymerize tubulins (Kinoshita *et al.*, 2005), phosphorylation of TACC by Aurora A might regulate the assembly of microtubules through the control of Msps/XMAP215 activity, in addition to the recruitment of TACC-Msps/XMAP215 complex (Barros *et al.*, 2005; Kinoshita *et al.*, 2005). Aurora A also phosphorylates many kinds of proteins that are involved in centrosome maturation and activation of the MAPK cascade regulating the cell cycle during the early embryogenesis of *Xenopus*, although the role of phosphorylation of these proteins is unknown (Table 12.1).

In *Xenopus*, TPX2, a microtubule-associated protein that localizes to centrosomes, is also a binding partner with Aurora A and is phosphorylated by that kinase in vitro (Kufer *et al.*, 2002; Eyers *et al.*, 2003; Tsai *et al.*, 2003). TPX2 appears to block the access of protein phosphatase 1 (PP1) to Aurora A, allowing the accumulation of active Aurora A by autophosphorylation (Eyers *et al.*, 2003; Tsai *et al.*, 2003). Depletion of TPX2 and TPXL1 (an ortholog of TPX2 in *C. elegans*) inhibits bipolar spindle assembly in mitotic *Xenopus* extracts and *C. elegans*, suggesting that TPX2 may regulate the activity of Aurora A, following the assembly of mitotic spindles (Tsai *et al.*, 2003; Ozlü *et al.*, 2005). Thus, Aurora A might function mainly in spindle formation through centrosome organization. Besides the regulation of the spindle, Aurora A seems to have a variety of roles.

12.3.1.2 Aurora B

Aurora B is a component of chromosomal passenger complexes that localizes first to the centromeres and then to the midzone of the central spindles and midbody in mitotic cells. It regulates chromosome segregation and cytokinesis. Many experiments show that the removal or disruption of the chromosomal passenger complexes causes defects in chromosome alignment due to the defects of attachment between kinetochores and spindle microtubules and failure in cytokinesis (Andrews *et al.*, 2003). Aurora B seems to regulate the spindle assembly at the centromeres opposite to the site of the centrosomes where Aurora A localizes and later modulates the stability of the central spindles. The chromosomal passenger complexes are composed of survivin, the inner centromere protein (INCENP), Borealin/DasraB, and Aurora B kinase, and they are conserved among different animal species (Glotzer, 2005). The function of Aurora B, i.e., activity, localization, and possibly the substrate specificity of Aurora B, depends on the interaction with three other nonenzymatic proteins: INCENP, survivin, and borealin (Table 12.1).

Many targets of Aurora B have been reported (Table 12.1), e.g., kinetochore proteins and cytokinesis-related proteins (Andrews *et al.*, 2003; Carmena and Earnshaw, 2003; Vader *et al.*, 2006b). One important role of Aurora B seems to be the regulation of chromosome condensation and chromosome bi-orientation from prometaphase to metaphase. The phosphorylation of histone H3 and H3 variant CENP-A by Aurora B, which is conserved from yeast to vertebrates, might help to drive mitotic chromatin condensation (Goto *et al.*, 1999, 2002; Hsu *et al.*, 2000; Adams *et al.*, 2001; Giet and Glover, 2001; Murnion *et al.*, 2001; Petersen *et al.*, 2001; Zeitlin *et al.*, 2001; Crosio *et al.*, 2002). Recent

Table 12.1 Substrate for Aurora kinases in animals

	Localization	Substrate	Possible function	References
Aurora A	Centrosome (spindle pole), mitotic spindles	D-TACC/Maskin	TACC loading to centrosomes, spindle microtubule assembly at the centrosome	Barros *et al.*, 2005; Kinoshita *et al.*, 2005; Peset *et al.*, 2005
		TPX2	Accumulation of active Aurora A (function of phosphorylation unknown)	Kufer *et al.*, 2002
		Eg5	Spindle assembly (function of phosphorylation unknown)	Giet *et al.*, 1999
		CPEB	Translational upregulation of Mos MAPKKK	Mendez *et al.*, 2000
		MBD3	Component of histone deacetylase, centrosome maturation	Sakai *et al.*, 2002
Aurora B	Centromere, midzone of the central spindle, midbody	INCENP	Conformational change in Aurora B, full activation of Aurora B	Kang *et al.*, 2001; Honda *et al.*, 2003; Sessa *et al.*, 2005
		Survivin	Regulation of localization of Aurora B complexes to centromeres	Speliotes *et al.*, 2000; Bolton *et al.*, 2002; Wheatley *et al.*, 2004
		Borealin	INCENP-Survivin interaction (function of phosphorylation unknown)	Romano *et al.*, 2003; Gassmann *et al.*, 2004; Vader *et al.*, 2006a
		Histone H3, CENP-A	Chromosome condensation and segregation of mitotic chromosomes	Goto *et al.*, 1999, 2002; Hsu *et al.*, 2000; Adam *et al.*, 2001; Giet and Glover, 2001; Murnion *et al.*, 2001; Petersen *et al.*, 2001; Zeitlin *et al.*, 2001; Crosio *et al.*, 2002
		DNA topoisomerase II	Chromosome condensation	Morrison *et al.*, 2002
		ISWI	Chromosome condensation	MacCallum *et al.*, 2002
		MCAK	Inhibition of the microtubule-depolymerizing activity, MCAK loading on the centromeres, microtubule–kinetocore attachment	Andrews *et al.*, 2004; Lan *et al.*, 2004
		MgcRacGAP/CYK4	Increase GAP activity toward RhoA, maintenance of the central spindle midzone	Minoshima *et al.*, 2003; Ban *et al.*, 2004
		Type III intermediate filaments	Intermediate filament formation and segregation, completion of cytokinesis	Goto *et al.*, 2003; Kawajiri *et al.*, 2003
		Myosin II regulatory light chain	Loading Myosin II to the spindle midzone, maintenance of midzone organization, furrow ingression	Straight *et al.*, 2003
Aurora C	Centromere, midzone of the central spindle, midbody	Unknown	Cooperation with Aurora B	Li *et al.*, 2004; Yan *et al.*, 2005

studies show another chromosomal passenger complex consisting of Aurora B and INCENP, and this complex seems to be responsible for modifying histone H3 (Gassmann *et al.*, 2004). In addition, it has been reported that Aurora B phosphorylates two nonhistone chromosome proteins, DNA topoisomerase II and the chromatin-remodeling factor ISWI (MacCallum *et al.*, 2002; Morrison *et al.*, 2002; Table 12.1). Aurora B might play a role in chromatin condensation and separation through chromatin modification by phosphorylation. To align and segregate chromosomes, Aurora B also appears to regulate kinetochore–microtubule interaction. INCENP, survivin and borealin that forms chromosomal passenger complex with Aurora B are all phosphorylated by Aurora B, thereby regulating its localization and activation (Honda *et al.*, 2003; Wheatley *et al.*, 2004; Sessa *et al.*, 2005). Such properly localized and activated Aurora B complexes phosphorylate the microtubule-destabilizing Kin I kinesin MCAK, and phosphorylation of MCAK by Aurora B inhibits its microtubule-depolymerizing activity (Andrews *et al.*, 2004; Lan *et al.*, 2004). Since MCAK localizes at both centromeres and centrosomes, there has been some discussion on how MCAK functions as depolymerizing motors to move chromosomes; however, Aurora B might play a key role in chromosome segregation through the spindle assembly.

Another role of Aurora B is the progression of cytokinesis. CYK4/MgcRacGAP, a member of the Rho family GTPase activating protein and a known component of centralspindlin complex along with a kinesin-like protein CeMKLP1/ZEN4, was identified as a substrate of Aurora B (Minoshima *et al.*, 2003). The centralspindlin complex is conserved in *C. elegans* and humans and functions to bundle antiparallel microtubules and stabilize the spindle midzone (Mishima and Glotzer, 2003). Phosphorylation of MgcRacGAP by Aurora B increased its latent GAP activity toward RhoA instead of Rac1/Cdc42, resulting in the completion of cytokinesis through the maintenance of the central spindle (Minoshima *et al.*, 2003; Ban *et al.*, 2004). It has been reported that Aurora B also phosphorylates type-III intermediate filaments such as vimentin, GFAP, and desmin, and their phosphorylation might promote filament formation and the segregation of cells at the final step of cytokinesis (Goto *et al.*, 2003; Kawajiri *et al.*, 2003; Table 12.1). The regulatory right chain of myosin II is also one of the substrates of Aurora B (Straight *et al.*, 2003; Table 12.1), which appears to drive and coordinate many aspects during cytokinesis, including the stability of the spindle midzone, maintenance of the central spindle through the regulation of the activity of small GTPase and intermediate filaments, and furrow ingression via the control of actomyosin force generation.

There are a few studies about Aurora C. During M phase, Aurora C is localized to chromosomes in prophase and metaphase and then transferred to the central spindle midzone and midbody as well as Aurora B (Li *et al.*, 2004; Yan *et al.*, 2005). Aurora C also interacts with INCENP (Li *et al.*, 2004) and can rescue the Aurora B-silenced cytokinesis-defective phenotype (Sasai *et al.*, 2004), suggesting that Aurora C cooperates with Aurora B to regulate the progression of chromosome segregation and cytokinesis.

12.3.2 A class of plant Aurora kinases

In plants, three members of the Aurora family were reported in *Arabidopsis* (Demidov *et al.*, 2005; Kawabe *et al.*, 2005). Three *Arabidopsis Aurora*-like genes conserve a catalytic domain of Aurora kinases in animals and yeast, and they are designated *AtAurora1* (At4g32830), *AtAurora2* (At2g258800), and *AtAurora3* (At2g45490). A search of plant databases showed the existence of putative Aurora-like genes in many plant species, and all members were grouped in a plant-specific clade divided into two major subgroups: plant Auroraα and plant Auroraβ (Demidov *et al.*, 2005). Subgroup Auroraα includes AtAurora1 and AtAurora2, and subgroup Auroraβ includes AtAurora3 in *Arabidopsis* (Demidov *et al.*, 2005). The transcripts and proteins of all three Auroras are accumulated in tissues containing dividing cells, e.g., in young roots, flowers, and flower buds (Demidov *et al.*, 2005).

Subcellular localization of AtAuroras was analyzed by GFP fusion in tobacco BY-2 cells (Table 12.2). At the onset of prophase, before nuclear envelope breakdown, GFP-AtAurora1 and GFP-AtAurora2 were located on the cytoplasmic side of the nuclear membrane, gradually migrating to the poles of the mitotic spindle as mitosis progressed. In the metaphase, they were located in the mitotic spindle (Demidov *et al.*, 2005; Kawabe *et al.*, 2005). GFP-AtAurora1 and GFP-AtAurora2 behaviors are similar until the end of anaphase. In telophase, GFP-AtAurora1 concentrated at the midzone of phragmoplasts and followed the expansion of cell plate until mother wall was reached (Van Damme *et al.*, 2004; Demidov *et al.*, 2005; Kawabe *et al.*, 2005). In contrast, the localization of GFP-AtAurora2 to the phragmoplast midzone was very faint. GFP-AtAurora3 is concentrated in the nucleus as dots in early prophase cells and then becomes evenly localized in chromosomes from metaphase to late anaphase (Demidov *et al.*, 2005; Kawabe *et al.*, 2005). Thus, AtAuroras during M phase exhibit several plant-specific distribution patterns, although they showed some common properties as chromosomal passenger proteins,

Table 12.2 Aurora kinases in *Arabidopsis thaliana*

	Localization	Substrate	Possible function	References
AtAurora1	Nuclear membrane, spindle poles, mitotic spindles, phragmoplast midzone	Histone H3	Unknown	Van Damme *et al.*, 2004; Demidov *et al.*, 2005; Kawabe *et al.*, 2005
AtAurora2	Nuclear membrane, spindle poles, mitotic spindles	Unknown	Unknown	Demidov *et al.*, 2005; Kawabe *et al.*, 2005
AtAurora3	Nuclear dots, spindle poles, centromeres, divided chromosomes	Histone H3	Chromosome condensation and segregation	Demidov *et al.*, 2005; Kawabe *et al.*, 2005; Kurihara *et al.*, 2006

including the dot-like localization in nuclear material and relocalization to the midzone in cytokinetic machinery. The accumulation patterns of AtAurora mRNAs and proteins and their subcellular localization suggest that plant Auroras have a conserved role in the signal transduction pathways to control the progression of mitosis.

12.3.3 Functions of plant Aurora kinases

Phosphorylation of Ser10 and Ser28 of histone H3 is carried out by Aurora B (Hsu *et al.*, 2000; Giet and Glover, 2001; Petersen *et al.*, 2001; Crosio *et al.*, 2002; Goto *et al.*, 2002; Hirota *et al.*, 2005). Although the phosphorylation of histone H3 during mitosis has been reported in plants (Houben *et al.*, 1999; Gernand *et al.*, 2003), the kinases involved in the process were not clear. Recently, it has been reported that Ser10 of histone H3 is phosphorylated by AtAurora1, AtAurora2, and AtAurora3 in vitro (Demidov *et al.*, 2005; Kawabe *et al.*, 2005). In tobacco BY-2 cells, the histone H3 phosphorylated at Ser10 or Ser28 is localized at the pericentromeric regions during mitosis, which was consistent with the localization of AtAurora3 (Kurihara *et al.*, 2006). In addition, the Aurora kinase inhibitor hesperadin inhibited the histone H3 phosphorylation at Ser10 and Ser28 during the M phase in BY-2 cells. Hesperadin treatment also increased the ratio of metaphase cells and decreased the ratio of anaphase/telophase cells, suggesting that it delays metaphase/anaphase transition (Kurihara *et al.*, 2006). Phosphorylation of histone H3, which might be involved in the condensation and segregation of chromosomes, might be carried out by Aurora-family kinases in plants as well as in yeast and animals. To further investigate the function of plant Auroras, identification and characterization of plant Aurora substrates is critical.

12.4 Cytokinesis modulated by the MAPK cascade

12.4.1 Plant cytokinesis

Cytokinesis in plant somatic cells occurs by cell plate formation through the fusion of new cell membranes and construction of new cross walls (designated cell walls) from the interior to the periphery of the cell (Nishihama and Machida, 2001; Jürgens, 2005). These dynamic events in cytokinesis are supported by a microtubule (MT)-based structure known as a phragmoplast, which consists of bundled antiparallel non-kinetochore MTs between the two daughter nuclei. At an early stage, the phragmoplast has a barrel-like shape, and a new cell plate is generated at the midzone of the phragmoplast. Once cell plate formation begins, the phragmoplast changes into a ring-like structure, and expands centrifugally while maintaining its localization at the edge of the growing cell plates. The cell plates are thought to originate from Golgi-derived transport vesicles (Yasuhara *et al.*, 1995, 2000; Nebenführ *et al.*, 2000; Seguí-Simarro *et al.*, 2004). The Golgi-derived vesicles are transported along the phragmoplast, continuously fused at the equatorial region, and the new cell plates expand as the cell walls mature.

Several factors involved in membrane fusion machinery in cytokinesis have been identified from *Arabidopsis* mutants (Nacry *et al.*, 2000; Jürgens, 2005). KNOLLE, a plant-specific syntaxin, and KEULE, a Sec1 homolog, are key regulators of vesicle trafficking. These two proteins interact and participate in the fusion of Golgi-derived vesicles, which results in the formation of cell plate during cytokinesis (Lukowitz *et al.*, 1996; Waizenegger *et al.*, 2000; Assaad *et al.*, 2001). Dynamin-related proteins (ADL1A and ADL1E) seem to be involved in vesicle fusion during cytokinesis (Kang *et al.*, 2003). Other factors including *KORRIGAN* and *CYT1* encode the enzymes required for synthesis and/or maturation of a new cell plate (Zuo *et al.*, 2000; Lukowitz *et al.*, 2001). Membrane dynamics, i.e., the formation of cell plates during cytokinesis, depend on the dynamics of microtubules and actin filaments in the phragmoplasts. It has been revealed that an MAPK cascade is essential for phragmoplast dynamics during the expansion of cell plates and progression of cytokinesis in plants (Fig. 12.1; Hülskamp *et al.*, 1997; Nishihama *et al.*, 1997, 2001, 2002; Spielman *et al.*, 1997; Strompen *et al.*, 2002; Soyano *et al.*, 2003; Yang *et al.*, 2003; Tanaka *et al.*, 2004). Chapter 4 of this book provides an in-depth account of various MAPK cascades in plants.

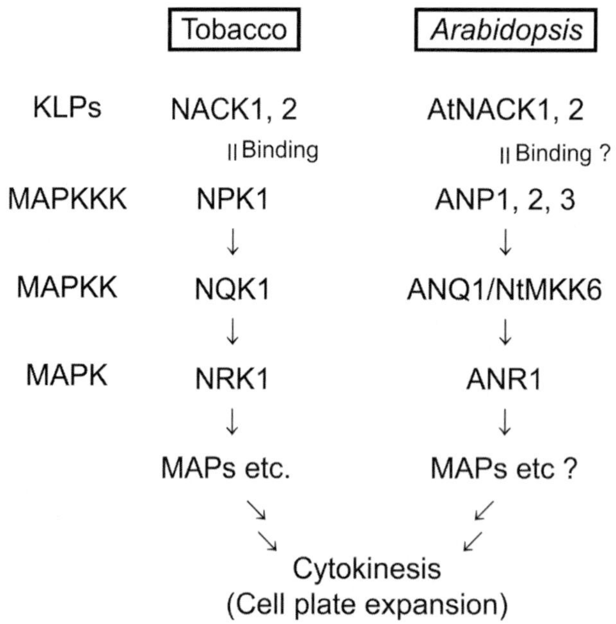

Figure 12.1 The NACK-PQR pathway that controls plant cytokinesis. The components of the NACK-PQR pathways in tobacco and *Arabidopsis* are shown.

12.4.2 NPK1: an MAPK kinase kinase required for cell plate expansion

NPK1 (nucleus- and phragmoplast-localized protein kinase 1) was identified as a gene that is specifically transcribed in actively dividing cells. The *NPK1* gene from tobacco encodes a member of the MAPKKK family, and the kinase domain of NPK1 can replace the functions of several yeast MAPKKKs (Banno *et al.*, 1993; Machida *et al.*, 1998). The *NPK1* gene is transcribed in meristematic tissues and immature organs, not in mature organs (Nakashima *et al.*, 1998). *Arabidopsis* homologs (*ANP1, ANP2*, and *ANP3*) of *NPK1* are also preferentially transcribed in organs that contain proliferating cells (Nishihama *et al.*, 1997). These results suggest that NPK1 MAPKKKs play a role in the signaling pathway regulating plant cell division (Nishihama *et al.*, 2001).

In BY-2 tobacco cells, the kinase activity of NPK1 specifically increases in the late M phase of the cell cycle, following transcription of the *NPK1* gene and accumulation of NPK1 protein at the initiation of M phase (Nishihama *et al.*, 2001). NPK1 is localized in the nucleus at the interphase and the prophase prior to the breakdown of the nuclear envelope, whereas it is localized in the cytoplasm at the metaphase (Nishihama *et al.*, 2001). During cytokinesis, when the kinase activity of NPK1 increases, NPK1 shifts to the leading edge of the equatorial zone of the phragmoplast (Fig. 12.2a, Color plate 21; Nishihama *et al.*, 2001). Overexpression of a kinase-defective mutant of NPK1 (NPK1:KW) in BY-2 cells caused the inhibition of the lateral expansion of the phragmoplast and the generation of multinucleated cells with incomplete cell plates (Fig. 12.2b, Color plate 21; Nishihama *et al.*, 2001).

Arabidopsis homologs of NPK1 (ANP1, ANP2, and ANP3) also appear to be involved in cytokinesis. Loss of function of two of the three homologs of NPK1 (ANP2 and ANP3) causes defects in cytokinesis, especially, the formation of multinucleated cells with incomplete cell walls (Krysan *et al.*, 2002). These results suggest that the MAPK cascade including the NPK1 MAPKKK family is necessary for the progression of cytokinesis in plants.

12.4.3 NACK1: activator of the NPK1 MAPKKK that is a kinesin-like protein

Animals and yeast have several proteins that can regulate the MAPKKKs via protein–protein interaction. To isolate the activators of NPK1 MAPKKK, we used a functional yeast genetic system based on the mating pheromone-responsive MAPK cascade, which consists of STE11 MAPKKK, STE7 MAPKK, and FUS3 MAPK (Irie *et al.*, 1994). We identified two KLPs in tobacco, which we designated as NPK1-activating kinesin-like proteins 1 and 2 (NACK1 and NACK2). Both these proteins interact with NPK1 and increase its protein kinase activity (Nishihama *et al.*, 2002). In tobacco BY-2 cells, the *NACK1* and *NACK2* mRNAs and NACK1 protein accumulate only at the M phase of the cell cycle, which is consistent with the increase of NPK1 kinase activity (Nishihama *et al.*, 2002).

Yeast two-hybrid and in vitro immunoprecipitation assays using recombinant proteins have shown that the stalk domain of NACK1 binds directly to the regulatory domain of NPK1 via these predicted coiled-coil structures (Ishikawa *et al.*, 2002). During late anaphase and telophase, NACK1 is consistently colocalized with NPK1 at the equatorial zone of the phragmoplast (Fig. 12.2a, Color plate 21), whereas the deletion of the regulatory domain of NPK1, which contains the NACK1-binding site, eliminates its localization to the equator of the phragmoplast (Nishihama *et al.*, 2002). Overexpression of a mutant NACK1 protein that lacks the putative motor region (NACK1:ST) in tobacco cells result in failure to accumulate NPK1 proteins at the phragmoplast equator and cytokinesis defects (Nishihama *et al.*, 2002). This suggests that NACK1 plays a role as a positive regulator in both the recruitment of NPK1 to the phragmoplast midzone and the activation of NPK1 MAPKKK during cytokinesis. Recently, Weingartner *et al.* (2004) reported that the overexpression of the constitutively active form of cyclin B1 disrupts the proper localization of NACK1 on the phragmoplast MTs during cytokinesis. It is interesting to know whether CDK/cyclin complexes regulate the localization and activation of NACK1/NPK1 complex.

The homologs of the *NACK1* and *NACK2* genes in *Arabidopsis* are designated *AtNACK1* and *AtNACK2* and are identical to *HINKEL* (*HIK*) and *STUD* (*STD*)/*TETRASPORE* (*TES*), respectively (Nishihama *et al.*, 2002; Strompen *et al.*, 2002; Yang *et al.*, 2003). Loss-of-function mutations in *AtNACK1/HIK* and *STD/TES/AtNACK2* result in the occasional failure of somatic and male-meiotic cytokinesis, respectively (Hülskamp *et al.*, 1997; Spielman *et al.*, 1997; Nishihama *et al.*, 2002; Strompen *et al.*, 2002; Yang *et al.*, 2003). Recently, it has been shown that these genes have redundant functions and are essential for cytokinesis during both male and female gametogenesis (Tanaka *et al.*, 2004).

12.4.4 The NACK-PQR pathway: an MAPK cascade that regulates the progression of cytokinesis

We identified the NQK1/NtMEK1 and NRK1 proteins of tobacco as a MAPKK and MAPK, respectively, which act downstream of NPK1 (Soyano *et al.*, 2003). To isolate downstream factors of NPK1, we used a yeast genetic system that is based on the osmosensing MAPK cascade of yeast (Brewster *et al.*, 1993; Maeda *et al.*, 1994, 1995). By screening the yeast cell's library under high osmotic conditions, we isolated *NQK1* cDNA. NRK1 was isolated as a binding partner of NQK1 using a yeast two-hybrid system. NPK1 phosphorylates and activates NQK1, which in turn phosphorylates and activates NRK1 (Soyano *et al.*, 2003). Although NPK1 and NACK1 proteins rapidly disappear after the M phase, NQK1 and NRK1 proteins accumulate throughout the cell cycle. Activities of NPK1, NQK1, and NRK1 in tobacco cells, however, increase at the late M phase of the cell cycle and decrease after the M phase, and the

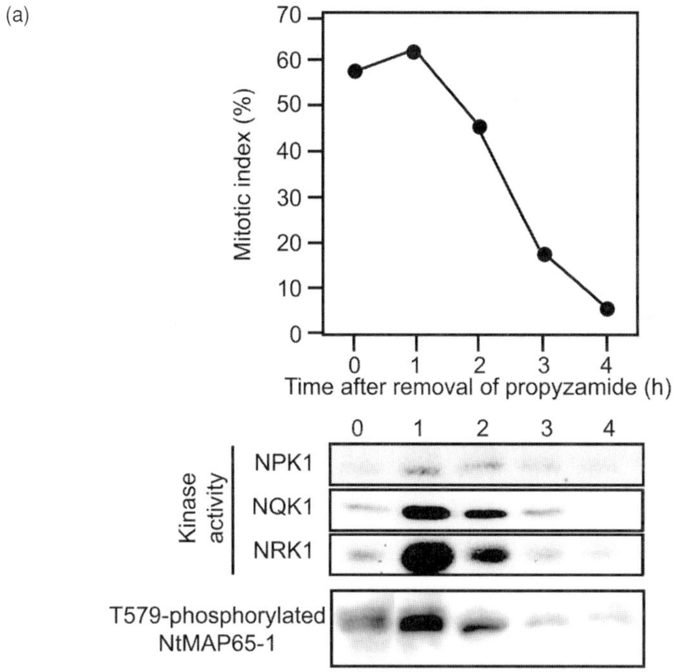

Figure 12.3 Specific activation of PQR MAPK cascade at the late M phase and phosphorylation of MAP65, a downstream factor of NRK1 MAPK. (a) The graph shows a plot of the mitotic indices of BY-2 cells synchronized at M phase. The cell cycle was arrested at prometaphase by propyzamide after release from an aphidicolin block. The protein kinase activities of NPK1, NQK1, and NRK1 were determined by immunocomplex kinase assays using recombinant kinase-negative NQK1, kinase-negative NRK1, and myelin basic proteins as substrates, respectively. Accumulation of NtMAP65-1 phosphorylated on Thr-579 was detected with rabbit antibodies against NtMAP65-1 phosphorylated on Thr-579 by Western blotting (bottom).

patterns of activation of these protein kinases are similar to the pattern of NACK1 accumulation (Fig. 12.3a; Nishihama *et al.*, 2001; Soyano *et al.*, 2003). Moreover, studies on the subcellular localization of NQK1 and NRK1 revealed that they were also localized at the equator of the phragmoplast at least during telophase (unpublished data). These results suggest that the activation of NPK1 MAPKKK by NACK1 binding causes the activation of NRK1 MAPK via the activation of NQK1 MAPKK at the equator of the phragmoplast during cytokinesis.

The role of NQK1 in cell division was demonstrated by overexpressing a kinase-defective mutant form of NQK1 (NQK1:KW) in BY-2 cells and by mutation of *ANQ1*, the *Arabidopsis* homolog of *NQK1*. The NQK1:KW expressing cells and *anq1* mutants were multinucleated and had incomplete cell plates (similar to Fig. 12.2b, Color plate 21). This indicates that NQK1 is required for the expansion of the phragmoplast and formation of the cell plates (Soyano *et al.*, 2003). Although the activation of NRK1 is tightly coupled to the activation of NPK1 and NQK1, involvement of NRK1 in the formation of cell plates has not yet been experimentally demonstrated. Recently, we identified a loss-of-function mutant of an *Arabidopsis* homolog of *NRK1* (*ANR1*) that is defective in cytokinesis (unpublished data). These results suggest that the MAPK cascade composed of NPK1 MAPKKK, NQK1 MAPKK, and NRK1 MAPK are activated by the binding of NACK1/2 KLP and promotes the formation of cell plate. We designated this cascade as the NACK-PQR pathway (Fig. 12.1; Soyano *et al.*, 2003).

The phenotypes generated by the overexpression of the dominant-negative mutants of NACK1, NPK1, and NQK1 are similar to those of cells treated with taxol, a compound that blocks the depolymerization of microtubules (Yasuhara *et al.*, 1993). This suggests that MT disassembly is required for phragmoplast expansion. In the cells from *atnack1/hik* mutant plants of *Arabidopsis*, phragmoplast microtubules persist in the center of the division plane, suggesting that the disassembly of phragmoplast microtubules is inhibited in these cells (Strompen *et al.*, 2002). Thus, the activation of the NACK-PQR pathway appears to be necessary for the reorganization of phragmoplast microtubules in the expansion of the phragmoplast during cell plate formation. Factors acting downstream of the NACK-PQR pathway may therefore control microtubule dynamics.

12.4.5 MAP65: an MAP that is a downstream factor of NRK1

Several MAPs purified from tobacco BY-2 cells are phosphorylated by active NRK1 in vitro. One of these candidate substrates is NtMAP65-1a, a protein belonging to the MAP65/Ase1/PRC1 family (Sasabe *et al.*, 2006; Sasabe and Machida, 2006). This family of proteins is conserved among a variety of organisms and includes Ase1p (anaphase spindle elongation factor) in yeast (Pellman *et al.*, 1995; Schuyler *et al.*, 2003), PRC1 (protein regulator of cytokinesis 1) in mammals (Jiang *et al.*, 1998), SPD1 (spindle defective1) in *C. elegans* (Verbrugghe and White, 2004), and Feo (Fascetto) in *Drosophila* (Vernì *et al.*, 2004). These MAPs localize to the cytokinetic apparatus, and most of them are involved in cytokinesis.

In vitro, NRK1 phosphorylates NtMAP65-1a at a single site, Thr-579, in the carboxy-terminal region. Specific antibodies against Thr-579-phosphorylated NtMAP65-1 have revealed that NtMAP65-1 is phosphorylated at this site in vivo. In synchronized BY-2 cells, NtMAP65-1 phosphorylated at Thr-579 accumulates at the late M phase (Fig. 12.3a, Color plate 22), although the total amount of NtMAP65-1 does not change. Such a pattern of phosphorylation is consistent with the pattern of NACK1 accumulation and NPK1, NQK1/NtMEK1, and NRK1/NTF6 activation. Immunostaining with NtMAP65-1 antibodies also revealed NtMAP65-1 on various MT structures throughout the cell cycle. Interestingly, NtMAP65-1 phosphorylated on Thr-579 is concentrated at the equator of the phragmoplast along with other components of the NACK-PQR pathway, although NtMAP65-1 can be found throughout the entire phragmoplast (Fig. 12.3b, Color plate 22; Sasabe *et al.*, 2006). Overexpression of NtMAP65-1 that is not phosphorylated by MAPK in tobacco cells confers the phragmoplast and cortical MTs resistance to depolymerization by propyzamide, and it delays the expansion of the phragmoplast (Sasabe *et al.*, 2006). These findings suggest that NtMAP65-1 is phosphorylated by MAPK, which localizes at the phragmoplast midzone in tobacco cells and regulates phragmoplast expansion by promoting the instability of midzone MTs. Putative MAPK phosphorylation sites are conserved in many members of the MAP65 protein family in plants and animals. It will be interesting to investigate whether the mechanism by which the MAP65 function is regulated by MAPKs is conserved among a variety of plant MAP65s.

12.5 Concluding remarks

Recent genetic and biochemical analyses have revealed the presence of regulatory systems for the progression of cell division including the entry of mitosis, segregation of chromosomes, and cytokinesis in plants. The controlling system for the progression of mitosis from the entry of mitosis to the exit of chromosome segregation in plants seems to be basically conserved in both plants and animals. These organisms, however, might have at least two distinct systems that operate for the progression of the cell cycle because processes of cytokinesis in these organisms appear to be different and a plant cell has a large number of diverged CDKs and cyclins. Cytokinesis is achieved by the formation of cell plates in plant cells and by the furrow constriction in animal cells. The formation of cell plate includes many different processes such as de novo synthesis of cell walls and fusion of cell membranes in addition to microtubule turnover. Such complex processes must be coordinated and regulated for the proper formation of cell plates. The plant cells must have evolved unique systems for the formation of cell plates. The NACK-PQR system might be one such unique system. It is interesting to examine whether the NACK-PQR system can regulate the above-mentioned multiple processes for cell plate formation. The NACK-PQR system is activated by the interaction between NACK1 KLP and NPK1 MAPKKK at the late M phase. The critical question may be how the interaction can be regulated during the progression of M phase. There might be a regulating system for the interaction because both NACK1 and NPK1 proteins are present from an early M phase in which NPK1 activity is not detected.

Another question for these distinct systems is related to the presence of a large number of diverged CDKs and cyclins—how the physical interaction between CDKs and cyclins can be regulated for their proper activation. In particular, it appears to be crucial to reveal the roles of plant-specific CDK, CDKB, and other diverged types of CDK proteins. It should also be important to elucidate how specific interactions between those CDKs and corresponding cyclins can be achieved during the progression of the cell cycle in a plant. In addition, further efforts by genetic and biochemical analyses may disclose molecular mechanisms that regulate systematic links among the pathways of CDK/cyclin, Aurora kinase, and MAPK.

Acknowledgments

This work was supported in part by a grant from the Program for the Promotion of Basic Research Activities for Innovative Biosciences, by Grants-in-Aid for Scientific Research on Priority Areas (no. 14036216) and the 21st Century COE Program (System Bioscience) from the Ministry of Education, Culture, Sports, Science, and Technology of Japan. M.S. is supported by a grant from the "Academic Frontier" Project for Private Universities (matching fund subsidy from Ministry of Education, Culture, Sports, Science and Technology (MEXT), 2005–2009).

References

Adams, R.R., Maiato, H., Earnshaw, W.C. and Carmena, M. (2001) Essential roles of *Drosophila* inner centromere protein (INCENP) and Aurora B in histone H3 phosphorylation, metaphase chromosome alignment, kinetochore disjunction, and chromosome segregation. *J Cell Biol*, **153**, 865–880.

Andrews, P.D., Knatko, E., Moore, W.J. and Swedlow, J.R. (2003) Mitotic mechanics: the auroras come into view. *Curr Opin Cell Biol*, **15**, 672–683.

Andrews, P.D., Ovechkina, Y., Morrice, N., Wagenbach, M., Duncan, K., Wordeman, L. and Swedlow, J.R. (2004) Aurora B regulates MCAK at the mitotic centromere. *Dev Cell*, **6**, 253–268.

Assaad, F.F., Huet, Y., Mayer, U. and Jürgens, G. (2001) The cytokinesis gene KEULE encodes a Sec1 protein that binds the syntaxin KNOLLE. *J Cell Biol*, **152**, 531–543.

Ban, R., Irino, Y., Fukami, K. and Tanaka, H. (2004) Human mitotic spindle-associated protein PRC1 inhibits MgcRacGAP activity toward Cdc42 during the metaphase. *J Biol Chem*, **279**, 16394–16402.

Banno, H., Hirano, K., Nakamura, T., Irie, K., Nomoto, S., Matsumoto, K. and Machida, Y. (1993) NPK1, a tobacco gene that encodes a protein with a domain homologous to yeast BCK1, STE11, and Byr2 protein kinases. *Mol Cell Biol*, **13**, 4745–4752.

Barros, T.P., Kinoshita, K., Hyman, A.A. and Raff, J.W. (2005) Aurora A activates D-TACC-Msps complexes exclusively at centrosomes to stabilize centrosomal microtubules. *J Cell Biol*, **170**, 1039–1046.

Bolton, M.A., Lan, W., Powers, S.E., McCleland, M.L., Kuang, J. and Stukenberg, P.T. (2002) Aurora B kinase exists in a complex with survivin and INCENP and its kinase activity is stimulated by survivin binding and phosphorylation. *Cell Mol Biol*, **13**, 3064–3077.

Boniotti, M.B. and Gutierrez, C. (2001) A cell-cycle-regulated kinase activity phosphorylates plant retinoblastoma protein and contains, in *Arabidopsis*, a CDKA/cyclin D complex. *Plant J*, **28**, 341–350.

Boudolf, V., Inze, D. and De Veylder, L. (2006) What if higher plants lack a CDC25 phosphatase? *Trends Plant Sci*, **11**, 474–479.

Brewster, J.L., de Valoir, T., Dwyer, N.D., Winter, E. and Gustin, M.C. (1993) An osmosensing signal transduction pathway in yeast. *Science*, **259**, 1760–1763.

Breyne, P., Dreesen, R., Vandepoele, K., De Veylder, L., Van Breusegem, F., Callewaert, L., Rombauts, S., Raes, J., Cannoot, B., Engler, G., Inzé, D. and Zabeau, M. (2002) Transcriptome analysis during cell division in plants. *Proc Natl Acad Sci USA*, **99**, 14825–14830.

Carmena, M. and Earnshaw, W.C. (2003) The cellular geography of Aurora kinases. *Nat Rev Mol Cell Biol*, **4**, 842–854.

Chan, C.S. and Botstein, D. (1993) Isolation and characterization of chromosome-gain and increase-in-ploidy mutants in yeast. *Genetics*, **135**, 677–691.

Cheeseman, I.M., Anderson, S., Jwa, M., Green, E.M., Kang, J., Yates, J.R., Chan, C.S., Drubin, D.G. and Barnes, G. (2002) Phospho-regulation of kinetochore-microtubule attachments by the Aurora kinase Ipl1p. *Cell*, **111**, 163–172.

Crosio, C., Fimia, G.M., Loury, R., Kimura, M., Okano, Y., Zhou, H., Sen, S., Allis, C.D. and Sassone-Corsi, P. (2002) Mitotic phosphorylation of histone H3: spatio-temporal regulation by mammalian Aurora kinases. *Mol Cell Biol*, **22**, 874–885.

Cullen, C.F. and Ohkura, H. (2001) Msps protein is localized to acentrosomal poles to ensure bipolarity of *Drosophila* meiotic spindles. *Nat Cell Biol*, **3**, 637–642.

Demidov, D., Van Damme, D., Geelen, D., Blattner, F.R. and Houben, A. (2005) Identification and dynamics of two classes of Aurora-like kinases in *Arabidopsis* and other plants. *Plant Cell*, **17**, 836–848.

Dewitte, W. and Murray, J.A. (2003) The plant cell cycle. *Annu Rev Plant Biol*, **54**, 235–264.

Doerner, P., Jørgensen, J.E., You, R., Steppuhn, J. and Lamb, C. (1996) Control of root growth and development by cyclin expression. *Nature*, **380**, 520–523.

Eyers, P.A., Erikson, E., Chen, L.G. and Maller, J.L. (2003) A novel mechanism for activation of the protein kinase Aurora A. *Curr Biol*, **13**, 691–697.

Ferreira, P.C.G., Hemerly, A.S., Villarroel, R., Van Montagu, M. and Inzé, D. (1991) The *Arabidopsis* functional homolog of the p34cdc2 protein kinase. *Plant Cell*, **3**, 531–540.

Fisher, R.P. and Morgan, D.O. (1994) A novel cyclin associates with MO15/CDK7 to form the CDK-activating kinase. *Cell*, **78**, 713–724.

Fobert, P.R., Gaudin, V., Lunness, P., Coen, E.S. and Doonan, J.H. (1996) Distinct classes of cdc2-related genes are differentially expressed during the cell division cycle in plants. *Plant Cell*, **8**, 1465–1476.

Francis, D. (2007) The plant cell cycle-15 years on. *New Phytol*, **174**, 261–278.

Gassmann, R., Carvalho, A., Henzing, A.J., Ruchaud, S., Hudson, D.F., Honda, R., Nigg, E.A., Gerloff, D.L. and Earnshaw, W.C. (2004) Borealin: a novel chromosomal passenger required for stability of the bipolar mitotic spindle. *J Cell Biol*, **166**, 179–191.

Gernand, D., Demidov, D. and Houben, A. (2003) The temporal and spatial pattern of histone H3 phosphorylation at serine 28 and serine 10 is similar in plants but differs between mono- and polycentric chromosomes. *Cytogenet Genome Res*, **101**, 172–176.

Giet, R. and Glover, D.M. (2001) *Drosophila* Aurora B kinase is required for histone H3 phosphorylation and condensin recruitment during chromosome condensation and to organize the central spindle during cytokinesis. *J Cell Biol*, **152**, 669–682.

Giet, R., Uzbekov, R., Cubizolles, F., Le Guellec, K. and Prigent, C. (1999) The Xenopus laevis aurora-related protein kinase pEg2 associates with and phosphorylates the kinesin-related protein XlEg5. *J Biol Chem*, **274**, 15005–15013.

Glotzer, M. (2005) The molecular requirements for cytokinesis. *Science*, **307**, 1735–1739.

Goto, H., Tomono, Y., Ajiro, K., Kosako, H., Fujita, M., Sakurai, M., Okawa, K., Iwamatsu, A., Okigaki, T., Takahashi, T. and Inagaki, M. (1999) Identification of a novel phosphorylation site on histone H3 coupled with mitotic chromosome condensation. *J Biol Chem*, **274**, 25543–25549.

Goto, H., Yasui, Y., Kawajiri, A., Nigg, E.A., Terada, Y., Tatsuka, M., Nagata, K. and Inagaki, M. (2003) Aurora-B regulates the cleavage furrow-specific vimentin phosphorylation in the cytokinetic process. *J Biol Chem*, **278**, 8526–8530.

Goto, H., Yasui, Y., Nigg, E.A. and Inagaki, M. (2002) Aurora-B phosphorylates Histone H3 at serine28 with regard to the mitotic chromosome condensation. *Genes Cells*, **7**, 11–17.

Harbour, J.W. and Dean, D.C. (2000) Chromatin remodeling and Rb activity. *Curr Opin Cell Biol*, **12**, 685–689.

Healy, J.M.S., Menges, M., Doonan, J.H. and Murray, J.A.H. (2001) The *Arabidopsis* D-type cyclins CycD2 and CycD3 both interact *in vivo* with the PSTAIRE cyclin-dependent kinase Cdc2a but are differentially controlled. *J Biol Chem*, **276**, 7041–7047.

Hemerly, A., Engler Jde, A., Bergounioux, C., Van Montagu, M., Engler, G., Inzé, D. and Ferreira, P. (1995) Dominant negative mutants of the Cdc2 kinase uncouple cell division from iterative plant development. *EMBO J*, **14**, 3925–3936.

Hirayama, T., Imajuku, Y., Anai, T., Matsui, M. and Oka, A. (1991) Identification of two cell-cycle-controlling cdc2 gene homologs in *Arabidopsis thaliana*. *Gene*, **105**, 159–165.

Hirota, T., Lipp, J.J., Toh, B.H. and Peters, J.M. (2005) Histone H3 serine 10 phosphorylation by Aurora B causes HP1 dissociation from heterochromatin. *Nature*, **438**, 1176–1180.

Honda, R., Körner, R. and Nigg, E.A. (2003) Exploring the functional interactions between Aurora B, INCENP, and survivin in mitosis. *Cell Mol Biol*, **14**, 3325–3341.

Houben, A., Wako, T., Furushima-Shimogawara, R., Presting, G., Kunzel, G., Schubert, I. and Fukui, K. (1999) Short communication: the cell cycle dependent phosphorylation of histone H3 is correlated with the condensation of plant mitotic chromosomes. *Plant J*, **18**, 675–679.

Hsu, J.Y., Sun, Z.W., Li, X., Reuben, M., Tatchell, K., Bishop, D.K., Grushcow, J.M., Brame, C.J., Caldwell, J.A., Hunt, D.F., Lin, R., Smith, M.M. and Allis, C.D. (2000) Mitotic phosphorylation of histone H3 is governed by Ipl1/Aurora kinase and Glc7/PP1 phosphatase in budding yeast and nematodes. *Cell*, **102**, 279–291.

Hülskamp, M., Parekh, N.S., Grini, P., Schneitz, K., Zimmermann, I., Lolle, S.J. and Pruitt, R.E. (1997) The STUD gene is required for male-specific cytokinesis after telophase II of meiosis in *Arabidopsis thaliana*. *Dev Biol*, **187**, 114–124.

Imajuku, Y., Hirayama, T., Endoh, H. and Oka, A. (1992) Exon–intron organization of the *Arabidopsis thaliana* protein kinase genes CDC2a and CDC2b. *FEBS Lett*, **304**, 73–77.

Inzé, D. and De Veylder, L. (2006) Cell cycle regulation in plant development. *Annu Rev Genet*, **40**, 77–105.

Irie, K., Gotoh, Y., Yashar, B.M., Errede, B., Nishida, E. and Matsumoto, K. (1994) Stimulatory effects of yeast and mammalian 14-3-3 proteins on the Raf protein kinase. *Science*, **265**, 1716–1719.

Ishikawa, M., Soyano, T., Nishihama, R. and Machida, Y. (2002) The NPK1 mitogen-activated protein kinase kinase contains a functional nuclear localization signal at the binding site for the NACK1 kinesin-like protein. *Plant J*, **32**, 789–798.

Iwakawa, H., Shinmyo, A. and Sekine, M. (2006) *Arabidopsis* CDKA;1, cdc2 homologue, controls proliferation of generative cells in male gametogenesis. *Plant J*, **45**, 819–831.

Jiang, W., Jimenez, G., Wells, N.J., Hope, T.J., Wahl, G.M., Hunter, T. and Fukunaga, R. (1998) PRC1: a human mitotic spindle-associated CDK substrate protein required for cytokinesis. *Mol Cell*, **2**, 877–885.

Joubès, J., Chevalier, C., Dudits, D., Heberle-Bors, E., Inzé, D., Umeda, M. and Renaudin, J.P. (2000) CDK-related protein kinases in plants. *Plant Mol Biol*, **43**, 607–620.

Jürgens, G. (2005) Plant cytokinesis: fission by fusion. *Trends Cell Biol*, **15**, 277–283.

Kang, B.H., Busse, J.S. and Bednarek, S.Y. (2003) Members of the *Arabidopsis* dynamin-like gene family, ADL1, are essential for plant cytokinesis and polarized cell growth. *Plant Cell*, **15**, 899–913.

Kang, J., Cheeseman, I.M., Kallstrom, G., Velmurugan, S., Barnes, G. and Chan, C.S. (2001) Functional cooperation of Dam1, Ipl1, and the inner centromere protein (INCENP)-related protein Sli15 during chromosome segregation. *J Cell Biol*, **155**, 763–774.

Karsenti, E. and Vernos, I. (2001) The mitotic spindle: a self-made machine. *Science*, **294**, 543–547.

Kawabe, A., Matsunaga, S., Nakagawa, K., Kurihara, D., Yoneda, A., Hasezawa, S., Uchiyama, S. and Fukui, K. (2005) Characterization of plant Aurora kinases during mitosis. *Plant Mol Biol*, **58**, 1–13.

Kawajiri, A., Yasui, Y., Goto, H., Tatsuka, M., Takahashi, M., Nagata, K. and Inagaki, M. (2003) Functional significance of the specific sites phosphorylated in desmin at cleavage furrow: Aurora-B may phosphorylate and regulate type III intermediate filaments during cytokinesis coordinatedly with Rho-kinase. *Cell Mol Biol*, **14**, 1489–1500.

Kinoshita, K., Habermann, B. and Hyman, A.A. (2002) XMAP215: a key component of the dynamic microtubule cytoskeleton. *Trends Cell Biol*, **12**, 267–273.

Kinoshita, K., Noetzel, T.L., Pelletier, L., Mechtler, K., Drechsel, D.N., Schwager, A., Lee, M., Raff, J.W. and Hyman, A.A. (2005) Aurora A phosphorylation of TACC3/maskin is required for centrosome-dependent microtubule assembly in mitosis. *J Cell Biol*, **170**, 1047–1055.

Kono, A., Umeda-Hara, C., Adachi, S., Nagata, N., Konomi, M., Nakagawa, T., Uchimiya, H. and Umeda, M. (2007) The *Arabidopsis* D-type cyclin CYCD4 controls cell division in the stomatal lineage of the hypocotyl epidermis. *Plant Cell*, **19**, 1265–1277.

Kono, A., Umeda-Hara, C., Lee, J., Ito, M., Uchimiya, H. and Umeda, M. (2003) *Arabidopsis* D-type cyclin CYCD4;1 is a novel cyclin partner of B2-type cyclin-dependent kinase. *Plant Physiol*, **132**, 1315–1321.

Krysan, P.J., Jester, P.J., Gottwald, J.R. and Sussman, M.R. (2002) An *Arabidopsis* mitogen-activated protein kinase kinase kinase gene family encodes essential positive regulators of cytokinesis. *Plant Cell*, **14**, 1109–1120.

Kufer, T.A., Silljé, H.H., Körner, R., Gruss, O.J., Meraldi, P. and Nigg, E.A. (2002) Human TPX2 is required for targeting Aurora-A kinase to the spindle. *J Cell Biol*, **158**, 617–623.

Kurihara, D., Matsunaga, S., Kawabe, A., Fujimoto, S., Noda, M., Uchiyama, S. and Fukui, K. (2006) Aurora kinase is required for chromosome segregation in tobacco BY-2 cells. *Plant J*, **48**, 572–580.

Labbé, J.C., Martinez, A.M., Fesquet, D., Capony, J.P., Darbon, J.M., Derancourt, J., Devault, A., Morin, N., Cavadore, J.C. and Dorée, M. (1994) p40MO15 associates with a p36 subunit and requires both nuclear translocation and Thr176 phosphorylation to generate cdk-activating kinase activity in Xenopus oocytes. *EMBO J*, **13**, 5155–5164.

Lan, W., Zhang, X., Kline-Smith, S.L., Rosasco, S.E., Barrett-Wilt, G.A., Shabanowitz, J., Hunt, D.F., Walczak, C.E. and Stukenberg, P.T. (2004) Aurora B phosphorylates centromeric MCAK and regulates its localization and microtubule depolymerization activity. *Curr Biol*, **14**, 273–286.

Lee, M.J., Gergely, F., Jeffers, K., Peak-Chew, S.Y. and Raff, J.W. (2001) Msps/XMAP215 interacts with the centrosomal protein D-TACC to regulate microtubule behaviour. *Nat Cell Biol*, **3**, 643–649.

Li, X., Sakashita, G., Matsuzaki, H., Sugimoto, K., Kimura, K., Hanaoka, F., Taniguchi, H., Furukawa, K. and Urano, T. (2004) Direct association with inner centromere protein (INCENP) activates the novel chromosomal passenger protein, Aurora-C. *J Biol Chem*, **279**, 47201–47211.

Lukowitz, W., Mayer, U. and Jürgens, G. (1996) Cytokinesis in the *Arabidopsis* embryo involves the syntaxin-related KNOLLE gene product. *Cell*, **84**, 61–71.

Lukowitz, W., Nickle, T.C., Meinke, D.W., Last, R.L., Conklin, P.L. and Somerville, C.R. (2001) *Arabidopsis* cyt1 mutants are deficient in a mannose-1-phosphate guanylyltransferase and point to a requirement of N-linked glycosylation for cellulose biosynthesis. *Proc Natl Acad Sci USA*, **98**, 2262–2267.

MacCallum, D.E., Losada, A., Kobayashi, R. and Hirano, T. (2002) ISWI remodeling complexes in Xenopus egg extracts: identification as major chromosomal components that are regulated by INCENP-aurora B. *Cell Mol Biol*, **13**, 25–39.

Machida, Y., Nakashima, M., Morikiyo, K., Soyano, T. and Nishihama, R. (1998) MAPKKK-related protein kinase NPK1: involvement in the regulation of the M phase of plant cell cycle. *J Plant Res*, **111**, 243–246.

Maeda, T., Takekawa, M. and Saito, H. (1995) Activation of yeast PBS2 MAPKK by MAPKKKs or binding of an SH3-containing osmosenser. *Science*, **269**, 554–558.

Maeda, T., Wurgler-Murphy, S.M. and Saito, H. (1994) A two-component system that regulates an osmosensing MAP kinase cascade in yeast. *Nature*, **369**, 242–245.

Magnard, J.L., Yang, M., Chen, Y.-C.S., Leary, M. and McCormick, S. (2001) The *Arabidopsis* gene *tardy asynchronous meiosis* is required for the normal pace and synchrony of cell division during male meiosis. *Plant Physiol*, **127**, 1157–1166.

Magyar, Z., Mészáros, T., Miskolczi, P., Deák, M., Fehér, A., Brown, S., Kondorosi, E., Athanasiadis, A., Pongor, S., Bilgin, M., Bakó, L., Koncz, C. and Dudits, D. (1997) Cell cycle phase specificity of putative cyclin-dependent kinase variants in synchronized alfalfa cells. *Plant Cell*, **9**, 223–235.

Mäkelä, T.P., Tassan, J.P., Nigg, E.A., Frutiger, S., Hughes, G.J. and Weinberg, R.A. (1994) A cyclin associated with the CDK-activating kinase MO15. *Nature*, **371**, 254–257.

Mendez, R., Hake, L.E., Andresson, T., Littlepage, L.E., Ruderman, J.V. and Richter, J.D. (2000) Phosphorylation of CPE binding factor by Eg2 regulates translation of c-mos mRNA. *Nature*, **404**, 302–307.

Menges, M., Samland, A.K., Planchais, S. and Murray, J.A. (2006) The D-type cyclin CYCD3;1 is limiting for the G1-to-S-phase transition in *Arabidopsis*. *Plant Cell*, **18**, 893–906.

Mészáros, T., Miskolczi, P., Ayaydin, F., Pettkó-Szandtner, A., Peres, A., Magyar, Z., Horváth, G.V., Bakó, L., Fehér, A. and Dudits, D. (2000) Multiple cyclin-dependent kinase complexes and phosphatases control G2/M progression in alfalfa cells. *Plant Mol Biol*, **43**, 595–605.

Minoshima, Y., Kawashima, T., Hirose, K., Tonozuka, Y., Kawajiri, A., Bao, Y.C., Deng, X., Tatsuka, M., Narumiya, S., May, W.S., Nosaka, T., Semba, K., Inoue, T., Satoh, T., Inagaki, M. and Kitamura, T. (2003) Phosphorylation by aurora B converts MgcRacGAP to a RhoGAP during cytokinesis. *Dev Cell*, **4**, 549–560.

Minshull, J., Sun, H., Tonks, N.K. and Murray, A.W. (1994) A MAP kinase-dependent spindle assembly checkpoint in Xenopus egg extracts. *Cell*, **79**, 475–486.

Mishima, M. and Glotzer, M. (2003) Cytokinesis: a logical GAP. *Curr Biol*, **13**, R589–R591.

Morgan, D.O. (1997) Cyclin-dependent kinases: engines, clocks, and microprocessors. *Annu Rev Cell Dev Biol*, **13**, 261–291.

Morrison, C., Henzing, A.J., Jensen, O.N., Osheroff, N., Dodson, H., Kandels-Lewis, S.E., Adams, R.R. and Earnshaw, W.C. (2002) Proteomic analysis of human metaphase chromosomes reveals topoisomerase II alpha as an Aurora B substrate. *Nucleic Acids Res* , **30**, 5318–5327.

Murnion, M.E., Adams, R.R., Callister, D.M., Allis, C.D., Earnshaw, W.C. and Swedlow, J.R. (2001) Chromatin-associated protein phosphatase 1 regulates aurora-B and histone H3 phosphorylation. *J Biol Chem*, **276**, 26656–26665.

Nacry, P., Mayer, U. and Jürgens, G. (2000) Genetic dissection of cytokinesis. *Plant Mol Biol*, **43**, 719–733.

Nakagami, H., Kawamura, K., Sugisaka, K., Sekine, M. and Shinmyo, A. (2002) Phosphorylation of retinoblastoma-related protein by the cyclin D/cyclin-dependent kinase complex is activated at the G1/S-phase transition in tobacco. *Plant Cell*, **14**, 1847–1857.

Nakagami, H., Sekine, M., Murakami, H. and Shinmyo, A. (1999) Tobacco retinoblastoma-related protein phosphorylated by a distinct cyclin-dependent kinase complex with Cdc2/cyclin D *in vitro*. *Plant J*, **18**, 243–252.

Nakashima, M., Hirano, K., Nakashima, S., Banno, H., Nishihama, R. and Machida, Y. (1998) The expression pattern of the gene for NPK1 protein kinase related to mitogen-activated protein kinase kinase kinase (MAPKKK) in a tobacco plant: correlation with cell proliferation. *Plant Cell Physiol.*, **3**, 690–700.

Nebenführ, A., Frohlick, J.A. and Staehelin, L.A. (2000) Redistribution of Golgi stacks and other organelles during mitosis and cytokinesis in plant cells. *Plant Physiol*, **124**, 135–151.

Nishihama, R., Banno, H., Kawahara, E., Irie, K. and Machida, Y. (1997) Possible involvement of differential splicing in regulation of the activity of *Arabidopsis* ANP1 that is related to mitogen-activated protein kinase kinase kinases (MAPKKKs). *Plant J*, **12**, 39–48.

Nishihama, R., Ishikawa, M., Araki, S., Soyano, T., Asada, T. and Machida, Y. (2001) The NPK1 mitogen-activated protein kinase kinase kinase is a regulator of cell-plate formation in plant cytokinesis. *Genes Dev*, **15**, 352–363.

Nishihama, R. and Machida, Y. (2001) Expansion of the phragmoplast during plant cytokinesis: a MAPK pathway may MAP it out. *Curr Opin Plant Biol*, **4**, 507–512.

Nishihama, R., Soyano, T., Ishikawa, M., Araki, S., Tanaka, H., Asada, T., Irie, K., Ito, M., Terada, M., Banno, H., Yamazaki, Y. and Machida, Y. (2002) Expansion of the cell plate in plant cytokinesis requires a kinesin-like protein/MAPKKK complex. *Cell*, **109**, 87–99.

Nowack, M.K., Grini, P.E., Jakoby, M.J., Lafos, M., Koncz, C. and Schnittger, A. (2006) A positive signal from the fertilization of the egg cell sets off endosperm proliferation in angiosperm embryogenesis. *Nat Genet*, **38**, 63–67.

Ozlü, N., Srayko, M., Kinoshita, K., Habermann, B., O'toole, E.T., Müller-Reichert, T., Schmalz, N., Desai, A. and Hyman, A.A. (2005) An essential function of the *C. elegans* ortholog of TPX2 is to localize activated Aurora A kinase to mitotic spindles. *Dev Cell*, **9**, 237–248.

Pagès, G., Lenormand, P., L'Allemain, G., Chambard, J.C., Meloche, S. and Pouysségur, J. (1993) Mitogen-activated protein kinases p42mapk and p44mapk are required for fibroblast proliferation. *Proc Natl Acad Sci USA*, **90**, 8319–8323.

Pellman, D., Bagget, M., Tu, Y.H., Fink, G.R. and Tu, H. (1995) Two microtubule-associated proteins required for anaphase spindle movement in *Saccharomyces cerevisiae*. *J Cell Biol*, **130**, 1373–1385.

Peset, I., Seiler, J., Sardon, T., Bejarano, L.A., Rybina, S. and Vernos, I. (2005) Function and regulation of Maskin, a TACC family protein, in microtubule growth during mitosis. *J Cell Biol*, **170**, 1057–1066.

Petersen, J., Paris, J., Willer, M., Philippe, M. and Hagan, I.M. (2001) The *S. pombe* Aurora-related kinase Ark1 associates with mitotic structures in a stage dependent manner and is required for chromosome segregation. *J Cell Sci*, **114**, 4371–4384.

Porceddu, A., Stals, H., Reichheld, J.P., Segers, G., De Veylder, L., De Pinho Barrôco, R., Casteels, P., Van Montagu, M., Inzé, D. and Mironov, V. (2001) A plant-specific cyclin-dependent kinase is involved in the control of G2/M progression in plants. *J Biol Chem*, **276**, 36354–36360.

Romano, A., Guse, A., Krascenicova, I., Schnabel, H., Schnabel, R. and Glotzer, M. (2003) CSC-1: a subunit of the Aurora B kinase complex that binds to the survivin-like protein BIR-1 and the INCENP-like protein ICP-1. *J Cell Biol*, **161**, 229–236.

Roudier, F., Fedorova, E., Gyorgyey, J., Feher, A., Brown, S., Kondorosi, A. and Kondorosi, E. (2000) Cell cycle function of a *Medicago sativa* A2-type cyclin interacting with a PSTAIRE-type cyclin-dependent kinase and a retinoblastoma protein. *Plant J*, **23**, 73–83.

Sakai, H., Urano, T., Ookata, K., Kim, M.H., Hirai, Y., Saito, M., Nojima, Y. and Ishikawa, F. (2002) MBD3 and HDAC1, two components of the NuRD complex, are localized at Aurora-A-positive centrosomes in M phase. *J Biol Chem*, **277**, 48714– 48723.

Sasabe, M. and Machida, Y. (2006) MAP65: a bridge linking a MAP kinase to microtubule turnover. *Curr Opin Plant Biol*, **9**, 563–570.

Sasabe, M., Soyano, T., Takahashi, Y., Sonobe, S., Igarashi, H., Itoh, T.J., Hidaka, M. and Machida, Y. (2006) Phosphorylation of NtMAP65-1 by a MAP kinase down-regulates its activity of microtubule bundling and stimulates progression of cytokinesis of tobacco cells. *Genes Dev*, **20**, 1004–1014.

Sasai, K., Katayama, H., Stenoien, D.L., Fujii, S., Honda, R., Kimura, M., Okano, Y., Tatsuka, M., Suzuki, F., Nigg, E.A., Earnshaw, W.C., Brinkley, W.R. and Sen, S. (2004) Aurora-C kinase is a novel chromosomal passenger protein that can complement Aurora-B kinase function in mitotic cells. *Cell Motil Cytoskeleton*, **59**, 249–263.

Scholey, J.M., Brust-Mascher, I. and Mogilner, A. (2003) Cell division. *Nature*, **422**, 746–752.

Schnittger, A., Schöbinger, U., Bouyer, D., Weinl, C., Stierhof, Y.D. and Hülskamp, M. (2002) Ectopic D-type cyclin expression induces not only DNA replication but also cell division in *Arabidopsis* trichomes. *Proc Natl Acad Sci USA*, **99**, 6410–6415.

Schuyler, S.C., Liu, J.Y. and Pellman, D. (2003) The molecular function of Ase1p: evidence for a MAP-dependent midzone-specific spindle matrix. *J Cell Biol*, **160**, 517–528.

Segers, G., Gadisseur, I., Bergounioux, C., De Almeida Engler, J., Jacqmard, A., Van Montagu, M. and Inzé, D. (1996) The *Arabidopsis* cyclin-dependent kinase gene cdc2bAt is preferentially expressed during S and G2 phases of the cell cycle. *Plant J*, **10**, 601–612.

Seguí-Simarro, J.M., Austin, J.R., White, E.A. and Staehelin, L.A. (2004) Electron tomographic analysis of somatic cell plate formation in meristematic cells of *Arabidopsis* preserved by high-pressure freezing. *Plant Cell*, **16**, 836–856.

Sessa, F., Mapelli, M., Ciferri, C., Tarricone, C., Areces, L.B., Schneider, T.R., Stukenberg, P.T. and Musacchio, A. (2005) Mechanism of Aurora B activation by INCENP and inhibition by hesperadin. *Mol Cell*, **18**, 379–391.

Shimotohno, A., Umeda-Hara, C., Bisova, K., Uchimiya, H. and Umeda, M. (2004) The plant speci?c kinase CDKF;1 is involved in activating phosphorylation of cyclin-dependent kinase-activating kinases in *Arabidopsis*. *Plant Cell*, **16**, 2954–2966.

Sorrell, D.A., Menges, M., Healy, J.M.S., Deveaux, Y., Amano, C., Kagami, H., Shinmo, A., Doonan, J.H., Sekine, M. and Murray, J.A.H. (2001) Cell cycle regulation of cyclin-dependent kinases in tobacco cultivar Bright Yellow-2 cells. *Plant Physiol*, **126**, 1214–1223.

Soyano, T., Nishihama, R., Morikiyo, K., Ishikawa, M. and Machida, Y. (2003) NQK1/NtMEK1 is a MAPKK that acts in the NPK1 MAPKKK-mediated MAPK cascade and is required for plant cytokinesis. *Genes Dev*, **17**, 1055–1067.

Speliotes, E.K., Uren, A., Vaux, D. and Horvitz, H.R. (2000) The survivin-like *C. elegans* BIR-1 protein acts with the Aurora-like kinase AIR-2 to affect chromosomes and the spindle midzone. *Mol Cell*, **6**, 211–223.

Spielman, M., Preuss, D., Li, F.L., Browne, W.E., Scott, R.J. and Dickinson, H.G. (1997) TETRASPORE is required for male meiotic cytokinesis in *Arabidopsis thaliana*. *Development*, **124**, 2645–2657.

Straight, A.F., Cheung, A., Limouze, J., Chen, I., Westwood, N.J., Sellers, J.R. and Mitchison, T.J. (2003) Dissecting temporal and spatial control of cytokinesis with a myosin II Inhibitor. *Science*, **299**, 1743–1747.

Strompen, G., El Kasmi, F., Richter, S., Lukowitz, W., Assaad, F.F., Jurgens, G. and Mayer, U. (2002) The *Arabidopsis* HINKEL gene encodes a kinesin-related protein involved in cytokinesis and is expressed in a cell cycle-dependent manner. *Curr Biol*, **12**, 153–158.

Takenaka, K., Moriguchi, T. and Nishida, E. (1998) Activation of the protein kinase p38 in the spindle assembly checkpoint and mitotic arrest. *Science*, **280**, 599–602.

Tanaka, H., Ishikawa, M., Kitamura, S., Takahashi, Y., Soyano, T., Machida, C. and Machida, Y. (2004) The AtNACK1/HINKEL and STUD/TETRASPORE/AtNACK2 genes, which encode functionally redundant kinesins, are essential for cytokinesis in *Arabidopsis*. *Genes Cells*, **9**, 1199–1211.

Tsai, M.Y., Wiese, C., Cao, K., Martin, O., Donovan, P., Ruderman, J., Prigent, C. and Zheng, Y. (2003) A Ran signalling pathway mediated by the mitotic kinase Aurora A in spindle assembly. *Nat Cell Biol*, **5**, 242–248.

Umeda, M., Bhalerao, R.P., Schell, J., Uchimiya, H. and Koncz, C. (1998) A distinct cyclin dependent kinase-activating kinase of *Arabidopsis thaliana*. *Proc Natl Acad Sci USA*, **95**, 5021–5026.

Umeda, M., Shimotohno, A. and Yamaguchi, M. (2005) Control of cell division and transcription by cyclin-dependent kinase-activating kinases in plants. *Plant Cell Physiol*, **46**, 1437–1442.

Umeda, M., Umeda-Hara, C., Yamaguchi, M., Hashimoto, J. and Uchimiya, H. (1999) Differential expression of genes for cyclindependent protein kinases in rice plants. *Plant Physiol*, **119**, 31–40.

Vader, G., Kauw, J.J., Medema, R.H. and Lens, S.M. (2006a) Survivin mediates targeting of the chromosomal passenger complex to the centromere and midbody. *EMBO Rep*, **7**, 85–92.

Vader, G., Medema, R.H. and Lens, S.M. (2006b). The chromosomal passenger complex: guiding Aurora-B through mitosis. *J Cell Biol*, **173**, 833–837.

Van Damme, D., Bouget, F.Y., Van Poucke, K., Inzé, D. and Geelen, D. (2004) Molecular dissection of plant cytokinesis and phragmoplast structure: a survey of GFP-tagged proteins. *Plant J*, **40**, 386–398.

Vandepoele, K., Raes, J., De Veylder, L., Rouzé, P., Rombauts, S. and Inzé, D. (2002) Genome-wide analysis of core cell cycle genes in *Arabidopsis*. *Plant Cell*, **14**, 903–916.

Verbrugghe, K.J. and White, J.G. (2004) SPD-1 is required for the formation of the spindle midzone but is not essential for the completion of cytokinesis in *C. elegans* embryos. *Curr Biol*, **14**, 1755–1760.

Vernì, F., Somma, M.P., Gunsalus, K.C., Bonaccorsi, S., Belloni, G., Goldberg, M.L. and Gatti, M. (2004) Feo, the *Drosophila* homolog of PRC1, is required for central-spindle formation and cytokinesis. *Curr Biol*, **14**, 1569–1575.

Waizenegger, I., Lukowitz, W., Assaad, F., Schwarz, H., Jürgens, G. and Mayer, U. (2000) The *Arabidopsis* KNOLLE and KEULE genes interact to promote vesicle fusion during cytokinesis. *Curr Biol*, **10**, 1371–1374.

Wang, Y., Magnard, J.L., McCormick, S. and Yang, M. (2004) Progression through meiosis I and meiosis II in *Arabidopsis* anthers is regulated by an A-type cyclin predominately expressed in prophase I. *Plant Physiol*, **136**, 4127–4135.

Weingartner, M., Criqui, M.C., Meszaros, T., Binarova, P., Schmit, A.C., Helfer, A., Derevier, A., Erhardt, M., Bogre, L. and Genschik, P. (2004) Expression of a nondegradable cyclin B1 affects plant development and leads to endomitosis by inhibiting the formation of a phragmoplast. *Plant Cell*, **16**, 643–657.

Wheatley, S.P., Henzing, A.J., Dodson, H., Khaled, W. and Earnshaw, W.C. (2004) Aurora-B phosphorylation *in vitro* identifies a residue of survivin that is essential for its localization and binding to inner centromere protein (INCENP) *in vivo*. *J Biol Chem*, **279**, 5655–5660.

Wright, J.H., Munar, E., Jameson, D.R., Andreassen, P.R., Margolis, R.L., Seger, R. and Krebs, E.G. (1999) Mitogen-activated protein kinase kinase activity is required for the G(2)/M transition of the cell cycle in mammalian fibroblasts. *Proc Natl Acad Sci USA*, **96**, 11335–11340.

Yan, X., Wu, Y., Li, Q., Cao, L., Liu, X., Saiyin, H. and Yu, L. (2005) Cloning and characterization of a novel human Aurora C splicing variant. *Biochem Biophys Res Commun*, **328**, 353–361.

Yang, C.Y., Spielman, M., Coles, J.P., Li, Y., Ghelani, S., Bourdon, V., Brown, R.C., Lemmon, B.E., Scott, R.J. and Dickinson, H.G. (2003) TETRASPORE encodes a kinesin required for male meiotic cytokinesis in *Arabidopsis*. *Plant J*, **34**, 229–240.

Yasuhara, H. and Shibaoka, H. (2000) Inhibition of cell-plate formation by brefeldin A inhibited the depolymerization of microtubules in the central region of the phragmoplast. *Plant Cell Physiol*, **41**, 300–310.

Yasuhara, H., Sonobe, S. and Shibaoka, H. (1993) Effects of taxol on the development of the cell plate and of the phragmoplast in tobacco BY-2 cells. *Plant Cell Physiol*, **34**, 21–29.

Yasuhara, H., Sonobe, S. and Shibaoka, H. (1995) Effects of brefeldin A on the formation of the cell plate in tobacco BY-2 cells. *Eur J Cell Biol*, **66**, 274–281.

Zeitlin, S.G., Shelby, R.D. and Sullivan, K.F. (2001) CENP-A is phosphorylated by Aurora B kinase and plays an unexpected role in completion of cytokinesis. *J Cell Biol*, **155**, 1147–1157.

Zuo, J., Niu, Q.W., Nishizawa, N., Wu, Y., Kost, B. and Chua, N.H. (2000) KORRIGAN, an *Arabidopsis* endo-1,4-beta-glucanase, localizes to the cell plate by polarized targeting and is essential for cytokinesis. *Plant Cell*, **12**, 1137–1152.

Annual Plant Reviews (2008) **33**, 362–387
doi: 10.1111/b.9781405160025.2008.00013.x
www.interscience.wiley.com

Chapter 13
GUARD CELL SIGNALING

Yan Wu

Key Laboratory of Ministry of Education for Plant Developmental Biology, College of Life Sciences, Wuhan University, Wuhan 430072, People's Republic of China

Abstract: A pair of guard cells forms a stomatal pore in the epidermis of plant aerial tissues. Stomatal pores can open and close to regulate gas exchange in response to environmental stimuli, such as the plant hormone abscisic acid (ABA), water status, temperature, and light conditions, CO_2 exchange between plants and the atmosphere, and bacterial invasion. For example, elevated ABA and CO_2 levels induce stomatal closure, whereas blue light causes stomatal opening. It is hypothesized that guard cells possess signaling mechanisms that integrate multiple environmental signals to persuade stomatal movement and to regulate plant survival under various conditions. Recent integrated approach has allowed us to gain a glimpse of the network of signal transduction mechanisms in the modulation of stomatal movement. Molecular genetic studies together with cell biological analysis have revealed many signaling components and pathways involved in the regulation of stomatal movement. Genomics- and systems-based approaches have led to the generation of models of the signaling network in guard cells. This chapter focuses on recent advances in our understanding of the mechanisms of guard cell signaling and the interaction of multiple signals to modulate guard cell movement.

Keywords: stomatal movement; abscisic acid (ABA); CO_2; cytosolic calcium; blue light

13.1 Introduction

A pair of specialized kidney-shaped cells, termed guard cells, forms a pore, named stomata, which are inhabitant in the epidermis of the aerial parts of higher plants. The opening and closing of stomatal pores are the result of movement of guard cells that requires turgor and volume changes, which are regulated by a number of signaling events (Allan *et al.*, 1994; Jacob *et al.*, 1999; Ng *et al.*, 2001; Schroeder *et al.*, 2001a,b; Israelsson *et al.*, 2006; Li *et al.*, 2006; MacRobbie, 2006). To regulate gas exchange between intercellular spaces within the plant tissue and the surrounding environment, the stomata opens to permit CO_2 entry for photosynthetic carbon fixation and closes to prevent loss of water evaporation during transpiration. Excessive loss of water results in drought stress, growth arrest, and even death of plants. Consequently, the balancing act of stomatal opening and closing to regulate photosynthesis and transpiration is critical for plant growth and survival (Hetherington and Woodward, 2003; Israelsson *et al.*, 2006).

The central role of guard cells in regulating gas exchange is of importance for ecological and biotechnological applications (Schroeder *et al.*, 2001b; Hetherington and Woodward, 2003). Stress conditions such as drought cause dramatic crop losses and freshwater consumption. Recent studies in *Arabidopsis* have demonstrated that stomatal responses can be manipulated by modifying guard cell signal transduction elements to reduce transpirational water loss and desiccation during drought periods (Schroeder *et al.*, 2001b; Hugouvieux *et al.*, 2002; Li *et al.*, 2002; Israelsson *et al.*, 2006). Stomatal movement is also an important mechanism for defense responses against pathogens that enter plant tissues through stomata (Melotto *et al.*, 2006). Many physiological signals, such as the hormones including abscisic acid (ABA), auxin, cytokinin, and gibberellins, and environmental cues, such as red and blue light, CO_2, plant pathogen (i.e., bacteria), influence stomatal opening and closing. Therefore, guard cells have become a popular single-cell system for exploring the signaling cascade of the genes and proteins and the mechanism for integrating various signals.

Guard cells have also been used as a model system for investigating the role and the regulation of ion channels (Schroeder *et al.*, 2001b; Hetherington and Woodward, 2003; Israelsson *et al.*, 2006; Li *et al.*, 2006). Stomatal movement is the result of guard cell turgor changes. An increase in turgor drives the expansion of the more elastic outer side of the guard cell wall, pulling open the inner sides that form the pore. The activity of both inward and outward ion channels ultimately plays a key role in the determination of turgor pressure in guard cells. Therefore, research on guard cell movement has revolved around guard cell ion channels and the mechanism regulating them. A patch clamping technique has been crucial in the identification and characterization of ion channels involved in guard cell movement (Schroeder *et al.*, 1987, 2001b; Allen *et al.*, 2001). Integration of this technique with other powerful approaches, including biochemistry, genetics, cell biology, genomics, and proteomics, has led to the wealth of knowledge of signaling mechanisms and networks underlying guard cell movement (Leonhardt *et al.*, 2004; Israelsson *et al.*, 2006; Li *et al.*, 2006). Consequently, the guard cell has become one of the most studied and better understood signaling systems in plants. Due to space limitation, this chapter is intended to highlight our recent advances in our understanding of the guard cell signaling network that integrates various signals, such as plant hormones, ABA, CO_2 concentration, blue light, and pathogen infections. A network of signals will be discussed (Fig. 13.1). Readers are referred to a number of excellent recent reviews for more details of elucidating the molecular mechanisms of guard cell signal transduction (Assmann and Shimazaki, 1999; Blatt, 2000; Hetherington, 2001; Schroeder

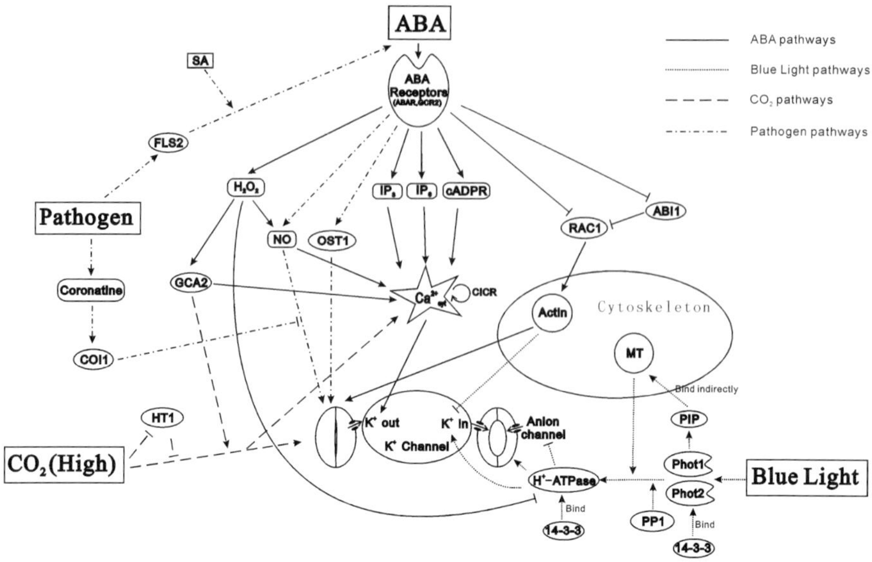

Figure 13.1 Guard cell signaling network. This figure shows four major signals, ABA, CO_2, blue light, and plant pathogen networking in guard cells. The solid line represents the ABA signaling pathways; the dashed line represents CO_2 signaling pathways; the dotted line represents blue light signaling pathways; and the dash-dotted line represents the pathogen signaling pathways. The full names of signaling components corresponding to each abbreviated label are shown below:
ABA, abscisic acid; ABAR, ABA receptor; ABI1, protein phosphatase 2C ABI1; Actin, actin cytoskeleton reorganization; Blue light, blue light signaling; cADPR, cyclic ADP-ribose; Ca^{2+}_{cyt}, cytosolic calcium; CICR, Ca^{2+}-induced Ca^{2+} release; CO_2, carbon dioxide signaling; COI1, coronatine-insensitive 1, a subunit of an E3 ubiquitin ligase; Coronatine, virulence factor secreted by *Pst DC3000*; FLS2, flagellin receptor; GCA2, ABA-insensitive mutant growth control by *aba2*; GCR2, G-protein-coupled receptor; H^+-ATPase, H^+-ATPase at the plasma membrane; H_2O_2, hydrogen peroxide; HT1, high leaf temperature 1, HT1 kinase; $InsP_3$, inositol-1,4,5-trisphosphate; $InsP_6$, *myo*-inositol hexakisphosphate; K^+_{in} / K^+_{out}, potassium (K^+) -inward / -outward channels; MT, microtubule; NO, nitric oxide; OST1, protein kinase open stomata 1; Pathogen, *Psuedomonas syringae* signaling model; Phot1 and Phot2, light-activated, FMN-binding receptor kinase, phototropin 1 and 2; PIP, Vfphot1a interacting protein; PP1, protein phosphatase 1; RAC1, small GTPase RAC1; SA, salicylic acid; 14-3-3, 14-3-3 protein.

et al., 2001a,b; Hetherington and Woodward, 2003; Israelsson *et al.*, 2006; Li *et al.*, 2006; Young *et al.*, 2006).

13.2 ABA-mediated guard cell signaling

ABA plays an important role in the regulation of development and responses to environmental stresses, such as drought, salinity, cold, as well as in the regulation of stomatal movement. To conserve water, ABA level builds up in the leaves upon drought stress, and ABA inhibits the opening of closed stomata and promotes the closing of open ones. These are two separate turgor-driven processes involving the coordinated activation and inhibition of inwardly and outwardly directed cation and anion channels present in the plasma and tonoplast membranes (MacRobbie, 1995, 1998; Blatt, 2000; Hetherington, 2001; Schroeder *et al.*, 2001b). Perhaps because of its importance in drought tolerance, ABA has received the most attention among various signals that regulate stomatal movement. Thus, ABA signaling mechanisms in guard cells are relatively well understood. Two recent advances have propelled our understanding of ABA signaling in guard cells yet to another level. ABA receptors had remained elusive until the recent consecutive reports of three distinct receptors (Razem *et al.*, 2006; Shen *et al.*, 2006; Liu *et al.*, 2007). Within the last two years, three *Arabidopsis* ABA receptors have been identified one after another: (a) FCA, a nucleus-localized RNA-binding protein that acts as an ABA receptor in complex with FY and specifically functions in flowering (Razem *et al.*, 2006), (b) ABAR/CHLH, the H subunit of Mg-chelatase (CHLH) that is localized to the chloroplast and acts as an ABA receptor in the regulation of seed germination, postgermination growth, and stomatal movement (Shen *et al.*, 2006), and (c) GCR2, a G-protein-coupled receptor that is localized to the plasma membrane (PM) and appears to mediate all known ABA responses in *Arabidopsis* (Liu *et al.*, 2007). However, the relative contribution for each of ABAR/CHLH and GCR2 to the regulation of stomatal movement has yet to be evaluated. The regulation of stomatal aperture is a highly dynamic complex process that requires a systems-level analysis of ABA receptors and their signaling components. A large number of cellular components in the ABA regulation of aperture changes are known (Fig. 13.2), but we were completely ignorant of their roles in stomatal movement at the quantitative level until a recent study from Assmann's group (Li *et al.*, 2006). Using mathematical modeling, Li *et al.* (2006) recapitulated a model that integrates more than 40 identified network components in the regulation of ABA-induced stomatal closure (Li *et al.*, 2006). The model reveals novel predictions regarding how strongest reduction in ABA responsiveness can be achieved by disrupting key nodes in the network, such as membrane depolarization, anion efflux, actin cytoskeleton reorganization, cytosolic pH increase, the phosphatidic acid pathway, or K^+ efflux through slowly activating K^+ channels at the PM (Li *et al.*, 2006) .

13.2.1 Ion channels

Stomatal closure requires massive ion efflux from guard cells (MacRobbie, 1995, 1998; Blatt, 2000; Schroeder *et al.*, 2001b), which involves coordinated activation and inhibition of inwardly and outwardly directed PM-localized cation and anion channels (Hamilton *et al.*, 2000; Hetherington, 2001; Schroeder *et al.*, 2001b; Maeser *et al.*, 2003). In guard cells, the initiation of K^+ and anions efflux, the removal of sucrose, and the conversion of malate to osmotically inactive starch can all result in the reduction of turgor pressure

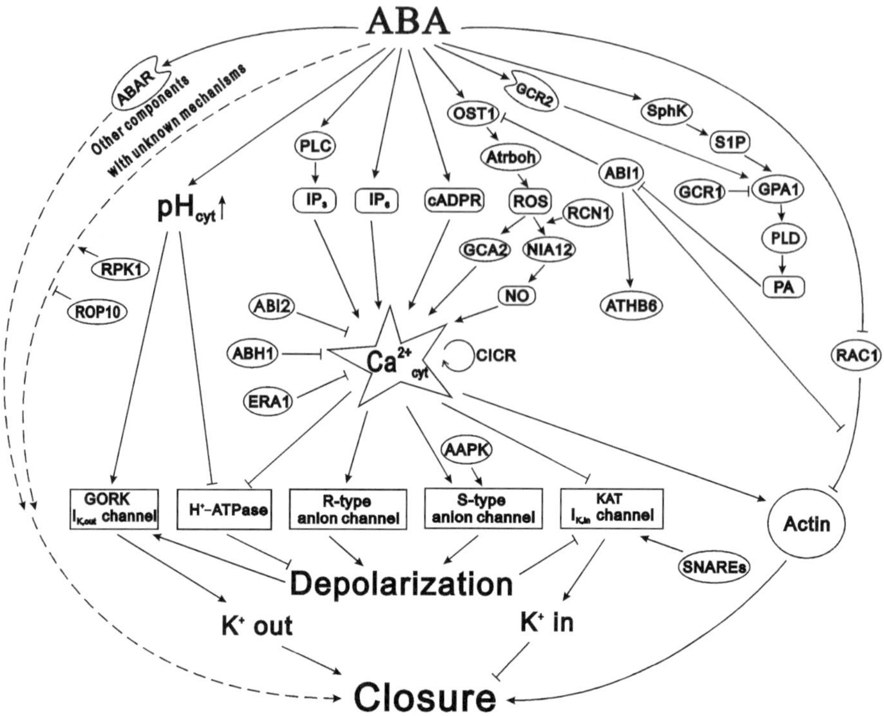

Figure 13.2 Guard cell ABA signaling. This figure summarizes most of the components involved in ABA-induced guard cell closure. The solid line represents the proposed working model for ABA signaling in guard cells. The dashed line represents the signaling pathways with unknown mechanisms. The full names of signaling components corresponding to each abbreviated label are shown below:
AAPK, an ABA-activated protein kinase; ABAR, ABA receptor; ABI1, protein phosphatase 2C ABI1; ABI2, protein phosphatase 2C ABI2; ABH1, mRNA cap-binding protein; Actin, actin cytoskeleton reorganization; ATHB6, *Arabidopsis* homeodomain-binding protein 6; Atrboh, NADPH oxidase; cADPR, cyclic ADP-ribose; Ca^{2+}_{cyt}, cytosolic calcium; CICR, Ca^{2+}-induced Ca^{2+} release; ERA1, farnesyl transferase ERA1; GCA2, ABA-insensitive mutant growth control by *aba2*; GCR1, putative G-protein-coupled receptor; GCR2, G-protein-coupled receptor; GORK, guard cell outward-rectifying K^+ channel; GPA1, heterotrimeric G protein α subunit; H^+-ATPase, H^+-ATPase at the plasma membrane; IP_3, inositol-1,4,5-trisphosphate; IP_6, *myo*-inositol hexakisphosphate; KAT1, inward-rectifying K^+channel; NIA12, nitrate reductase; NO, nitric oxide; OST1, protein kinase open stomata 1; PA, phosphatidic acid; pH_{cyt}, cytosolic pH; PLC, phospholipase C; PLD, phospholipase D; RAC1, small GTPase RAC1; RCN1, protein phosphatase 2A; ROS, reactive oxygen species; ROP10, small GTPase ROP10; RPK1, receptor-like protein kinase1; R-type anion channel, rapid transient type of anion channel; SphK, sphingosine kinase; S1P, sphingosine-1-phosphate; SNAREs, soluble *N*-ethylmaleimide-sensitive factor attachment protein receptors; S-type anion channel, slow transient type of anion channel.

of guard cells, leading to stomatal closure (Schroeder *et al.*, 1987; MacRobbie, 1995, 1998). As depicted in Fig. 13.2, ABA regulates three cellular activities that coordinately inhibit inward K^+ (K^+_{in}) channels and activate outward K^+ channels (K^+_{out}). First of all, ABA elevates cytosolic Ca^{2+} concentrations ($[Ca^{2+}]_{cyt}$)

(McAinsh *et al.*, 1990). The elevated $[Ca^{2+}]_{cyt}$ inhibits PM-localized proton pumps (Kinoshita *et al.*, 1995; Brault *et al.*, 2004) and K^+_{in} channels and activate two different types of anion channels: slow-activating sustained (S-type) and rapid transient (R-type) anion channels (Schroeder and Hagiwara, 1989; Hedrich *et al.*, 1990; Schroeder and Keller 1992). Secondly, activation of these anion channels and inhibition of the proton pump cause PM depolarization, leading to K^+_{in} inhibition and K^+_{out} activation. Thirdly, ABA also elevates cytosolic pH (pH_{cyt}) by an unknown mechanism, although a pH_{cyt} increase is also dependent upon $[Ca^{2+}]_{cyt}$ elevation (Irving *et al.*, 1992; Brault *et al.*, 2004; Jia and Davies, 2007). Increases in pH_{cyt} levels directly enhance the opening of voltage-activated $K^+_{out}$and promote the opening of anion efflux channels (Blatt and Armstrong, 1993; Hosy *et al.*, 2003; Wang *et al.*, 2005). The persistent efflux of both anions and K^+ contributes to the loss of guard cell turgor (Homann and Thiel, 2002; Hosy *et al.*, 2003), leading to stomatal closing.

In the guard cell, over 90% of ions exported from the guard cells during stomatal closing must first be transported from vacuoles into the cytosol (Ward and Schroeder, 1994; MacRobbie, 1995; Ward *et al.*, 1995; Allen and Sanders, 1996; MacRobbie, 1998, 2002, 2006; Allen *et al.*, 1999b, 2001). The $[Ca^{2+}]_{cyt}$ elevation also activates vacuolar K^+(VK) channels to release K^+ from the vacuole (Ward and Schroeder, 1994; MacRobbie, 2002). The signaling mechanism for $[Ca^{2+}]_{cyt}$ activation of VK remains obscure. Evidence suggests protein tyrosine dephosporylation may be involved, but its molecular basis is unknown (MacRobbie, 2002).

13.2.2 Ca^{2+} elevation and oscillation

As discussed above, it is well established that ABA-induced stomatal closing is dependent upon Ca^{2+} (DeSilva *et al.*, 1985; Schwartz, 1985; Schroeder and Hagiwara, 1989; Gilroy *et al.*, 1990; Webb *et al.*, 2001; Marten *et al.*, 2006), although a Ca^{2+}-independent ABA signaling pathway also appears to exist (Allan *et al.*, 1994; Israelsson *et al.*, 2006; Marten *et al.*, 2006). It seems that both $[Ca^{2+}]_{cyt}$ elevation and oscillation signal to ABA-mediated stomatal closure. Molecular and genetic studies support the importance of ABA-induced $[Ca^{2+}]_{cyt}$ elevations in guard cells (McAinsh *et al.*, 1990; Allen *et al.*, 1999a, 2001, 2002; Wood *et al.*, 2000; Schroeder *et al.*, 2001b; Han *et al.*, 2003; Tang *et al.*, 2007). In *Arabidopsis*, the ABA-insensitive mutants, *abi1-1* and *abi2-1*, in which ABA induction of stomatal closure is suppressed, an ABA-induced $[Ca^{2+}]_{cyt}$ increase is greatly compromised (Allen *et al.*, 1999a). By using viable stomata in epidermal strips of a transgenic line of tobacco expressing aequorin (the proteinous luminescent reporter of Ca^{2+}), it was shown that the ABA-induced stomatal closure resulted from $[Ca^{2+}]_{cyt}$ increases in epidermal strips (Wood *et al.*, 2000). ABA-induced $[Ca^{2+}]_{cyt}$ increases were also observed in the study of the ABA-hypersensitive mutant *ear1-2*. At low ABA concentrations, increases of $[Ca^{2+}]_{cyt}$ in guard cells and stomatal closure were greatly enhanced in the *ear1-2* mutant compared with the wild type (Allen *et al.*, 2002).

Oscillations in free $[Ca^{2+}]_{cyt}$ are important for Ca^{2+}-based signaling (McAinsh *et al.*, 1995; Staxén *et al.*, 1999; Allen *et al.*, 2000, 2001; Schroeder *et al.*, 2001b). Different mechanisms may contribute to the generation of $[Ca^{2+}]_{cyt}$ oscillations (Staxén *et al.*, 1999; Allen *et al.*, 2001; Schroeder *et al.*, 2001b). A defined window of guard cell Ca^{2+} oscillation parameters was varied by using a "calcium clamp" (Allen *et al.*, 2000, 2001), and it was shown that the long-term steady-state stomatal closure was "calcium programmed" by guard cell $[Ca^{2+}]_{cyt}$ oscillations (Allen *et al.*, 2001). During stomatal closure, Ca^{2+} oscillations can be induced by an increase in external (apoplastic) Ca^{2+}, including the repetitive Ca^{2+} influx across the PM coupled to Ca^{2+} release from intracellular stores for each separate Ca^{2+} transient (McAinsh *et al.*, 1995; Grabov and Blatt, 1998; Allen *et al.*, 2000, 2001). ABA induces oscillations in the guard cell $[Ca^{2+}]_{cyt}$, and the pattern of the oscillations depends on ABA concentrations and correlates with the final stomatal aperture (Staxén *et al.*, 1999). In the *Arabidopsis* V-ATPase mutant *de-etiolated 3* (*det3*), external Ca^{2+} and oxidative stress elicited prolonged Ca^{2+} increases, which did not oscillate, and stomatal closure was abolished. Moreover, in *det3* guard cells, experimentally imposing external Ca^{2+}-induced oscillations rescued stomatal closure. These data provide genetic evidence that stimulus-specific Ca^{2+} oscillations are necessary for stomatal closure (Allen *et al.*, 2000).

13.2.3 Regulation of PM-localized Ca^{2+} influxes

It is not clear what are the upstream mechanisms by which the sensitivity of $[Ca^{2+}]_{cyt}$ is regulated. Genetic and cell biological studies in *Arabidopsis* have implicated a role for reactive oxygen species (ROS) and H_2O_2-induced $[Ca^{2+}]_{cyt}$ elevations in ABA signaling in guard cells (Pei *et al.*, 2000). Further studies confirm that there is a close relationship between ABA and the spatial and temporal pattern of NO and H_2O_2 production in guard cells. ABA-induced stomatal closure is dependent upon the endogenous H_2O_2 production. H_2O_2 in turn induces NO generation (Figs. 13.1 and 13.2). Both ABA and H_2O_2 stimulate NO synthesis (Garcia-Mata and Lamattina, 2002; Bright *et al.*, 2006). Moreover, in *Arabidopsis* and *Vicia* guard cells, it has been shown that NO selectively regulates Ca^{2+}-sensitive ion channels by promoting Ca^{2+} release from intracellular stores to raise $[Ca^{2+}]_{cyt}$, and this NO-sensitive Ca^{2+} release might act via a cGMP-dependent cascade (Garcia-Mata *et al.*, 2003) (Fig. 13.2).

13.2.4 Regulation of intracellular Ca^{2+} stores

Several intracellular Ca^{2+} stores have been implicated in ABA and $[Ca^{2+}]_{cyt}$ signaling in guard cells. There are three second messengers implicated in the regulation of $[Ca^{2+}]_{cyt}$ stores in animal cells: (a) cyclic ADP-ribose (cADPR), (b) nicotinic acid adenine dinucleotide phosphate (NAADP), and (c) inositol-1,4,5-trisphosphate ($InsP_3$) (Lee, 2001). In light of the function of intracellular Ca^{2+} stores in the ABA signal transduction pathways, second messengers of

Ca^{2+} stores, cADPR, $InsP_3$ and $InsP_6$ are summarized here (Figs. 13.1 and 13.2). cADPR is produced from NAD through action of the enzyme ADP-ribosyl cyclase, and mobilizes Ca^{2+} from intracellular stores by activating an endomembrane ion channel known as the ryanodine receptor (RYR) (Lee, 2001). In plants, the vacuolar Ca^{2+} release can be stimulated by nanomolar concentrations of cADPR (Allen *et al.*, 1995). Microinjected cADPR activated transient expression of two ABA-responsive genes in tomato hypocotyls cells (Wu *et al.*, 1997). Furthermore, overexpression of the *Aplysia ADPR cyclase* gene in *Arabidopsis* resulted in an increase in ADPR cyclase activity and cADPR levels, as well as elevated expression of a set of ABA-responsive genes (Sánchez *et al.*, 2004). Taken together, these studies implicated cADPR as a second messenger in ABA signaling to the regulation of Ca^{2+}elevation. In *Commelina* guard cells, ABA-induced $[Ca^{2+}]_{cyt}$ increases and stomatal closure are partially mediated by cADPR (Leckie *et al.*, 1998), suggesting that additional parallel $[Ca^{2+}]_{cyt}$ elevation mechanisms exist in the ABA signaling cascade.

Recent studies have suggested that at least two inositol phosphates can act as second messengers releasing Ca^{2+} in the ABA modulation of stomatal closure (Hunt *et al.*, 2003; Lemtiri-Chlieh *et al.*, 2000, 2003). $InsP_3$ is generated from phosphatidylinositol-4,5-bisphosphate [PtdIns(4,5)P_2] by the action of phospholipase C (PLC) (see Chapter 8). Immunolocalization suggests the presence of a Ca^{2+}-activated PLC in tobacco guard cells (Hunt *et al.*, 2003). In transgenic tobacco plants with reduced levels of PLC (due to cosupression induced by the full-length NrPLC2 cDNA expressed under the control of a guard cell-specific promoter), the guard cells appeared partially defective in ABA induction of aperture closing. These results imply the involvement of PLC in the amplification of the Ca^{2+} signal responsible for reduction of stomatal aperture in response to ABA (Hunt *et al.*, 2003). Other inositol phosphates may also act as second messengers in ABA signal transduction pathways. The *myo*-inositol hexakisphosphate ($InsP_6$) is the most abundant inositol phosphate in cells. It can mobilize an endomembrane store of calcium in guard cells. ABA elevates $InsP_6$ levels, and $InsP_6$triggers the release of Ca^{2+} from endomembrane stores, which in turn inactivates the PM inward K^+ conductance in a cytosolic Ca^{2+}-dependent manner (Lemtiri-Chlieh *et al.*, 2000, 2003).

13.2.5 Ca^{2+} sensing

A recent study has identified Ca^{2+} sensors that function to transduce the ABA signal in guard cells (Mori *et al.*, 2006). The functions of two *Arabidopsis* guard cell-expressed *CDPK* genes, *CPK3* and *CPK6*, were studied. Double mutant alleles *cpk3–cpk6* showed impairment in ABA- and Ca^{2+}-activation of S-type anion channels. Surprisingly, they also impair ABA activation of Ca^{2+}-permeable channels. Furthermore, ABA- and Ca^{2+}-induced stomatal closing were partially impaired in these *CDPK* mutants. These findings showed important functions of CDPKs (CPK3 and CPK6) in guard cell ion channel regulation and provided genetic evidence for Ca^{2+} sensors involved in guard cell signaling (Israelsson *et al.*, 2006; Mori *et al.*, 2006).

13.2.6 The actin cytoskeleton

Pharmacological studies support a role for the reorganization of actin filaments in the regulation of stomatal movement (Hwang *et al.*, 1997; Galatis and Apostolakos, 2004). Treatment with actin-depolymerizing drug cytochalasin D activated the K^+_{in} channels and enhanced light-induced stomatal opening, whereas phalloidin (an actin filament stabilizer) inhibited K^+_{in} currents and light-induced stomatal opening (Hwang *et al.*, 1997). In ABA-treated guard cells, the actin structure was altered from a radial pattern to a randomly oriented and short-fragmented pattern, raising the possibility that ABA could regulate stomatal movement through its action on the actin cytoskeleton (Eun and Lee, 1997). This notion is further supported by the involvement of ROP/RAC GTPases in the regulation of guard cell movement and ABA signaling (Lemichez *et al.*, 2001). ROP/RAC GTPases are the plant-specific subfamily of conserved Rho-family GTPases that are known to modulate actin organization (Yang, 2002) (see Chapter 3). *Arabidopsis* AtRac1/ROP6 was reported to act as a negative regulator in the ABA-induced actin reorganization that promotes stomatal closure (Lemichez *et al.*, 2001). ABA treatment induced AtRac1/ROP6 inactivation and disruption of the actin cytoskeleton in guard cells. In the ABA-insensitive mutant *abi1-1*, which is impaired in stomatal closure, neither AtRac1/ROP6 inactivation nor actin disruption was observed upon ABA treatment. These observations indicate that inactivation of AtRac1/ROP6 by ABA is essential for stabilization of actin architecture and for stomatal closure (Lemichez *et al.*, 2001).

ABA-induced actin filaments reorganization has also been observed by a pharmacological study in guard cells of *Commelina communis*. ABA-induced actin changes were mediated by cytosolic calcium levels and by protein kinase and phosphatase activities (Hwang and Lee, 2001). When the guard cells were treated with $CaCl_2$, the reorganization of actin filaments was similar to that induced by ABA. A protein kinase inhibitor, staurosporine, could inhibit actin reorganization induced by ABA or $CaCl_2$. However, a protein phosphatase inhibitor, calyculin A, could mimic ABA treatment and induced long radial cortical actin filaments in ABA- or $CaCl_2$-treated guard cells (Hwang and Lee, 2001). Taken together, these studies reveal the significance of actin reorganization and small GTP-binding proteins in stomatal movement (Figs. 13.1 and 13.2).

13.2.7 Genetic and genomic dissection of ABA signaling components in guard cells

Genetic approach has been powerful for the discovery of new ABA signaling components in guard cells. The *ABA-INSENSITIVE* 1 (*ABI1*) and *ABI2* genes, which encode protein phosphatase 2C(PP2C), have been shown to negatively regulate ABA signaling (Saez *et al.*, 2004, 2006; Yoshida *et al.*, 2006a,b). ABI1, but not ABI2, has been demonstrated to interact with OPEN STOMATA 1 (OST1/SnRK2E), a serine–threonine protein kinase that acts as a positive regulator in ABA signaling (Mustilli *et al.*, 2002; Yoshida *et al.*, 2006b). In contrast to *abi1-1*, *abi2-1* did not abolish ABA activation of OST1 kinase activity; this suggests ABI2 functions downstream of OST1 (Yoshida *et al.*, 2006b), which is consistent with the differential effects of *abi1-1* and *abi2-1* in the guard cell ABA signaling network (Murata *et al.*, 2001). In *Vicia faba*, ABA-activated protein kinase (AAPK), a homolog of OST1, interacts with mRNA-binding protein AAPK-INTERACTING PROTEIN1 (AKIP1) (Li *et al.*, 2002). The *Arabidopsis* homolog to AKIP1, UBP1-INTERACTING PROTEIN 2a (UBA2a) showed ABA-induced relocalization to nuclear speckles, but it did not appear to interact with the OST1 kinase (Riera *et al.*, 2006).

Combined genetic and biochemical analyses have revealed a connection between PHOSPHOLIPASE Dα1 (PLDα1) and ABI1 in response to ABA (Zhang *et al.*, 2005; Mishra *et al.*, 2006). PLDα1-generated PA binds to ABI1, abolishing ABI1 inhibition of ABA-induced stomatal closing (Zhang *et al.*, 2005; Mishra *et al.*, 2006). In the knockout mutant *pldDα1*, ABA-induced stomatal closure was abolished but the ABA response was rescued in the *pldDα1abi1* double mutant. G-PROTEIN ALPHA SUBUNIT 1 (GPA1) was shown to interact with PLDα1 and PA to mediate ABA inhibition of stomatal opening (Mishra *et al.*, 2006) (Fig. 13.2). This finding agrees with the previous observation that the mutant *gpa1* is insensitive to ABA inhibition of both stomatal opening and K^+_{in} channel regulation in guard cells (Wang *et al.*, 2001).

An mRNA cap-binding protein, ABH1, was identified from the isolation of a recessive *ABA- hypersensitive* mutant, *abh1*. *ABH1* encodes a nuclear mRNA cap-binding protein and functions in a heterodimeric complex to bind the mRNA cap structure (Hugouvieux *et al.*, 2001, 2002). *abh1* mutations enhance ABA induction of guard cell closure (guard cell closure was induced by 0.5 μM ABA) and increase cytosolic calcium levels, suggesting that ABH1 is involved in the amplification of early ABA signaling. It was suggested that ABH1 represents a modulator of ABA signaling by affecting transcriptional regulation of early ABA signaling elements (Hugouvieux *et al.*, 2001, 2002). Two guard-cell-specific MYB transcription factors that control stomatal aperture were identified almost simultaneously, *AtMYB60* and *AtMYB61*, both of which are in the *Arabidopsis* R2R3-MYB family of transcription factors (Cominelli *et al.*, 2005; Liang *et al.*, 2005).

ABA regulates K^+ channel activity at the PM of guard cells, leading to stomatal closure. Analysis of a knockout mutant for GUARD CELL OUTWARD RECTIFIYING K^+ CHANNEL (GORK) indicates that additional types of K^+ transporters may act in parallel with guard cell K^+ channels (Hosy *et al.*, 2003). By expressing a dominant-negative fragment, it was shown that the impairment of a SOLUBLE *N*-ETHYLMALEIMIDE-SENSITIVE-FACTOR ATTACHMENT PROTEIN RECEPTOR (SNARE) in tobacco prevented ABA-induced enhancement of K^+_{out} activity and reduction of K^+_{in} rectifying channel activity (Leyman *et al.*, 1999). Further study indicates a key role for SNAREs in trafficking and positional anchoring of K^+ channel proteins in the PM (Sutter *et al.*, 2006). The inward-rectifying *Arabidopsis* K^+ channel (KAT1) is dependent on SNAREs for delivery to the PM, and KAT1 is localized within microdomain clusters at the PM (Sutter *et al.*, 2006).

A recent study has identified TWO PORE CHANNEL 1 (TPC1), a gene that encodes essential subunits of the ubiquitous slow vacuolar channels. TPC1 functions in Ca^{2+}-induced stomatal closing. Its regulation of stomatal movement, however, is independent of ABA (Peiter *et al.*, 2005).

A role of ROS as a second messenger in guard cell ABA signaling was demonstrated by characterizing the *Arabidopsis* double mutant of AtrbohD (*Arabidopsis thaliana* RESPIRATORY BURST OXIDASE PROTEIN D) and AtrbohF (Kwak *et al.*, 2003). In *atrbohD atrbohF*, ABA-induced production of ROS was disrupted. ABA-induced Ca^{2+}-permeable channel activation and stomatal closing were abolished, but the impairment was restored by H_2O_2 application (Kwak *et al.*, 2003). Consistent with a role of ROS in guard cell signaling, ozone can stimulate stomatal closing. Ozone can induce oxidative burst, which is dependent on heterotrimeric G proteins (Joo *et al.*, 2005). ROS signaling in guard cells has been discussed intensively in a recent review (Kwak *et al.*, 2006).

A genetic screen for mutants that have altered responses to reduced relative humidity (RH) identified two genes involved in guard cell ABA signaling. *OST1* encodes a protein kinase involved in ABA-induced stomatal closure, and *ABA2* encodes an enzyme involved in ABA biosynthesis. Both *aba2* and *ost1* are impaired in the reduction of stomatal conductance in response to low humidity transitions (Xie *et al.*, 2006; Yoshida *et al.*, 2006b). This suggests that OST1 protein is a required component in stomatal RH and ABA signaling. The participation of ABA signaling in RH responses supports an organized networking in guard cell signaling (Xie *et al.*, 2006).

Genetic screens for the negative regulators of ABA responses identified the *Arabidopsis* ENHANCED RESPONSE TO ABSCISIC ACID 1 (ERA1) gene that encodes the β-subunit of farnesyl transferase (FTase). Knockout mutations of ERA1 cause ABA-hypersensitive anion and Ca^{2+} channel regulation, stomatal closing, and reduced transpirational water loss (Culter *et al.*, 1996; Pei *et al.*, 1998; Allen *et al.*, 2002). ERA1 target proteins have not been clearly demonstrated yet. One putative target is ROP10, a member of the ROP/RAC GTPase family (Zheng *et al.*, 2002). The *rop10-1* knockout mutant exhibited hypersensitivity to ABA stimulation of stomatal closure. The PM localization of ROP10 is dependent on ERA1, leading to the suggestion that the ROP10 may be an ERA1 target. α-Carboxyl methyltransferase, an enzyme that methylates the C-terminus of farnesylated proteins, may be important in ABA responses. Inhibitors of α-carboxyl methyltransferase cause increased ABA sensitivity and stomatal closure, suggesting that α-carboxyl methylation of prenylation proteins may be necessary for negative regulation of ABA signal transduction in *Arabidopsis* (Chary *et al.*, 2002; Zheng *et al.*, 2002). Therefore, stress-stimulated inhibition of farnesylation provides an approach for engineering drought avoidance in crop plants. *Brassica napus* transgenic plants expressing the antisense construct for the α- or β-subunit of farnesyl transferase under the drought-inducible promoter *rd29A* showed enhancement of ABA response and drought tolerance. The results showed the reduction of stomatal conductance under stress conditions and enhanced yields (Wang *et al.*, 2005).

Genomic approaches have led to a more systematic understanding of ABA signaling in guard cells (Leonhardt *et al.*, 2004). Microarray analysis of the *Arabidopsis* guard cell transcriptome identified 1309 genes that are significantly expressed in guard cells (Leonhardt *et al.*, 2004). In comparison to that of mesophyll cells, only 64 transcripts were detectable solely in guard cells. These genes encode transcription factors, signal transduction proteins such as protein kinases, receptor protein kinases, and metabolic pathway proteins. Many known guard cell ABA signaling components are modulated at the transcription level. Further analysis reveals that 69 ABA-inducible genes and 64 ABA-repressed genes are preferentially present in guard cells. The ABA-upregulated genes can be classified, with respect to their putative functions, into two major classes: (1) genes implicated in cell protection and in production of important metabolic proteins, and (2) genes implicated in signal transduction, such as protein kinases, protein phosphatases, receptor kinases, transcription factors, 14-3-3 proteins, reactive oxygen turnover enzymes (NADPH oxidases, catalase, and glutathione *S*-transferase), and protein ubiquitination components. Among the ABA-downregulated genes, except for the photosynthesis-related genes, several genes that function in signal transduction and osmoregulation, such as protein kinases and the K^+ channels *KAT1*, *KAT2*, and *AKT2*, are also repressed by ABA. The functional characterization of these regulatory proteins will clearly contribute to more comprehensive understanding of the ABA signaling network in plants (Leonhardt *et al.*, 2004). For example, the analysis of *Arabidopsis* mutants for ABA-induced PP2C genes in guard cells has been carried out. Recessive mutation of *AtP2C-HA* (homolog to ABI1/ABI2) induces ABA-hypersensitive stomatal closing and seed germination (Saez *et al.*, 2006). *AtPP2CA*, a different ABA-regulated PP2CA gene, was also identified from forward genetic screening. PP2CA mutation causes ABA hypersensitivity during seed germination (Kuhn *et al.*, 2006; Yoshida *et al.*, 2006a). ABI1 and ABI2 display differential binding to different members of the SALT OVERLY SENSITIVE 2 (SOS2)-like protein kinase in *Arabidopsis* (Ohta *et al.*, 2003). Therefore, combined genomic and genetic analyses of genes expressed in *Arabidopsis* guard cells provide a robust tool for identifying new signaling mechanisms and for a systems-level understanding of signaling networks in guard cells.

13.3 CO_2 signaling in guard cells

As pores for CO_2 exchange between the intracellular spaces and the atmosphere, stomata can adjust their aperture by sensing the ambient CO_2 concentrations. When the CO_2 concentration is high, the stomata close, whereas low

CO_2 concentrations induce their opening (Assmann, 1999; Hashimoto *et al.*, 2006; Israelsson *et al.*, 2006) (Fig. 13.1). Increased CO_2 concentrations activate outward rectifying K^+ channels and induce Cl^-efflux via the anion channels and enhance malate production. Malate has been shown to play a role in regulating R-type anion channels. Stomatal opening is mediated by ion and organic solute concentrations (Hanstein and Felle, 2002). High CO_2 causes guard cell depolarization, which in turn diminishes the $[Ca^{2+}]_{cyt}$ transient rate. However, evidence suggests that elevated CO_2 concentrations trigger rises in $[Ca^{2+}]_{cyt}$ (Webb *et al.*, 1996). Recent studies show that Ca^{2+} regulation in stomatal movement may be more complicated. The *Arabidopsis gca2* mutant (*growth controlled by abscisic acid*) exhibits ABA insensitivity. Interestingly, it shows no marked change in the average $[Ca^{2+}]_{cyt}$ transient rate when the ambient CO_2 concentration changes. As a result, high CO_2-induced stomatal closure is not observed in this mutant suggesting that *gca2* is insensitive to high CO_2 (Young *et al.*, 2006). A model has been proposed in which CO_2 concentrations prime Ca^{2+} sensors that could mediate the specificity of Ca^{2+}signaling (Young *et al.*, 2006). These observations also imply a crosstalk between ABA and CO_2 signaling pathways (Assmann, 1999; Vavasseur and Raghavendra, 2005) at the level of GCA2 and both these signals regulate many ion-transport mechanisms (Assmann, 1999; Vavasseur and Raghavendra, 2005; Young *et al.*, 2006) (Fig. 13.1).

Two allelic *Arabidopsis* mutants, *ht1-1* and *ht1-2*, have been obtained from the genetic screen for mutants with leaf temperature changes (Hashimoto *et al.*, 2006). They both exhibit higher leaf temperature at low CO_2 concentrations. The *ht1-1* mutant has a reduction of stomatal aperture at low CO_2 concentrations, reflecting a CO_2-hypersensitive response, but the strong allele *ht1-2* shows severe impairments in stomatal movements with constitutive high CO_2 response. In response to ABA, blue light and fusicoccin (FC), however, *ht1-2* appears to be normal, indicating that HT1 (HIGH LEAF TEMPERATURE 1), which encodes a protein kinase that is expressed in guard cells, does not have a general function in the regulation of all stomatal responses. These data suggest that the HT1 kinase is a negative regulator of CO_2 signal transduction (Fig. 13.1). Furthermore, *ht1* mutants appear to exhibit normal responses to ABA, blue light, and fungal phytotoxin FC. These findings indicate that the HT1 kinase is important for regulation of stomatal movements and its function is more pronounced in response to CO_2 than other signals such as ABA and light (Hashimoto *et al.*, 2006).

Evidence suggests a crosstalk between photosynthesis and stomata response to CO_2 (Roelfsema *et al.*, 2002; Messinger *et al.*, 2006). Stomatal opening in response to red light has been suggested to link to a decrease in intercellular CO_2 concentrations (Olsen *et al.*, 2002; Roelfsema *et al.*, 2002). The action spectra for mesophyll photosynthesis and for the stomatal response to red light are similar (Olsen *et al.*, 2002). In addition, stomatal opening is normal in response to red light, while the CO_2 concentration is held by adjusting ambient CO_2 concentration, suggesting that a CO_2 concentration decrease is independent pathway of red light-induced stomatal opening (Olsen *et al.*, 2002; Messinger *et al.*, 2006).

13.4 Light signaling in guard cells

Stomatal opening can be stimulated by both red and blue light (Zeiger and Zhu, 1998; Assmann and Shimazaki, 1999; Kinoshita and Shimazaki, 1999; Kinoshita *et al.*, 2001; Roelfsema *et al.*, 2001; Taylor and Assmann, 2001; Olsen *et al.*, 2002). In most species, guard cells are the only epidermal cells that contain chloroplasts, and chlorophylls in guard cells were implicated as a photoreceptor in the light responses of stomata. Guard cell chlorophyll absorbs red and blue wavelengths of light, and guard cell sensitivity to red and blue light is consistent with a role for the opening of stomata for photosynthesis (Assmann and Shimazaki, 1999). It was suggested that guard cells might possess a specific blue light photoreceptor, which agrees with the generation of greater quantum efficiency of blue light over red light in stimulating stomatal opening (Zeiger and Zhu, 1998; Assmann and Shimazaki, 1999). The stomatal response to blue light may be particularly important under the blue-enriched illumination that triumphs in the early morning light and in sunspecks. As discussed in this section, fundamental questions concern the cellular signals bridging light perception to downstream regulators in guard cells.

The blue light receptors, PHOTOTROPINS 1 and 2 (PHOT1 and PHOT2) (Kinoshita *et al.*, 2001), are responsible for blue light induction of stomatal opening (Kinoshita *et al.*, 2001; Briggs and Christie, 2002). Although the blue light-induced stomatal opening is only slightly affected in either *phot1* and *phot2* single mutants, the *phot1phot2* double mutants are completely defective in blue light induction of stomatal opening, indicating that PHOT1 and PHOT2 are functionally redundant in guard cells (Kinoshita *et al.*, 2001).

Phototropins are PM-localized light-dependant protein kinases containing two LOV (light, oxygen, or voltage) domains and a Ser/Thr kinase domain. The LOV domains bind chromophore FMN (FLAVIN MONONUCLEOTIDE) and produce a flavin-C cysteinyl adduct after blue light illumination, resulting in the C-terminus kinase activation. Subsequently, 14-3-3 proteins bind to the phototropins and initiate signaling. The 14-3-3 protein binding to phototropins is mediated by phosphorylation in guard cells of broad bean, and the Vfphot-14-3-3 complex is able to interact with other components in *V. faba* (Kinoshita and Shimazaki, 1999; Emi *et al.*, 2001; Kinoshita *et al.*, 2003; Emi *et al.*, 2005).

Light mediates stomatal movement through its regulation of the PM H^+-ATPase activity (Kinoshita and Shimazaki, 1999; Emi *et al.*, 2001; Zhang *et al.*, 2004), which is also subject to the regulation by other pathways, such as ABA-induced H_2O_2 activity (Zhang *et al.*, 2004), K^+ influx (Roelfsema *et al.*, 2001; Taylor and Assmann, 2001), Ca^{2+} elevation (Kinoshita *et al.*, 1995; Assmann and Shimazaki, 1999; Schroeder *et al.*, 2001a), binding of 14-3-3 (Kinoshita *et al.*, 2003), and protein phosphorylation activities (Kinoshita and Shimazaki, 1999; Takemiya *et al.*, 2006). The blue light-activated signal is transmitted to the PM H^+-ATPase, resulting in phosphorylation of the H^+-ATPase C-terminus (Kinoshita and Shimazaki, 1999). The 14-3-3 protein binds to the PM H^+-ATPase when it is activated by blue light and FC (Kinoshita *et al.*, 2003); in turn, the activated H^+-ATPase elevates the negative electrical potential gradient inside of the PM. The hyperpolarized PM drives a K^+_{in} channel. As a result, the accumulation of potassium salt and increase in turgor pressure in guard cells lead to stomatal opening (Kinoshita and Shimazaki, 1999). PP1 (PROTEIN PHOSPHATASE 1) has been found to play a role in blue light-induced signaling in guard cells (Takemiya *et al.*, 2006). PP1 inhibitor, tautomysin, inhibited blue light-induced phosphorylation of H^+-ATPase in the PM and stomatal opening. In transgenic plants carrying dominant-negative PP1c or inhibitor-2 (a protein that specifically binds and inhibits PP1c), guard cells are impaired in blue light-induced stomatal opening. These data suggest that PP1 functions as a positive regulator downstream of phototropins and upstream of H^+-ATPase in the blue light signaling pathway of guard cells (Takemiya *et al.*, 2006). It is unclear how PHOT1 and 2 transmit light signals to PP1 and how PP1 regulate H^+-ATPase.

VfPIP, a protein that interacts with Vfphot1a, has been isolated from *V. faba* guard cells. VfPIP localized on cortical microtubules (MT), suggesting that the function of Vfphoto1a may be mediated by cortical MT. Stomata treated with MT-depolymerizing compounds failed to open in response to blue light, further supporting an important role for cortical MT in blue light induction of stomatal opening (Emi *et al.*, 2005). ABA inhibits blue light-induced H^+ pumping by reducing the phosphorylation of the PM H^+-ATPase and disrupting the binding of 14-3-3 protein to H^+-ATPase. The inhibitory effect results from ABA-induced H_2O_2 production and exogenous H_2O_2 inhibits H^+-ATPase in the same fashion as ABA. However, H_2O_2 has no effect on H^+-ATPase in the microsomal fraction, suggesting that H_2O_2 reduces phosphorylation of H^+-ATPase via inhibition of signaling components that transmit the blue light signal to the PM H^+-ATPase (Zhang *et al.*, 2004) (Fig. 13.1). A recent experiment tested the working model of two independent light signaling pathways in the modulation of guard cell movement (Marten *et al.*, 2007). It was shown that the inhibitory effect of blue light on the guard cell PM anion channels is also involved in phototropin signaling. This pathway is independent of blue light stimulation of the PM H^+-ATPases (Marten *et al.*, 2007).

Earlier studies suggest the presence of carotenoid zeaxanthin-mediated blue light signaling in guard cells. DTT, a reducing agent that effectively inhibits zeaxanthin formation, can completely inhibit blue light-induced stomatal opening. The *Arabidopsis* mutant *npq1* (Frechijia *et al.*, 1999), which failed to convert violaxanthin into zeaxanthin, exhibits impaired blue light-induced stomatal opening. Based on these data, zeaxanthin was hypothesized to be a candidate photoreceptor in blue light signaling of guard cells (Frechijia *et al.*, 1999; Lascève *et al.*, 1999), but its direct role as a photoreceptor was put in doubt after the results from the analysis of the *phot1phot2* double mutant (Kinoshita *et al.*, 2001). This led to the hypothesis that zeaxanthin may play a modulating role in blue light-induced stomatal opening.

In response to blue light, the loss of H^+ from guard cells contributes to a hyperpolarization of the PM, which creates an electrical gradient driving force for the uptake of K^+ and activates K^+-sensitive ion channels in the guard cell membrane. The H^+extrusion by an H^+-ATPase is also presumed to be crucial for anion uptake (Assmann and Shimazaki, 1999). In addition to K^+ accumulation driven by membrane hyperpolarization, blue light also stimulates an increase in intracellular concentrations of the organic acid malate as well as sucrose accumulation in guard cells under certain conditions (Talbott and Zeiger, 1998; Assmann and Shimazaki, 1999). It was observed that K^+ and sucrose act on osmoticum in guard cells at different day times. The increase in the osmotic concentration driven by the uptake of K^+ and accumulation of sucrose results in guard cell swelling and stomatal aperture widening. The K^+concentration in guard cells increases at the beginning of a daily cycle, but it decreases in the early afternoon (Talbott and Zeiger, 1998). Then the sucrose becomes predominantly active solute after a slow increase in the morning, and the stomatal apertures continue increasing until the sucrose content decreases at the end of the day (Ritte *et al.*, 1999; Outlaw and De Vlieghere-He, 2001).

13.5 Innate immunity in guard cells

When pathogens such as *Pseudomonas syringae* (*P. syringae*, or other plant and human pathogens) attack plant leaves, the stomata close rapidly to prevent their entrance into its interior (Melotto *et al.*, 2006). To assure a successful invasion, *P. syringae* cells tend to surround the open stomata specifically and evolve specific virulence factors to instigate stomatal reopening, such as coronatine (COR). The mutant strain of *P. syringae* that lacks COR fails to initiate disease and reopening of the stomata. Although inoculation of *E. coli* could also stimulate stomatal closure, it could not make the stomata reopen (Melotto *et al.*, 2006). PAMPs (PATHOGEN-ASSOCIATED MOLECULAR PATTERNs), which are specific bacterial components, such as flg22 and LPS, play a role in triggering stomatal closure (Melotto *et al.*, 2006). Experiments have shown that plants perceive flg22 through the FLS2 receptor. The flg22 peptide fails to induce stomatal closure in the *fls2* mutant. Further study shows that PAMP-induced stomatal closure is part of the SA-regulated innate immune system. The ability of stomata to close in response to bacteria was compromised in SA-deficient mutants (Melotto *et al.*, 2006). Interestingly, ABA is involved in PAMP-induced stomatal closure. The *Arabidopsis* guard cell mutant *ost1* and ABA-deficient mutant *aba3-1* both show the inability to close stomata in response to bacterial invasion, suggesting that pathogen attack may trigger ABA synthesis to promote stomatal closure (Melotto *et al.*, 2006).

NO is also found to be required for PAMP-induced stomatal closure. A NO synthase inhibitor effectively prevents stomatal closure triggered by flg22 or LPS. COI1 is a subunit of an E3 ubiquitin ligase and is necessary for the phytotoxin COR to inhibit PAMP-induced ABA signaling. *P. syringae* fails to reopen stomata in the COR-insensitive *coi1* mutant. Either PAMP- or ABA-induced production of NO is normal after treatment with COR, suggesting that COR acts downstream of NO production to counteract stomatal closure (Melotto *et al.*, 2006) (Fig. 13.1).

In the *Plasmopara viticola*-infected grapevine leaves, stomata do not close in response to dehydration and exogenously added ABA. The induction of stomatal closure and decrease of transcription was inhibited. In the epidermal peels, which were separated from infected leaves, ABA-induced stomatal closure was similar to that in healthy leaves. The author hypothesized that the *P. viticola*-induced stomatal abnormality might result from a nonsystemic compound or a reduction of the backpressure exerted by surrounding epidermal cells (Allègre *et al.*, 2007).

13.6 Extracellular Ca^{2+} sensing in guard cells

Studies have shown that extracellular Ca^{2+}, $[Ca^{2+}]_{ext}$, is required for various physiological and developmental processes in plants (Sanders *et al.*, 2002; Han *et al.*, 2003; Chen *et al.*, 2004). A high proportion of the total Ca^{2+} is often located in the cell wall and at the exterior surface of the PM (Sanders *et al.*, 2002; Han *et al.*, 2003). As described in this chapter, multiple factors such as ABA, CO_2, and light can regulate stomatal movements and cause guard cell $[Ca^{2+}]_{cyt}$ changes and subsequently stomatal closure (McAinsh *et al.*, 1990; Assmann and Shimazaki, 1999; Schroeder *et al.*, 2001b). A PM-localized extracellular Ca^{2+}-sensing receptor CAS has been shown to regulate the guard cell $[Ca^{2+}]_{cyt}$. Suppression of CAS expression could disrupt $[Ca^{2+}]_{ext}$ signaling in guard cells. Elevated $[Ca^{2+}]_{ext}$ activates CAS, which mediates stomatal closure, preventing excessive Ca^{2+} unloading (see Chapter 5). Thus, $[Ca^{2+}]_{ext}$-induced stomatal closure could function as a feedback mechanism of Ca^{2+} supply (Han *et al.*, 2003). Further study on the molecular mechanisms of CAS revealed that $[Ca^{2+}]_{ext}$-regulated CAS activity in turn directs release of Ca^{2+} from internal stores via the IP_3 pathway in *Arabidopsis*. The changes in $[Ca^{2+}]_{cyt}$ are synchronized to extracellular Ca^{2+} concentration oscillations through CAS. These results suggest that $[Ca^{2+}]_{cyt}$ oscillations are coupled to the $[Ca^{2+}]_{ext}$ oscillations–CAS–IP_3 pathway. The phase and period of oscillations are likely determined by the stomatal conductance (Tang *et al.*, 2007).

Externally applied camodulin (ExtCaM), a calcium-binding protein, was shown to stimulate a cascade of intracellular signaling events to regulate stomatal movement (Chen *et al.*, 2004). The ExtCaM-induced stomatal closure is mediated by the activation of heterotrimeric G protein, generation of H_2O_2, and changes in $[Ca^{2+}]_{cyt}$. This finding implicates that in guard cells, heterotrimeric G protein may transmit the ExtCaM signal, generating H_2O_2 to induce $[Ca^{2+}]_{cyt}$ elevation, leading to stomatal closure (Chen *et al.*, 2004). Therefore, ExtCaM could be another extracellular Ca^{2+} sensor in the guard cell signaling and stomatal movements.

13.7 Conclusions and prospects

Tremendous progress has been made in our understanding of signaling mechanisms regulating guard cell movement, which not only has provided us with important insights into the mechanisms modulating the adaptive guard cell responses to various external and physiological stimuli, but also has enhanced our general knowledge of signal transduction in plants. A linear signaling cascade is clearly not adequate to explain all signaling mechanisms. As a matter of fact, the proposed signaling mechanisms discussed in this chapter involve the interaction networking of signals such as ABA, CO_2, cytoskeleton, light, and pathogen attack (Figs. 13.1 and 13.2). Many signaling components, e.g., Ca^{2+} and Ca^{2+} stores, K^+-related ion channels, ROS and NO, protein kinases, receptors for certain signals, and their signaling pathways, have been revealed. It is also clear that any given signal does not merely regulate a linear cascade of signaling molecules to achieve the modulation of guard cell movement, but rather signaling networks are required. Perhaps such networks facilitate the interaction among different signals and their integration to allow dynamic regulation of stomatal movement under the complex environmental setting in nature. However, how the specificity of different signaling pathways that may involve some common components such as Ca^{2+} could be achieved and how different signals interact with each other and are integrated to control stomatal movement remain poorly understood and present major challenges for future research. With the combination of molecular genetic, biochemistry, and cell biology approaches, and the ability to manipulate guard cell signaling genes in *Arabidopsis*, future research in this area will endow us with new insights into the mechanisms underlying dynamic guard cell movement, which may ultimately be useful for engineering crop plants with improved gas exchange in response to drought, atmospheric CO_2 variation, and other environmental stresses (Schroeder *et al.*, 2001b; Hetherington and Woodward, 2003).

Acknowledgments

The author gratefully thanks Yuxuan Wu for the help in preparing this chapter. Other members of the Wu lab are acknowledged for comments on the manuscript. Because of space limitation, the author apologizes for not being able to include all the literature on guard cell signaling. Research in the author's laboratory was supported by NSFC (grant #30640021) and Wuhan University in People's Republic of China.

References

Allan, A.C., Fricker, M.D., Ward, J.L., Beale, M.H. and Trewavas, A.J. (1994) Two transduction pathways mediate rapid effects of abscisic acid in Commelina guard cells. *Plant Cell*, **6**, 1319–1328.

Allègre, M., Daire, X., Héloir, M.-C., Trouvelot, S., Mercier, L., Adrian, M. and Pugin, A. (2007) Stomata deregulation in *Plasmopara viticola*-infected grapevine leaves. *New Phytol*, **173**, 832–840.

Allen, G.J., Chu, S.P., Harrington, C.L., Schumacher, K., Hoffmann, T., Tang, Y.Y., Grill, E. and Schroeder, J.I. (2001) A defined range of guard cell calcium oscillation parameters encodes stomatal movements. *Nature*, **411**, 1053–1057.

Allen, G.J., Chu, S.P., Schumacher, K., Shimazaki, C.T., Vafeados, D., Kemper, A., Hawke, S.D., Tallman, G., Tsien, R.Y., Harper, J.F., Chory, J. and Schroeder, J.I. (2000) Alteration of stimulus-specific guard cell calcium oscillations and stomatal closing in *Arabidopsis det3* mutant. *Science*, **289**, 2338–2342.

Allen, G.J., Kuchitsu, K., Chu, S.P., Murata, Y. and Schroeder, J.I. (1999a) *Arabidopsis abi1-1* and *abi2-1* phosphatase mutations reduce abscisic acid-induced cytoplasmic calcium rises in guard cells. *Plant Cell*, **11**, 1785–1798.

Allen, G.J., Kwak, J.M., Chu, S.P., Llopis, J., Tsien, R.Y., Harper, J.F. and Schroeder, J.I. (1999b) Cameleon calcium indicator reports cytoplasmic calcium dynamics in *Arabidopsis* guard cells. *Plant J*, **19**, 735–747.

Allen, G.J., Muir, S.R. and Sanders, D. (1995) Release of $Ca^{2+}$from individual plant vacuoles by both InsP3 and cyclic ADP-ribose. *Science*, **268**, 735–737.

Allen, G.J., Murata, Y., Chu, S.P., Nafisi, M. and Schroeder, J.I. (2002) Hypersensitivity of abscisic acid-induced cytosolic calcium increases in *Arabidopsis* farnesyltransferase mutant *era1-2*. *Plant Cell*, **14**, 1649–1662.

Allen, G.J. and Sanders, D. (1996) Control of ionic currents guard cell vacuoles by cytosolic and luminal calcium. *Plant J*, **10**, 1055–1069.

Assmann, S.M. (1999) The cellular basis of guard cell sensing of rising CO_2. *Plant Cell Environ*, **22**, 629–637.

Assmann, S.M. and Shimazaki, K.-I. (1999) The multisensory guard cell. Stomatal responses to blue light and abscisic acid. *Plant Physiol*, **119**, 809–816.

Blatt, M.R. (2000) Cellular signaling and volume control in stomatal movements in plants. *Ann Rev Cell and Devel Biol*, **16**, 221–241.

Blatt, M.R. and Armstrong, F. (1993) K^+ channels of stomatal guard cells: abscisic acid-evoked control of the outward-rectifier mediated by cytoplasmic pH. *Planta*, **191**, 330–341.

Brault, M., Amiar, Z., Pennarun, A.-M., Monestiez, M., Zhang, Z., Cornel, D., Dellis, O., Knight, H., Bouteau, F. and Rona, J.-P. (2004) Plasma membrane depolarization induced by abscisic acid in *Arabidopsis* suspension cells involves reduction of proton pumping in addition to anion channel activation, which are both Ca^{2+} dependent. *Plant Physiol*, **135**, 231–243.

Briggs, W.R. and Christie, J.M. (2002) Phototropsins 1 and 2: versatile plant blue-light receptors. *Trends Plant Sci*, **7**, 204–210.

Bright, J., Desikan, R., Hancock, J.T., Weir, I.S. and Neill, S.J. (2006) ABA-induced NO generation and stomatal closure in *Arabidopsis* are dependent on H_2O_2 synthesis. *Plant J*, **45**, 113–122.

Chary, S.N., Bultema, R.L., Packard, C.E. and Crowell, D.N. (2002) Prenylcysteine α-carboxyl methyltransferease expression and function in *Arabidopsis thaliana*. *Plant J*, **32**, 735–747.

Chen, Y.-L., Huang, R., Xiao, Y.-M., Lü, P., Chen, J. and Wang, X.-C. (2004) Extracellular calmodulin-induced stomatal closure is mediated by heterotrimeric G protein and H_2O_2. *Plant Physiol*, **136**, 4096–4103.

Cominelli, E., Galbiati, M., Vavasseur, A., Conti, L., Sala, T., Vuylsteke, M., Leonhardt, N., Dellaporta, S.L. and Tonelli, C. (2005) A guard-cell-specific MYB transcription factor regulates stomatal movements and plant drought tolerance. *Curr Biol*, **15**, 1196–1200.

Culter, S., Ghassemian, M., Bonetta, D., Cooney, S. and McCourt, P. (1996) A protein farnesyl transferase involved in abscisic acid signal transduction in *Arabidopsis*. *Science*, **273**, 1239–1241.

DeSilva, D.L.R., Hetherington, A.M. and Mansfield, T.A. (1985) Synergism between calcium ions and abscisic acid in preventing stomatal opening. *New Phytol*, **100**, 473–482.

Emi, T., Kinoshita, T., Sakamoto, K., Mineyuki, Y. and Shimazaki, K.-I. (2005) Isolation of a protein interacting with Vfphot1a in guard cells of *Vicia faba*. *Plant Physiol*, **138**, 1615–1626.

Emi, T., Kinoshita, T. and Shimazaki, K.-I. (2001) Specific binding of vf14-3-3a isoform to the plasma membrane H^+-ATPase in response to blue light and fusicoccin in guard cells of broad bean. *Plant Physiol*, **125**, 1115–1125.

Eun, S.O. and Lee, Y. (1997) Actin filaments of guard cells are reorganized in response to light and abscisic acid. *Plant Physiol*, **115**, 1491–1498.

Frechijia, S., Zhu, J., Talbott, L.D. and Zeiger, E. (1999) Stomata from *npq1*, a zeaxanthin-less *Arabidopsis* mutant, lack a specific response to blue light. *Plant Cell Physiol*, **40**, 949–954.

Galatis, B. and Apostolakos, P. (2004) The role of the cytoskeleton in morphogenesis and function of stomatal complexes. *New Phytol*, **161**, 613–639.

Garcia-Mata, C., Gay, R., Sokolovski, S., Hills, A., Lamattina, L. and Blatt, M.R. (2003) Nitric oxide regulates K^+ and Cl^- channels in guard cells through a subset of abscisic acid-evoked signaling pathways. *Proc Natl Acad Sci USA*, **100**, 11116–11121.

Garcia-Mata, C. and Lamattina, L. (2002) Nitric oxide and abscisic acid cross talk in guard cells. *Plant Physiol*, **128**, 790–792.

Gilroy, S., Read, N.D. and Trewavas, A.J. (1990) Elevation of cytosolic calcium by caged calcium or caged inositol triphosphate initiates stomatal closure. *Nature*, **346**, 769–771.

Grabov, A. and Blatt, M.R. (1998) Membrane voltage initiates Ca^{2+} waves and potentiates Ca^{2+} increases with abscisic acid in stomatal guard cells. *Proc Natl Acad Sci USA*, **95**, 4778–4783.

Hamilton, D.W.A., Hills, A., Koehler, B. and Blatt, M.R. (2000) Ca^{2+} channels at the plasma membrane of stomatal guard cells are activated by hyperpolarization and abscisic acid. *Proc Natl Acad Sci USA*, **97**, 4967–4972.

Han, S., Tang, R., Anderson, L.K., Woerner, T.E. and Pei, Z.-M. (2003) A cell surface receptor mediates extracellular Ca^{2+} sensing in guard cells. *Nature*, **425**, 196–200.

Hanstein, S.M. and Felle, H.H. (2002) CO_2-triggered chloride release from guard cells in intact fava bean leaves. Kinetics of the onset of stomatal closure. *Plant Physiol*, **130**, 940–950.

Hashimoto, M., Negi, J., Young, J., Israelsson, M., Schroeder J.I. and Iba, K. (2006) *Arabidopsis* HT1 kinase controls stomatal movements in response to CO_2. *Nature Cell Biol*, **8**, 391–397.

Hedrich, R., Busch, H. and Raschke, K. (1990) Ca^{2+} and nucleotide dependent regulation of voltage dependent anion channels in the plasma membrane of guard cells. *EMBO J*, **9**, 3889–3892.

Hetherington, A.M. (2001) Guard cell signaling. *Cell*, **107**, 711–714.

Hetherington, A.M. and Woodward, F.I. (2003) The role of stomata in sensing and driving environmental change. *Nature*, **424**, 901–908.

Homann, U. and Thiel, G. (2002) The number of K^+ channels in the plasma membrane of guard cell protoplasts changes in parallel with the surface area. *Proc Natl Acad Sci USA*, **99**, 10215–10220.

Hosy, E., Vavasseur, A., Mouline, K., Dreyer, I., Gaymard, F., Porée, F., Boucherez, J., Lebaudy, A., Bouchez, D., Véry, A.-A., Simonneau, T., Thibaud, J.-B. and Sentenac, H. (2003) The *Arabidopsis* outward K^+ channel *GORK* is involved in regulation of stomatal movements and plant transpiration. *Proc Natl Acad Sci USA*, **100**, 5549–5554.

Hugouvieux, V., Murata, Y., Young, J.J., Kwak, J.M., Mackesy, D.Z. and Schroeder, J.I. (2002) Localization, ion channel regulation, and genetic interactions during abscisic acid signaling of the nuclear mRNA cap-binding protein, ABH1. *Plant Physiol*, **130**, 1276–1287.

Hugouvieux, V., Kwak, J.M. and Schroeder, J.I. (2001) An mRNA cap binding protein, ABH1, modulates early abscisic acid signal transduction in *Arabidopsis*. *Cell*, **106**, 477–487.

Hunt, L., Mills, L.N., Pical, C., Leckie, C.P., Aitken, F.L., Kopka, J., Mueller-Roeber, B., McAinsh, M.R., Hetherington, A.M. and Gray, J.E. (2003) Phospholipase C is required for the control of stomatal aperture by ABA. *Plant J*, **34**, 47–55.

Hwang, J.U. and Lee, Y. (2001) Abscisic acid-induced actin reorganization in guard cells of dayflower is mediated by cytosolic calcium levels and by protein kinase and protein phosphatase activities. *Plant Physiol*, **125**, 2120–2128.

Hwang, J.-U., Suh, S., Yi, H., Kim, J. and Lee, Y. (1997) Actin filaments modulate both stomatal opening and inward K^+-channel activities in guard cells of *Vicia faba* L. *Plant Physiol*, **115**, 335–342.

Irving, H.R., Gehring, C.A. and Parish, R.W. (1992) Changes in cytosolic pH and calcium of guard cells precede stomatal movements. *Proc Natl Acad Sci USA*, **89**, 1790–1794.

Israelsson, M., Siegel, R.S., Young, J., Hashimoto, M., Iba, K. and Schroeder, J.I. (2006) Guard cell ABA and CO_2 signaling network updates and Ca^{2+} sensor priming hypothesis. *Curr Opin Plant Biol*, **9**, 654–663.

Jacob, T., Ritchie, S., Assmann, S.M. and Gilroy, S. (1999) Abscisic acid signal transduction in guard cells is mediated by phospholipase D activity. *Proc Natl Acad Sci USA*, **96**, 12192–12197.

Jia, W. and Davies, W.J. (2007) Modification of leaf apoplastic pH in relation to stomatal sensitivity to root-sourced abscisic acid signals. *Plant Physiol*, **143**, 68–77.

Joo, J.H., Wang, S., Chen, J.G., Jones, A.M. and Fedoroff, N.V. (2005) Different signaling and cell death roles of heterotrimeric G protein α and β subunits in the *Arabidopsis* oxidative stress response to ozone. *Plant Cell*, **17**, 957–970.

Kinoshita, T., Doi, M., Suetsugu, N., Kagawa, T., Wada, M. and Shimazaki, K.-I. (2001) phot1 and phot2 mediate blue light regulation of stomatal opening. *Nature*, **414**, 656–660.

Kinoshita, T., Emi, T., Tominaga, M., Sakamoto, K., Shigenaga, A., Doi, M. and Shimazaki, K.-I. (2003) Blue-light- and phosphorylation-dependent binding of a 14-3-3 protein to phototropins in stomatal guard cells of broad bean. *Plant Physiol*, **133**, 1453–1463.

Kinoshita, T., Nishimura, M. and Shimazaki, K.-I. (1995) Cytosolic concentration of Ca^{2+}regulates the plasma membrane H^+-ATPase in guard cells of fava bean. *Plant Cell*, **7**, 1333–1342.

Kinoshita, T. and Shimazaki, K.-I. (1999) Blue light activates the plasma membrane H^+-ATPase by phosphorylation of the C-terminus in stomatal guard cells. *EMBO J*, **18**, 5548–5558.

Kuhn, J.M., Boisson-Dernier, A., Dizon, M.B., Maktabi, M.H. and Schroeder, J.I. (2006) The protein phosphatase *AtPP2CA* negatively regulates abscisic acid signal transduction in *Arabidopsis*, and effects of *abh1* on *AtPP2CA* mRNA. *Plant Physiol*, **140**, 127–139.

Kwak, J.M., Mori, I.C., Pei, Z.-M., Leonhardt, N., Torres, M.A., Dangl, J.L., Bloom, R.E., Bodde, S., Jones, J.D.G. and Schroeder, J.I. (2003) NADPH oxidase *AtrbohD* and *AtrbohF* genes function in ROS-dependent ABA signaling in *Arabidopsis*. *EMBO J*, **22**, 2623–2633.

Kwak, J.M., Nguyen, V. and Schroeder, J.I. (2006) The role of reactive oxygen species in hormonal responses. *Plant Physiol*, **141**, 323–329.

Lascève, G., Leymarie, J., Olney, M.A., Liscum, E., Christie, J.M., Vavasseur, A. and Briggs, W.R. (1999) *Arabidopsis* contains at least four independent blue-light-activated signal transduction pathways. *Plant Physiol*, **120**, 605–614.

Leckie, C.P., McAinsh, M.R., Allen, G.J., Sanders, D. and Hetherington, A.M. (1998) Abscisic acid-induced stomatal closure mediated by cyclic ADP-ribose. *Proc Natl Acad Sci USA*, **95**, 15837–15842.

Lee, H.C. (2001) Physiological functions of cyclic ADP-ribose and NAADP as calcium messengers. *Annu Rev Pharmacol Toxicol*, **41**, 317–345.

Lemichez, E., Wu, Y., Sanchez, J.-P., Mettouchi, A., Mathur, J. and Chua, N.-H. (2001) Inactivation of AtRac1 by abscisic acid is essential for stomatal closure. *Genes Dev*, **15**, 1808–1816.

Lemtiri-Chlieh, F., MacRobbie, E.A.C. and Brearley, C.A. (2000) Inositol hexakisphosphate is a physiological signal regulating the K^+—inward rectifying conductance in guard cells. *Proc Natl Acad Sci USA*, **97**, 8687–8692.

Lemtiri-Chlieh, F., MacRobbie, E.A., Webb, A.A.R., Manison, N.F., Brownlee, C., Skepper, J.N., Chen, J., Prestwich, G.D. and Brearley, C.A. (2003) Inositol hexakisphosphate mobilizes an endomembrane store of calcium in guard cells. *Proc Natl Acad Sci USA*, **100**, 10091–10095.

Leonhardt, N., Kwak, J.M., Robert, N., Waner, D., Leonhardt, G. and Schroeder, J.I. (2004) Microarray expression analyses of *Arabidopsis* guard cells and isolation of a recessive abscisic acid hypersensitive protein phosphatase 2C mutant. *Plant Cell*, **16**, 596–615.

Leyman, B., Geelen, D., Quintero, F.J. and Blatt, M.R. (1999) A tobacco syntaxin with a role in hormonal control of guard cell ion channels. *Science*, **283**, 537–540.

Li, J., Kinoshita, T., Pandey, S., Ng, C.K.-Y., Gygi, S.P., Shimazaki, K.-I. and Assmann, S.M. (2002) Modulation of an RNA-binding protein by abscisic-acid-activated protein kinase. *Nature*, **418**, 793–797.

Li, S., Assmann, S.M. and Albert, R. (2006) Predicting essential components of signal transduction networks: a dynamic model of guard cell abscisic acid signaling. *PLoS Biology*, **4**, e312.

Liang, Y.-K., Dubos, C., Dodd, I.C., Holroyd, G.H., Hetherington, A.M. and Campbell, M.M. (2005) *AtMYB61*, an R2R3-MYB transcription factor controlling stomatal aperture in *Arabidopsis thaliana*. *Curr Biol*, **15**, 1201–1206.

Liu, X., Yue, Y., Nie, Y., Li, W., Wu, W.-H. and Ma, L. (2007) A G protein-coupled receptor is a plasma membrane receptor for the plant hormone abscisic acid. *Science*, **315**, 1712–1716.

MacRobbie, E.A.C. (1995) ABA-induced ion efflux in stomatal guard cells: multiple actions of ABA inside and outside the cell. *Plant J*, **7**, 565–576.

MacRobbie, E.A.C. (1998) Signal transduction and ion channels in guard cells. *Phil Trans Roy Soc Lond Ser B*, **353**, 1475–1488.

MacRobbie, E.A.C. (2002) Evidence for a role for protein tyrosine phosphatase in the control of ion release from the guard cell vacuole in stomatal closure. *Proc Natl Acad Sci USA*, **99**, 11963–11968.

MacRobbie, E.A.C. (2006) Osmotic effects on vacuolar release in guard cells. *Proc Natl Acad Sci USA*, **103**, 1135–1140.

Maeser, P., Leonhardt, N. and Schroeder, J.I. (2003) "The clickable guard cell: electronically linked model of guard cell signal transduction pathways." In: *The Arabidopsis Book*. American Society of Plant Biologists, Rockville, MD.

Marten, H., Hedrich, R. and Roelfsema, M.R.G. (2007) Blue light inhibits guard cell plasma membrane anion channels in a phototropin-dependent manner. *Plant J*, **50**, 29–39.

Marten, H., Konrad, K.R., Dietrich, P., Roelfsema, M.R.G. and Hedrich, R. (2006) Ca^{2+}-dependent and -independent abscisic acid activation of plasma membrane anion channels in guard cells of *Nicotinana tabacum*. *Plant Physiol*, **143**, 28–37.

McAinsh, M.R., Brownlee, C. and Hetherington, A.M. (1990) Abscisic acid-induced elevation of guard cell cytosolic Ca^{2+} precedes stomatal closure. *Nature*, **343**, 186–188.

McAinsh, M.R., Webb, A.A.R., Taylor, J.E. and Hetherington, A.M. (1995) Stimulus-induced oscillations in guard cell cytosolic free calcium. *Plant Cell*, **7**, 1207–1219.

Melotto, M., Underwood, W., Koczan, J., Nomura, K. and He, S.Y. (2006) Plant stomata function in innate immunity against bacterial invasion. *Cell*, **126**, 969–980.

Messinger, S.M., Buckley, T.N. and Mott, K.A. (2006) Evidence for involvement of photosynthetic processes in the stomatal response to CO_2. *Plant Physiol*, **140**, 771–778.

Mishra, G., Zhang, W., Deng, F., Zhao, J. and Wang, X. (2006) A bifurcating pathway directs abscisic acid effects on stomatal closure and opening in *Arabidopsis*. *Science*, **312**, 264–266.

Mori, I.C., Murata, Y., Yang, Y., Munemasa, S., Wang, Y.-F., Andreoli, S., Tiriac, H., Alonso, J.M., Harper, J.F., Ecker, J.R., Kwak, J.M. and Schroeder, J.I. (2006) CDPKs CPK6 and CPK3 function in ABA regulation of guard cell S-type anion- and Ca^{2+} -permeable channels and stomatal closure. *PLoS Biology*, **4**, e327.

Murata, Y., Pei, Z.-M., Mori, I.C. and Schroeder, J.I. (2001) Abscisic acid activation of plasma membrane Ca^{2+} channels in guard cells requires cytosolic NAD(P)H and is differentially disrupted upstream and downstream of reactive oxygen species production in *abi1-1* and *abi2-1* protein phosphatase 2C mutants. *Plant Cell*, **13**, 2513–2523.

Mustilli, A.-C., Merlot, S., Vavasseur, A., Fenzi, F. and Giraudat, J. (2002) *Arabidopsis* OST1 protein kinase mediates the regulation of stomatal aperture by abscisic acid and acts upstream of reactive oxygen species production. *Plant Cell*, **14**, 3089–3099.

Ng, C.K.-Y., Carr, K., McAinsh, M.R., Powell, B. and Hetherington, A.M. (2001) Drought-induced guard cell signal transduction involves sphingosine-1-phosphate. *Nature*, **410**, 596–599.

Ohta, M., Guo, Y., Halfter, U. and Zhu, J.-K. (2003) A novel domain in the protein kinase SOS2 mediates interaction with the protein phosphatase 2C ABI2. *Proc Natl Acad Sci USA*, **100**, 11771–11776.

Olsen, R.L., Pratt, R.B., Gump, P., Kemper, A. and Tallman, G. (2002) Red light activates a chloroplast-dependent ion uptake mechanism for stomatal opening under reduced CO_2 concentrations in *Vicia* spp. *New Phytol*, **153**, 497–508.

Outlaw, W.H. and De Vlieghere-He, X. (2001) Transpiration rate. An important factor controlling the sucrose content of the guard cell apoplast of broad bean. *Plant Physiol*, **126**, 1716–1724.

Pei, Z.-M., Ghassemian, M., Kwak, C.M., McCourt, P. and Schroeder, J.I. (1998) Role of farnesyltransferase in ABA regulation of guard cell anion channels and plant water loss. *Science*, **282**, 287–290.

Pei, Z.-M., Murata, Y., Benning, G., Thomine, S., Kluesener, B., Allen, G.J., Grill, E. and Schroeder, J.I. (2000) Calcium channels activated by hydrogen peroxide mediate abscisic acid signaling in guard cells. *Nature*, **406**, 731–734.

Peiter, E., Maathuis, F.J.M., Mills, L.N., Knight, H., Pelloux, J., Hetherington, A.M. and Sanders, D. (2005) The vacuolar Ca^{2+}-activated channel TPC1 regulates germination and stomatal movement. *Nature*, **434**, 404–408.

Razem, F.A., El-Kereamy, A., Abrams, S.R. and Hill, R.D. (2006) The RNA-binding protein FCA is an abscisic acid receptor. *Nature*, **439**, 290–294.

Riera, M., Redko, Y. and Leung, J. (2006) *Arabidopsis* RNA-binding protein UBA2a relocalizes into nuclear speckles in response to abscisic acid. *FEBS Lett*, **580**, 4160–4165.

Ritte, G., Rosenfeld, J., Rohrig, K. and Raschke, K. (1999) Rates of sugar uptake by guard cell protoplasts of *Pisum sativum* L. related to the solute requirement for stomatal opening. *Plant Physiol*, **121**, 647–656.

Roelfsema, M.R.G., Steinmeyer, R., Staal, M. and Hedrich, R. (2001) Single guard cell recordings in intact plants: light-induced hyperpolarization of the plasma membrane. *Plant J*, **26**, 1–13.

Roelfsema, M.R.G., Hanstein, S., Felle, H.H. and Hedrich, R. (2002) CO_2 provides an intermediate link in the red light response of guard cells. *Plant J*, **32**, 65–75.

Saez, A., Apostolova, N., Gonzalez-Guzman, M., Gonzalez-Garcia, M.P., Nicolas, C., Lorenzo, O. and Rodriguez, P.L. (2004) Gain-of-function and loss-of-function phenotypes of the protein phosphatase 2C *HAB1* reveal its role as a negative regulator of abscisic acid signalling. *Plant J*, **37**, 354–369.

Saez, A., Robert, N., Maktabi, M.H., Schroeder, J.I., Serrano, R. and Rodriguez, P.L. (2006) Enhancement of abscisic acid sensitivity and reduction of water consumption in *Arabidopsis* by combined inactivation of the protein phophatases type 2C ABI1 and HAB1. *Plant Physiol*, **141**, 1389–1399.

Sánchez, J.-P., Duque, P. and Chua, N.-H. (2004) ABA activates ADPR cyclase and cADPR induces a subset of ABA-responsive genes in *Arabidopsis*. *Plant J*, **38**, 381–395.

Sanders, D., Pelloux, J., Brownlee, C. and Harper, J.F. (2002) Calcium at the crossroads of signaling. *Plant Cell*, **14**, S401–S417.

Schroeder, J.I., Allen, G.J., Hugouvieux, V., Kwak, J.M. and Waner, D. (2001b) Guard cell signal transduction. *Ann Rev Plant Physiol Plant Mol Biol*, **52**, 627–658.

Schroeder, J.I. and Hagiwara, S. (1989) Cytosolic calcium regulates ion channels in the plasma membrane of *Vicia faba* guard cells. *Nature*, **338**, 427–430.

Schroeder, J.I. and Keller, B.U. (1992) Two types of anion channels currents in guard cells with distinct voltage regulation. *Proc Natl Acad Sci USA*, **89**, 5025–5029.

Schroeder, J.I., Kwak, J.M. and Allen, G.J. (2001a) Guard cell abscisic acid signalling and engineering drought hardiness in plants. *Nature*, **410**, 327–330.

Schroeder, J.I., Raschke, K. and Neher, E. (1987) Voltage dependence of K^+channels in guard cell protoplasts. *Proc Natl Acad Sci USA*, **84**, 4108–4112.

Schwartz, A. (1985) Role of Ca^{2+} and EGTA on stomatal movements in *Commelina communis* L. *Plant Physiol*, **79**, 1003–1005.

Shen, Y.-Y., Wang, X.-F., Wu, F.-Q., Du, S.-Y., Cao, Z., Shang, Y., Wang, X.-L., Peng, C.-C., Yu, X.-C., Zhu, S.-Y., Fan, R.-C., Xu, Y.-H. and Zhang, D.-P. (2006) The Mg-chelatase H subunit is an abscisic acid receptor. *Nature*, **443**, 823–826.

Staxén, I., Pical, C., Montgomery, L.T., Gray, J.E., Hetherington, A.M. and McAinsh, M.R. (1999) Abscisic acid induces oscillations in guard-cell cytosolic free calcium that involve phosphoinositide-specific phospholipase C. *Proc Natl Acad Sci USA*, **96**, 1779–1784.

Sutter, J.-U., Campanoni, P., Tyrrell, M. and Blatt, M.R. (2006) Selective mobility and sensitivity to SNAREs is exhibited by the *Arabidopsis* KAT1 K^+ channel at the plasma membrane. *Plant Cell*, **18**, 935–954.

Takemiya, A., Kinoshita, T., Asanuma, M. and Shimazaki, K.-I. (2006) Protein phosphatase 1 positively regulates stomatal opening in response to blue light in *Vicia faba*. *Proc Natl Acad Sci USA*, **103**, 13549–13554.

Talbott, L. and Zeiger, E. (1998) The role of sucrose in guard cell osmoregulation. *J Exp Bot*, **49**, 329–337.

Tang, R.-H., Han, S., Zheng, H., Cook, C.W., Choi, C.S., Woerner, T.E., Jackson, R.B. and Pei, Z.-M. (2007) Coupling diurnal cytosolic Ca^{2+} oscillations to the CAS-IP_3 pathway in *Arabidopsis*. *Science*, **315**, 1423–1426.

Taylor, A.R. and Assmann, S.M. (2001) Apparent absence of a redox requirement for blue light activation of pump current in broad bean guard cells. *Plant Physiol*, **125**, 329–338.

Vavasseur, A. and Raghavendra, A.S. (2005) Guard cell metabolism and CO_2 sensing. *New Phytol*, **165**, 665–682.

Wang, X.-Q., Ullah, H., Jones, A.M. and Assmann, S.M. (2001) G protein regulation of ion channels and abscisic acid signaling in *Arabidopsis* guard cells. *Science*, **292**, 2070–2072.

Wang, Y., Ying, J., Kuzma, M., Chalifoux, M., Sample, A., MaArthur, C., Uchacz, T., Sarvas, C., Wan, J., Dennis, D.T., McCourt, P. and Huang, Y. (2005) Molecular trailoring of farnesylation for plant drought tolerance and yield protection. *Plant J*, **43**, 413–424.

Ward, J.M., Pei, Z.-M. and Schroeder, J.I. (1995) Roles of ion channels in initiation of signal transduction in higher plants. *Plant Cell*, **7**, 833–844.

Ward, J.M. and Schroeder, J.I. (1994) Calcium-activated K^+channels and calcium-induced calcium release by slow vacuolar ion channels in guard cell vacuoles implicated in the control of stomatal closure. *Plant Cell*, **6**, 669–683.

Webb, A.A.R., Larman, M.G., Montgomery, L.T., Taylor, J. E. and Hetherington, A.M. (2001) The role of calcium in ABA-induced gene expression and stomatal movements. *Plant J*, **26**, 351–362.

Webb, A.A.R., McAinshi, M.R., Mansfield, T.A. and Hetherington, A.M. (1996) Carbon dioxide induces increases in guard cell cytosolic free calcium. *Plant J*, **9**, 297–304.

Wood, N.T., Allan, A.C., Haley, A., Viry-Moussaïd, M. and Trewavas, A.J. (2000) The characterization of differential calcium signalling in tobacco guard cells. *Plant J*, **24**, 335–344.

Wu, Y., Kuzma, J., Marechal, E., Graeff, R., Lee, H.C., Foster, R. and Chua, N.-H. (1997) Abscisic acid signaling through cyclic ADP-ribose in plants. *Science*, **278**, 2126–2130.

Xie, X., Wang, Y., Williamson, L., Holroyd, G.H., Tagliavia, C., Murchie, E., Theobald, J., Knight, M.R., Davies, W.J., Ottoline Leyser, H.M. and Hetherington, A.M. (2006) The identification of genes involved in the stomatal response to reduce atomospheric relative humidity. *Curr Biol*, **16**, 882–887.

Yang, Z. (2002) Small GTPases: versatile signaling switches in plants. *Plant Cell*, **14**, S375–S388.

Yoshida, T., Nishimura, N., Kitahata, N., Kuromori, T., Ito, T., Asami, T., Shinozaki, K. and Hirayama, T. (2006a) *ABA-HYPERSENSITIVE GERMINATION3* encodes a protein phosphatase 2C (AtPP2CA) that strongly regulates abscisic acid signaling during germination among *Arabidopsis* protein phosphatase 2Cs. *Plant Physiol*, **140**, 115–126.

Yoshida, R., Umezawa, T., Mizoguchi, T., Takahashi, S., Takahashi, F. and Shinozaki, K. (2006b) The regulatory domain of SRK2E/OST1/SnRK2.6 interacts with ABI1 and integrates abscisic acid (ABA) and osmotic stress signals controlling stomatal closure in *Arabidopsis*. *J Biol Chem*, **281**, 5310–5318.

Young, J.J., Mehta, S., Israelsson, M., Godoski, J., Grill, E. and Schroeder, J.I. (2006) CO_2 signaling in guard cells: calcium sensitivity response modulation, a Ca^{2+}-independent phase, and CO_2 insensitivity of the *gca2* mutant. *Proc Natl Acad Sci USA*, **103**, 7506–7511.

Zeiger, E. and Zhu, J. (1998) Role of zeaxanthin in blue light photoreception and the modulation of light-CO_2 interactions in guard cells. *J Exp Bot*, **49**, 433–442.

Zhang, W., Yu, L., Zhang, Y. and Wang, X. (2005) Phospholipase D in the signaling networks of plant response to abscisic acid and reactive oxygen species. *Biochim Biophys Acta*, **1736**, 1–9.

Zhang, X., Wang, H., Takemiya, A., Song, C.-p., Kinoshita, T. and Shimazaki, K.-I. (2004) Inhibition of blue light-dependent H^+ pumping by abscisic acid through hydrogen peroxide-induced dephosphorylation of the plasma membrane H^+-ATPase in guard cell protoplasts. *Plant Physiol*, **136**, 4150–4158.

Zheng, Z.-L., Nafisi, M., Tam, A., Li, H., Crowell, D.N., Chary, S.N., Schroeder, J.I., Shen, J. and Yang, Z. (2002) Plasma membrane-associated ROP10 small GTPase is a specific negative regulator of abscisic acid responses in *Arabidopsis*. *Plant Cell*, **14**, 2787–2797.

Annual Plant Reviews (2008) **33**, 388–416
doi: 10.1111/b.9781405160025.2008.00014.x
www.interscience.wiley.com

Chapter 14

THE MOLECULAR NETWORKS OF ABIOTIC STRESS SIGNALING

Zhizhong Gong,[1] Viswanathan Chinnusamy,[2] and Jian-Kang Zhu[3]

[1] *State Key Laboratory of Plant Physiology and Biochemistry, College of Biological Sciences, National Center for Plant Gene Research (Beijing), China Agricultural University, Beijing 100094, People's Republic of China*
[2] *Water Technology Centre, Indian Agricultural Research Institute, New Delhi 110 012, India*
[3] *Department of Botany and Plant Sciences, Institute for Integrative Genome Biology, 2150 Batchelor Hall, University of California, Riverside, CA 92521, USA*

Abstract: In response to harsh changing environmental conditions, sessile plants mobilize complex regulatory networks to modulate physiological changes, allowing them to tolerate and adapt to these conditions. Both abiotic stresses and diurnal rhythm regulate biosynthesis and metabolism of abscisic acid (ABA), which plays a crucial role in integrating various environmental as well as cellular signals. ABA is perceived by several receptor systems including flowering time control protein A (FCA), ABA receptor (ABAR/CHLH, H subunit of Mg-chelatase), and G-protein-coupled receptor 2 (GCR2). These receptors regulate ABA responses in seed dormancy, germination, root growth, and stomatal responses. Cold stress signaling mechanism consists of both CBF (CRT/DRE-binding factor)-dependent and -independent pathways in *Arabidopsis*. The CBF regulon is conserved across many plant species. The MYC-type bHLH transcription factor ICE1 (inducer of CBF expression 1) regulates freezing tolerance by activating the transcription of *CBFs* and some other cold-induced regulons. ICE1 is negatively regulated by the HOS1 (high expression of osmotically responsive gene 1)-mediated ubiquitination and proteosomal degradation and positively regulated by SIZ1 [*S*AP (scaffold attachment factor) and M*iz*1 (Msx2-interacting zinc finger)] mediated sumoylation under cold stress. Salt stress tolerance is primarily conferred by the SOS (salt overly sensitive) pathway, in which a calcium-responsive SOS3–SOS2 protein kinase complex controls the activity of ion transporters such as SOS1. The SOS pathway is conserved between *Arabidopsis* and rice. Several transcription factors and their target genes induced under various abiotic stresses have been identified. Cross talks in environmental stress signaling networks appear to occur at various levels ranging from second messengers and secondary signals to phosphorelay and transcriptional regulators. Furthermore, RNA metabolism especially small RNA-mediated mRNA cleavage translation repression, and chromatin remodeling are emerging hot topics in abiotic stress studies.

Keywords: abiotic stress; *Arabidopsis*; ABA; salt stress; drought; low temperature

14.1 Introduction

Crop production and natural plant distribution are greatly limited by abiotic stresses such as drought, salt, and low- or high-temperature stresses, all of which can lead to dehydration and oxidative stress in plant cells (Zhu, 2002). In addition to their common effects, different stresses exert specific effects on plant growth and development. Drought stress constrains photosynthesis and nutrient uptake from water-limited soil (Schroeder *et al.*, 2001; Shinozaki *et al.*, 2003; Shinozaki and Yamaguchi-Shinozaki, 2007). Ion toxicity is the major culprit for inhibiting plants growth in saline soils (Hasegawa *et al.*, 2000; Zhu, 2001; Xiong *et al.*, 2002). Low temperatures directly affect membrane fluidity and the activities of various enzymes and membrane transporters (Thomashow, 1999, 2001; Zhu, 2001; Xiong *et al.*, 2002). Under severe stress conditions, the growth of plants can be retarded, or completely stopped. In the natural environment, plants usually experience multiple stresses at the same time. In response to each of these different stresses, plants have adopted complex mechanisms to achieve stress tolerance. Different plant genotypes or the same plants at different developmental stages may utilize different response mechanisms. The input of different stress signals might lead to both common and specific outputs through a complex network signal transduction pathways. In this chapter, we summarize various abiotic stress signaling pathways and their cross talks.

14.2 Abscisic acid

Abscisic acid (ABA), a plant hormone identified in 1960s, plays critical roles in plant growth and development including seed maturation, seed dormancy, germination, vegetative growth, flowering, and plant adaptation to environmental stresses (Milborrow, 2001; Xiong and Zhu, 2003; Nambara and Marion-Poll, 2005). Due to its diverse functions, ABA biosynthesis, catabolism, perception, and signal transduction have been extensively studied. Under normal conditions, ABA contents are kept at low levels, and the low levels of ABA are important for maintaining cell vigor. ABA-deficient mutants are usually smaller with poor growth. Osmotic stress caused by drought and salt stress rapidly activates the production of ABA, which triggers stomatal closure and reduce transpiration. ABA also regulates the expression of many stress-inducible genes such as late-embryogenesis-abundant- (LEA-) like proteins that are critical for cell survival under dehydration stress.

14.2.1 ABA biosynthesis and metabolism

Cellular ABA is primarily synthesized by a de novo pathway, which involves different enzymes encoded by genes whose expression can be induced under different stress conditions, mainly drought, salt, and, to some extent, low temperatures (Xiong and Zhu, 2003). When the environmental conditions become favorable, excess ABA is inactivated by hydroxylation or conjugation to glucose (Lee *et al.*, 2006b). Through various genetic screening methods, ABA-deficient mutants were isolated from different plant species such as *Nicotiana plumbaginifolia* (Marin *et al.*, 1996), *Zea mays* (Schwartz *et al.*, 1997; Tan *et al.*, 1997), *Lycopersicon esculentum* (Sagi *et al.*, 2002; Thompson *et al.*, 2000a,b), and *Arabidopsis* (Leon-Kloosterziel *et al.*, 1996). ABA-deficient mutants usually show precocious germination and severe wilty phenotypes. The first ABA biosynthesis gene *ABA2* was isolated from *N. plumbaginifolia* by transposon tagging using the maize activator transposon (Marin *et al.*, 1996). *ABA2* gene encodes a zeaxanthin epoxidase (ZEP) which catalyzes the epoxidation of zeaxanthin and antheraxanthinto violaxanthin, the first step in ABA biosynthesis in plastids (Marin *et al.*, 1996). The *Arabidopsis* orthologous mutant *aba1* was restored to wild-type phenotype by transforming the *ABA2* cDNA under the control of the CaMV 35S promoter (Marin *et al.*, 1996).

The second gene *VP14* encoding a 9-*cis*-epoxycarotenoid dioxygenase (NCED) was isolated from the maize *vp14* mutant by transposon tagging (Tan *et al.*, 1997). NCED catalyzes the oxidative cleavage of the major epoxycarotenoid 9-*cis*-neoxanthin to yield xanthoxin, the first committed step of ABA biosynthesis in plastids (Tan *et al.*, 1997). Xanthoxin is exported from plastids to the cytosol where it is converted to ABA aldehyde by a short-chain alcohol dehydrogenase/reductase (SDR) encoded by the *AtABA2/GIN1* (*GLUCOSE INSENSITIVE1*) gene in *Arabidopsis* (Rook *et al.*, 2001; Cheng *et al.*, 2002; Gonzalez-Guzman *et al.*, 2002). *LOS5/ABA3* encodes a molybdenum cofactor (MoCo) sulfurase which catalyzes the synthesis of sulfurylated form of MoCo, the cofactor of ABA aldehyde oxidase (AAO3). AAO3 catalyzes the conversion of ABA aldehyde to ABA (Seo *et al.*, 2000; Bittner *et al.*, 2001; Xiong *et al.*, 2001b). The expression of both *ZEP* and *NCED* in tomato shows a diurnal rhythm in leaves (Thompson *et al.*, 2000a). Earlier studies suggest that *SDR1* expression is uniquely upregulated by sugar (Cheng *et al.*, 2002), while the expression of all the other ABA biosynthetic genes is induced by drought and salt stress (Xiong *et al.*, 2001b). However, more recent analysis of the *ABA2* promoter-driven expression of the *GUS* reporter in transgenic *Arabidopsis* revealed that *ABA2* expression is enhanced by prolonged periods of drought, salt, cold, and flooding, although it is not regulated by short-term stresses (Lin *et al.*, 2006). Besides the upregulation by drought and salt stress, the expression of most ABA biosynthetic genes are also transcriptionally induced by ABA, suggesting self-regualtion in ABA accumulation under abiotic stress conditions (Xiong and Zhu, 2003).

When stresses are relieved, the transcripts of ABA biosynthetic genes are reduced to basic levels, and ABA is either conjugated to inactive form by glycosylation or hydroxylated by ABA 8′-hydroxylases encoded by the cytochrome P450 *CYP707A* gene subfamily (Kushiro *et al.*, 2004; Saito *et al.*, 2004). *CYP707A3* is highly induced by both dehydration and rehydration of *Arabidopsis* seedlings (Umezawa *et al.*, 2006), and hence may play a crucial role in controlling the threshold level of ABA in cells during dehydration and rehydration. CYP707A1 and CYP707A2 are likely important in seed development and postgermination growth at different stages (Millar *et al.*, 2006; Okamoto *et al.*, 2006). ABA glycosylation is catalyzed by the ABA-glucosyltransferase, which is induced by ABA and water-deficit stress (Xu *et al.*, 2002). Glycosylated ABA constitutes an inactive pool of ABA stored in vaculoes or the apoplastic space (Dietz *et al.*, 2000). In *Arabidopsis*, *AtBG1* encoding a β-glucosidase homolog localized to the ER is responsible for hydrolyzing glucose-conjugated ABA (Lee *et al.*, 2006b) to produce active ABA. *AtBG1* is induced by various abiotic stresses and exogenous ABA. Dehydration induces rapid polymerization of AtBG1, which increases AtBG1 enzyme activity by fourfolds, leading to rapid accumulation of active ABA. AtBG1 also functions in day/night cycles to regulate ABA fluctuation for plants to finely adjust to the physiological and environmental changes (Lee *et al.*, 2006b).

14.2.2 ABA receptors

The first step for ABA responses is the recognition of ABA by receptor(s). Physiological studies using impermeable ABA derivatives or microinjection of ABA into cells suggested that ABA receptors exist in both intracellular and extracellular locations (Leung and Giraudat, 1998; Finkelstein *et al.*, 2002). Forward genetics thus far has failed to identify ABA receptors. However, recently, three groups have succeeded in isolating three different ABA-binding proteins using biochemical approaches and have shown that these proteins act as ABA receptors (Razem *et al.*, 2004; Shen *et al.*, 2006; Liu *et al.*, 2007).

By using ABA-mimicking anti-idiotypic antibodies, Razem *et al.* (2004) purified an ABA-binding protein (ABAP) from the barley aleurone. Based on its sequence, an *Arabidopsis* ABAP homolog FCA (flowering time control protein A) was identified (Razem *et al.*, 2006). FCA was initially identified for its role in promoting flowering in *Arabidopsis* and is a nuclear protein with two conserved RNA-binding motifs (Macknight *et al.*, 1997). FCA interacts with FY (flowering locus Y) to form a complex that inhibits the transcript accumulation of the flowering suppressor FLC (flowering locus C) (Simpson *et al.*, 2003). As a receptor, FCA stereospecifically binds (+)-ABA with high affinity, and this binding inhibits the formation of FCA–FY complex, leading to the accumulation of FLC transcripts and delaying of flowering (Razem *et al.*, 2006). The ABA-binding region is localized in the C-terminal side of FCA, although the specific binding residues are not identified yet. FCA lesions also affect ABA inhibition of lateral root formation, but are not required for ABA inhibition of seed germination or stomatal regulation by ABA (Razem *et al.*, 2006). The results further suggest that other ABA receptor(s) must exist (Finkelstein, 2006).

Zhang *et al.* (2002) purified a 42-kDa protein, ABAR (ABA receptor), from broad bean leaves by using ABA-linked AEH-Sepharose 4B as an affinity resin (Zhang *et al.*, 2002). A partial sequence of ABAR shows high homology with the H subunit of Mg-chelatase (CHLH) (Shen *et al.*, 2006). The 42-kDa protein purified from broad bean leaves is complementary only to the C-terminal part of CHLH, suggesting that CHLH is unstable, and the C-terimnal region contains the ABA-binding residues. CHLH is encoded by *GUN5* (genomes uncoupled 5), which was initially found to play vital roles in plastid-to-nucleus retrograde signaling (Mochizuki *et al.*, 2001). ABAR/CHLH expressed in yeast shows high specific binding affinity only for the active form, (+)-ABA, but not for two inactive ABA isomers, (−)-ABA and *trans*-ABA. ABAR has an equilibrium dissociation constant (K_d) of 32 nM, which precisely matches the ABA level in living cells (Shen *et al.*, 2006). Downregulation of *CHLH* by RNAi or antisense suppression reduced the sensitivity of transgenic plants to ABA inhibition of seed germination, seedling growth, and to ABA promotion of stomatal closure. In contrast, transgenic plants with overexpression of *CHLH* showed ABA-hypersensitive phenotypes, and were more tolerant to drought stress. *CHLH* appears to be an essential gene since a homozygous T-DNA insertion mutation in *CHLH* is lethal. ABAR positively regulates the transcripts of ABA responsive genes such as *RD29A*, *MYB2*, *MYC2*, *ABI4*, *ABI5*, and *OST1* or seed-specific genes like *ABI3* and *ABI5*, and negatively controls the expression of *ABI1*, *ABI2*, and *CIPK15*. However, unlike FCA, ABAR is not involved in the regulation of flowering. It is interesting that *gun2-5* mutant plants did not show any ABA-related phenotypes, whereas the *cch* mutant, another allele of *gun5*, had ABA-insensitive phenotypes. The *cch* but not the *gun5* mutation impaired the ABA-binding capacity of CHLH. The mutated residue in *cch* might be involved in ABA binding. The results suggest that ABAR-mediated ABA signaling is distinct from chlorophyll metabolism or plastid signalling mediated by GUN5 (Shen *et al.*, 2006). ABI2 interacts with a prefibrillin protein, which is a plastid-associated lipid-binding protein probably functioning in protection of PSII from photoinhibition (Yang *et al.*, 2006). The expression of fibrillin is regulated at transcriptional and posttranscriptional levels by ABI1 and ABI2, respectively (Yang *et al.*, 2006). Because the ABA receptor ABAR is also localized in plastids, it would be interesting to dissect the network of ABA signaling mediated by ABI2, ABAR, fibrillin, and other proteins in plastids.

Analysis of T-DNA mutation in the *Arabidopsis GPA1* gene, which encodes the α-subunit of heterotrimeric GTP-binding (G) protein, revealed a role for GPA1 in guard cell ABA signaling. *gpa1* mutants show ABA insensitivity in

ABA inhibition of stomatal opening, and thus rate of water loss from *gpa1* mutants is greater than that from wild-type plants (Wang *et al.*, 2001). Recently, a G-protein-coupled receptor, GCR2, was identified as a plasma membrane ABA receptor that regulates ABA signaling through GPA1 in *Arabidopsis* (Liu *et al.*, 2007). Transmembrane structure prediction and GCR2-YFP localization showed that GCR2 is an integral membrane protein with seven transmembrane helices. Surface plasmon resonance spectroscopy, split-ubiquitin system, bimolecular fluorescence complementation, and coimmunoprecipitation studies showed that carboxy terminal (C290-401) of GCR2 interact with heteromeric G-protein α (GPA1). *Arabidopsis* seeds of *gcr2* mutants exhibited ABA insensitivity, while seeds of transgenic *Arabidopsis* overexpressing *GCR2* showed hypersensitivity to ABA during germination as compared to the wild-type plants. ABA-induced expression of ABA responsive genes such as *RD29A*, *KIN1*, and *ABI5* was also significantly reduced in *gcr2* mutant. Both ABA-induced stomata closing and ABA-inhibited stomata opening were insensitive in *gcr2*, while *GCR2* overexpressing plants exhibited hypersensitive stomatal closure response as compared to wild-type plants. Thus, *gcr2* mutations result in ABA insensitivity, while *GCR2* overexpression results in ABA hypersensitivity. Furthermore, GCR2 binds ABA with high affinity at physiological concentrations (equilibrium dissociation constant, K_d, is 20.1 nM) with expected kinetics and stereospecificity. It is expected that ligand binding to GCR2 should induce dissociation of G protein from GCR2–G protein complex to activate signal transduction. In this study it was shown that the binding of physiologically active form of ABA to the GCR2 leads to the dissociation of the GCR2–GPA1 complex in yeast. These results demonstrate that GCR2 is a plasma membrane ABA receptor (Liu *et al.*, 2007). GPA1 has been shown to be a positive regulator of auxin-mediated cell division (Ullah *et al.*, 2001) and brassinosteroids-mediated germination (Ullah *et al.*, 2002). Thus, GPA1 is involved in the cross talk between plant hormones ABA, auxin, and brassinosteroids. Identification of downstream components of these ABA receptors will further enhance our knowledge on ABA signaling cross talk and specificity.

14.2.3 ABA signaling and stress responsive gene expression

The molecular mechanisms of ABA signaling have been studied extensively (Giraudat, 1995; Grill and Himmelbach, 1998; Fedoroff, 2002; Finkelstein *et al.*, 2002; Zhu, 2002) (see Chapter 13). An early genetic screen conducted in *Arabidopsis* using ABA-resistant germination identified several ABA-insensitive mutants including *abi1*, *abi2*, *abi3*, *abi4*, *abi5*, and *abi8* (Finkelstein and Somerville, 1990; Giraudat *et al.*, 1992; Leung *et al.*, 1994, 1997; Meyer *et al.*, 1994; Finkelstein *et al.*, 1998; Rodriguez *et al.*, 1998; Finkelstein and Lynch, 2000; Brocard-Gifford *et al.*, 2004). Among them, ABI1 and ABI2 are homologous members of the protein phosphatase 2C family, and function as key negative regulators in ABA signaling (Leung *et al.*, 1994, 1997; Meyer *et al.*, 1994; Gosti *et al.*, 1999; Merlot *et al.*, 2001).

Phospholipid metabolism plays a pivotal role in ABA signaling. Phosphatidylinositol 4,5-bisphosphate (PIP_2), diacylglycerol, and inositol 1,4,5-trisphosphate (IP_3), the products of phospholipids, are three important signal molecules (see Chapter 8). PIP_2 is hydrolyzed by phosphoinositide-specific phospholipase C (PI-PLC) to produce IP_3 and diacylglycerol (Xiong *et al.*, 2002). Phosphatidylinositol-4-phosphate 5-kinase (PIP5K) phosphorylates phosphatidylinositol 4-phosphate to produce PIP_2. The expression of *PIP5K* genes was induced by drought, salt, and ABA treatments, supporting an important role for PIP5K in water stress signal transduction (Mikami *et al.*, 1998). Drought stress increased the expression of some *PI-PLC* genes in potato (Kopka *et al.*, 1998) and mung bean (Kim *et al.*, 2004). The increase in PI-PLC could in turn enhance the cleavage of PIP_2 to generate more IP_3 and diacylglycerol. PIP_2 signaling is terminated through the action of inositol polyphosphate phosphatases and phosphoinositide phosphatases. Mutation in a plant unique SAC domain phosphoinositide phosphatase gene *SAC9* caused *Arabidopsis* to accumulate high levels of both PIP_2 and IP_3. The *sac9* mutants constitutively express higher levels of stress-inducible genes and accumulate more reactive-oxygen species than wild-type plants, and show closed stomata even under nonstress conditions (Williams *et al.*, 2005). Transient increases in PIP_2 and IP_3 were observed in plant cells during ABA and other abiotic stress treatments (DeWald *et al.*, 2001; Xiong *et al.*, 2002). It has been shown in many cell types that IP_3 is an important signal for triggering Ca^{2+} release from internal stores. IP_3-mediated signaling is terminated in *Arabidopsis* by the inositol 5-phosphatase At5PTase1, which hydrolyzes IP_3 (Burnette *et al.*, 2003). Transgenic plants overexpressing *At5PTase1* were insensitive to ABA and exhibited a decrease in ABA-induced gene expression (Sanchez and Chua, 2001). In *Arabidopsis*, the *FIERY1* gene encodes an inositol polyphosphate-1-phosphatase, which functions in the metabolism of IP_3 (Xiong *et al.*, 2001c). *fiery1* mutant plants accumulated more IP_3 under ABA treatment, and showed a hypersensitive response to ABA in both seed germination and ABA-induced expression of stress responsive genes (Xiong *et al.*, 2001c). Thus, ABA signaling to induce abiotic stress responsive genes is mediated by IP_3.

Phosphatidic acid, a hydrolytic product of phospholipids produced by phospholipase D alpha1, is another important lipid signal in ABA signaling. Phospholipase D alpha1 and phosphatidic acid interact with the α-subunit of heterotrimeric G protein (Gα) to positively mediate ABA inhibition of stomatal opening (Zhao and Wang, 2004; Mishra *et al.*, 2006) (see Chapters 2 and 8). Gα appears to also interact with *Arabidopsis* GCR1 (G-protein-coupled receptor 1) in ABA signaling. In contrast to *gpa1* and *gcr2* mutants, *gcr1* mutants exhibit ABA hypersensitivity in inhibition of root growth, gene expression, and stomatal response, and show more drought tolerance than wild-type plants. Thus, GCR1 may act as a negative regulator of GPA1-mediated ABA signaling (Pandey and Assmann, 2004).

Two small guanosine triphosphatases (GTPases), ROP6/AtRac1 and ROP10, members of the *Arabidopsis* Rho GTPase family, are important mediators of ABA signaling (Lemichez *et al.*, 2001; Zheng *et al.*, 2002) (see Chapters 3 and 13). ABA induces stomatal closure through inactivating ROP6/AtRac1, leading to the disruption of the guard cell actin cytoskeleton (Lemichez *et al.*, 2001). ROP10, a plasmalemma Rho GTPase, has also been shown to be a negative regulator in ABA inhibition of seed germination, root elongation, and ABA promotion of stomatal closure (Zheng *et al.*, 2002). ROP2 has also been implicated in the negative regulation of ABA promotion of seed dormancy and of ABA inhibition of seed germination (Li *et al.*, 2001). It is unknown whether these GTPases act downstream of any of the three known ABA receptors described above. These GTPases are localized to the plasma membrane. The plasma membrane-localized leucine-rich repeat receptor kinase 1 (AtRPK1) is also involved in ABA signaling *Arabidopsis* plants (Osakabe *et al.*, 2005). It would be interesting to see whether AtRPK1 is linked to the ROP small GTPases in ABA signaling.

14.3 The molecular mechanisms of salt tolerance

High salt stress leads to Na^+ toxicity as well as hyperosmotic and oxidative stresses, inhibiting the growth and development of plants. Under high saline conditions, plant cells can reestablish ion homeostasis by extruding toxic sodium out of cells or compartmentalizing sodium into vacuoles. Genetic studies using a root-bending assay in *Arabidopsis* have identified three important salt overly sensitive (SOS) genes, named *SOS1*, *SOS2*, and *SOS3*, which constitute the main salt tolerance pathway in plants (Chinnusamy and Zhu, 2003, 2006b) (Fig. 14.1). SOS1 is a Na^+/H^+ antiporter with 10–12 transmembrane domains and an unusually long cytoplasmic C-terminal tail (Shi *et al.*, 2000). SOS1 might also function as a sodium sensor (Chinnusamy and Zhu, 2003). SOS2 is a serine/threonine protein kinase with a unique regulatory domain in the C-terminal part and a catalytic domain in the N-terminal region (Liu *et al.*, 2000). SOS3 is a myristoylated calcium-binding protein that is recruited to the plasma membrane. Salt shock transiently increases cytosolic Ca^{2+}, which can be sensed by SOS3 (Liu and Zhu, 1998; Ishitani *et al.*, 2000). SOS3 then interacts with and activates SOS2 (Halfter *et al.*, 2000). The activated SOS3–SOS2 complex phosphorylates and activates SOS1 to efficiently transport cytosolic sodium out of the cell (Zhu, 2003). Mutations in any of the SOS genes render plants hypersensitive to salt stress. Salt stress leads to ABA accumulation and part of the ABA signal might be transduced through SOS2. In yeast two-hybrid assays, SOS2 physically interacts with ABI2 and the interaction is abolished by the *abi2-1* mutation. The *abi2-1* mutant seedlings show more salt tolerance, but are ABA-insensitive (Ohta *et al.*, 2003). These results connect the SOS pathway with ABA signal transduction through ABI2 (Ohta *et al.*, 2003). Salt stress also causes oxidative stress. It was

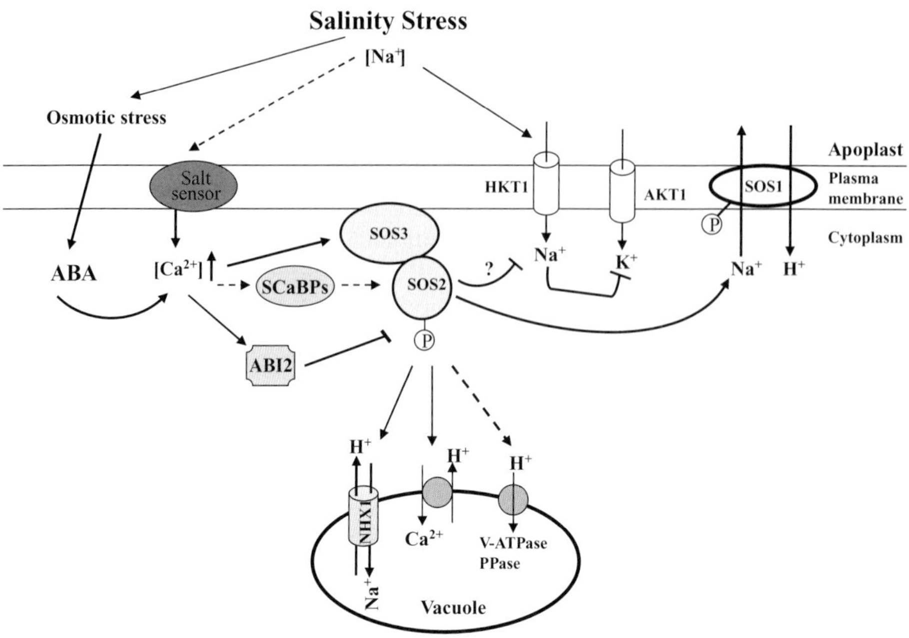

Figure 14.1 Regulation of ion homeostasis under salt stress by SOS pathway. The SOS3 calcium sensor perceives the salt-stress-induced Ca^{2+} signals and then activates SOS2 kinase. Activated SOS2 kinase activates SOS1, a plasma membrane Na^+/H^+ antiporter, by phosphorylation. Activated SOS1 then pumps Na^+ out of cytosol. SOS2 kinase also activates tonoplast Na^+/H^+ antiporter that sequesters Na^+ into the vacuole. Na^+entry into cytosol through Na^+ transporter HKT1 activity may also be restricted by SOS2. Activation of NHX1 and VCX1 by SOS2 are SOS3-independent and probably regulated through SOS3-like Ca^{2+} binding proteins (SCaBPs). ABI2 interact with SOS2 and negatively regulate ion homeostasis by inhibiting either the SOS2 kinase activity or the activities of SOS2 targets.

found that the predicted cytoplasmic tail of SOS1 physically interacts with RCD1 (radical-induced cell death 1), a transcriptional regulator of oxidative stress responses in *Arabidopsis* (Ahlfors *et al.*, 2004; Katiyar-Agarwal *et al.*, 2006). Under unstressed conditions, RCD1 is localized in the nucleus. Under high salt or oxidative stress, the localization of RCD1 is found both in the nucleus and in the cytoplasm. Both *sos1* and *rcd1* mutants are sensitive to salt and apoplastic oxidative stresses (Katiyar-Agarwal *et al.*, 2006). These results are of great interest because it provides direct evidence to connect the SOS ion homeostasis pathway with ABA and oxidative stress signaling in plant salt tolerance. The SOS pathway has been characterized in detail in the dicot *Arabidopsis*. A recent study in rice indicated that the SOS salt tolerance pathway is highly conserved in monocots (Martinez-Atienza *et al.*, 2007). Overexpression of SOS1 or activated SOS2 improves the salt tolerance of transgenic plants (Shi *et al.*, 2003; Guo *et al.*, 2004).

Salt stress causes Na^+–K^+ disequilibrium due to the inhibitory effect of Na^+ on K^+ acquisition and nutrition. The *sos1*, *sos2*, and *sos3* mutants all are deficient in K^+ under salt stress. A genetic screen for suppressors of *sos3* mutation identified the *AtHKT1* gene, which encodes a Na^+ transporter (Izhaki *et al.*, 2001) (Fig. 14.1). In fact, mutations in *AtHKT1* suppress not only *sos3*, but also *sos1* and *sos2*. The expression of *AtHKT1* is restricted to phloem tissues; and *athkt1* mutations block Na^+ recirculation through the phloem from shoots to roots and cause overaccumulation of Na^+ in the leaves (Izhaki *et al.*, 2001; Berthomieu *et al.*, 2003; Bethke *et al.*, 2004). In the two coastal *Arabidopsis* accessions Ts-1 and Tsu-1, a deletion in the upstream of *AtHKT1* gene was found to be responsible for the low expression of *AtHKT1* in roots, resulting in higher accumulation of Na^+ in the shoots (Rus *et al.*, 2006). However, in contrast to null *athkt1* mutants, which are sensitive to salt stress, it seems that this novel version of *AtKHT1* is genetically correlated with high salt tolerance (Rus *et al.*, 2006). In rice, a quantitative trait locus *SKC1* encoding a sodium transporter *OsHKT8* has been isolated from a salt-tolerant variety (Ren *et al.*, 2005). The expression of *SKC1* is observed mainly in parenchyma cells surrounding the xylem vessels, and voltage clamp analysis showed a selective Na^+ transport activity for SKC1. SKC1 in the salt-tolerant rice variety is important for K^+ homeostasis. These results suggest that *HKT1* regulates Na^+ homeostasis, which in turn affects K^+ acquisition.

Potassium uptake in low-potassium soils is mainly mediated by the K^+ channel AKT1 in *Arabidopsis*. In a genetic screen for low-K^+-sensitive mutants (*lks*), Wu and colleagues (Xu *et al.*, 2006) isolated the *lks1* mutant that showed chlorotic leaves when grown on low-potassium medium. *LKS1* encodes a CBL-interacting protein kinase 23 (CIPK23, a SOS2 homolog), which was shown to physically interact with CBL1 and CBL9 (calcineurin B-like proteins, SOS3 homologs) that are localized in the plasma membrane (Xu *et al.*, 2006). *cbl1cbl9* double but not single mutants showed a chlorotic phenotype similar to *lks1* under low-potassium conditions, suggesting that CBL1 and CBL9 function redundantly in the low-potassium response pathway. Transgenic plants overexpressing these genes show improved K^+ acquisition. Genetic and voltage patch-clamp analyses indicated that the CBL1/9–CIPK23 complex regulates AKT1 activity for K^+ uptake through phosphorylation of AKT1 (Xu *et al.*, 2006).

Sodium sequestration into the vacuole is a very important strategy both for osmotic adjustment and for reducing sodium concentration in the cytosol of plant cells. Na^+/H^+ antiporters as well as V-type ATPasess and H^+-pyrophosphatases (H^+-PPase) play critical roles in Na^+ sequestration into the vacuole. Energy-dependent Na^+ transport across vacuolar membranes depends on the proton gradient generated by these H^+ pumps (Fig. 14.1). Overexpression of the H^+-PPase gene *AVP1* in *Arabidopsis* increased cation uptake by vacuolar membrane vesicles and enhanced solute accumulation and water retention, and the transgenic plants were more salt- and drought-tolerant than wild-type plants (Gaxiola *et al.*, 2001). *AVP1* overexpression also resulted in increased cell division by facilitating auxin fluxes, and the transgenic plants became bigger with more roots. In addition, the transgenic tomato plants showed increased drought tolerance (Gaxiola *et al.*, 2001; Li *et al.*, 2005; Park *et al.*, 2005). In *Arabidopsis*, there exist six endosomal Na^+/H^+ antiporters with different localization and expression patterns (Yokoi *et al.*, 2002). AtNHX1 and AtNHX2 localize to the tonoplast membrane, and their gene expression was induced by high salt stress through an ABA-dependent pathway (Yokoi *et al.*, 2002). Expression of *AtNHX1*, *AtNHX2*, and *AtNHX5* in a yeast endosomal/vacuolar Na^+/H^+ antiporter *nhx1* mutant complemented its salt-sensitive phenotype, suggesting that these three genes might function in sodium compartmentation into the plant vacuole (Gaxiola *et al.*, 1999; Quintero *et al.*, 2000; Yokoi *et al.*, 2002).

Salinity inhibits the growth of plants, whereas the phytohormone gibberellin (GA) promotes growth by regulating the degradation of DELLA proteins that restrain cell proliferation and expansion. Salt stress reduces the production of bioactive GAs, and interestingly the growth of quadruple-DELLA mutant seedlings lacking GAI, RGA, RGL1, and RGL2 (four of five DELLA proteins) is less inhibited by moderate salt stress compared to wild-type plants. The study suggests that the DELLA proteins are negative regulators of plant growth under salt stress. A reduction in GA level under salt stress results in more DELLA protein accumulation in the nucleus (Achard *et al.*, 2006). The *ga1-3* mutants that are deficient in GA biosynthesis or *gai* (GA-insensitive) dominant mutants (Peng *et al.*, 1997) are more salt-tolerant than wild type or the quadruple-DELLA mutant. These results suggest that stabilized DELLA proteins enhance plant survival under high salt conditions, but inhibit plant growth in moderate saline environments (Achard *et al.*, 2006).

14.4 The transcriptional regulation of cold- and drought-inducible genes

Transcripts of many genes with different functions are induced by various environmental stresses. The inducible genes can be divided into several groups according to their functions. One group includes genes for signal transduction pathways as we have discussed above. The other genes encode effector proteins such as LEA proteins, enzymes for osmolyte biosynthesis, and detoxification enzymes. Transcription factors regulate the expression of stress-inducible genes through direct binding to conserved *cis*-elements in the promoter regions of the inducible genes. These *cis*-elements include the ABA responsive element (ABRE), the dehydration responsive element (DRE)/C-repeat (CRT), MYCRS (MYC recognition sequence)/MYBRS (MYB recognition sequence), and other sequences. Earlier studies using yeast one-hybrid assays identified several important transcription factors that bind to stress responsive *cis*-elements (Stockinger *et al.*, 1997; Liu *et al.*, 1998; Choi *et al.*,

2000; Uno *et al.*, 2000). CBF/DREB1 (C-repeat-binding factor/DRE binding protein 1) and DREB2 are AP2 domain transcription activators that bind to the DRE/CRT sequence (Stockinger *et al.*, 1997; Liu *et al.*, 1998), while the basic leucine zipper-type ABF (ABRE-binding factors)/AREB (ABA responsive element binding protein) transcription factors bind to the ABRE element (Choi *et al.*, 2000; Uno *et al.*, 2000). The transcription factors AtMYC2 and AtMYB2 bind the MYCRS and MYBRS element, respectively (Abe *et al.*, 2003).

In *Arabidopsis*, CBFs/DREB1s are major regulators of cold stress responsive gene expression. There are three cold-inducible CBF/DREB1 genes named *CBF1/DREB1B*, *CBF3/DERB1A*, and *CBF2/DREB1C* (Stockinger *et al.*, 1997; Liu *et al.*, 1998), and these are also induced by ABA (Knight *et al.*, 2004; Chinnusamy *et al.*, 2006a) (Fig. 14.2). Overexpression of *CBF1*, *CBF2*, or *CBF3*

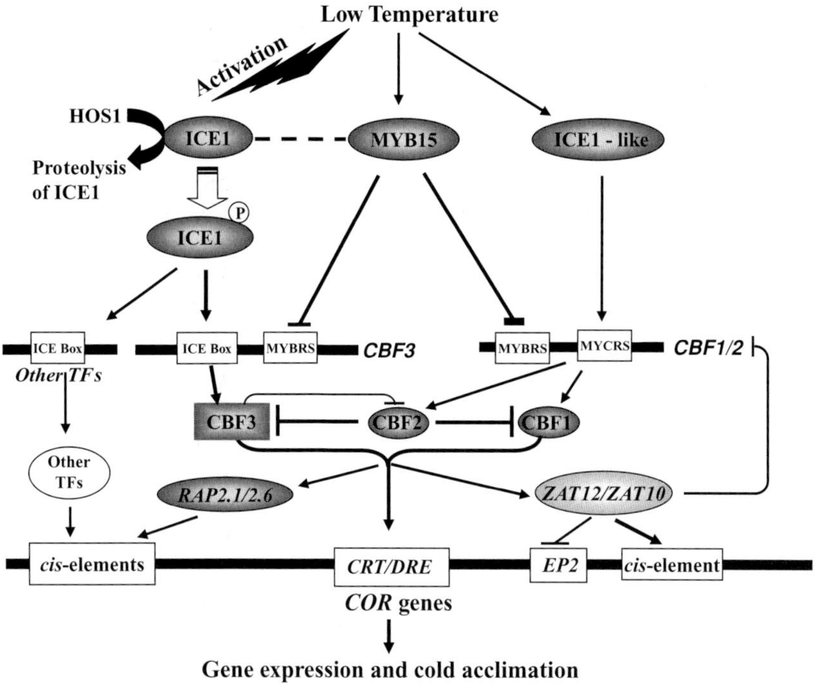

Figure 14.2 Transcriptional networks and their regulation under cold stress. Cold stress activates the ICE1 and ICE1-like proteins which induce transcription of *CBFs* and other transcription factors. CBF2 negatively regulate the transcription level of CBF1/3, while CBF3 regulate the expression level of CBF2. CBFs induce the expression of C2H2 zinc finger transcription factors such as ZAT12 and ZAT10, which in turn either positively or negatively regulate the expression of target *COR* genes. The level of ICE1 protein is tagged with ubiquitin for proteolysis under cold stress by RING finger E3 ubiquitin ligase HOS1. MYB15 transcriptional regulator negatively regulates the expression of CBFs during cold stress. The negative regulators, namely, HOS1 and MYB15, may be necessary to maintaining cold responsive transcriptome homeostasis. Line arrow, induction of expression; line ending with bar, repression of gene expression.

improved the salt, drought, and freezing tolerance of transgenic plants by activating the transcription of downstream stress-inducible genes (Jaglo-Ottosen *et al.*, 1998; Liu *et al.*, 1998; Vogel *et al.*, 2005). The expression of *CBF1* and *CBF3* appears to be negatively regulated by CBF2 (Novillo *et al.*, 2004). Using a *CBF3* promoter-luciferase (*CBF3-LUC*) reporter gene, the *ice1* (*inducer of CBF expression 1*) mutant with impaired expression of CBF3-LUC under cold stress was isolated from an ethyl methane sulfonate-mutagenized *Arabidopsis* population (Chinnusamy *et al.*, 2003). Map-based cloning identified ICE1 as a MYC-like bHLH protein, which is localized in the nucleus and can recognize MYC-binding sequences in the *CBF3* promoter. It seems that ICE1 mainly regulates the expression of *CBF3*, but only slightly regulates the expression of *CBF1* and *CBF2* under cold stress. The expression of *CBF3* and its downstream target genes is greatly reduced in the dominant *ice1* mutant, and the mutant is hypersensitive to both chilling and freezing stress. On the other hand, overexpression of *ICE1* increased the freezing tolerance of transgenic plants (Chinnusamy *et al.*, 2003; Lee *et al.*, 2005) (Fig. 14.2). Overexpression of *ICE1* increased the *CBF3* transcript level only under cold treatment, but not in normal warm conditions, suggesting that ICE1 is regulated at the posttranslational level (Chinnusamy *et al.*, 2003). Analysis of the CBF promoters also identified MYB recognition sequences, and the MYB protein AtMYB15, which binds to MYBRS, was identified as a negative regulator of CBF expression. The cold upregulated AtMYB15 interacts with ICE1. There is a reduction in cold-induced *CBF* expression in transgenic plants overexpressing *AtMYB15* and increased expression of *CBFs* in *myb15* loss-of-function mutants (Agarwal *et al.*, 2006).

In another genetic screen for mutants with altered expression of *RD29A-LUC*, many loci important for stress responsive gene expression were identified (Xiong *et al.*, 2002; Zhu, 2002). Among these, *HOS1* (high expression of osmotically responsive genes) is a negative regulator of cold responses. *hos1* mutation increased the expression of *RD29A* and other stress-inducible genes including *CBFs* under low temperatures. *HOS1* encodes a RING-finger protein that has an ubiquitin E3 ligase activity. The protein is localized in the cytoplasm at normal conditions but is imported into the nucleus under low-temperature stress. HOS1 physically interacts with ICE1 and mediates the degradation of ICE1 under cold stress through the ubiquitination–proteosomal pathway (Dong *et al.*, 2006a). Recently, a role for sumoylation in cold acclimation was found through studies on the *Arabidopsis* SUMO (small ubiquitin-related modifier) E3 ligase SIZ1 (SAP and Miz1) (Miura *et al.*, 2007). Sumoylation is a posttranslational protein modification by which SUMO proteins are conjugated to protein substrates by SUMO E3 ligases. Desumoylation is the removal of SUMO proteins from their target proteins by SUMO proteases. Sumoylation and desumoylation of proteins play a crucial role in hormonal, abiotic, and biotic stress responses in plants. The SUMO E3 ligase *siz1* null mutant is impaired in the accumulation of SUMO conjugates during cold stress and is hypersensitive to chilling and freezing stresses. The *siz1* mutant exhibits significant reduction in cold induction of *CBFs* and its target *COR* genes (*COR15A*, *COR47*, and *KIN1*). Conversely, *siz1* mutation enhances the cold induction of *AtMYB15*, a negative regulator of *CBFs*. SIZ1 mediates SUMO conjugation to K393 of ICE1 during cold acclimation, and this reduces polyubiquitination of ICE1. A K393R substitution in ICE1 [ICE1(K393R)] blocks SIZ1-mediated sumoylation. Transgenic *Arabidopsis* plants overexpressing *ICE1* but not *ICE1(K393R)* exhibit an enhanced cold induction of *CBFs* and enhanced freezing tolerance, and *ICE1(K393R)* overexpressing transgenic plants exhibit a moderate increase in *MYB15* expression under cold stress. These results show that SIZ1-mediated sumoylation positively regulates ICE1 stability and activity to fine-tune the transcription of *COR* genes during cold acclimation (Miura *et al.*, 2007).

Besides the CBF regulon, there also exist CBF-independent regulons and pathways for plant cold acclimation (Fig. 14.2). *HOS9* encodes a homeodomain transcription factor (Zhu *et al.*, 2004). A lesion in *HOS9* increased the expression of *RD29A* and some other stress-inducible genes independent of CBFs (Zhu *et al.*, 2004). *HOS10* encodes a R2R3-type MYB transcription factor that is localized in the nucleus (Zhu *et al.*, 2005). The *hos10* mutation impairs cold acclimation and the mutant plants are hypersensitive to freezing temperatures as well as to NaCl. The induction of *NCED3* by dehydration and ABA accumulation are reduced by this mutation (Zhu *et al.*, 2005). The *Arabidopsis eskimo1* (*esk1*) mutants accumulate high levels of proline and are constitutively freezing-tolerant, but the expression of cold-regulated genes with the DRE/CRT *cis*-element is not affected in the mutants, suggesting that *ESK1*-mediated freezing tolerance pathway is distinct from the CBF pathway (Xin and Browse, 1998). *ESK1* encodes a DUF (domain of unknown function) 231 protein. Microarray analysis identified 312 genes with altered expression in *esk1* mutant, of which only 12 genes are commonly upregulated in *esk1* mutants and *CBF2* overexpressing *Arabidopsis* plants. These results suggest that freezing tolerance imparted by the recessive *esk1* mutation is distinct from that of ICE1–CBF pathway (Xin *et al.*, 2007). The *sfr6* (sensitivity to freezing) mutation reduces the expression of genes such as *KIN1*, *COR15A*, and *RD29A* that carry the CRT/DRE element, but the mutation does not affect the expression of *CBFs* (Knight *et al.*, 1999; Boyce *et al.*, 2003). Thus, SFR6 may be required for CBF function under cold stress.

Although the expression of *DREB2A* and *DREB2B* is induced by drought and salt stress, transgenic plants overexpressing these two genes did not show improved stress tolerance, indicating that posttranscriptional modifications might be required for their activities (Liu *et al.*, 1998). Domain deletion analysis indicated that *DREB2A* contains both activation and inhibitory domains. Overexpression of a constitutively active *DREB2A* (*DREB2A CA*) increased drought tolerance in transgenic *Arabidopsis* plants (Sakuma *et al.*, 2006b). Interestingly, the upregulated genes in plants overexpressing *DREB2A CA* include both dehydration responsive genes and heat shock-related genes (Sakuma *et al.*, 2006a). The expression of *DREB2A* is also rapidly induced by heat shock.

DREB2A knockout mutants are hypersensitive to heat stress, whereas transgenic plants overexpressing *DREB2A CA* showed increased thermotolerance (Sakuma *et al.*, 2006a). These results suggest that DREB2A mediates the cross talk between drought and heat stress.

The expression of *ABF1*, *ABF2/AREB1*, *ABF3*, and *ABF4/AREB2* is induced by ABA (Choi *et al.*, 2000; Uno *et al.*, 2000). Constitutive overexpression of *ABF3* and *ABF4* enhanced the ABA sensitivity of seed germination and seedling growth, and the transgenic *Arabidopsis* plants had reduced transpiration and were more drought-tolerant than wild-type plants (Kang *et al.*, 2002). However, the constitutive overexpression of full-length *ABF2/AREB1* did not induce the downstream genes. Like *DREB2A*, *ABF2/AREB1* may also require posttranslational modifications for its activity. Indeed, an ABA-activated 42-kDa kinase is found to phosphorylate and activate *AREB1* upon exogenous ABA application. Transgenic plants overexpressing an active form of AREB1 containing substitution of the phosphorylatable Ser/Thr residues to Asp induced the expression of many ABA responsive genes even without ABA treatment (Furihata *et al.*, 2006).

Recently, other transcriptional factors were identified, which bind to non-ABRE/DRE *cis*-elements in promoters of ABA responsive genes (Soderman *et al.*, 1999; Himmelbach *et al.*, 2002; Fujita *et al.*, 2004; Lu *et al.*, 2007). *ATHB6* is a negative regulator, which physically interacts with ABI1 and acts downstream of ABI1 in the ABA pathway. ATHB6 recognizes a *cis*-element with a core motif of CAATTATTA and mediates the expression of *ATHB6*- and ABA-dependent genes (Himmelbach *et al.*, 2002). AtERF7, an APETALA2/EREBP-type transcription factor, binds to the GCC box to repress the expression of target genes. AtERF7 interacts with and may be regulated by the protein kinase PKS3, a global regulator of ABA response. The transcriptional repression activity of *AtERF7* is enhanced through interaction with AtSin3, the *Arabidopsis* homolog of a human global corepressor of transcription, which in turn may interact with HDA19, a histone deacetylase. Consistent with the transcriptional repression activity, overexpression of *AtERF7* in *Arabidopsis* reduced the sensitivity of guard cells to ABA, whereas reducing the expression of *AtERF7* and *AtSin3* in RNAi lines increased the sensitivity to ABA during seed germination (Song *et al.*, 2005). These results suggest that histone modifications and chromatin remodeling are involved in ABA-regulated gene expression, which is further supported by the finding that AtHD2C, a histone deacetylase, functions in ABA response (Sridha and Wu, 2006). *AtHD2C* expression is constitutive but is downregulated by ABA. Transgenic plants overexpressing *AtHD2C* displayed reduced transpiration and enhanced tolerance to salt and drought stresses.

14.5 Oxidative stress management

Reactive oxygen species (ROS), namely superoxide radicals ($O_2^{\cdot -}$), hydrogen peroxide (H_2O_2), and hydroxyl radicals ($OH^{\cdot}$), are produced in aerobic cellular processes and the production/accumulation of these radicals reaches toxic levels in cells under various abiotic stresses (see Chapter 7). Since ROS causes oxidative damage to membrane lipids, proteins, and nucleic acids, ROS detoxification is a crucial defense against abiotic stresses. Plants employ antioxidant metabolites and enzymes to detoxify ROS.

Osmotic stresses, extreme temperatures, and ABA induce enhanced accumulation of ROS, such as H_2O_2. ROS may be perceived by cells through oxidation of proteins. In plants, redox-sensitive receptor-like kinases and two-component histidine kinases may act as potential sensors of ROS (Vranová *et al.*, 2002). These sensory kinases likely activate a mitogen-activated protein kinase (MAPK) module that regulates ROS management (see Chapter 4). The AtMEKK1/ANP1 (=MAPKKK)–AtMEK1/AtMKK2 (=MAPKK)–AtMPK3/6 (=MAPK) cascade may transduce ROS signals in *Arabidopsis* (Kovtun *et al.*, 2000; Chinnusamy *et al.*, 2004; Teige *et al.*, 2004). Transgenic tobacco or maize plants overexpressing a constitutively active tobacco *NPK1* (ortholog of ANP1) showed enhanced tolerance to drought and other abiotic stresses through induction of some stress responsive genes (Kovtun *et al.*, 2000; Shou *et al.*, 2004a,b). In rice, *OsMAPK5* expression and its kinase activity is induced by ABA, biotic, and abiotic stresses including salt, drought, wounding, and cold. Transgenic overexpression of *OsMAPK5* in rice increased tolerance to several abiotic stresses, including drought stress, probably by enhancing ROS detoxification (Xiong and Yang, 2003). These observations suggest that ROS signaling under various abiotic stresses may converge at MAPK cascades.

14.6 Posttranscriptionl regulation of gene expression

Besides transcriptional regulation, gene expression can also be regulated at RNA levels during transcript elongation, splicing, maturation, export, translation, and degradation.

14.6.1 mRNA metabolism under abiotic stresses

Genetic screening for drought response mutants identified *abo1* (ABA overly sensitive 1), which showed a drought resistance phenotype when grown in soil (Chen *et al.*, 2006). *abo1* mutations enhance ABA sensitivity in both stomatal closing and inhibition of seedling growth. The expression of some drought- or ABA-inducible genes is reduced to a lower level comparable to that of wild-type plants. Map-based cloning revealed ABO1 as the largest subunit of the holo-Elongator, which is highly conserved from yeast to animals and plays critical roles in assisting transcriptional elongation, secretion, and tRNA modification (Chen *et al.*, 2006). The *sta1-1* (*stabilized1-1*) mutation in *Arabidopsis* increases the stability of normally unstable luciferase transgene transcript and some endogenous gene transcripts. *STA1* encodes a nuclear protein similar to the human U5 snRNP-associated 102-kDa protein and the yeast pre-mRNA splicing factors Prp1p and Prp6p. The expression of *STA1* is induced by cold stress, and the splicing of cold-induced *COR15A* gene was impaired in *sta1-1* mutant. *sta1-1* mutant is more sensitive to ABA inhibition of germination and chilling stress. These results suggest that the pre-mRNA splicing factor STA1 is required for both splicing and the turnover of unstable transcripts under abiotic stresses (Lee *et al.*, 2006a).

ABH1 (ABA hypersensitive 1) encodes the CBP80 subunit of the dimeric mRNA cap-binding complex (CBC), which binds to the 7-methylated cap structure of mRNA, and plays important roles in mRNA processing. Plants with lesions in *ABH1* display increased ABA sensitivity for seed germination, seedling growth, and stomatal closing. *abh1* mutants transpire less and are more drought-tolerant than wild-type plants (Hugouvieux *et al.*, 2001). Plants with mutations in another subunit (CBP20) of CBC show similar drought-resistant phenotypes as *abh1* (Papp *et al.*, 2004). The *Arabidopsis sad1* (supersensitive to ABA and drought) mutation increases the expression of *RD29A-LUC* reporter gene and other stress responsive genes under ABA or hyperosmotic stress treatment (Xiong *et al.*, 2001a), and enhances ABA and osmotic stress sensitivity of seed germination and vegetative growth. Interestingly, *sad1* mutation also impairs drought-induced ABA accumulation by blocking the positive feedback regulation of ABA biosynthesis. Although stomatal closing in *sad1* is hypersensitive to ABA, *sad1* plants are hypersensitive to drought stress due to the defective ABA biosynthesis. *SAD1* encodes a protein similar to multifunctional Sm-like snRNP proteins, which are known to participate in RNA metabolism including splicing, export, and degradation (Xiong *et al.*, 2001a).

14.6.2 mRNA export

Import of signaling proteins into the nucleus and the export of mRNAs and small RNAs to the cytoplasm through the nuclear pore complex of the nuclear envelop plays a pivotal role in gene regulation in eukaryotes. Export-competent mRNP consists of mRNA cargo and nucleocytoplasmic shuttling nuclear proteins such as the RNA export factors, DEAD-box protein 5 and nucleoporins (Cole and Scarcelli, 2006). The *los4-1* (for low expression of osmotically responsive genes) mutation reduces or delays the expression of *CBF* genes and their downstream target genes. *los4-1* mutant plants are hypersensitive to chilling especially in the dark. Overexpression of *CBF3* in *los4-1* mutant plants reverses the chilling sensitivity (Gong *et al.*, 2002). The *cryophyte* mutant (*los4-2*) is allelic to *los4-1*. Interestingly, in contrast to *los4-1*, *los4-2* mutant plants show enhanced cold induction of *CBF2* and more chilling and freezing tolerance than the wild-type *Arabidopsis*. *LOS4* encodes a DEAD box RNA helicase essential for mRNA export in plant cells. mRNA export from the nucleus is blocked in the *cryophyte/los4-2* mutant only at warm temperatures, while the *los4-1* mutation weakens mRNA export at both cold and warm temperatures (Gong *et al.*, 2002, 2005).

Nuclear pore complexes are made up of several proteins collectively called nucleoporins (NUPs). Molecular analysis of a NUP from *Arabidopsis*, AtNUP160, revealed that NUPs regulating mRNA export play a critical role in cold acclimation. In *atnup160-1* mutant, poly(A) mRNA export is dramatically reduced under cold stress. This results in impairment of cold induction of *CBFs* and hypersensitivity to chilling and freezing stresses. These results suggest that mRNA export under cold stress is critical for chilling and freezing tolerance (Dong *et al.*, 2006b). Protein nuclear transport is also involved in abiotic stress responses (Verslues *et al.*, 2006). *SAD2* encodes an importin beta family protein, which might function in the nuclear transport of proteins. The expression of some stress-inducible genes was higher in *sad2-1* mutant than in the wild type, and *sad2-1* mutant exhibits ABA hypersensitivity in seed germination and seedling growth. The results thus suggest a role for protein nuclear transport in ABA signaling (Verslues *et al.*, 2006).

14.6.3 Small RNAs

Posttranscriptionl regulation by noncoding RNAs, namely, micro-RNAs (miRNAs) and short interfering RNAs (siRNAs), plays a crucial role in abiotic stress responsive gene expression (Borsani *et al.*, 2005; Mallory and Vaucheret, 2006). miRNAs are single-stranded noncoding RNAs of ~20–24 nucleotides (nt) in length. miRNAs are produced by DICER-like (ribonuclease III-like) enzymes from imperfect hairpin RNAs, which originate from longer pri-miRNAs transcribed from *MIR* genes. siRNAs are double-stranded 21–26-nt small RNAs derived from long double-stranded RNAs. miRNAs and siRNAs regulate gene expression through mRNA cleavage, translation repression, or chromatin remodeling (Borsani *et al.*, 2005; Mallory and Vaucheret, 2006). Under abiotic stresses, stress-upregulated small RNAs may downregulate their target genes, which are likely negative regulators of stress tolerance, while stress-downregulated small RNAs may result in accumulation of their target gene mRNAs, which may positively regulate stress tolerance.

Screening of small RNAs from plants treated with various stresses led to the identification of several miRNA and siRNAs that are regulated by cold, salt, or drought stress (Sunkar and Zhu, 2004). These small RNAs play important roles in regulating their target gene expression under different environmental conditions (Sunkar and Zhu, 2004). In *Arabidopsis*, some antisense overlapping gene pairs generate a new type of siRNAs called nat-siRNAs. The founding member of nat-siRNAs is derived from a *cis*-NAT gene pair of *SRO5* (a stress-inducible gene encoding a RCD1-like protein) and *P5CDH* (Δ^1-pyrroline-5-carboxylate dehydrogenase). Studies on this nat-siRNA demonstrated a crucial role of nat-siRNAs in osmoprotection and oxidative stress management under salt stress in *Arabidopsis*. Salt stress-induced *SRO5* mRNA and the *P5CDH* mRNA produce a dsRNA, which is processed by an siRNA biogenesis pathway to produce a 24-nt SRO5-P5CDH nat-siRNA. The 24-nt nat-siRNA guides the cleavage of the *P5CDH* transcript to further produce

21-nt SRO5–P5CDH nat-siRNAs. The induction of *SRO5* is required to initiate the production of the nat-siRNAs, which direct the cleavage of *P5CDH* transcripts to allow proline accumulation. The SRO5 protein is also required to manage ROS under salt stress (Borsani *et al.*, 2005).

Oxidative stress is a common secondary stress induced in cells by various abiotic stresses. Superoxide dismutases (SODs) form the first line of defense against oxidative stress as it rapidly converts superoxide to hydrogen peroxide (H_2O_2) and molecular oxygen. Overexpression of *SODs* in transgenic plants was shown to result in enhanced stress tolerance (Alscher *et al.*, 2002). Recently, a role for miRNAs in defense against oxidative stress was found in *Arabidopsis*. Oxidative stress downregulates miR398, which targets the transcripts of two Cu–Zn SOD genes namely *CSD1* (a cytosolic SOD) and *CSD2* (a chloroplastic SOD) (Sunkar and Zhu, 2004; Sunkar *et al.*, 2006). The expression patterns of miR398 and its target genes show an inverse correlation in various developmental stages and oxidative stress conditions. A transient coexpression assay with miR398 with its target genes in *Nicotiana* showed that miR398 directs the degradation of *CSD1* and *CSD2* mRNAs in vivo. Furthermore, the role of miR398 was examined in transgenic *Arabidopsis* overexpressing the normal *CSD2* gene or an miR398-resistant form of *CSD2* gene, *mCSD2* (miR398 target site mutated without modifying amino acid sequence). The *mCSD2* transgenic plants showed highest *CSD2* mRNA levels and oxidative stress tolerance as compared with *CSD2* transgenics and wild-type plants (Sunkar *et al.*, 2006).

14.7 Future perspectives

Although substantial progress has been made in our understanding of abiotic stress signaling and tolerance mechanisms, the picture of stress signaling network is still quite incomplete. The isolation of the stress-related ABA receptor ABAR will be helpful in future investigation of ABA signal transduction. Forward and reverse genetic studies using the model plants *Arabidopsis* and rice continue to be critical for identifying unknown signaling components. Searching for superior alleles of key stress tolerance determinants such as *SOS1* from naturally stress-tolerant plants will be very useful for engineering stress hardier crops. With more and more stress-relevant genes cloned, the big challenge for molecular biologists and crop breeders is how to sort out the more important genes and combine them in transgenic plants or use marker-assisted breeding to improve crop production under stress conditions. Small RNA studies will shed more light on the molecular mechanisms of stress regulation. Overexpression of miRNA-resistant target genes will help overcome posttranscriptional gene silencing, and thus may lead to better expression of engineered trait in transgenic plants. Understanding the roles of small RNAs and chromatin regulation in transcriptome homeostasis, cellular tolerance, phenological and developmental plasticity of plants under abiotic stress, and recovery will be important for effective genetic engineering of high-level abiotic stress resistance in crop plants.

References

Abe, H., Urao, T., Ito, T., Seki, M., Shinozaki, K. and Yamaguchi-Shinozaki, K. (2003) *Arabidopsis* AtMYC2 (bHLH) and AtMYB2 (MYB) function as transcriptional activators in abscisic acid signaling. *Plant Cell*, **15**, 63–78.

Achard, P., Cheng, H., De Grauwe, L., Decat, J., Schoutteten, H., Moritz, T., Van Der Straeten, D., Peng, J. and Harberd, N.P. (2006) Integration of plant responses to environmentally activated phytohormonal signals. *Science*, **311**, 91–94.

Agarwal, M., Hao, Y., Kapoor, A., Dong, C.H., Fujii, H., Zheng, X. and Zhu, J.K. (2006) A R2R3 type MYB transcription factor is involved in the cold regulation of *CBF* genes and in acquired freezing tolerance. *J Biol Chem*, **281**, 37636–37645.

Ahlfors, R., Lang, S., Overmyer, K., Jaspers, P., Brosche, M., Tauriainen, A., Kollist, H., Tuominen, H., Belles-Boix, E., Piippo, M., Inze, D., Palva, E.T. and Kangasjarvi, J. (2004) *Arabidopsis* RADICAL-INDUCED CELL DEATH1 belongs to the WWE protein–protein interaction domain protein family and modulates abscisic acid, ethylene, and methyl jasmonate responses. *Plant Cell*, **16**, 1925–1937.

Alscher, R.G., Erturk, N. and Heath, L.S. (2002) Role of superoxide dismutases (SODs) in controlling oxidative stress in plants. *J Exp Bot*, **53**, 1331–1341.

Berthomieu, P., Conejero, G., Nublat, A., Brackenbury, W.J., Lambert, C., Savio, C., Uozumi, N., Oiki, S., Yamada, K., Cellier, F., Gosti, F., Simonneau, T., Essah, P.A., Tester, M., Very, A.A., Sentenac, H. and Casse, F. (2003) Functional analysis of AtHKT1 in *Arabidopsis* shows that Na(+) recirculation by the phloem is crucial for salt tolerance. *EMBO J*, **22**, 2004–2014.

Bethke, P.C., Gubler, F., Jacobsen, J.V. and Jones, R.L. (2004) Dormancy of *Arabidopsis* seeds and barley grains can be broken by nitric oxide. *Planta*, **219**, 847–855.

Bittner, F., Oreb, M. and Mendel, R.R. (2001) ABA3 is a molybdenum cofactor sulfurase required for activation of aldehyde oxidase and xanthine dehydrogenase in *Arabidopsis thaliana*. *J Biol Chem*, **276**, 40381–40384.

Borsani, O., Zhu, J., Verslues, P.E., Sunkar, R. and Zhu, J.K. (2005) Endogenous siRNAs derived from a pair of natural cis-antisense transcripts regulate salt tolerance in *Arabidopsis*. *Cell*, **123**, 1279–1291.

Boyce, J.M., Knight, H., Deyholos, M., Openshaw, M.R., Galbraith, D.W., Warren, G. and Knight, M.R. (2003) The *sfr6* mutant of *Arabidopsis* is defective in transcriptional activation via CBF/DREB1 and DREB2 and shows sensitivity to osmotic stress. *Plant J*, **34**, 395–406.

Brocard-Gifford, I., Lynch, T.J., Garcia, M.E., Malhotra, B. and Finkelstein, R.R. (2004) The *Arabidopsis thaliana ABSCISIC ACID-INSENSITIVE8* encodes a novel protein mediating abscisic acid and sugar responses essential for growth. *Plant Cell*, **16**, 406–421.

Burnette, R.N., Gunesekera, B.M. and Gillaspy, G.E. (2003) An *Arabidopsis* inositol 5-phosphatase gain-of-function alters abscisic acid signaling. *Plant Physiol*, **132**, 1011–1019.

Chen, Z., Zhang, H., Jablonowski, D., Zhou, X., Ren, X., Hong, X., Schaffrath, R., Zhu, J.K. and Gong, Z. (2006) Mutations in *ABO1/ELO2*, a subunit of holo-Elongator, increase abscisic acid sensitivity and drought tolerance in *Arabidopsis thaliana*. *Mol Cell Biol*, **26**, 6902–6912.

Cheng, W.H., Endo, A., Zhou, L., Penney, J., Chen, H.C., Arroyo, A., Leon, P., Nambara, E., Asami, T., Seo, M., Koshiba, T. and Sheen, J. (2002) A unique short-chain dehydrogenase/reductase in *Arabidopsis* glucose signaling and abscisic acid biosynthesis and functions. *Plant Cell*, **14**, 2723–2743.

Chinnusamy, V., Ohta, M., Kanrar, S., Lee, B.H., Hong, X., Agarwal, M. and Zhu, J.K. (2003) ICE1: a regulator of cold-induced transcriptome and freezing tolerance in *Arabidopsis*. *Genes Dev*, **17**, 1043–1054.

Chinnusamy, V., Schumaker, K. and Zhu, J.K. (2004) Molecular genetic perspectives on cross-talk and specificity in abiotic stress signalling in plants. *J Exp Bot*, **55**, 225–236.

Chinnusamy, V., Zhu, J. and Zhu, J.K. (2006a) Gene regulation during cold acclimation in plants. *Physiol Plant*, **126**, 52–61.

Chinnusamy, V., Zhu, J. and Zhu, J.K. (2006b) Salt stress signaling and mechanisms of plant salt tolerance. *Genet Eng (N Y)*, **27**, 141–177.

Chinnusamy, V. and Zhu, J.K. (2003) Plant salt tolerance. *Top Curr Genet*, **4**, 241–270.

Choi, H., Hong, J., Ha, J., Kang, J. and Kim, S.Y. (2000) ABFs, a family of ABA-responsive element binding factors. *J Biol Chem*, **275**, 1723–1730.

Cole, C.N. and Scarcelli, J.J. (2006) Transport of messenger RNA from the nucleus to the cytoplasm. *Curr Opin Cell Biol*, **18**, 299–306.

DeWald, D.B., Torabinejad, J., Jones, C.A., Shope, J.C., Cangelosi, A.R., Thompson, J.E., Prestwich, G.D. and Hama, H. (2001) Rapid accumulation of phosphatidylinositol 4,5-bisphosphate and inositol 1,4,5-trisphosphate correlates with calcium mobilization in salt-stressed *Arabidopsis*. *Plant Physiol*, **126**, 759–769.

Dietz, K.J., Sauter, A., Wichert, K., Messdaghi, D. and Hartung, W. (2000) Extracellular beta-glucosidase activity in barley involved in the hydrolysis of ABA glucose conjugate in leaves. *J Exp Bot*, **51**, 937–944.

Dong, C.H., Agarwal, M., Zhang, Y., Xie, Q. and Zhu, J.K. (2006a) The negative regulator of plant cold responses, HOS1, is a RING E3 ligase that mediates the ubiquitination and degradation of ICE1. *Proc Natl Acad Sci USA*, **103**, 8281–8286.

Dong, C.H., Hu, X.Y., Tang, W., Zheng, X., Kim, Y.S., Lee, B.H. and Zhu, J.K. (2006b) A putative *Arabidopsis* nucleoporin AtNUP160 is critical for RNA export and required for plant tolerance to cold stress. *Mol Cell Biol*, **26**, 9533–9543.

Fedoroff, N.V. (2002) Cross-talk in abscisic acid signaling. *Sci STKE*, **2002**, RE10.

Finkelstein, R.R. (2006) Studies of abscisic acid perception finally flower. *Plant Cell*, **18**, 786–791.

Finkelstein, R.R., Gampala, S.S. and Rock, C.D. (2002) Abscisic acid signaling in seeds and seedlings. *Plant Cell*, **14** (Suppl), S15–S45.

Finkelstein, R.R. and Lynch, T.J. (2000) The *Arabidopsis* abscisic acid response gene *ABI5* encodes a basic leucine zipper transcription factor. *Plant Cell*, **12**, 599–609.

Finkelstein, R.R. and Somerville, C.R. (1990) Three classes of abscisic acid (ABA)-insensitive mutations of *Arabidopsis* define genes that control overlapping subsets of ABA responses. *Plant Physiol*, **94**, 1172–1179.

Finkelstein, R.R., Wang, M.L., Lynch, T.J., Rao, S. and Goodman, H.M. (1998) The *Arabidopsis* abscisic acid response locus *ABI4* encodes an APETALA 2 domain protein. *Plant Cell*, **10**, 1043–1054.

Fujita, M., Fujita, Y., Maruyama, K., Seki, M., Hiratsu, K., Ohme-Takagi, M., Tran, L.S., Yamaguchi-Shinozaki, K. and Shinozaki, K. (2004) A dehydration-induced NAC protein, RD26, is involved in a novel ABA-dependent stress-signaling pathway. *Plant J*, **39**, 863–876.

Furihata, T., Maruyama, K., Fujita, Y., Umezawa, T., Yoshida, R., Shinozaki, K. and Yamaguchi-Shinozaki, K. (2006) Abscisic acid-dependent multisite phosphorylation regulates the activity of a transcription activator AREB1. *Proc Natl Acad Sci USA*, **103**, 1988–1993.

Gaxiola, R.A., Li, J., Undurraga, S., Dang, L.M., Allen, G.J., Alper, S.L. and Fink, G.R. (2001) Drought- and salt-tolerant plants result from overexpression of the AVP1 H^+-pump. *Proc Natl Acad Sci USA*, **98**, 11444–11449.

Gaxiola, R.A., Rao, R., Sherman, A., Grisafi, P., Alper, S.L. and Fink, G.R. (1999) The *Arabidopsis thaliana* proton transporters, AtNhx1 and Avp1, can function in cation detoxification in yeast. *Proc Natl Acad Sci USA*, **96**, 1480–1485.

Giraudat, J. (1995). Abscisic acid signaling. *Curr Opin Cell Biol*, **7**, 232–238.

Giraudat, J., Hauge, B.M., Valon, C., Smalle, J., Parcy, F. and Goodman, H.M. (1992) Isolation of the *Arabidopsis ABI3* gene by positional cloning. *Plant Cell*, **4**, 1251–1261.

Gong, Z., Dong, C.H., Lee, H., Zhu, J., Xiong, L., Gong, D., Stevenson, B. and Zhu, J.K. (2005) A DEAD box RNA helicase is essential for mRNA export and important for development and stress responses in *Arabidopsis*. *Plant Cell*, **17**, 256–267.

Gong, Z., Lee, H., Xiong, L., Jagendorf, A., Stevenson, B. and Zhu, J.K. (2002) RNA helicase-like protein as an early regulator of transcription factors for plant chilling and freezing tolerance. *Proc Natl Acad Sci USA*, **99**, 11507–11512.

Gonzalez-Guzman, M., Apostolova, N., Belles, J.M., Barrero, J.M., Piqueras, P., Ponce, M.R., Micol, J.L., Serrano, R. and Rodriguez, P.L. (2002) The short-chain alcohol dehydrogenase ABA2 catalyzes the conversion of xanthoxin to abscisic aldehyde. *Plant Cell*, **14**, 1833–1846.

Gosti, F., Beaudoin, N., Serizet, C., Webb, A.A., Vartanian, N. and Giraudat, J. (1999) ABI1 protein phosphatase 2C is a negative regulator of abscisic acid signaling. *Plant Cell*, **11**, 1897–1910.

Grill, E. and Himmelbach, A. (1998) ABA signal transduction. *Curr Opin Plant Biol*, **1**, 412–418.

Guo, Y., Qiu, Q.S., Quintero, F.J., Pardo, J.M., Ohta, M., Zhang, C., Schumaker, K.S. and Zhu, J.K. (2004) Transgenic evaluation of activated mutant alleles of SOS2 reveals a critical requirement for its kinase activity and C-terminal regulatory domain for salt tolerance in *Arabidopsis thaliana*. *Plant Cell*, **16**, 435–449.

Halfter, U., Ishitani, M. and Zhu, J.K. (2000) The *Arabidopsis* SOS2 protein kinase physically interacts with and is activated by the calcium-binding protein SOS3. *Proc Natl Acad Sci USA*, **97**, 3735–3740.

Hasegawa, P.M., Bressan, R.A., Zhu, J.K. and Bohnert, H.J. (2000) Plant cellular and molecular responses to high salinity. *Annu Rev Plant Physiol Plant Mol Biol*, **51**, 463–499.

Himmelbach, A., Hoffmann, T., Leube, M., Hohener, B. and Grill, E. (2002) Homeodomain protein ATHB6 is a target of the protein phosphatase ABI1 and regulates hormone responses in *Arabidopsis*. *EMBO J*, **21**, 3029–3038.

Hugouvieux, V., Kwak, J.M. and Schroeder, J.I. (2001) An mRNA cap binding protein, ABH1, modulates early abscisic acid signal transduction in *Arabidopsis*. *Cell*, **106**, 477–487.

Ishitani, M., Liu, J., Halfter, U., Kim, C.S., Shi, W. and Zhu, J.K. (2000) SOS3 function in plant salt tolerance requires N-myristoylation and calcium-binding. *Plant Cell*, **12**, 1667–1677.

Izhaki, A., Swain, S.M., Tseng, T.S., Borochov, A., Olszewski, N.E. and Weiss, D. (2001) The role of SPY and its TPR domain in the regulation of gibberellin action throughout the life cycle of *Petunia hybrida* plants. *Plant J*, **28**, 181–190.

Jaglo-Ottosen, K.R., Gilmour, S.J., Zarka, D.G., Schabenberger, O. and Thomashow, M.F. (1998) *Arabidopsis CBF1* overexpression induces *COR* genes and enhances freezing tolerance. *Science*, **280**, 104–106.

Kang, J.Y., Choi, H.I., Im, M.Y. and Kim, S.Y. (2002) *Arabidopsis* basic leucine zipper proteins that mediate stress-responsive abscisic acid signaling. *Plant Cell*, **14**, 343–357.

Katiyar-Agarwal, S., Zhu, J., Kim, K., Agarwal, M., Fu, X., Huang, A. and Zhu, J.K. (2006) The plasma membrane Na^+/H^+ antiporter SOS1 interacts with RCD1 and functions in oxidative stress tolerance in *Arabidopsis*. *Proc Natl Acad Sci USA*, **103**, 18816–18821.

Kim, Y.J., Kim, J.E., Lee, J.H., Lee, M.H., Jung, H.W., Bahk, Y.Y., Hwang, B.K., Hwang, I. and Kim, W.T. (2004) The *Vr-PLC3* gene encodes a putative plasma membrane-localized phosphoinositide-specific phospholipase C whose expression is induced by abiotic stress in mung bean (*Vigna radiata* L.). *FEBS Lett*, **556**, 127–136.

Knight, H., Veale, E.L., Warren, G.J. and Knight, M.R. (1999) The *sfr6* mutation in *Arabidopsis* suppresses low-temperature induction of genes dependent on the *CRT/DRE* sequence motif. *Plant Cell*, **11**, 875–886.

Knight, H., Zarka, D.G., Okamoto, H., Thomashow, M.F. and Knight, M.R. (2004) Abscisic acid induces *CBF* gene transcription and subsequent induction of cold-regulated genes via the *CRT* promoter element. *Plant Physiol*, **135**, 1710–1717.

Kopka, J., Pical, C., Gray, J.E. and Muller-Rober, B. (1998) Molecular and enzymatic characterization of three phosphoinositide-specific phospholipase C isoforms from potato. *Plant Physiol*, **116**, 239–250.

Kovtun, Y., Chiu, W.L., Tena, G. and Sheen, J. (2000) Functional analysis of oxidative stress-activated mitogen-activated protein kinase cascade in plants. *Proc Natl Acad Sci USA*, **97**, 2940–2945.

Kushiro, T., Okamoto, M., Nakabayashi, K., Yamagishi, K., Kitamura, S., Asami, T., Hirai, N., Koshiba, T., Kamiya, Y. and Nambara, E. (2004) The *Arabidopsis* cytochrome P450 *CYP707A* encodes ABA 8′-hydroxylases: key enzymes in ABA catabolism. *EMBO J*, **23**, 1647–1656.

Lee, B.H., Henderson, D.A. and Zhu, J.K. (2005) The *Arabidopsis* cold-responsive transcriptome and its regulation by ICE1. *Plant Cell*, **17**, 3155–3175.

Lee, B.H., Kapoor, A., Zhu, J. and Zhu, J.K. (2006a) STABILIZED1, a stress-upregulated nuclear protein, is required for pre-mRNA splicing, mRNA turnover, and stress tolerance in *Arabidopsis*. *Plant Cell*, **18**, 1736–1749.

Lee, K., Piao, H., Kim, H., Choi, S., Jiang, F., Hartung, W., Hwang, I., Kwak, J., Lee, I. and Hwang, I. (2006b) Activation of glucosidase via stress-induced polymerization rapidly increases active pools of abscisic acid. *Cell*, **126**, 1109–1120.

Lemichez, E., Wu, Y., Sanchez, J.P., Mettouchi, A., Mathur, J. and Chua, N.H. (2001) Inactivation of AtRac1 by abscisic acid is essential for stomatal closure. *Genes Dev*, **15**, 1808–1816.

Leon-Kloosterziel, K.M., Gil, M.A., Ruijs, G.J., Jacobsen, S.E., Olszewski, N.E., Schwartz, S.H., Zeevaart, J.A. and Koornneef, M. (1996) Isolation and characterization of abscisic acid-deficient *Arabidopsis* mutants at two new loci. *Plant J*, **10**, 655–661.

Leung, J., Bouvier-Durand, M., Morris, P.C., Guerrier, D., Chefdor, F. and Giraudat, J. (1994) *Arabidopsis* ABA response gene *ABI1*: features of a calcium-modulated protein phosphatase. *Science*, **264**, 1448–1452.

Leung, J. and Giraudat, J. (1998) Abscisic acid signal transduction. *Annu Rev Plant Physiol Plant Mol Biol*, **49**, 199–222.

Leung, J., Merlot, S. and Giraudat, J. (1997) The *Arabidopsis* ABSCISIC ACID-INSENSITIVE2 (*ABI2*) and *ABI1* genes encode homologous protein phosphatases 2C involved in abscisic acid signal transduction. *Plant Cell*, **9**, 759–771.

Li, H., Shen, J.J., Zheng, Z.L., Lin, Y. and Yang, Z. (2001) The Rop GTPase switch controls multiple developmental processes in *Arabidopsis*. *Plant Physiol*, **126**, 670–684.

Li, J., Yang, H., Peer, W.A., Richter, G., Blakeslee, J., Bandyopadhyay, A., Titapiwantakun, B., Undurraga, S., Khodakovskaya, M., Richards, E.L., Krizek, B., Murphy, A.S., Gilroy, S. and Gaxiola, R. (2005) *Arabidopsis* H^+-PPase AVP1 regulates auxin-mediated organ development. *Science*, **310**, 121–125.

Lin, P.C., Hwang, S.G., Endo, A., Okamoto, M., Koshiba, T. and Cheng, W.H. (2006) Ectopic expression of *ABA2/GIN1* in *Arabidopsis* promotes seed dormancy and stress tolerance. *Plant Physiol*, **143**, 745–758.

Liu, J., Ishitani, M., Halfter, U., Kim, C.S. and Zhu, J.K. (2000) The *Arabidopsis thaliana SOS2* gene encodes a protein kinase that is required for salt tolerance. *Proc Natl Acad Sci USA*, **97**, 3730–3734.

Liu, J. and Zhu, J.K. (1998) A calcium sensor homolog required for plant salt tolerance. *Science*, **280**, 1943–1945.

Liu, Q., Kasuga, M., Sakuma, Y., Abe, H., Miura, S., Yamaguchi-Shinozaki, K. and Shinozaki, K. (1998) Two transcription factors, DREB1 and DREB2, with an EREBP/AP2 DNA binding domain separate two cellular signal transduction pathways in drought- and low-temperature-responsive gene expression, respectively, in *Arabidopsis*. *Plant Cell*, **10**, 1391–1406.

Liu, X., Yue, Y., Li, B., Nie, Y., Li, W., Wu, W.H. and Ma, L. (2007) A G protein-coupled receptor is a plasma membrane receptor for the plant hormone abscisic acid. *Science*, **315**, 1712–1716.

Lu, P.L., Chen, N.Z., An, R., Su, Z., Qi, B.S., Ren, F., Chen, J. and Wang, X.C. (2007) A novel drought-inducible gene, ATAF1, encodes a NAC family protein that negatively regulates the expression of stress-responsive genes in *Arabidopsis*. *Plant Mol Biol*, **63**, 289–305.

Macknight, R., Bancroft, I., Page, T., Lister, C., Schmidt, R., Love, K., Westphal, L., Murphy, G., Sherson, S., Cobbett, C. and Dean, C. (1997) FCA, a gene controlling flowering time in *Arabidopsis*, encodes a protein containing RNA-binding domains. *Cell*, **89**, 737–745.

Mallory, A.C. and Vaucheret, H. (2006) Functions of microRNAs and related small RNAs in plants. *Nat Genet*, **38** (Suppl), S31–S36.

Marin, E., Nussaume, L., Quesada, A., Gonneau, M., Sotta, B., Hugueney, P., Frey, A. and Marion-Poll, A. (1996) Molecular identification of zeaxanthin epoxidase of *Nicotiana plumbaginifolia*, a gene involved in abscisic acid biosynthesis and corresponding to the ABA locus of *Arabidopsis thaliana*. *EMBO J*, **15**, 2331–2342.

Martinez-Atienza, J., Jiang, X., Garciadeblas, B., Mendoza, I., Zhu, J.K., Pardo, J.M. and Quintero, F.J. (2007) Conservation of the SOS salt tolerance pathway in rice. *Plant Physiol*, **143**, 1001–1012.

Merlot, S., Gosti, F., Guerrier, D., Vavasseur, A. and Giraudat, J. (2001) The ABI1 and ABI2 protein phosphatases 2C act in a negative feedback regulatory loop of the abscisic acid signalling pathway. *Plant J*, **25**, 295–303.

Meyer, K., Leube, M.P. and Grill, E. (1994) A protein phosphatase 2C involved in ABA signal transduction in *Arabidopsis thaliana*. *Science*, **264**, 1452–1455.

Mikami, K., Katagiri, T., Iuchi, S., Yamaguchi-Shinozaki, K. and Shinozaki, K. (1998) A gene encoding phosphatidylinositol-4-phosphate 5-kinase is induced by water stress and abscisic acid in *Arabidopsis thaliana*. *Plant J*, **15**, 563–568.

Milborrow, B.V. (2001) The pathway of biosynthesis of abscisic acid in vascular plants: a review of the present state of knowledge of ABA biosynthesis. *J Exp Bot*, **52**, 1145–1164.

Millar, A.A., Jacobsen, J.V., Ross, J.J., Helliwell, C.A., Poole, A.T., Scofield, G., Reid, J.B. and Gubler, F. (2006) Seed dormancy and ABA metabolism in *Arabidopsis* and barley: the role of ABA 8′-hydroxylase. *Plant J*, **45**, 942–954.

Mishra, G., Zhang, W., Deng, F., Zhao, J. and Wang, X. (2006) A bifurcating pathway directs abscisic acid effects on stomatal closure and opening in *Arabidopsis*. *Science*, **312**, 264–266.

Miura, K., Jin, J.B., Lee, J., Yoo, C.Y., Stirm, V., Miura, T., Ashworth, E.N., Bressan, R.A., Yun, D.J. and Hasegawa, P.M. (2007) SIZ1-mediated sumoylation of ICE1 controls *CBF3/DREB1A* expression and freezing tolerance in *Arabidopsis*. *Plant Cell*, **19**, 1403–1414.

Mochizuki, N., Brusslan, J.A., Larkin, R., Nagatani, A. and Chory, J. (2001) *Arabidopsis* genomes uncoupled 5 (GUN5) mutant reveals the involvement of Mg-chelatase H subunit in plastid-to-nucleus signal transduction. *Proc Natl Acad Sci USA*, **98**, 2053–2058.

Nambara, E. and Marion-Poll, A. (2005) Abscisic acid biosynthesis and catabolism. *Annu Rev Plant Biol*, **56**, 165–185.

Novillo, F., Alonso, J.M., Ecker, J.R. and Salinas, J. (2004) CBF2/DREB1C is a negative regulator of *CBF1/DREB1B* and *CBF3/DREB1A* expression and plays a central role in stress tolerance in *Arabidopsis*. *Proc Natl Acad Sci USA*, **101**, 3985–3990.

Ohta, M., Guo, Y., Halfter, U. and Zhu, J.K. (2003) A novel domain in the protein kinase SOS2 mediates interaction with the protein phosphatase 2C ABI2. *Proc Natl Acad Sci USA*, **100**, 11771–11776.

Okamoto, M., Kuwahara, A., Seo, M., Kushiro, T., Asami, T., Hirai, N., Kamiya, Y., Koshiba, T. and Nambara, E. (2006) *CYP707A1* and *CYP707A2*, which encode abscisic acid 8′-hydroxylases, are indispensable for proper control of seed dormancy and germination in *Arabidopsis*. *Plant Physiol*, **141**, 97–107.

Osakabe, Y., Maruyama, K., Seki, M., Satou, M., Shinozaki, K. and Yamaguchi-Shinozaki, K. (2005) Leucine-rich repeat receptor-like kinase1 is a key membrane-bound regulator of abscisic acid early signaling in *Arabidopsis*. *Plant Cell*, **17**, 1105–1119.

Pandey, S. and Assmann, S.M. (2004) The *Arabidopsis* putative G protein-coupled receptor GCR1 interacts with the G protein α subunit GPA1 and regulates abscisic acid signaling. *Plant Cell*, **16**, 1616–1632.

Papp, I., Mur, L.A., Dalmadi, A., Dulai, S. and Koncz, C. (2004) A mutation in the Cap Binding Protein 20 gene confers drought tolerance to *Arabidopsis*. *Plant Mol Biol*, **55**, 679–686.

Park, S., Li, J., Pittman, J.K., Berkowitz, G.A., Yang, H., Undurraga, S., Morris, J., Hirschi, K.D. and Gaxiola, R.A. (2005) Up-regulation of a H^+-pyrophosphatase

(H^+-PPase) as a strategy to engineer drought-resistant crop plants. *Proc Natl Acad Sci USA*, **102**, 18830–18835.

Peng, J., Carol, P., Richards, D.E., King, K.E., Cowling, R.J., Murphy, G.P. and Harberd, N.P. (1997) The *Arabidopsis GAI* gene defines a signaling pathway that negatively regulates gibberellin responses. *Genes Dev*, **11**, 3194–3205.

Quintero, F.J., Blatt, M.R. and Pardo, J.M. (2000) Functional conservation between yeast and plant endosomal Na($^+$)/H($^+$) antiporters. *FEBS Lett*, **471**, 224–228.

Razem, F., Luo, M., Liu, J.-H., Abrams, S. and Hill, R. (2004) Purification and characterization of a barley aleurone abscisic acid binding protein. *J Biol Chem*, **279**, 9922–9929.

Razem, F.A., El-Kereamy, A., Abrams, S.R. and Hill, R.D. (2006) The RNA-binding protein FCA is an abscisic acid receptor. *Nature*, **439**, 290–294.

Ren, Z.H., Gao, J.P., Li, L.G., Cai, X.L., Huang, W., Chao, D.Y., Zhu, M.Z., Wang, Z.Y., Luan, S. and Lin, H.X. (2005) A rice quantitative trait locus for salt tolerance encodes a sodium transporter. *Nat Genet*, **37**, 1141–1146.

Rodriguez, P.L., Benning, G. and Grill, E. (1998) ABI2, a second protein phosphatase 2C involved in abscisic acid signal transduction in *Arabidopsis*. *FEBS Lett*, **421**, 185–190.

Rook, F., Corke, F., Card, R., Munz, G., Smith, C. and Bevan, M.W. (2001) Impaired sucrose-induction mutants reveal the modulation of sugar-induced starch biosynthetic gene expression by abscisic acid signalling. *Plant J*, **26**, 421–433.

Rus, A., Baxter, I., Muthukumar, B., Gustin, J., Lahner, B., Yakubova, E. and Salt, D.E. (2006) Natural variants of At*HKT1* enhance Na^+ accumulation in two wild populations of *Arabidopsis*. *PLoS Genet* **2** (12), e210.

Sagi, M., Scazzocchio, C. and Fluhr, R. (2002) The absence of molybdenum cofactor sulfuration is the primary cause of the *flacca* phenotype in tomato plants. *Plant J*, **31**, 305–317.

Saito, S., Hirai, N., Matsumoto, C., Ohigashi, H., Ohta, D., Sakata, K. and Mizutani, M. (2004) *Arabidopsis CYP707As* encode (+)-abscisic acid 8′-hydroxylase, a key enzyme in the oxidative catabolism of abscisic acid. *Plant Physiol*, **134**, 1439–1449.

Sakuma, Y., Maruyama, K., Osakabe, Y., Qin, F., Seki, M., Shinozaki, K. and Yamaguchi-Shinozaki, K. (2006b) Functional analysis of an *Arabidopsis* transcription factor, DREB2A, involved in drought-responsive gene expression. *Plant Cell*, **18**, 1292–1309.

Sakuma, Y., Maruyama, K., Qin, F., Osakabe, Y., Shinozaki, K. and Yamaguchi-Shinozaki, K. (2006a) Dual function of an *Arabidopsis* transcription factor DREB2A in water-stress-responsive and heat-stress-responsive gene expression. *Proc Natl Acad Sci USA*, **103**, 18822–18827.

Sanchez, J.P. and Chua, N.H. (2001) *Arabidopsis* PLC1 is required for secondary responses to abscisic acid signals. *Plant Cell*, **13**, 1143–1154.

Schroeder, J.I., Kwak, J.M. and Allen, G.J. (2001) Guard cell abscisic acid signalling and engineering drought hardiness in plants. *Nature*, **410**, 327–330.

Schwartz, S.H., Tan, B.C., Gage, D.A., Zeevaart, J.A. and McCarty, D.R. (1997) Specific oxidative cleavage of carotenoids by VP14 of maize. *Science*, **276**, 1872–1874.

Seo, M., Peeters, A.J., Koiwai, H., Oritani, T., Marion-Poll, A., Zeevaart, J.A., Koornneef, M., Kamiya, Y. and Koshiba, T. (2000) The *Arabidopsis* aldehyde oxidase 3 (*AAO3*) gene product catalyzes the final step in abscisic acid biosynthesis in leaves. *Proc Natl Acad Sci USA*, **97**, 12908–12913.

Shen, Y.Y., Wang, X.F., Wu, F.Q., Du, S.Y., Cao, Z., Shang, Y., Wang, X.L., Peng, C.C., Yu, X.C., Zhu, S.Y., Fan, R.C., Xu, Y.H. and Zhang, D.P. (2006) The Mg-chelatase H subunit is an abscisic acid receptor. *Nature*, **443**, 823–826.

Shi, H., Ishitani, M., Kim, C.S. and Zhu, J.K. (2000) The *Arabidopsis thaliana* salt tolerance gene *SOS1* encodes a putative Na^+/H^+ antiporter. *Proc Natl Acad Sci USA*, **97**, 6896–6901.

Shi, H., Lee, B.H., Wu, S.J. and Zhu, J.K. (2003) Overexpression of a plasma membrane Na^+/H^+ antiporter gene improves salt tolerance in *Arabidopsis thaliana*. *Nat Biotechnol*, **21**, 81–85.

Shinozaki, K. and Yamaguchi-Shinozaki, K. (2007) Gene networks involved in drought stress response and tolerance. *J Exp Bot*, **58**, 221–227.

Shinozaki, K., Yamaguchi-Shinozaki, K. and Seki, M. (2003) Regulatory network of gene expression in the drought and cold stress responses. *Curr Opin Plant Biol*, **6**, 410–417.

Shou, H., Bordallo, P., Fan, J.B., Yeakley, J.M., Bibikova, M., Sheen, J. and Wang, K. (2004b) Expression of an active tobacco mitogen-activated protein kinase kinase kinase enhances freezing tolerance in transgenic maize. *Proc Natl Acad Sci USA*, **101**, 3298–3303.

Shou, H., Bordallo, P. and Wang, K. (2004a) Expression of the *Nicotiana* protein kinase (*NPK1*) enhanced drought tolerance in transgenic maize. *J Exp Bot*, **55**, 1013–1019.

Simpson, G.G., Dijkwel, P.P., Quesada, V., Henderson, I. and Dean, C. (2003) FY is an RNA 3′ end-processing factor that interacts with FCA to control the *Arabidopsis* floral transition. *Cell*, **113**, 777–787.

Soderman, E., Hjellstrom, M., Fahleson, J. and Engstrom, P. (1999) The HD-Zip gene *ATHB6* in *Arabidopsis* is expressed in developing leaves, roots and carpels and up-regulated by water deficit conditions. *Plant Mol Biol*, **40**, 1073–1083.

Song, C.P., Agarwal, M., Ohta, M., Guo, Y., Halfter, U., Wang, P. and Zhu, J.K. (2005) Role of an *Arabidopsis* AP2/EREBP-type transcriptional repressor in abscisic acid and drought stress responses. *Plant Cell*, **17**, 2384–2396.

Sridha, S. and Wu, K. (2006) Identification of AtHD2C as a novel regulator of abscisic acid responses in *Arabidopsis*. *Plant J*, **46**, 124–133.

Stockinger, E.J., Gilmour, S.J. and Thomashow, M.F. (1997) *Arabidopsis thaliana CBF1* encodes an AP2 domain-containing transcriptional activator that binds to the C-repeat/DRE, a cis-acting DNA regulatory element that stimulates transcription in response to low temperature and water deficit. *Proc Natl Acad Sci USA*, **94**, 1035–1040.

Sunkar, R., Kapoor, A. and Zhu, J.K. (2006) Posttranscriptional induction of two Cu/Zn superoxide dismutase genes in *Arabidopsis* is mediated by downregulation of miR398 and important for oxidative stress tolerance. *Plant Cell*, **18**, 2051–2065.

Sunkar, R. and Zhu, J.K. (2004) Novel and stress-regulated microRNAs and other small RNAs from *Arabidopsis*. *Plant Cell*, **16**, 2001–2019.

Tan, B.C., Schwartz, S.H., Zeevaart, J.A. and McCarty, D.R. (1997) Genetic control of abscisic acid biosynthesis in maize. *Proc Natl Acad Sci USA*, **94**, 12235–12240.

Teige, M., Scheikl, E., Eulgem, T., Doczi, R., Ichimura, K., Shinozaki, K., Dangl, J.L. and Hirt, H. (2004) The MKK2 pathway mediates cold and salt stress signaling in *Arabidopsis*. *Mol Cell*, **15**, 141–152.

Thomashow, M.F. (1999) Plant cold acclimation: freezing tolerance genes and regulatory mechanisms. *Annu Rev Plant Physiol Plant Mol Biol*, **50**, 571–599.

Thomashow, M.F. (2001) So what's new in the field of plant cold acclimation? Lots! *Plant Physiol*, **125**, 89–93.

Thompson, A.J., Jackson, A.C., Parker, R.A., Morpeth, D.R., Burbidge, A. and Taylor, I.B. (2000a). Abscisic acid biosynthesis in tomato: regulation of zeaxanthin epoxidase and 9-*cis*-epoxycarotenoid dioxygenase mRNAs by light/dark cycles, water stress and abscisic acid. *Plant Mol Biol*, **42**, 833–845.

Thompson, A.J., Jackson, A.C., Symonds, R.C., Mulholland, B.J., Dadswell, A.R., Blake, P.S., Burbidge, A. and Taylor, I.B. (2000b). Ectopic expression of a tomato 9-*cis*-epoxycarotenoid dioxygenase gene causes over-production of abscisic acid. *Plant J*, **23**, 363–374.

Ullah, H., Chen, J.-G., Wang, S. and Jones, A.M. (2002) Role of a heterotrimeric G protein in regulation of *Arabidopsis* seed germination. *Plant Physiol*, **129**, 897–907.

Ullah, H., Chen, J.-G., Young, J., Im, K.-H., Sussman, M. and Jones, A. (2001) Modulation of cell proliferation by heterotrimeric G protein in *Arabidopsis*. *Science*, **292**, 2066–2069.

Umezawa, T., Okamoto, M., Kushiro, T., Nambara, E., Oono, Y., Seki, M., Kobayashi, M., Koshiba, T., Kamiya, Y. and Shinozaki, K. (2006) CYP707A3, a major ABA 8′-hydroxylase involved in dehydration and rehydration response in *Arabidopsis thaliana*. *Plant J*, **46**, 171–182.

Uno, Y., Furihata, T., Abe, H., Yoshida, R., Shinozaki, K. and Yamaguchi-Shinozaki, K. (2000) *Arabidopsis* basic leucine zipper transcription factors involved in an abscisic acid-dependent signal transduction pathway under drought and high-salinity conditions. *Proc Natl Acad Sci USA*, **97**, 11632–11637.

Verslues, P.E., Guo, Y., Dong, C.H., Ma, W. and Zhu, J.K. (2006) Mutation of SAD2, an importin beta-domain protein in *Arabidopsis*, alters abscisic acid sensitivity. *Plant J*, **47**, 776–787.

Vogel, J.T., Zarka, D.G., Van Buskirk, H.A., Fowler, S.G. and Thomashow, M.F. (2005) Roles of the CBF2 and ZAT12 transcription factors in configuring the low temperature transcriptome of *Arabidopsis*. *Plant J*, **41**, 195–211.

Vranová, E., Inzé, D. and Van Breusegem, F. (2002) Signal transduction during oxidative stress. *J Exp Bot*, **53**, 1227–1236.

Wang, X.Q., Ullah, H., Jones, A.M. and Assmann, S.M. (2001) G protein regulation of ion channels and abscisic acid signaling in *Arabidopsis* guard cells. *Science*, **292**, 2070–2072.

Williams, M.E., Torabinejad, J., Cohick, E., Parker, K., Drake, E.J., Thompson, J.E., Hortter, M. and Dewald, D.B. (2005) Mutations in the *Arabidopsis* phosphoinositide phosphatase gene *SAC9* lead to overaccumulation of PtdIns (4,5)P2 and constitutive expression of the stress-response pathway. *Plant Physiol*, **138**, 686–700.

Xin, Z. and Browse, J. (1998) *Eskimo1* mutants of *Arabidopsis* are constitutively freezing-tolerant. *Proc Natl Acad Sci USA*, **95**, 7799–7804.

Xin, Z., Mandaokar, A., Chen, J., Last, R.L. and Browse, J. (2007) *Arabidopsis ESK1* encodes a novel regulator of freezing tolerance. *Plant J*, **49**, 786–799.

Xiong, L., Gong, Z., Rock, C.D., Subramanian, S., Guo, Y., Xu, W., Galbraith, D. and Zhu, J.K. (2001a) Modulation of abscisic acid signal transduction and biosynthesis by an Sm-like protein in *Arabidopsis*. *Dev Cell*, **1**, 771–781.

Xiong, L., Ishitani, M., Lee, H. and Zhu, J.K. (2001b) The *Arabidopsis* LOS5/ABA3 locus encodes a molybdenum cofactor sulfurase and modulates cold stress- and osmotic stress-responsive gene expression. *Plant Cell*, **13**, 2063–2083.

Xiong, L., Lee, B., Ishitani, M., Lee, H., Zhang, C. and Zhu, J.K. (2001c) *FIERY1* encoding an inositol polyphosphate 1-phosphatase is a negative regulator of abscisic acid and stress signaling in *Arabidopsis*. *Genes Dev*, **15**, 1971–1984.

Xiong, L., Schumaker, K.S. and Zhu, J.K. (2002) Cell signaling during cold, drought, and salt stress. *Plant Cell*, **14**, S165–S183.

Xiong, L. and Yang, Y. (2003) Disease resistance and abiotic stress tolerance in rice are inversely modulated by an abscisic acid-inducible mitogen-activated protein kinase. *Plant Cell*, **15**, 745–759.

Xiong, L. and Zhu, J.K. (2003) Regulation of abscisic acid biosynthesis. *Plant Physiol*, **133**, 29–36.

Xu, J., Li, H., Chen, L., Wang, Y., Liu, L., He, L. and Wu, W. (2006) A protein kinase, interacting with two calcineurin B-like proteins, regulates K^+ transporter AKT1 in *Arabidopsis*. *Cell*, **125**, 1347–1360.

Xu, Z.J., Nakajima, M., Suzuki, Y. and Yamaguchi, I. (2002) Cloning and characterization of the abscisic acid-specific glucosyltransferase gene from adzuki bean seedlings. *Plant Physiol*, **129**, 1285–1295.

Yang, Y., Sulpice, R., Himmelbach, A., Meinhard, M., Christmann, A. and Grill, E. (2006) Fibrillin expression is regulated by abscisic acid response regulators and is involved in abscisic acid-mediated photoprotection. *Proc Natl Acad Sci USA*, **103**, 6061–6066.

Yokoi, S., Quintero, F.J., Cubero, B., Ruiz, M.T., Bressan, R.A., Hasegawa, P.M. and Pardo, J.M. (2002) Differential expression and function of *Arabidopsis thaliana* NHX Na^+/H^+ antiporters in the salt stress response. *Plant J*, **30**, 529–539.

Zhang, D.P., Wu, Z.Y., Li, X.Y. and Zhao, Z.X. (2002) Purification and identification of a 42-kilodalton abscisic acid-specific-binding protein from epidermis of broad bean leaves. *Plant Physiol*, **128**, 714–725.

Zhao, J. and Wang, X. (2004) *Arabidopsis* phospholipase Dalpha1 interacts with the heterotrimeric G-protein α-subunit through a motif analogous to the DRY motif in G-protein-coupled receptors. *J Biol Chem*, **279**, 1794–1800.

Zheng, Z.L., Nafisi, M., Tam, A., Li, H., Crowell, D.N., Chary, S.N., Schroeder, J.I., Shen, J. and Yang, Z. (2002) Plasma membrane-associated ROP10 small GTPase is a specific negative regulator of abscisic acid responses in *Arabidopsis*. *Plant Cell*, **14**, 2787–2797.

Zhu, J., Shi, H., Lee, B.H., Damsz, B., Cheng, S., Stirm, V., Zhu, J.K., Hasegawa, P.M. and Bressan, R.A. (2004) An *Arabidopsis* homeodomain transcription factor gene, *HOS9*, mediates cold tolerance through a CBF-independent pathway. *Proc Natl Acad Sci USA*, **101**, 9873–9878.

Zhu, J., Verslues, P.E., Zheng, X., Lee, B.H., Zhan, X., Manabe, Y., Sokolchik, I., Zhu, Y., Dong, C.H., Zhu, J.K., Hasegawa, P.M. and Bressan, R.A. (2005) *HOS10* encodes an R2R3-type MYB transcription factor essential for cold acclimation in plants. *Proc Natl Acad Sci USA*, **102**, 9966–9971.

Zhu, J.K. (2001). Cell signaling under salt, water and cold stresses. *Curr Opin Plant Biol*, **4**, 401–406.

Zhu, J.K. (2002) Salt and drought stress signal transduction in plants. *Annu Rev Plant Biol*, **53**, 247–273.

Zhu, J.K. (2003) Regulation of ion homeostasis under salt stress. *Curr Opin Plant Biol*, **6**, 441–445.

INDEX